MTP International Review of Science

Aliphatic Compounds

MTP International Review of Science

Publisher's Note

The MTP International Review of Science is an important new venture in scientific publishing, which we present in association with MTP Medical and Technical Publishing Co. Ltd. and University Park Press, Baltimore. The basic concept of the Review is to provide regular authoritative reviews of entire disciplines. We are starting with chemistry because the problems of literature survey are probably more acute in this subject than in any other. As a matter of policy, the authorship of the MTP Review of Chemistry is international and distinguished; the subject coverage is extensive, systematic and critical; and most important of all, new issues of the Review will be published every two years.

In the MTP Review of Chemistry (Series One), Inorganic, Physical and Organic Chemistry are comprehensively reviewed in 33 text volumes and 3 index volumes, details of which are shown opposite. In general, the reviews cover the period 1967 to 1971. In 1974, it is planned to issue the MTP Review of Chemistry (Series Two), consisting of a similar set of volumes covering the period 1971 to 1973. Series Three is planned for 1976, and so on.

The MTP Review of Chemistry has been conceived within a carefully organised editorial framework. The over-all plan was drawn up, and the volume editors were appointed, by three consultant editors. In turn, each volume editor planned the coverage of his field and appointed authors to write on subjects which were within the area of their own research experience. No geographical restriction was imposed. Hence, the 300 or so contributions to the MTP Review of Chemistry come from many countries of the world and provide an authoritative account of progress in chemistry.

To facilitate rapid production, individual volumes do not have an index. Instead, each chapter has been prefaced with a detailed list of contents, and an index to the 10 volumes of the MTP Review of Organic Chemistry (Series One) will appear, as a separate volume, after publication of the final volume. Similar arrangements will apply to the MTP Review of subsequent series.

Organic Chemistry
Series One
Consultant Editor
D. H. Hey, F.R.S.
Department of Chemistry
King's College, University of London

Volume titles and Editors

1 STRUCTURE DETERMINATION IN ORGANIC CHEMISTRY
Professor W. D. Ollis, F.R.S.,
University of Sheffield

2 ALIPHATIC COMPOUNDS
Professor N. B. Chapman,
Hull University

3 AROMATIC COMPOUNDS
Professor H. Zollinger, *Swiss Federal Institute of Technology*

4 HETEROCYCLIC COMPOUNDS
Dr. K. Schofield, *University of Exeter*

5 ALICYCLIC COMPOUNDS
Professor W. Parker, *University of Stirling*

6 AMINO ACIDS, PEPTIDES AND RELATED COMPOUNDS
Professor D. H. Hey, F.R.S. and
Dr. D. I. John,
King's College, University of London

7 CARBOHYDRATES
Professor G. O. Aspinall, *Trent University, Ontario*

8 STEROIDS
Dr. W. F. Johns, *G. D. Searle & Co., Chicago*

9 ALKALOIDS
Professor K. Wiesner, F.R.S.,
University of New Brunswick

10 FREE RADICAL REACTIONS
Professor W. A. Waters, F.R.S.,
University of Oxford

INDEX VOLUME

Butterworth & Co. (Publishers) Ltd.

Organic Chemistry
Series One

Consultant Editor
D. H. Hey, F.R.S.

MTP International Review of Science

Volume 2

Aliphatic Compounds

Edited by **N. B. Chapman**
University of Hull

Butterworths · London
University Park Press · Baltimore

THE BUTTERWORTH GROUP

ENGLAND
Butterworth & Co (Publishers) Ltd
London: 88 Kingsway, WC2B 6AB

AUSTRALIA
Butterworths Pty Ltd
Sydney: 586 Pacific Highway 2067
Melbourne: 343 Little Collins Street, 3000
Brisbane: 240 Queen Street, 4000

NEW ZEALAND
Butterworths of New Zealand Ltd
Wellington: 26–28 Waring Taylor Street, 1

SOUTH AFRICA
Butterworth & Co (South Africa) (Pty) Ltd
Durban: 152–154 Gale Street

ISBN 0 408 70276 1

UNIVERSITY PARK PRESS

U.S.A. and CANADA
University Park Press
Chamber of Commerce Building
Baltimore, Maryland, 21202

Library of Congress Cataloging in Publication Data

Chapman, Norman Bellamy, 1916–
  Aliphatic compounds.

  (Organic chemistry, series one, v. 2) (MTP international review of science)
  1. Aliphatic compounds. I. Title.
QD251.2.074 vol. 2 [QD305.H5]   547'.008 [547'.4]
ISBN 0–8391–1030–8   73–5666

First Published 1973 and © 1973
MTP MEDICAL AND TECHNICAL PUBLISHING CO. LTD.
St Leonard's House
St Leonardgate
Lancaster, Lancs.
and
BUTTERWORTH & CO. (PUBLISHERS) LTD.

Filmset by Photoprint Plates Ltd., Rayleigh, Essex
Printed in England by Redwood Press Ltd., Trowbridge, Wilts
and bound by R. J. Acford Ltd., Chichester, Sussex

Consultant Editor's Note

The subject of Organic Chemistry is in a rapidly changing state. At the one extreme it is becoming more and more closely involved with biology and living processes and at the other it is deriving a new impetus from the extending implications of modern theoretical developments. At the same time the study of the subject at the practical level is being subjected to the introduction of new techniques and advancements in instrumentation at an unprecedented level. One consequence of these changes is an enormous increase in the rate of accumulation of new knowledge. The need for authoritative documentation at regular intervals on a world-wide basis is therefore self-evident.

The ten volumes in Organic Chemistry in this First Series of biennial reviews in the MTP International Review of Science attempt to place on record the published achievements of the years 1970 and 1971 together with some earlier material found desirable to assist the initiation of the new venture. In order to do this on an international basis Volume Editors and Authors have been drawn from many parts of the world.

There are many alternative ways in which the subject of Organic Chemistry can be subdivided into areas for more or less self-contained reviews. No single system can avoid some overlapping and many such systems can leave gaps unfilled. In the present series the subject matter in eight volumes is defined mainly on a structural basis on conventional lines. In addition, one volume has been specially devoted to methods of structure determination, which include developments in new techniques and instrumental methods. A further separate volume has been devoted to Free Radical Reactions, which is justified by the rapidly expanding interest in this field. If there prove to be any major omissions it is hoped that these can be remedied in the Second Series.

It is my pleasure to thank the Volume Editors who have made the publication of these volumes possible.

London   D. H. Hey

Preface

There are those who would readily sound the death-knell of Organic Chemistry. The fallaciousness of this attitude is vividly demonstrated by the contents of the present volume on Aliphatic Chemistry, which might *a priori* be expected to be the least lively part of a supposedly moribund subject.

Nowhere is this more obvious than in the first chapter on hydrocarbons: new methods jostle with deeper analyses of old problems, and in so doing proclaim the unitary character of modern chemistry. Wherever else one looks that mastery over subtle chemical transformation and potent determination of intricate molecular structure for which aliphatic chemistry is rightly renowned is richly exemplified. Nitrogen compounds have long occupied a major position in this field and continue to do so: the chapter on phosphorus chemistry reveals the competition for eminence from that element. Sulphur compounds and halogeno-compounds are also close contenders. Possibly some of the most novel and exciting chemistry is to be found in the short review on boron compounds, a topic now essential to a modern representative treatment of organic chemistry whether aliphatic or aromatic. In the field of oxygen-functional compounds, the paths may be well trodden, but new vistas appear and useful short-cuts are established. In relation to naturally occurring compounds carboxylic acids occupy pride of place: there is a continuing splendour about aliphatic organic chemistry in this area. Some subjects have been excluded because they are dealt with elsewhere in the series as a whole, e.g. amino acids and proteins, and compounds closely related to simple carbohydrates. One or two minor topics, e.g. carbonic acid derivatives are only treated incidentally, and systematic treatment of organometallic compounds is also yielded to other volumes in the series.

It is hoped that the present review will provide an up-to-date, critical and stimulating account of what is manifestly a vital feature of a still vigorously growing and fascinating subject.

Hull   N. B. Chapman

Contents

1
Hydrocarbons

D. E. WEBSTER
University of Hull

1.1   INTRODUCTION

This chapter is a review of part of the recent literature concerning aliphatic hydrocarbons. A search of the 1970 and 1971 journals revealed *c.* 1600 papers and patents that would be appropriate for this section. The space available will not allow discussion of more than *c.* 25% of these, so drastic selection has been necessary.

The criteria used were as follows. Two major topics, namely heterogeneous catalysis involving aliphatic hydrocarbons (about 200 references) and relevant patents (about 500), have been entirely omitted; there is insufficient space to include all the work, and selection of parts of either would be most arbitrary.

The remaining 900 papers were reduced by omitting those devoted to the physical chemistry of aliphatic hydrocarbons. The remaining papers, those of clear interest to organic chemists, form the basis of this chapter.

Although some mechanistic aspects of aliphatic hydrocarbon chemistry are discussed, the emphasis of the chapter is on compounds rather than mechanism. Reactions of aliphatic hydrocarbons that are catalysed homogeneously in solution by transition-metal complexes are included. A more complete review of such reactions is in another part of this series[1]. There are no recent reviews that cover the same area as the subject matter of this chapter.

1.2   THEORETICAL STUDIES

1.2.1   Electronic structures

There have been rapid advances during the past few years in the methods available for treating the electronic structures of organic molecules[2], and self-consistent field (SCF) methods are now widely used. A survey[3] of molecular orbital methods that treat all the valence electrons, and can be used for three-dimensional molecules or complexes in general, that discusses both semi-empirical and *ab initio* methods, and gives the results for methane, ethane, ethylene, acetylene, and propyne has appeared. Two more detailed papers having the same senior author[4, 5] use a set of Gaussian fitted Slater-type orbitals (STO-3G)[6] and two extended sets of orbitals (4-31G)[7] and (6-31G)[5], to obtain the geometries and energies of the neutral C_1, C_2 and C_3 hydrocarbons, and of the related C_1 and C_2 positive ions. The agreement between calculated and experimentally determined bond lengths and bond angles is good, as can be seen in Table 1.1, which lists the values for the neutral C_1 and C_2 molecules studied. Equally good agreement is obtained for the bond angles of propyne, allene, cyclopropene, cyclopropane and propane[5]. Structures of protonated methane and ethane[4, 8] derived by the same methods are of interest, as these intermediates are found during the electrophilic substitution of alkanes (Section 1.3.2.2). CH_5^+ has been analysed by using three models. That of lowest energy [Figure 1.1 (a) with $r_1 = 0.1098$, $r_2 = 0.1106$, $r_3 = 0.1370$ and $r_4 = 0.1367$ nm, $\alpha = 140$, $\beta = 83.8$, $\theta = 37.2$ and $\zeta = 117.7$ degrees] is of C_s symmetry and is equivalent to a loose complex

Table 1.1

(From Pople, Lathan and Hehre[4], by courtesy of the American Chemical Society)

Molecule	Parameter	Calc. value (STO-3G)	Calc. value (4-31G)	Exptl. value
CH_4	r_{CH}	0.1083 nm	0.1081 nm	0.1085 nm
C_2H_2	r_{CC}	0.1168 nm	0.1190 nm	0.1203 nm
	r_{CH}	0.1065 nm	0.1051 nm	0.1061 nm
C_2H_4	r_{CC}	0.1306 nm	0.1316 nm	0.1330 nm
	r_{CH}	0.1082 nm	0.1073 nm	0.1076 nm
	∠HCH	115.6 degrees	116.0 degrees	116.6 degrees
C_2H_6	r_{CC}	0.1538 nm	0.1529 nm	0.1548 nm
	r_{CH}	0.1086 nm	0.1083 nm	0.1086 nm
	∠HCH	108.2 degrees	107.7 degrees	107.8 degrees

of CH_3^+ and H_2. The analogous model for $C_2H_7^+$ [Figure 1.1(b)], which is equivalent to a loose complex between $C_2H_5^+$ and H_2, is not that of lowest energy; the complex with a bridging proton [Figure 1.1 (c) with $r_1 = 0.2362$, $r_2 = 0.1251$, $r_3 = 0.1097$ and $r_4 = 0.1094$ nm, $\alpha = 52.7$, $\beta = 89.5$ and $\theta = 115.6$ degrees] is 11 kcal mol^{-1} lower in energy. In this complex the bridging proton is at the vertex of a very flat isosceles triangle with a long C—C distance, indicating that protonation of alkanes can easily lead to C—C cleavage. That $C_2H_5^+$ has a smaller affinity than CH_3^+ for H_2 is presumably because the additional methyl group is donating electrons to the carbenium* ionic centre.

The energy of the ethyl cation [Figure 1.1(d)] is found to be 11.42 and 6.76 kcal mol^{-1} lower than that of the bridged structure [Figure 1.1(e)] by using the STO-3G and the 4-31G method respectively. Corresponding values obtained[9] by using two non-empirical LCAO–MO–SCF techniques are 5.16 and 3.39 kcal mol^{-1}.

*Cf. p.12.

The main characteristics of the electronic spectra of small saturated hydrocarbons have been studied by using the CNDO (complete neglect of differential overlap), the INDO (intermediate neglect of differential overlap) and the RCNDO [CNDO including higher (Rydberg) atomic orbitals] methods[10]. All three correctly interpret the bathochromic shift in the electronic

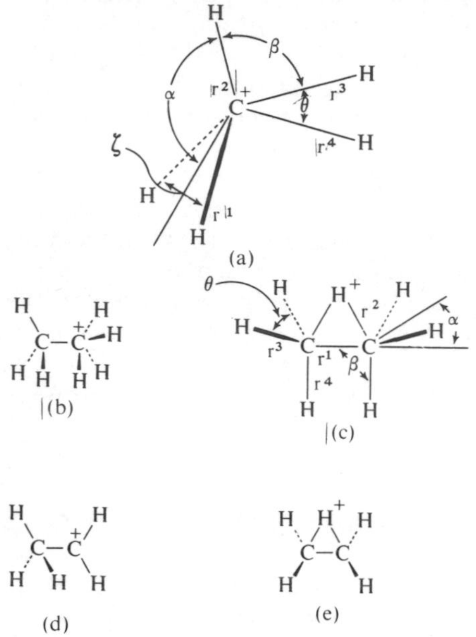

(a)

(b)

(c)

(d)

(e)

Figure 1.1   Models for protonated alkanes and alkenes
(From Pople, Lathan and Hehre[4], by courtesy of the American Chemical Society)

spectra of methane, ethane, propane, butane, n-pentane, isobutane, isopentane and neopentane. The RCNDO calculations show that electrons giving the first singlet–singlet and singlet–triplet transitions in the electronic spectra are almost entirely valence shell electrons, that the singlet–singlet transition involves principally C—H electrons for CH_4, C_2H_6, and C_3H_8, and principally C—C electrons for C_4H_{10} and C_5H_{12}, and that the singlet–triplet transition involves C—H electrons for all linear alkanes and C—C and C—H electrons for the branched alkanes.

Despite the view[2] that 'there no longer seems any point in carrying out calculations by less refined procedures (than SCF–LCAO–MO methods)' an interesting application of the Hückel method to π-resonance energies has classified single and double bonds into eight types according to the number of attached hydrogen atoms[11]. All acyclic and cyclic polyenes can be constructed by combinations of these eight types of bonds. The π-energies and resonance energies of cyclic polyenes, which can be readily calculated, correlate well with chemical behaviour, and may be compared with the values obtained by the more complex CNDO method[2].

The continuing controversy as to whether or not C—C bond lengths depend on π-bond order or the degree of hybridisation (s-character) of the orbitals forming the σ-bond in unsaturated and conjugated hydrocarbons, has now been extended to saturated hydrocarbons[12–14]. There is a strong correlation between experimental bond lengths and the amount of s-character calculated for hybrid orbitals forming single bonds[12], and, in contrast, it has been shown that if π-overlap is neglected, the calculated bond lengths of ethylene and acetylene are essentially identical with those calculated for ethane, implying that π-bonding is entirely responsible for the shortening of the C—C bond in ethylene and acetylene[13]. Calculations[14] using the INDO method with ethane as the model indicate that the predominant effect is due to π-bonding, and a pseudo-π description[15] with the H orbitals combined as a group (σ, $\pi_{symmetric}$, and $\pi_{antisymmetric}$) directly relates the s-character with the pseudo-π overlap of the hydrogen π-orbitals with the p_y and p_z orbitals of the carbon atoms.

The pseudo-π model has also been used in an 'explanation' of the long-standing problem of the origin of the barrier to internal rotation in ethane and other alkanes[16]. The behaviour of two occupied pairs of orbitals during the rotation from the staggered to the eclipsed form is suggested as the origin of the energy barrier. However, the problem is more complex. Two sets of SCF calculations[17] [19] show how elusive is the source of this energy barrier. Comparison of the components making up the barrier energy from these two treatments (Table 1.2) shows that although the calculated barriers agree

Table 1.2

(From Epstein and Lipscomb[19], by courtesy of the American Chemical Society)

| | Energy components (in atomic units) for ethane | |
	Ref. 17	Ref. 18
Kinetic	0.020 14	0.009 94
Nuclear repulsion	0.007 49	−0.118 48
Nuclear attraction	−0.048 98	0.201 13
Electron repulsion	0.026 57	−0.087 36
Barrier	*0.005 22	*0.005 23

*Equal to 3.3 kcal mol^{-1}.

to within 10^{-5} of an atomic unit, three of the four contributing terms have different signs in the two methods. The total energies of the functions studied are large and energy decompositions for such functions are generally erratic. It is suggested that there are essentially two effects to be considered, namely the rotation of the methyl groups and the adjustment of the molecular geometry to the new configuration after rotation. For ethane the second point negates the first, but for cases where no change in the molecular geometry occurs (other than the rotation being studied) the explanation based on the pseudo-π model[16] may be useful. Other recent *ab initio* LCAO–MO–SCF calculations[20-24] of the rotational energy barrier in ethane give respectively values of 2.58, 3.07, 3.17, 3.331 and 2.521 kcal mol^{-1}. (The experimental value[25] is 2.928 ± 0.025 kcal mol^{-1}). The origin of the barrier has been further discussed in terms of attractive (nuclear–electron), repulsive (nuclear–nuclear, electron–electron) and kinetic terms[20, 24, 26]. For ethane the barrier is due to the repulsive terms, and arises from ordinary chemical bonding by the action of the Pauli principle, although there are very substantial inherent difficulties in splitting up a molecular wave function clearly to bring out the

workings of the Pauli principle. Some insight is obtained by carrying out a
charge density analysis of the barrier[24]. This clearly requires some care as
the energy difference between the eclipsed and staggered forms of ethane is
only 1/20 000 of the total molecular energy. The total molecular electron
distributions for eclipsed and staggered ethane in the xy plane perpendicular
to the centre of the C—C bond are shown in Figure 1.2(a) and (b). The
three-fold symmetry of the eclipsed form is clear, but the six-fold symmetry

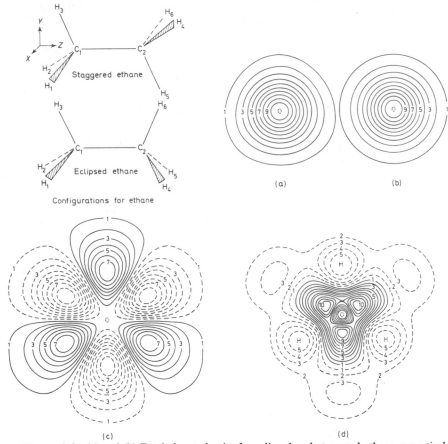

(a)   (b)

(c)   (d)

Figure 1.2 (a) and (b) Total charge density for eclipsed and staggered ethane respectively,
perpendicular to the mid-point of the C—C bond. (c) Difference plot of the eclipsed ethane
electron density minus the staggered ethane electron density at the same location as in (a) and (b).
(d) As for (c) on the C—C bond 0.0263 nm from C-1. For (a) and (b) contour 1 is at 0.018 au
and the contour interval is 0.018 au. For (c) and (d) contour 1 is at $\pm 0.000\,02$ au and the contour
interval is $\pm 0.000\,02$ au
(From Jorgensen and Allen[24], by courtesy of the American Chemical Society)

of the staggered form is obscure, and it is not clear from a comparison of these
total molecular charge densities that there is a greater repulsion between the
methyl groups in the eclipsed than in the staggered form. The difference in
the electron densities (eclipsed minus staggered) [Figure 1.2(c)] shows that
there is less electron density, hence greater repulsion, between the opposing

C–H bonds in the eclipsed than the staggered conformer since the negative contours (dotted lines) encompass more area than the equivalent positive contours (solid lines). An equivalent difference plot in an xy plane between the carbon atoms, but close to one of them [Figure 1.2(d)] clearly shows that there is an electron density decrease in the eclipsed relative to the staggered conformer in front of the C—H bonds, i.e. that an increase in the repulsion between eclipsing C—H bonds is the dominant factor in determining the nature of the ethane barrier. Barriers to rotation of methyl groups and C—C bonds in other molecules have also been calculated (Table 1.3)[23, 27–29].

Table 1.3

(From Random and Pople[23], by courtesy of the American Chemical Society, except for * Ref. 27, † Ref. 28 and ‡ Ref. 29)

Hydrocarbon	Barrier (calc.) /kcal mol^{-1}	Barrier (exptl.) /kcal mol^{-1}
But-1-yne	3.46	—
But-2-yne	0.006	—
Propene	1.54	1.95–2.04
Propene*	1.48	1.95–2.04
Propene*	1.25	1.95–2.04
Buta-1,2-diene	1.40	1.59
Propane	3.45	3.40
2-Methylpropene	1.705	2.12–2.35
cis-But-2-ene	0.42	0.73
trans-But-2-ene	1.54	1.95
2-Methylpropane	3.88	3.90
Ethylene	138.6	65
Allene	91.9	—
Allene†	72.7	
Butatriene	73.9	—
Vinylacetylene	137.7	—
Buta-1,3-diene	6.73	5.0
Buta-1,3-diene‡	5.15	5.0
n-Butane	3.40	—
But-1-ene (skew)	3.46	3.16
But-2-ene (cis)	4.96	3.99

A comparison of the electronic structure of ethylene and diborane by has recently appeared[30].

A comparison of the electronic structure of ethylene and diborane by molecular S.C.F. calculations and high-resolution photoelectron spectroscopy[31] substantiates Pitzer's[32] view that diborane is very similar to ethylene with the double bond protonated, there being some changes in the ordering of the molecular orbitals, in particular the π-MO in B_2H_6 is the fourth to be filled, whereas in ethylene it is of highest energy. There have been further studies of the photoelectron spectra of alkanes[33–36] and the spectra have been analysed by means of molecular orbital theory[35].

A detailed description of the electronic structure of acetylene, obtained by using an extended set of S.C.F. wave functions[37], shows that the σ-electronic charge in the C—C region is c. 2 as expected, but that the π-electronic charge is 1.2 electrons per π-bond, i.e. about 40% of the π-charge of the C—C bond is not in the region between the carbon nuclei (Figure 1.3). It is, therefore, somewhat misleading to say that there is a triple bond between the carbon atoms: relative to the σ-bond each π-bond is more like half a bond. This result agrees well with an earlier semi-empirical calculation[38] and puts the

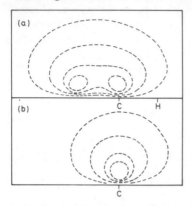

Figure 1.3 (a) Charge density contours of the π-electrons in acetylene. (b) Charge density contours of a 2p π-orbital of a free carbon atom. The contours correspond to the same charge densities as in (a)
(From Politzer and Harris[37], by courtesy of Pergamon Press)

conclusion 'that there is a breakdown of the traditional point of view according to which there are four π-electrons in the C—C bond of acetylene'[38], on a much firmer theoretical basis. The π-electrons outside the C—C region will help to strengthen the C—H bonds, hence an explanation of the shorter and stronger C—H bond in acetylene compared with ethane (0.1058 nm and 0.589 mdyn nm^{-1}, cf. 0.1093 nm and 0.479 mdyn nm^{-1}).

The intermediates formed in electrophilic additions of H^+, F^+, Cl^+, Br^+, SR^+, HgX^+ or Ag^+ to ethylene, have been studied by extended Hückel calculations[39]. The extensive results correlate well with experimental data and chemical intuition. The mode of addition is influenced by the amount of energy required to convert the symmetrical onium ion (1) into the unsymmetrical form (2). Perturbational molecular orbital theory has been success-

fully applied to the photocyclo-addition reactions of trimethylethylene (2-methylbut-2-ene) with benzaldehyde and of ethylene with formaldehyde[40]. The calculated lowest energy pathway involves electrophilic attack on the alkene by the lone non-bonding electron in the n,π^* state of the aldehyde to generate a biradical intermediate, in agreement with the mechanism postulated on the basis of experiment.

1.1.2.2   Woodward-Hoffmann orbital symmetry rules

Although the application of the Woodward–Hoffmann orbital symmetry rules to aliphatic hydrocarbons[41] has not developed further during the past 2 years, there has been considerable interest in the effect of transition metals as catalysts for symmetry-forbidden reactions. Since 1967 there have been a number of papers applying the ideas of symmetry conservation to a number of transition-metal catalysed reactions. Of particular interest are cyclo-addition (equation (1.1)) and alkene metathesis (equation (1.2)).

$$2CH_2{:}CH_2 \quad \rightleftharpoons \quad \text{cyclobutane} \tag{1.1}$$

$$\tag{1.2}$$

There is a continuing controversy as to whether or not such reactions are concerted, and whether, therefore, the Woodward–Hoffmann rules are applicable to them. It has been shown that the valence isomerism of cubane[42], an alicyclic cyclo-addition[43], and the nickel-catalysed cyclo-addition of buta-1,3-diene[44] proceed by non-concerted routes involving oxidative-addition. In contrast the metathesis of propene, labelled[45] with either carbon-14 or deuterium[46] in which all the ^{14}C or D is found in the butene produced (equation (1.3) and (1.4)), indicates a concerted mechanism with a cyclo-butane type of intermediate when a heterogeneous rhenium oxide–alumina catalyst is used.

$$2\ H_2C{:}^{14}CH{\cdot}Me \rightarrow H_2C{:}CH_2 + Me{\cdot}^{14}CH{:}^{14}CH{\cdot}Me \tag{1.3}$$

$$2\ H_2C{:}CD{\cdot}Me \rightarrow H_2C{:}CH_2 + Me{\cdot}CD{:}CD{\cdot}Me \tag{1.4}$$

Clearly the observation that a symmetry-forbidden transformation takes place with a transition-metal catalyst, does not necessarily mean that the metal has allowed a previously forbidden process. Nevertheless the requirements of a potential catalyst that will allow such a process have been outlined[47]. Cyclo-addition (equation (1.1)) will involve the exchange of electron pairs between the metal and the transforming ligands. This will redistribute the metal's electrons and can introduce energy barriers due to the ligand fields of the non-reacting ligands. Even small energy barriers might prevent the cyclo-addition and this could be so for simple alkenes such as ethylene or propene, which do not undergo such reactions. However the alkene metathesis reaction (equation (1.2)) is now well-known; simple alkenes are readily interconverted, and cyclobutane has been suggested as a model for a transition state, or as a short-lived intermediate[48]. The suggested types of catalyst are complexes that show high lability to interconversion of polytopal isomers[49], particularly those that are 6-coordinate trigonal prismatic, or preferably 7-coordinate monocapped trigonal prismatic. It has also been suggested[50, 51] that favourable catalysts for the alkene metathesis (equation (1.2)) will have high-spin d^4 or d^6 electronic configurations, and catalysts for the cyclo-addition (equation (1.1)) will be low-spin d^8. Again trigonal-prismatic complexes are favoured. Another view that has been expressed is that the introduction of energy levels of the metal between the highest-occupied levels and the lowest-unoccupied levels of the alkene is sufficient explanation of the catalytic activity[52], a view that has been strongly contested[53]. A mechanism that does not involve a normal cyclobutane ring as the intermediate for the metathesis has also been proposed[54]. It is suggested that a bis-ethylene π-complex is converted into a multicentred organometallic system, most conveniently described as metal orbitals associated with four methylenic units. A review which discusses the catalysis of symmetry-forbidden reactions in terms of valence-bond theory has recently been published[55]. Application of the Zimmerman topological approach[404] to the oxidation of alkenes by inorganic ions has shown that selection rules play a significant part in determining the mechanisms in many such reactions[405].

1.3 ALKANES

1.3.1 Preparation and isolation

There is little to report. Isobutane has been prepared as one of the products of the acid-catalysed dehydration of butanols[56], in a 10:90 v/v sulphuric acid–85% phosphoric acid mixture at 164 °C. The product yields are very variable, presumably because the alkenes formed will rearrange under the

Table 1.4
(From Warkentin and Hine[56], by courtesy of the National Research Council of Canada)

Reactant alcohol	Isobutane	Products(%)		
		But-1-ene and isobutene	trans-But-2-ene	cis-But-2-ene
Bu^{n}OH	5–19	20–51	22–47	16–27
Bu^{s}OH	2–6	31–46	29–36	23–27
Bu^{i}OH	25–43	42–66	7–11	2–5
Bu^{t}OH	0.5–2.0	98–99	trace	trace

reaction conditions. Isobutyl alcohol (3-methylpropan-1-ol) gives the highest yield of alkane (Table 1.4) partly because it can form an intermediate carbenium ion $[Me_3C^+]$ readily, but also because the isobutane arises by ready hydride transfer from the alcohol to the carbenium ion.

2,2,3,5-Tetramethyl-, 5-ethyl-2,2,3-trimethyl- and 5-ethyl-2,3-dimethyl-heptane, each with a rigid conformation, have been prepared by a long series of reactions[57]. The C_{18} hydrocarbons, 6-,7- and 8-methylheptadecane have been shown, by g.l.c. and mass spectrometry, to be present in a 1:4.5:4.5 ratio in blue-green algae[58].

1.3.2 Reactions

1.3.2.1 Radical processes

The chlorination of isopentane in the gas phase under u.v. radiation gives four monochlorinated products and their dichlorinated derivatives[59]. The reaction is clearly a radical process and the relative proportions of the isomers formed shows the following increase in the reactivity of the hydrogen atoms: primary < secondary < tertiary. The same order of reactivity has been observed for the chlorination of pentane, hexane and heptane under similar conditions[60]. The relative rate of substitution of the secondary hydrogen compared with that for the primary decreases with increase in temperature. The chlorine atoms attack the C—H bond with the highest electron availability, the strength of the C—H bond is not important. Iodine monochloride in daylight has also been used as the source of chlorine atoms, and relative rates of hydrogen abstraction (per available hydrogen) from methane, ethane, propane (primary) and propane (secondary) are 2.0, 100, 110 and 480 respectively[61]. The reactions required to explain the results are given in equations (1.5–1.8). Chloramine, under the influence of u.v. radiation or heat also gives

$$ICl + h\nu \rightarrow I^{\bullet} + Cl^{\bullet} \qquad (1.5)$$

$$Cl^{\bullet} + RH \rightarrow R^{\bullet} + HCl \qquad (1.6)$$

$$Cl^{\bullet} + ICl \rightarrow Cl_2 + I^{\bullet} \qquad (1.7)$$

$$R^{\bullet} + ICl \rightarrow RCl + I^{\bullet} \qquad (1.8)$$

chlorinated derivatives of isobutane and neopentane[62]. The reaction proceeds as given by equations (1.9–1.12), the amino radicals having a strong tendency to abstract hydrogen from the alkane. The major product (3) from isobutane is that resulting from substitution at the primary carbon atom, and the amount of tertiary carbon substitution product (4) is c. 25% of the total. There are no chloroaminated products.

$$NH_2Cl + h\nu \rightarrow NH_2^{\bullet} + Cl^{\bullet} \qquad 1.9)$$

$$RH + NH_2^{\bullet} \rightarrow R^{\bullet} + NH_3 \qquad (1.10)$$

$$R^{\bullet} + NH_2Cl \rightarrow RCl + NH_2^{\bullet} \qquad (1.11)$$

$$\text{or} \quad R^{\bullet} + Cl^{\bullet} \rightarrow RCl \qquad (1.12)$$

Sulphochlorination of n-butane with chlorine and sulphur dioxide under u.v. irradiation gives a complex mixture; both mono- and di-sulphonyl chlorides, and mono- and poly-chloro compounds are produced as well as mixed

$$Me_2 \cdot CH \cdot CH_2Cl \qquad\qquad\qquad Me_3CCl$$
$$(3) \qquad\qquad\qquad\qquad (4)$$

chlorination and sulphochlorination products[63]. Increase in the butane concentration, at a constant ratio of chlorine and sulphur dioxide concentration, favours sulphochlorination. Photodifluoroamination of ethane and n-butane by N_2F_4 gives, for ethane, both ethyldifluoroamine and acetonitrile, and for n-butane both the terminal product, NN-difluoro-n-butylamine and the product of substitution at the secondary carbon atom, NN-difluoro-s-butylamine[64]. The reaction proceeds as in equations (1.13–1.17). Earlier in

$$N_2F_4 \rightarrow 2NF_2 \qquad (1.13)$$

$$NF_2 + h\nu \rightarrow F^{\bullet} + NF^{\bullet} \qquad (1.14)$$

$$RH + F^{\bullet} \rightarrow HF + R^{\bullet} \qquad (1.15)$$

$$R^{\bullet} + NF_2 \rightarrow RNF_2 \qquad (1.16)$$

$$R^{\bullet} + N_2F_4 \rightarrow RNF_2 + NF_2 \qquad (1.17)$$

this section it is reported that secondary carbon–hydrogen bonds are more reactive than primary in photo-oxidation (by chlorine). Photo-oxidation of methyl groups by a mixture of chlorine and nitric oxide has been studied by using 2,2,3,3-tetramethylbutane, as this contains only primary carbon–hydrogen bonds. In carbon tetrachloride solution the major product is the aldehyde oxime hydrochloride (5) at low hydrocarbon concentrations, and the aldehyde oxime at high hydrocarbon concentrations[65]. With benzene as the solvent the product yields are considerably reduced.

$$EtCH:NOH,HCl$$

(5)

cis- and trans-Hexadec-2-ene are produced in the radiolysis of n-hexadecane by γ-rays from ^{60}Co [66]. This is in contrast to the alkenic product formed from polyethylene under similar conditions, when only the corresponding trans-isomer was reported. It is found that cis-hexadec-2-ene reacts very rapidly with air (the concentration was below the limit detectable by g.l.c. after c. 12 h) and it is suggested that the cis-product formed from polyethylene might have been lost by aerial oxidation. In addition to the hexadec-2-ene produced, there are a large number of linear (probably terminal) alkenes produced by scission of C—C bonds. These account for 20–25% of the product. The reactions in an electrical discharge of ethane and propane, in the presence of argon or helium, produce much hydrogen, simple hydrocarbons and polymeric products[67]. The identified simple hydrocarbons include ethane, ethylene, acetylene, propane, propene, butane, pentane and hexane. Although these complex reactions are similar to those taking place in photolysis and radiolysis, for example propene is formed in highest yield from propane, there are significant differences with ethane as the reactant. This also gives high yields of propene.

1.3.2.2 Electrophilic substitutions

The past 2 years have seen the beginnings of significant developments in the activation of alkanes whereby they can undergo non-radical reactions. This 'activation' can be achieved by superacids[68-71, 406] or by transition-metal complexes[72-77]: the first of these is discussed here, the second in Section 1.3.2.4.

It has been realised for some time that electrophilic reactions of hydrocarbons (alkenes, alkynes, and aromatic compounds) involve hydrocarbon ions in which the carbon is trivalent (formerly called carbonium ions but preferably called carbenium ions as suggested by Olah[71]). The reaction depends on the ability of the hydrocarbon to act as a π-donor to the electrophile. The donor ability of unshared electron pairs of heteroatoms gives rise to a second group of electrophilic reactions. The direct observation of alkylcarbenium ions has only recently become possible[78, 79] and has led to the development of a much wider concept of hydrocarbon cations (the suggested general name being carbocations[71]), there being both carbenium ions and hydrocarbon ions in which the carbon is pentavalent (the suggested name being carbonium ions[71]). Study of these carbonium ions has resolved the long-standing controversy concerning classical and non-classical carbenium ions and they are the key to electrophilic reactions of alkanes. Covalent C—C and C—H bonds exhibit electrophilic reactivity by a third major type of electron-donor ability, namely the donation of the shared electron-pairs of single bonds to form two-electron three-centre bonds with the electrophile. This σ-basicity is shown by primary, secondary, and tertiary C—C and C—H bonds and its recognition opens up a new area of chemistry in which alkanes can undergo a wide variety of electrophilic substitutions. Acid-catalysed fragmentation, isomerisation, and alkylation are known to involve trivalent carbenium ions, the mechanism involving intermolecular hydride transfer to a carbenium ion and rearrangement of hydrogen atoms or alkyl groups in the carbenium ion so produced. The understanding of hydrogen–deuterium exchange in hydrogen gas[80] and alkanes[68] in the superacids DF–SbF$_5$ or FSO$_3$D–SbF$_5$ mixtures, follows from the realisation that protolytic attack takes place at the H—H, C—H, or C—C bond, (6) and (7), and not at the hydrogen or carbon atoms themselves. Reaction of isobutane with the deuteriated superacids DSO$_3$F–SbF$_5$ or DF–SbF$_5$ at $-78\,°C$ and atmospheric pressure, illustrates the reactions involved (Figure 1.4).

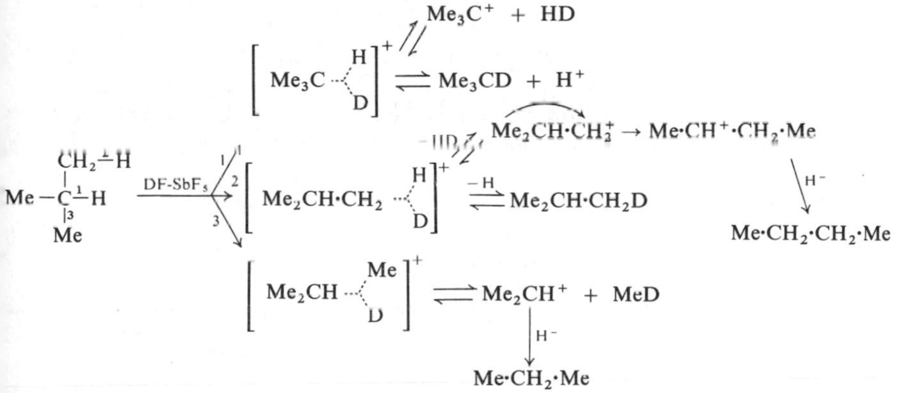

Figure 1.4 Hydrogen–deuterium exchange in isobutane in DF–SbF$_5$ solution (From Olah, Halpern, Shen, and Mo[68], by courtesy of the American Chemical Society)

There is substantial exchange of the methine proton (path 1 in Figure 1.4), and the t-butyl cation that is formed by the same route is a relatively

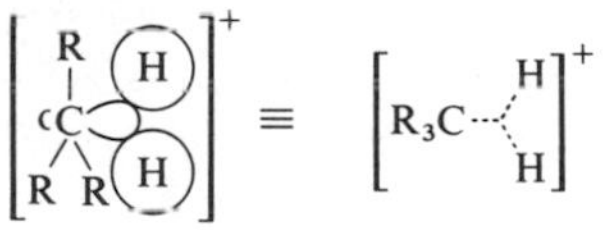

(6)   (7)

stable species. There is negligible exchange in the methyl group (path 2 in Figure 1.4) but this path allows some isomerisation to n-butane. Cleavage of the C—C bond to give methane and propane also occurs (path 3 in Figure 1.4). Cleavage of the C—C bond in ethane takes preference over C—H bond cleavage, as shown by the product (CH$_4$:H$_2$) ratio of 8:1. *Ab initio* molecular orbital calculations discussed earlier[4] give the structure of C$_2$H$_5^+$ of lowest energy as that in which the proton bridges the C—C bond (Figure 1.1(e)). Polymerisation can also occur. For example, when methane is the reactant, the reactions shown in equation (1.18) occur, leading to higher homologues. The reactivity order of the bonds in alkanes is tertiary C—H > C—C >

secondary C—H > primary C—H, but this order may be affected by steric

$$CH_4 + H^+ \rightleftharpoons H_2 + Me^+ \xrightarrow{CH_4} C_2H_7^+ \rightleftharpoons H_2 + C_2H_5^+ \xrightarrow{CH_4} C_4H_9^+ \text{ etc.} \qquad (1.18)$$

hindrance. The alkylation (methylation of methane) just discussed can be extended to alkylation in general by treating stable alkylcarbenium hexa-fluoroantimonates with alkanes in SO_2ClF solution[69]. For example 2,2,3-trimethylbutane (8) is formed in up to 12% yield when isobutane is treated with isopropyl hexafluoroantimonate, or when propane is treated with t-butyl hexafluoroantimonate. In the latter case intermolecular hydrogen transfer occurs, and it is generally much faster than alkylation, and the product is formed by propylation of isobutane (equations (1.19) and (1.20)).

$$Me_2CH_2 + \overset{+}{C}Me_3 \rightleftharpoons Me_2CH^+ + HCMe_3 \qquad (1.19)$$

$$Me_3CH + H\overset{+}{C}Me_2 \rightleftharpoons Me_3C \cdot CHMe_2 + H^+ \qquad (1.20)$$
$$(8)$$

Since such intermolecular hydrogen transfers will almost always occur, products of alkylation reactions are often complex. In a similar manner the first electrophilic aliphatic nitration has been achieved by using stable nitronium salts. The hexafluorophosphate ($NO_2^+ PF_6^-$) or hexafluoroanti-monate ($NO_2^+ SbF_6^-$) or tetrafluoroborate ($NO_2^+ BF_4^-$) have been used in methylene chloride–sulpholane as solvent[70]. The alkanes studied and

Table 1.5   Nitration and nitronium ion cleavage of alkanes with $NO_2^+PF_6^-$ in CH_2Cl_2–sulpholane at 25°C
(From Olah and Lin[70], by courtesy of the American Chemical Society)

Alkane	Nitroalkane products and their molar ratios
CH_4	$MeNO_2$
C_2H_6	$MeNO_2 > EtNO_2$ (2.9:1)
C_3H_8	$MeNO_2 > EtNO_2 > Pr^iNO_2 > Pr^nNO_2$ (2.8:1:0.5:0.1)
iso-C_4H_{10}	$Bu^tNO_2 > MeNO_2$ (3:1)
n-C_4H_{10}	$MeNO_2 > EtNO_2 > Bu^nNO_2 \approx Bu^tNO_2$ (5:4:1.5:1)
neo-C_5H_{12}	$MeNO_2 > Bu^tNO_2$ (3.3:1)

Figure 1.5   Nitration and nitronium ion cleavage of propane by $NO_2^+PF_6^-$ in methylene chloride–sulpholane solution at 25 °C
(From Olah and Lin[70], by courtesy of the American Chemical Society)

the products are given in Table 1.5; these products result from reactions as illustrated for propane in Figure 1.5.

Tertiary C—H bonds show the highest reactivity, and C—C bonds are usually more reactive than secondary or primary C—H bonds, giving preferential nitronium-ion cleavage of n-alkanes (Table 1.5).

Anodic oxidation of alkanes in anhydrous FSO_3H solution gives α,β-unsaturated ketones[407]. Skeletal rearrangement takes place and unidentified ketones have been prepared from n-hexane, n-heptane, n-pentane and isopentane.

1.3.2.3 Reactions in a plasma jet

There is continuing interest in reactions of alkanes in a plasma jet at temperatures up to 15 000 K. The major product from these reactions is acetylene. With methane as the reactant[81–84], work has concentrated on optimising the acetylene yield; increase in the temperature of the jet increases the yield, 80 % of the methane being converted into acetylene in an argon plasma jet[81, 82] at 1800 K, or in a hydrogen[81], or methane[83] plasma jet, whilst at lower temperatures vinylacetylene, but-1-ene, and buta-1,3-diene are also formed[81]. At higher temperatures the acetylene decomposes. A theoretical study supports the experimental observations[84]. For propane in a hydrogen jet at 3000–15 000 K the products are acetylene, ethylene, and propene[85], and methane[86]. The same products are obtained from a propane–butane mixture[87] and hexane (and other hydrocarbons) also give acetylene as the major reaction product[88].

1.3.2.4 Homogeneous catalysis

There are extensive reports of reactions of organic compounds catalysed by complexes of transition elements in solution; in these the organic compounds are invariably alkenes, alkynes, or aromatic compounds. These are able to form π-bonds with the transition metal. In contrast, alkanes, which have no π-electrons, have not been studied. For example, the recent review by Davidson[1], in which over 400 papers are surveyed, does not include a single example in which an alkane is the reactant (except for Ref. 73, when the significant feature, that the reactant is an alkane, is omitted). Alkanes do interact with platinum(II) in solution, and hydrogen–deuterium exchange[72–77] in a large range of alkanes[73, 74, 76, 77] has been reported. The mechanism of this alkane activation is not yet fully understood, but a correlation of the rate of exchange of the alkane or of an aromatic compound with its ionisation potential[73, 74] and a similar correlation for halogen-substituted hydrocarbons[75] (Figure 1.6) supports the suggestion that coordination of an electron-pair from the alkane to the platinum(II) is the rate-determining step (A in Figure 1.7). Subsequent steps in the reaction involve oxidative addition of the alkane to the platinum (B in Figure 1.7), and loss of HCl to and gain of DCl from the $CH_3 \cdot CO_2D–D_2O$ solvent (C in Figure 1.7). The order of reactivity of the carbon–hydrogen bonds is primary > secondary > tertiary, and branched-chain alkanes are less reactive than the normal isomers because of steric effects. The rates for a range of alkanes are shown on the

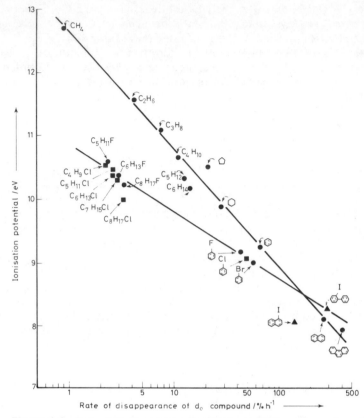

Figure 1.6 Rate of H–D exchange in alkanes, aromatic compounds and halogen derivatives plotted against ionisation potential

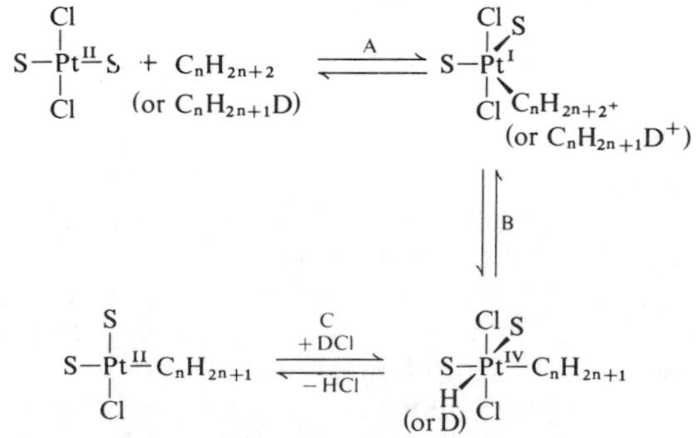

Figure 1.7 Reaction scheme for hydrogen–deuterium exchange in alkanes, catalysed by Pt^{II} in $CH_3 \cdot CO_2D$–D_2O solution at 80–100 °C
(From Hodges, Webster and Wells[74], by courtesy of The Chemical Society)

abscissa of Figure 1.6. Platinum complexes with ligands of low *trans*-effect alone are catalysts for this reaction. Replacement of the chloride ligands of $PtCl_4^{2-}$ by solvent molecules increases the activity[76] but various phosphine complexes have been shown to be inactive[75]. Perhaps in the next 2 years the extension of transition-metal catalysis to other alkane reactions will be revealed, and so will open up a new area of alkane chemistry.

1.3.2.5 Enzyme catalysed oxidations

There is continuing interest and activity in the oxidation of alkanes (and alkenes) by enzymes in micro-organisms (yeasts, fungi, and bacteria). An excellent review[89] covers many aspects of these oxidations, particularly the developments of the last 6 years. Studies on a variety of new micro-organisms include a marine bacterium[90], a new variant of the much studied *Candida* yeast[91], and other yeasts[92, 93]. The broad features of the oxidation, namely initial hydroxylation and subsequent further oxidations are well known; the detailed mechanism is much less certain. The fatty acids produced become esterified inside the cell, and there is some interest in carrying out the oxidation of alkanes with the active enzymes as cell-free extracts[94-96]. Both bacterial[94] and yeast[95, 96] extracts have been used; molecular oxygen is always the oxygen source. The active enzymes have been partially identified and are apparently located in the cell membrane. A detailed discussion of the economics of producing fatty acids from n-alkanes[97] concludes that it may not be possible to produce even the shorter-chain fatty acids (i.e. the most expensive) at a commercially viable price, but the much more expensive cocoa butter might one day be replaced by a fat produced by yeast. The range of fatty acids produced by different micro-organisms from long-chain n-alkanes is also discussed[97, 98] and the fatty acids present in the triglycerides of yeasts fed on n-alkanes have been determined[99]. Further studies on the bacterial oxidation of smaller n-alkanes (ethane, propane and n-butane) have been reported[100, 101].

1.4 ALKENES

1.4.1 Preparation

A new general synthesis of tetrasubstituted ethylenes from aliphatic nitro-compounds gives high yields of pure products[102]. Aliphatic dinitro-compounds react with sodium sulphide in NN-dimethylformamide under fluorescent light to give alkenes by a radical-anion chain mechanism (equation (1.21)). The dinitro-derivatives are readily prepared from their mononitro-analogues

$$\begin{array}{ccc} \underset{R^2}{\overset{R^1}{>}}C\underset{NO_2}{\overset{H}{<}} & \longrightarrow & \underset{R^2}{\overset{R^1}{>}}C\underset{O_2N}{-}\underset{NO_2}{C}\overset{R^1}{\underset{R^2}{<}} & \longrightarrow & \underset{R^2}{\overset{R^1}{>}}C:C\overset{R^1}{\underset{R^2}{<}} \end{array} \qquad (1.21)$$

by treatment with lithium and bromine. Typically 3,4-dimethylhex-3-ene is formed in 82% yield. The unsymmetrical alkenes (equation (1.22)) cannot be prepared pure as symmetrical compounds are also formed. Alk-1-enes are formed when primary alcohols are dehydrated by $N^1N^2N^3$-hexamethyl-

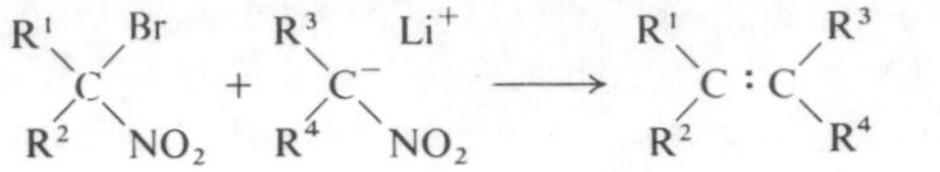

$$\begin{array}{c}\overset{R^1}{\underset{R^2}{>}}\!\!C\!\!\overset{Br}{\underset{NO_2}{<}} + \overset{R^3}{\underset{R^4}{>}}\!\!C^{-}\!\!\overset{Li^+}{\underset{NO_2}{<}} \longrightarrow \overset{R^1}{\underset{R^2}{>}}\!\!C:C\!\!\overset{R^3}{\underset{R^4}{<}}\end{array} \qquad (1.22)$$

phosphoric triamide[103]: octan-1-ol gives oct-1-ene together with an approximately equal amount of 1-dimethylamino-octane. Isoprene may be prepared by treating isobutene with formaldehyde[104]. The yield is improved by first converting the CH_2O into $ClCH_2OMe$ by reaction with methanolic hydrogen chloride. The 3-chloro-3-methylbutyl methyl ether then formed is efficiently decomposed thermally to give isoprene, methanol and hydrogen

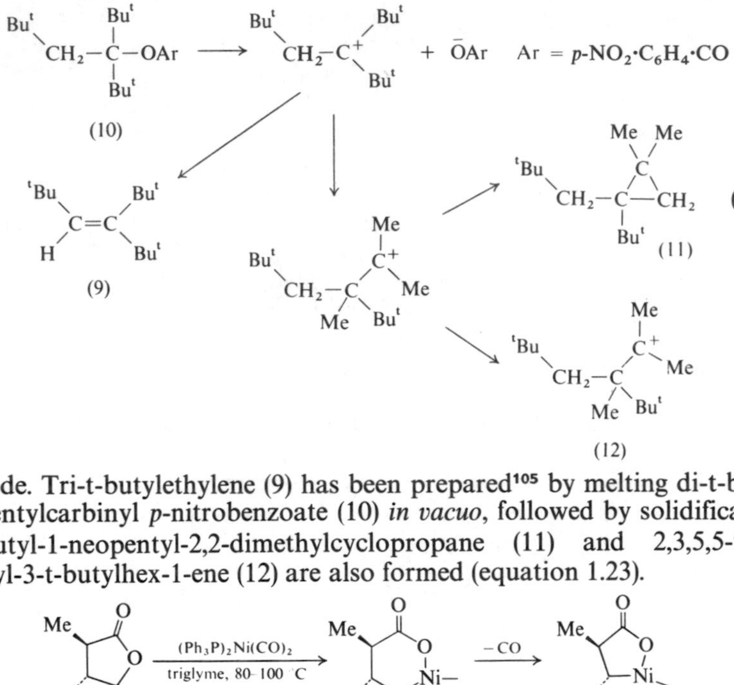

$$(1.23)$$

chloride. Tri-t-butylethylene (9) has been prepared[105] by melting di-t-butyl-neopentylcarbinyl p-nitrobenzoate (10) in vacuo, followed by solidification. 1-t-Butyl-1-neopentyl-2,2-dimethylcyclopropane (11) and 2,3,5,5-tetra-methyl-3-t-butylhex-1-ene (12) are also formed (equation 1.23).

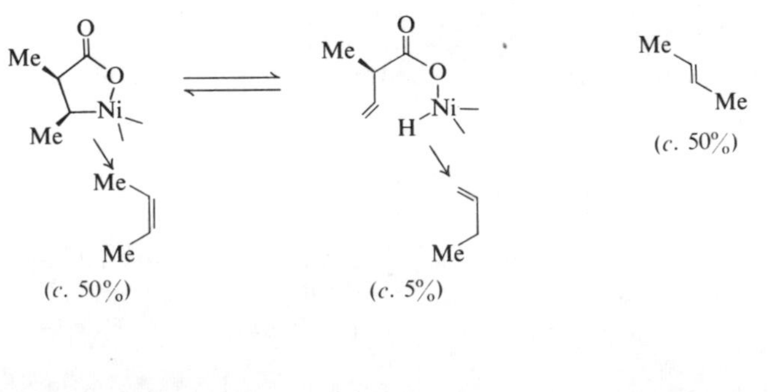

$$(1.24)$$

Magnesium bromide with a little magnesium amalgam will reduce epoxides to alkenes[106], by a reaction analogous to the well-known reductive elimination reactions of β-substituted alkyl halides. A new method of converting dicarboxylic acid anhydrides into alkenes involves the use of nickel, iron, or rhodium complexes[107]. If the anhydride has abstractable β-hydrogens, e.g. 2,3-dimethylsuccinic anhydride (13), then more than one product is formed (equation (1.24)). Thioanhydrides may also be used. Stereoselective and stereospecific alkene syntheses have been reviewed[108].

1.4.2 Reactions

1.4.2.1 Isomerisation

Double-bond migrations in alkenes may be catalysed by acids or bases, metals, metal complexes and boron compounds. They may be photochemically induced or initiated in other ways. All aspects of such alkene rearrangements are discussed in three excellent reviews[109–111]. The cis–trans isomerisation of cis-but-2-ene catalysed by NO_2 [112], also of trans-but-2-ene and but-1-ene in $(Me_2N)_3PO$ solution in the presence of EtLi [113], and of long-chain alkenes in dimethyl sulphoxide containing potassium t butoxide[114] have been studied and reaction mechanisms have been proposed. The base-catalysed isomerisation of trans-oct-4-ene (14) gives cis-oct-3-ene (15), and that of cis-oct-4-ene (16) gives trans-oct-3-ene (17)[115]. This is explained by the relative stabilities of the anions formed during the isomerisation. N.M.R. spectra show that the trans-1,cis-3 (18) (or cis-1, trans-3 (19)) anion is more stable than the trans-1,trans-3 (20) or cis-1,cis-3 (21) anion (equations (1.25) and (1.26)).

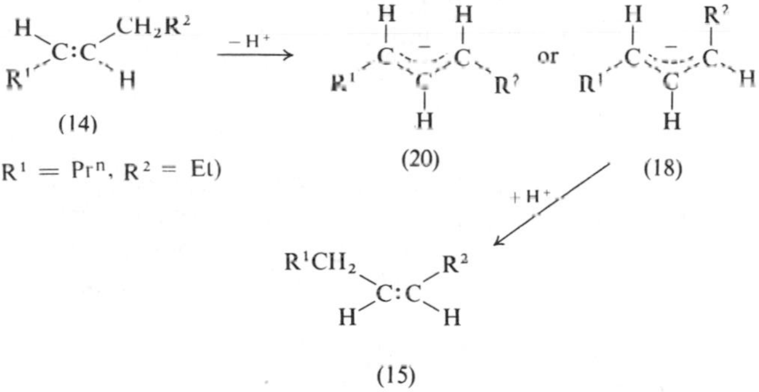

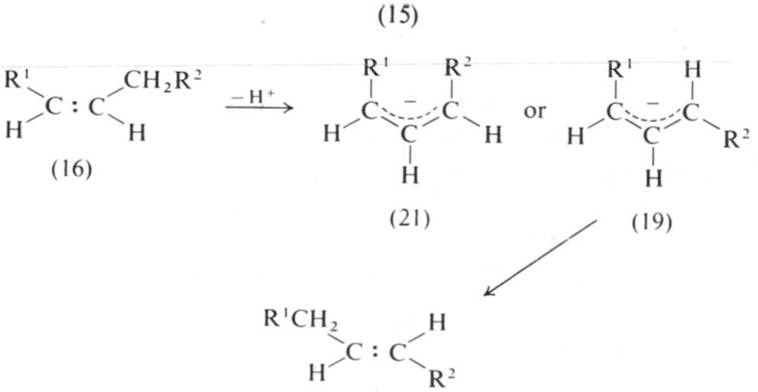

$$(1.25)$$

$$(1.26)$$

There are continuing reports of isomerisations of alkenes catalysed by solutions of transition-metal complexes. The mechanisms of these reactions are of two types. They may involve alkyl intermediates, with hydrogen addition to the olefin followed by hydrogen abstraction, and require a metal hydride as the catalyst, or they may involve allylic intermediates. One or both of these routes is available to complexes of all of the Group VIII and of several other transition metals. Recent papers discuss isomerisation by Ru, Rh and Pd chlorides[116], $RhH(CO)(PPh_3)_2$ [117], $RhCl(CO)(PPh_3)_2$ [118], $CoN_2(PPh_3)_3$ [119, 120], $RhCl(PPh_3)_3$ [121], $RuCl_2(PPh_3)_3$ [122, 123], $RuHCl(PPh_3)_3$[123] $Ni[P(OEt)_3]_4$ [124], $OsHCl(CO)(PPh_3)_3$ [124], $FeCl_3–AlEt_3$ [125], $CoCl_2–AlEt_3$ [125], $NiCl_2–AlEt_3$ [125], cyclo-octa-1,5-diene $W(CO)_4$ [126], and $Ni[PPh_2(CH_2)_4PPh_2]_2$ + HCN [127]. $RuCl_3$ is reported to be reduced to Ru^{II} before catalytic activity develops[116], in agreement with studies using other ruthenium complexes where the Ru^{II} hydride[123] and carbonyl hydride (formed by added hydroperoxide[122]) complexes are the catalysts. Catalysis by rhodium complexes also occurs by the alkyl route; kinetic studies[117] and the necessity of a hydrogen atmosphere[118] provide added proof. For the cobalt–nitrogen complex an intermolecular hydrogen transfer at a dimeric cobalt complex is suggested as the prime route for isomerisation[119]. This is essentially the alkyl route and is supported by the extensive hydrogen–deuterium scrambling that occurs when a deuteriated olefin is used as the substrate[128]. With the nickel complex, trifluoroacetic acid must be present, and it has been shown that the $NiH[P(OEt)_3]_4^+$ cation is the catalyst[124]. For this and for the osmium complex[124], alkyl intermediates are involved in alkene isomerisation. Organo-aluminium compounds[129] (and benzophenone[203]) have also been used as photosensitisers for the initiation by u.v. light or γ-radiation of the radical cis–trans isomerisation of alkenes.

1.4.2.2   Disproportionation

Theoretical studies of the application of the Woodward–Hoffmann orbital symmetry rules to alkene disproportionation (alkene metathesis) (equation (1.2)) have been reviewed in Section 1.2.2. One proposed intermediate, a multicentred complex with four methylenic units[54], has received experimental support from the study of alkene metathesis using the homogeneous catalysts $W(CO)_6$, $Mo(CO)_6$ and arene–$W(CO)_3$ [130]. Related homogeneous catalysts, formed by the reaction of nitrosyl-tungsten or -molybdenum complexes with organoaluminium compounds are very effective metathesis catalysts[131, 132]. A mixture of $MoCl_2(NO)_2(PPh_3)_2$ and $Me_3Al_2Cl_3$ in chlorobenzene will convert pent-1-ene into ethylene and oct-4-ene at 0 °C, and give deca-1,5,9-triene from ethylene and cyclo-octa-1,5-diene (equation (1.27)). In contrast, the catalyst based on WCl_6 will cause disproportionation of internal alkenes only[131]. It is found that $Me_3Al_2Cl_3$ gives a more active catalyst than either $Et_3Al_2Cl_3$ or $EtAlCl_2$ [132].

$$\begin{array}{c} CH_2 \\ \parallel \\ CH_2 \end{array} + \begin{array}{c} CH \\ \parallel \\ CH \end{array} \bigcirc \longrightarrow \begin{array}{c} CH_2{=}CH \\ CH_2{=}CH \end{array} \bigcirc \qquad (1.27)$$

A homogeneous catalyst prepared by treating $ReCl_5$ with Bu_4^nSn has been used to study the metathesis of pent-2-ene to but-2-ene and hex-3-ene[133], and a catalyst prepared from Wpy_2Cl_4 (py = pyridine) and $EtAlCl_2$ is found to increase in metathetical activity when carbon monoxide is added to the system[134].

1.4.2.3 Reduction

The reduction of alkenes to alkanes may be carried out with a conventional inorganic or organic reducing agent, or with hydrogen gas and a transition-metal complex as a catalyst. An efficient inorganic reducing agent is sodium hexamethylphosphoramide in t-butyl alcohol[135]: this will reduce even tetra-alkylethylenes. At room temperature, hex-1-ene and trans-hex-3-ene are converted into n-hexane in $> 98\%$ yield after 6 h.

Reduction using transition-metal complexes as catalysts is the subject of two reviews[136, 137] and recent work is also reviewed in another part of this series[1]. Further studies on the reduction of alkenes catalysed by Rh-(CO)-$(PPh_3)_3$ have shown that the observed gradual deactivation of the catalyst is due to dimerisation leading to an inactive species[138]. Interesting new catalysts for the hydrogenation of alkenes include $CoH_3(PPh_3)_3$ [139], $[Ru(C_6H_6)-Cl_2]_n$ [140], $IrCl(PPh_3)_2$ [141], Ru and Rh acetate[142, 143], and 1-methallyldicyclo-pentadienyltitanium(III)[144]. The iridium complex, $IrCl(PPh_3)_2$, is formed in situ from added $IrCl(PPh_3)_3$, and is about 10 times more active, for the reduction of hex-1-ene, than the corresponding complex of rhodium. Rhodium and ruthenium acetate appear to be active catalysts in both neutral[143] and acidic solution[142], in which they exist in a protonated form. In methanol, $Rh_2(OAc)_4$ or $Ru_2(OAc)_4$ with added HBF_4 and PPh_3, readily catalyse the reduction of hex-1-ene, hex-2-ene, cis-hept-2-ene, cis-pent-2-ene, hexa-1,5-diene and hex-1-yne[142]. The same two complexes in ethanol, NN-dimethylformamide, NN-dimethylacetamide, tetrahydrofuran, dioxan or dimethyl sulphoxide, without added acid or phosphine, also catalyse the reduction of hex-1-ene, dec-1-ene, and other alkenes[143]. When 1-methallyl-dicyclopentadienyltitanium(III) is used as the catalyst for the reduction of alk-1-enes, a rapid isomerisation to alk-2-enes takes place, followed by the hydrogenation[144]. If homogeneous catalysts are used for alkene reduction, a serious technical problem is the separation of the alkane produced from the catalyst. One way of overcoming the problem is to support the catalyst, in a manner similar to that used for heterogeneous catalysts. The complex $RhCl(PPh_3)_3$ impregnated on polystyrene beads, and rhodium complexes prepared from phosphines that have been polymerised into the polystyrene have both been successfully used for alkene reduction[145]. Similarly $PdCl_2$ supported on an ion-exchange resin is an active hydrogenation catalyst[146]. It is probable that it is in this way that homogeneous catalysts will be industrially useful for reactions of alkenes, and we may expect an increasing research effort in this field in the next few years.

Alkenes and alkynes can also be reduced by formic acid and alkali metal formates, in the presence of transition-metal complexes[147]. It is likely that, as in all the reactions discussed above, a transition-metal hydride is the key intermediate, the difference here being that formic acid, rather than hydrogen gas, is the hydrogen source. By using this system at 100 °C with a variety of Group VIII-metal complexes, oct-1-ene, oct-4-ene, hept-1-ene and hex-3-yne have all been reduced.

1.4.2.4   Oligomerisation and polymerisation

Alkene polymerisation giving polyalkenes is, of course, of considerable industrial importance. Except for reactions at high pressure, such poly-merisations are catalysed reactions, the best known catalysts being those of the Ziegler–Natta type. Not only polymers, however, are of industrial value, but dimers and other low oligomers of alkenes have numerous industrial applications. The high selectivity which can be obtained in polymerisation by Ziegler–Natta catalysts is also of interest in dimerisation reactions[148].

Oligomerisation of ethylene, propene, n-butenes and mixtures of these alkenes, catalysed by Pd^{II} complexes gives a mixture of products (equations (1.28–1.30))[149].

$$C_2H_4 \xrightarrow{\text{(PhCN)}_2\text{PdCl}_2,\ 100\,°\text{C, 24 h, 55 atm}} C_4H_8(90\%) + C_6H_{12}(10\%) + C_8H_{16}(<1\%) \quad (1.28)$$

$$C_3H_6 \xrightarrow{\text{(PhCN)}_2\text{PdCl}_2,\ 100\,°\text{C, 24 h, 55 atm. (35\% conversion)}} n\text{-}C_6H_{12}(92\%) + MeC_5H_9(8\%) \quad (1.29)$$

$$C_2H_4 + C_3H_6 \xrightarrow{\text{(PhCN)}_2\text{PdCl}_2,\ 100\,°\text{C, 72 h, 55 atm.}} C_4H_8(61\%) + C_5H_{10}(28\%) + C_6H_{12}(61\%) + \\ C_7H_{14}(1\%) \quad (1.30)$$

(Product yields in mol %)

By using mixtures of an aluminium halide with nickel, cobalt or iron di-t-butylbenzoates, ethylene and propene are oligomerised mainly to butenes and hexenes[150]. The catalytic effect increases in the order Ni < Co < Fe, and is always enhanced by added tertiary phosphines. Aluminium alkyls with nickel β-diketonates give 75–85 % selectivity to linear dimerisation products from alk-1-enes[151]. The high yield of hex-2-ene in a typical hexene mixture from the dimerisation of propene (Table 1.6) shows the high selectivity for the formation of linear dimers by head-to-head addition of propene, pre-

Table 1.6   Hexenes from propene dimerisation
(From Jones and Symes[151], by courtesy of the Chemical Society)

	Yield/%		Yield/%
Hex-1-ene	4.2	cis-4-Methylpent-2-ene	2.0
cis-Hex-2-ene	32.7	trans-4-Methylpent-2-ene	1.8
trans-Hex-2-ene	28.9	2-Methylpent-2-ene	10.7
cis-Hex-3-ene	6.0	2-Methylpent-1-ene	5.6
trans-Hex-3-ene	5.8	2,3-Dimethylbut-1-ene	0.5
4-Methylpent-1-ene	1.7	2,3-Dimethylbut-2-ene	0.1

sumably by an insertion type of mechanism. Other highly selective dimerisa-tion catalysts based on nickel are nickel oleate–di-isobutylaluminium chloride[152, 153], NiBr·R(PPh₃)₂ where R = 1-naphthyl or o-tolyl[154], $NiCl_3$

(PPr_3^i) [155], nickel halide–aluminium alkyl[156], and $Ni(hal)_2(PPh_3)_2$ with a Grignard reagent[157]. The Nioleate–Bu_2^iAlCl system[153] gives high yields of n-pentenes from the co-dimerisation of ethylene and propene. However, if a tertiary phosphine is added then methylbutenes are the preferred products; the electron-donor phosphine increases the degree of ionic character of the Ni—C σ-bond in the intermediate and promotes Markovnikov addition. The nickel halide–aluminium alkyl system[156] and related cobalt halide catalysts are also highly selective in giving dimers as the products. The suggested mechanism, which explains the predominance of 4-methylpent-2-ene when propene is dimerised, is given in equation (1.31), where $Cat^+ An^-$ represents the (undefined) cationic and anionic parts of the catalyst.

$$Cat^+An^- + \overset{\delta^-}{CH_2}=\overset{\delta^+}{CH\cdot Me} \longrightarrow Cat-CH_2\cdot CH^+\cdot Me \quad An^-$$

$$\xrightarrow{CH_2=CH\cdot Me} \underset{Me}{\overset{Cat-CH_2}{\diagdown}}CH\cdot CH_2\cdot \overset{+}{CH}\cdot Me \longrightarrow \left[\underset{Me}{\overset{Cat-CH_2}{\diagdown}}CH\cdot CH:CH\cdot Me + H^+An^-\right]$$

$$\longrightarrow Me_2CH\cdot CH:CH\cdot Me + Cat^+An^- \qquad (1.31)$$

Higher oligomers (polymers) can be prepared if other catalysts are used. A catalyst of $(EtO)_3TiCl + EtAlCl_2$ in toluene at $-20°C$ and 12 atm. pressure of ethylene gives C_6 to C_{40} alkenes with predominantly vinylic end groups[158]; and C_7 to C_{21} alkenes are obtained from hex-1-ene, hept-1-ene, and oct-1-ene by using a WCl_6–$EtAlCl_2$ catalyst[159]. The coupling step in these reactions involves alkene insertion into metal–alkyl bonds. Such reactions have been studied for compounds of Et_2AlCl or Et_3Al with V [160], Fe, Co or Ni salts[161]. The results indicate that for the metal alkenyls formed the stability decreases as follows: V $\gg$ Ni > Co > Fe [161]. Polyketones $(H(CH_2\cdot CH_2\cdot CO)_nR, R = Et,$ MeO and EtO) are produced by the co-oligomerisation of ethylene and carbon monoxide[162]. For example, solutions of rhodium oxide in 1:1 acetic acid–methanol at 130 °C for 24 h give octane-3,6-dione (26%), undecane-3,6,9-trione (6%), tetradecane-3,6,9,12-tetraone (1%), and methyl homolaevulinate (15%), together with higher homologues.

A review of polymers containing double bonds has recently appeared[163].

1.4.2.5 Reactions with organic compounds

The base-catalysed alkylations of alkylpyridines by alkenes give normal addition products for the 2- and 4-alkylpyridines[164], but, because of the high electrophilicity of the 2-position of the pyridine ring, a novel cyclo-alkylation product (22) is obtained when 3-ethyl- or 3-s-butyl-pyridine reacts with ethylene in the presence of sodium or potassium[165]. The normal alkylation product, 3-s-butylpyridine (23) is formed together with 6,7-dihydro-7,7-

(22)   (23)   (24)

diethyl-5-methyl-5H-1-pyrindine (22). With longer reaction times further ethylation occurs to give 6,7-dihydro-5,7,7-triethyl-5-methyl-5H-1-pyrindine (24).

The acid-catalysed alkylation of o-, m- or p-cymene with isobutene gives a complex mixture of over thirty products, as polymerisation, rearrangement, and fragmentation all occur[166]. Cyclopropanes are formed from alkenes, in more controlled reactions than with carbenes, if lithium carbenoids are used[167]. At $-50\,°C$ in pentane, $LiCH_2Br$ reacts with oct-1-ene to give n-hexylcyclopropane, the intermediate having a 'butterfly' structure (25).

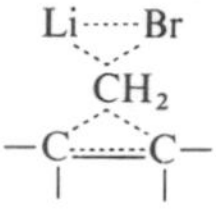

(25)

Tetrafluorodehydrobenzene (26) reacts with alkenes as shown in equation (1.32)[168]. The 1-(2,3,4,5-tetrafluorophenyl)hex-2-ene (27) is predominantly the *trans*-isomer.

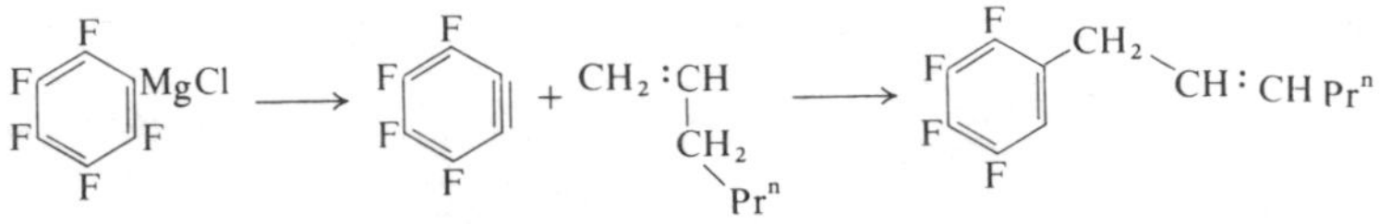

(1.32)

(26)   (27)

The acid-catalysed reaction of alk-1-enes with formaldehyde can give a number of condensation products[169, 170]. Typically hex-1-ene will give 4-butyl-1,3-dioxan (28) and 3-propyltetrahydropyran-4-ol (29) in approximately equal yields, when the reaction is carried out at high temperatures (Prins reaction) (equation (1.33))[169] but at low temperatures, in the presence

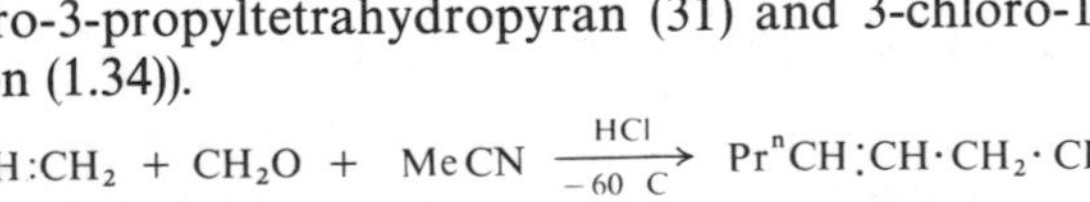

$$Bu^nCH{:}CH_2 \;+\; 2HCHO \xrightarrow[175\ C]{HCl}$$

(1.33)

(28)   (29)

of acetonitrile homoallylic acetates are formed[170]. Hex-1-ene, for example, gives *cis*- and *trans*-hept-3-enyl acetate (30) and small amounts of *cis*- and *trans*-4-chloro-3-propyltetrahydropyran (31) and 3-chloro-1-heptyl acetate (32) (equation (1.34)).

$$Bu^nCH{:}CH_2 \;+\; CH_2O \;+\; MeCN \xrightarrow[-60\ C]{HCl} Pr^nCH{:}CH{\cdot}CH_2{\cdot}CH_2{\cdot}O{\cdot}Ac$$

(30)

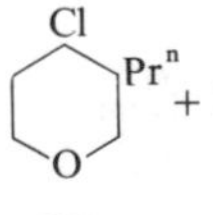

(1.34)

$+ \; Pr^nCH_2{\cdot}CHCl{\cdot}CH_2{\cdot}CH_2{\cdot}OAc$

(32)   (Ac = MeCO)

(31)

Tetrafluoroboric acid may be used as the catalyst for the addition of phenols or acetic acid to alkenes[171]; the reaction is of the same type as under photochemical conditions. Ethylene reacts with propionic acid (or methyl propionate) in the presence of di-t-butyl ether to give the insertion products (33) to (35); also (36) by rearrangement of (35). The yields are 44%, 25%, 8% and 23% respectively[172]. This reaction may also be initiated by γ-radiation[173]. Similarly with ethyl acetylacetoate in the presence of dicyclohexyl peroxycarbonate, ethylene reacts to give the insertion products (37) to (40) in 59%, 16%, 18% and 7% yield respectively[172].

$CHMeEt \cdot CO_2H$
(33)

$Me(CH_2)_3 \cdot CHMe \cdot CO_2H$
(34)

$Me(CH_2)_5 \cdot CHMe \cdot CO_2H$
(35)

$Me(CH_2)_3 \cdot CMeEt \cdot CO_2H$
(36)

$MeCO \cdot CHEt \cdot CO_2Et$
(37)

$MeCO \cdot CHBu^n \cdot CO_2Et$
(38)

$MeCO \cdot CEtBu^n \cdot CO_2Et$
(39)

$MeCO \cdot CEt_2 \cdot CO_2Et$
(40)

Some experimental evidence relating to the application of the Woodward–Hoffmann rules to cyclo-addition reactions, is provided by the addition of dimethylketene to alkenes[174]. The rate constant for addition to cis-but-2-ene to give cyclobutanones (equation (1.35)) is at least three times that for the trans-isomer, and of the pent-2-ene isomers only the cis-isomer reacts. This

$$Me_2C:C:O + \underset{H}{\overset{Mc}{\diagdown}}C:C\underset{H}{\overset{Me}{\diagup}} \longrightarrow \text{(structure)} \tag{1.35}$$

suggests that, as predicted by Woodward and Hoffmann, the interacting double bonds are at right angles as the two molecules approach each other (41). The trans-alkenes will present a bulky substituent towards the ketene and the cis- will not, whereas if the approach is planar there will be no difference in the steric effects of the cis- and the trans-isomer. The addition to alkenes of

(41)   (42)   (43)

the elusive ethoxyketene is also correlated with the Woodward Hoffmann rules–(intermediate (41))[175]. Ethoxyketene, prepared either by photolysis of ethyl diazoacetate or dehydrochlorination of ethoxyacetyl chloride with a trialkylamine, gives ethoxycyclobutanone derivatives with cis-but-2-ene (42; $R^1 = R^3 = H$, $R^2 = R^4 = Me$), trans-but-2-ene (42; $R^2 = R^3 = H$, $R^1 = R^4 = Me$), isobutene (42; $R^1 = R^2 = H$, $R^3 = R^4 = Me$), and 2,3-dimethylbut-2-ene (42; $R^1 = R^2 = R^3 = R^4 = Me$). The mode of reaction of diphenylcarbene (43) depends on the structure of the alkene, steric factors controlling the mechanism. Hept-1-ene[176] and the but-2-enes[177] react by abstraction–addition, but 3-methylbut-1-ene, propene and isobutene form cyclo-addition products[177].

Derivatives of 3-isoxazolidines are formed by the reaction of $(NO_2)_4C$ with alkenes[178]. For example, 2,4-dimethylpent-2-ene with $(NO_2)_4C$ in ether for 30 days gives 4-isopropyl-5,5-dimethyl-2-(1,1,3-trimethyl-2-nitrobutoxy)-3,3-dinitroisoxazolidine (44). Imidazoles (45) are prepared by the

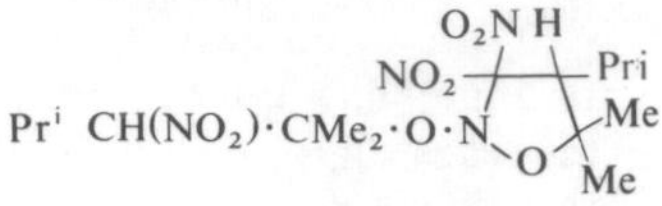

ALIPHATIC COMPOUNDS

$$Pr^i\;CH(NO_2)\cdot CMe_2\cdot O\cdot N \underset{O}{\overset{O_2N\;H}{\underset{\displaystyle Me}{\overset{\displaystyle NO_2}{\bigg|}}}}\substack{Pri\\Me}$$

(44)

reaction of nitrosonium tetrafluoroborate and alkyl cyanides with alkenes (equation (1.36))[179].

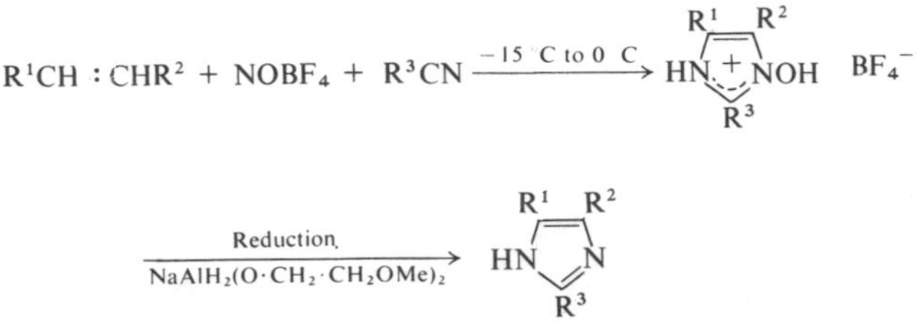

$$R^1CH : CHR^2 + NOBF_4 + R^3CN \xrightarrow{-15\;C\;to\;0\;C} \underset{R^3}{\overset{R^1\quad R^2}{HN\overset{+}{\cdot}NOH}}\;BF_4^-$$

(1.36)

$$\xrightarrow[NaAlH_2(O\cdot CH_2\cdot CH_2OMe)_2]{Reduction} \underset{R^3}{\overset{R^1\quad R^2}{HN\;\;N}}$$

(45)

An unusual reaction of ethylene is that with the products of the explosion of silver acetylide[180]. The carbon radicals formed react as C_2 and abstract hydrogen from ethylene giving acetylene, in yields of c. 10%.

One of the principal and rapidly developing areas of alkene chemistry is the numerous types of photochemical reactions. A comprehensive review[181] and a review of the specific area of cyclo-addition of α,β-unsaturated ketones[182] have been published recently. A unique tricyclic system (46) is formed when carbostyril is added to alkenes (equation (1.37), $R^1 = R^2 = R^3 = R^4 = Me$, or $R^1 = R^2 = Me$, $R^3 = R^4 = H$, or $R^1 = R^2 = Et$, $R^3 = R^4 = H$) under photochemical conditions[183]; some carbostyril dimer is also produced.

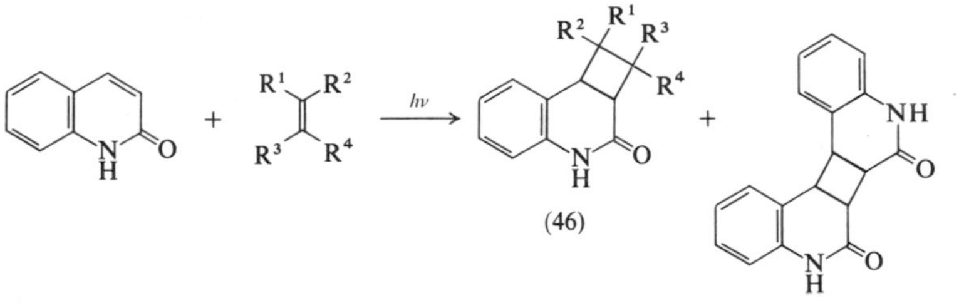

(46)

(1.37)

Isocarbostyril gives a similar product (47), although the reaction is much less efficient[184]. The photochemical reaction of butane-2,3-dione (48) with

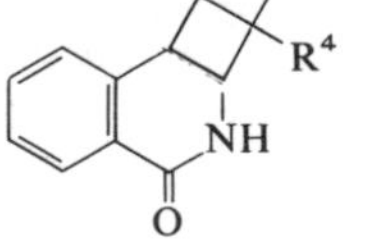

(47)

2-methylbut-2-ene gives the acyclic ether (49) and the oxetane (50) in a 2:1 ratio (equation (1.38))[185].

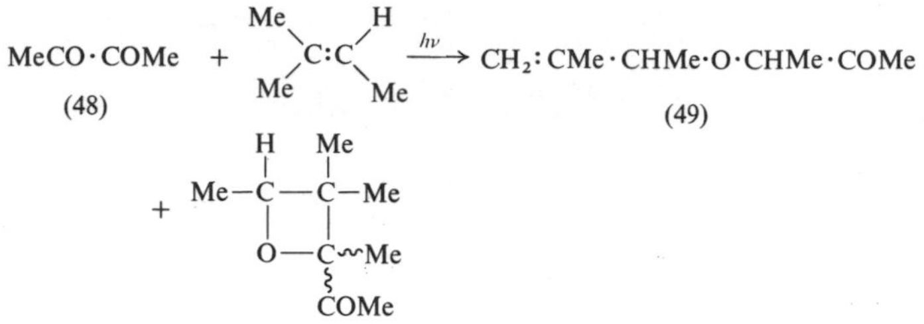

$$\text{MeCO·COMe} + \underset{\underset{(48)}{}}{\overset{\displaystyle \underset{Me}{\overset{Me}{}}\diagdown C:C \diagup \underset{Me}{\overset{H}{}}}{}} \xrightarrow{h\nu} \text{CH}_2\text{: CMe·CHMe·O·CHMe·COMe}$$

(48)   (49)

$$+ \quad \underset{\substack{| \\ \text{O}-\text{C}\sim\text{Me} \\ \\ \text{COMe}}}{\overset{\substack{\text{H} \quad \text{Me} \\ | \quad\quad |}}{\text{Me}-\text{C}-\text{C}-\text{Me}}} \qquad\qquad (1.38)$$

(50)

A mechanistic study of the photochemical addition of nitro-compounds to alkenes confirms earlier results that the addition takes place through the nitro-group to give 1,3,2-dioxazolidines (e.g. (51) from $PhNO_2$ and 2-methyl-but-2-ene)[186].

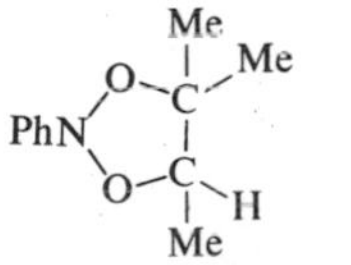

$$\text{PhN} \underset{\substack{\diagdown \\ \text{O}-\underset{|}{\overset{|}{\text{C}}}\diagdown \text{H} \\ \text{Me}}}{\overset{\substack{\diagup \\ \text{O}-\underset{|}{\overset{|}{\text{C}}}\diagup \text{Me} \\ }}{}}$$

(51)

A new route to 1,3-dithiolan derivatives (52) is provided by the photochemical reaction of 5-phenyl-1,2,4-dithiazole-3-thione (53) with alkenes (equation (1.39))[187].

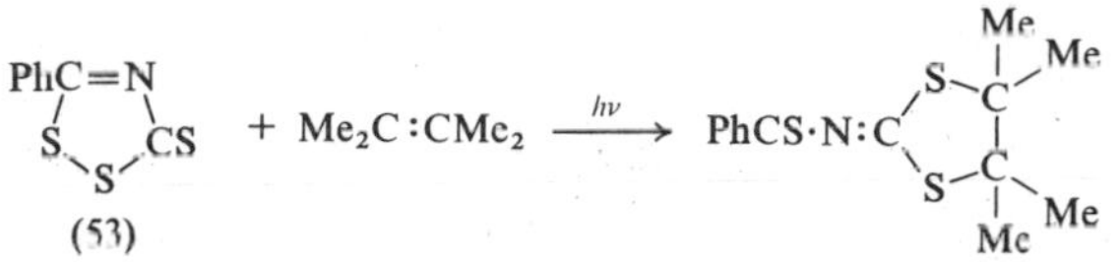

$$\underset{(53)}{\overset{\substack{\text{PhC}=\text{N} \\ \diagup \quad \diagdown \\ \text{S} \quad\quad \text{CS} \\ \diagdown_{\text{S}}\diagup}}{}} + \text{Me}_2\text{C:CMe}_2 \xrightarrow{h\nu} \underset{(52)}{\overset{\substack{\text{Me} \\ | \\ \text{PhCS·N:C}\overset{\substack{\text{S}-\overset{|}{\text{C}}\diagdown\text{Me} \\ | \\ \text{S}-\overset{|}{\text{C}}\diagdown\text{Me}}}{\diagup} \\ \text{Mc}\diagup\text{Me}}}{}} \qquad (1.39)$$

The photochemical cyclo-addition of benzene to cis- and to trans-but-2-ene gives the stereoisomeric adducts 6,7-dimethyltricyclo[3.3.0.0^{2,8}]oct-3-ene (54), with retention of configuration of the butene residue[188]. If the addition is to the tetrasubstituted ethylenic system of 2,3-dimethylbut-2-ene, the analogue of (54) is produced in only moderate amounts, the major product being 3-(1,1,2-trimethylallyl)cyclohexa-1,4-diene (55)[189].

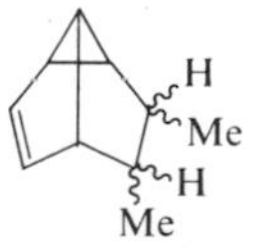

(54) 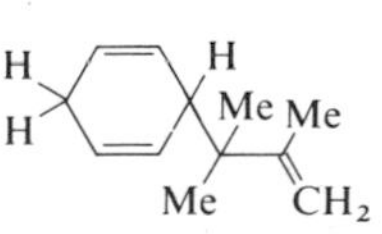   (55)

The arylation of alkenes by using organopalladium compounds as intermediates (equation (1.40)), as first reported by Heck[190] and Fujiwara et al.[191], is the subject of continuing study. Strong evidence for the suggested mechanism, involving a σ-bonded alkene–Pd^{II} intermediate, is provided by the reaction of a preformed complex involving 2,2-dichlorovinyl palladium(II) with benzene in the presence of silver acetate (equation (1.41))[192, 193].

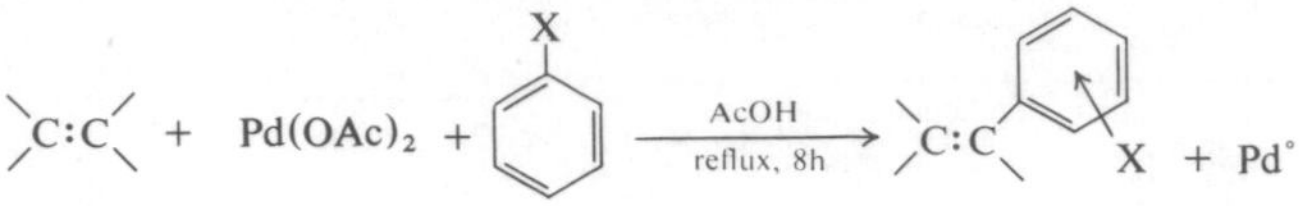

$$\text{>C:C<} + Pd(OAc)_2 + \text{[PhX]} \xrightarrow[\text{reflux, 8h}]{\text{AcOH}} \text{>C:C<} \cdot X + Pd° \qquad (1.40)$$

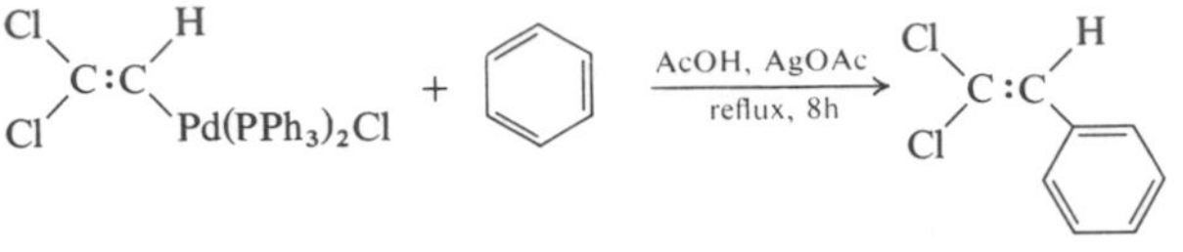

$$\underset{Cl}{\overset{Cl}{>}}C:C\underset{Pd(PPh_3)_2Cl}{\overset{H}{<}} + \text{[Ph]} \xrightarrow[\text{reflux, 8h}]{\text{AcOH, AgOAc}} \underset{Cl}{\overset{Cl}{>}}C:C\overset{H}{<}\text{[Ph]} \qquad (1.41)$$

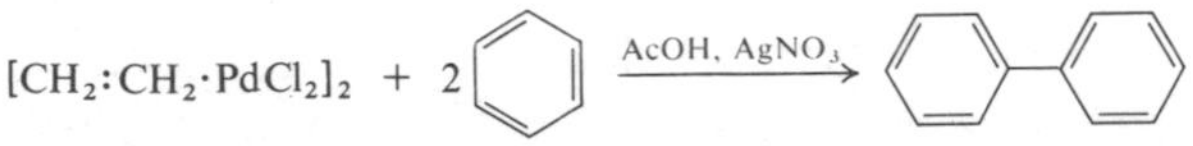

$$[CH_2:CH_2 \cdot PdCl_2]_2 + 2\,\text{[Ph]} \xrightarrow{\text{AcOH, AgNO}_3} \text{[Ph-Ph]} \qquad (1.42)$$

If π-alkene–palladium complexes are used, biphenyls are formed (e.g. equation (1.42))[194]. Ferrocene can also be used as the aromatic compound[195]. A significant advance in this area is the report that in the presence of oxygen the reaction becomes catalytic with respect to Pd^{II} [196]; the oxygen oxidises the palladium metal formed and so takes it back into solution. Arylation of alkenes with aryl iodides, using $MeCO_2K$ as the HI acceptor is also catalytic with respect to Pd^{II} (equation (1.43), X = H or Me)[197]. For the above[194] reactions hot acetic acid is used as the solvent. Methanol at room temperature can be used if organocobalt compounds are the source of the aryl or an alkyl group[198]. The transfer of the aryl or alkyl group to the alkene

$$PhI + CH_2:CHX + MeCO_2K \xrightarrow{PdCl_2} PhCH:CHX + MeCO_2H + KI \qquad (1.43)$$

probably occurs, as above, through a σ-bonded organopalladium complex. Two types of organocobalt complex have been used, organic derivatives of cobaloximes, e.g. methyl-pyridine-bis(dimethylglyoximato)cobalt(III), methyl pyridine cobaloxime' (56) and bis(salicylidenato)ethylenediamine cobalt(III), with Li_2PdCl_4 at 20–25 °C.

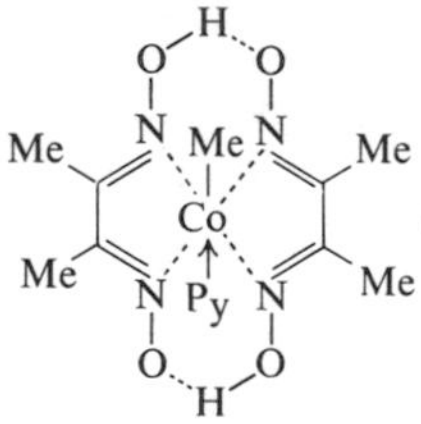

(56)

Propene has been phenylated with benzene to give isopropylbenzene by using $BF_3–D_2O$ as the catalyst[199], and alkenes have been phenylated by using Ph_3Al followed by hydrolysis[200].

Rhodium and iridium complexes catalyse the addition of secondary aliphatic amines to ethylene (equation (1.44), R^1 and R^2 = alkyl), N-ethyl-piperidine for example, being formed after 3 h at 180 °C from piperidine and ethylene in tetrahydrofuran in the presence of $RhCl_3 \cdot 3H_2O$ [201]. Higher alkenes than ethylene, and primary amines are unreactive. Both steric effects and basicity are important since amines of similar basicity give differing yields (Me_2NH, Et_2NH and Bu^n_2NH give 54%, 4% and 3% of product respectively), and reducing the basic strength reduces the yield (piperidine, pyrrolidine, and morpholine give 70%, 36%, and < 2% of product respectively).

$$CH_2:CH_2 + \begin{array}{c} R^1 \quad R^2 \\ \diagdown \quad \diagup \\ N \\ | \\ H \end{array} \longrightarrow \begin{array}{c} R^1 \quad R^2 \\ \diagdown \quad \diagup \\ N \\ | \\ Et \end{array} \qquad (1.44)$$

Radiolysis of solutions of ethylene in liquid nitrogen by γ-rays from ^{60}Co forms nitrogen atoms and ethyl radicals, which react with the ethylene to give HCN and CH_3CN in 10% yield[202]. The radiolysis of alkenes has been reviewed[204].

1.4.2.6 Carbonylation and hydroformylation

Metal carbonyls are widely used as catalysts for reactions between an alkene and carbon monoxide, typically the OXO and Reppe processes. Addition of CO and CCl_4 to alkenes to form 2-alkyl-4,4,4-trichlorobutanoyl chlorides (equation (1.45)) are catalysed by many carbonyls. Studies using $[Co(CO)_4]_2$, $[C_5H_5Fe(CO)_2]_2$ and $[C_5H_5Mo(CO)_3]_2$ with a wide range of alkenes, show the mechanism to be a succession of CO insertion and CCl_4 cleavage reactions[279]. This and similar reactions have been reviewed[205]. Perhaps the best known reaction of this type is hydroformylation (the addition of H_2 and CO)[206-209].

$$RCH:CH_2 + CO + CCl_4 \xrightarrow{\text{metal carbonyl}} Cl_3C \cdot CH_2 \cdot CHR \cdot CO \cdot Cl + Cl_3C \cdot CH_2 \cdot CHRCl \quad (1.45)$$

Hydroformylation by using $RhH(CO)(PPh_3)_3$ has been studied in depth[206, 207], and many intermediate species have been identified[206]. Preference for the formation of straight-chain aldehydes, which usually occurs, is enhanced if molten PPh_3 is used as the solvent[207]. The efficiency of the cobalt-group carbonyls as hydroformylation catalysts decreases in the order $Co_2(CO)_8 >$ $Rh(CO)_4 > Ir(CO)_4$, and the extent of formation of straight-chain aldehydes decreases in the same order[208]. The mechanism of hydroformylation of alkenes in the vapour phase has been studied; $HCo(CO)_4$ adds to propene mainly in a Markovnikov manner[209].

Carboxylic acids are produced by the addition of CO and H_2O to alkenes by using an acid catalyst[210-215]. 2-Ethylhex-1-ene[210, 211], isobutene[212], di-isobutene[213], pent-1-ene[213], 3-methylbut-1-ene[214], and hex-1-ene have been used with BF_3[212, 214], $BF_3-H_3PO_4$[210, 211, 213], or $HCO_2H-H_3PO_4$[215] as the catalyst. 2-Ethylhex-1-ene, for example, gives 2-ethyl-2-methylhexanoic acid (equation (1.46)). If methanol is present during the reaction, the expected methyl esters are formed[215].

$$Bu^nCEt:CH_2 \xrightarrow{CO+H_2O} Bu^nCMeEt \cdot CO_2H \qquad (1.46)$$

1.4.2.7 Hydrosilylation

The very important synthetic method for producing organosilicon compounds, by catalytic addition of R_3Si—H across the double bond of an alkene has been reviewed[216]. Although rhodium[217] and iridium complexes as catalysts have been much studied, hydrosilylation with nickel complexes has received little attention. A mixture of $NiCl_2$ and a ditertiary phosphine is found to be a very effective catalyst and unique in that exchange of H and Cl on silicon takes place during reaction[218]. Because they are easily reduced, palladium complexes have also been largely ignored but palladium phosphine complexes can be used[219]. Studies using aminosilanes have been reported[220].

1.4.2.8   Hydration

The extensive studies by Brown and his co-workers have shown that hydroboration is a highly convenient method of anti-Markovnikov hydration of

Table 1.7   Alcohols from alkenes by oxymercuration–demercuration
(From Brown and Geoghegan[221], by courtesy of the American Chemical Society)

Alkene	Alcohol	Yield %
Terminal		
Pent-1-ene	Pentan-2-ol	97
Hex-1-ene	Hexan-2-ol	95
Octadec-1-ene	Octadecan-2-ol	93
3,3-Dimethylbut-1-ene	3,3-Dimethylbutan-2-ol	94
Disubstituted terminal		
2-Methylbut-1-ene	2-Methylbutan-1-ol	92
2,4,4-Trimethylpent-1-ene	2,4,4-Trimethylpentan-2-ol	87
Disubstituted internal		
cis-Pent-2-ene	Pentan-2-ol (64 %)	93
	Pentan-3-ol (36 %)	
trans-Pent-2-ene	Pentan-2-ol (56 %)	91
	Pentan-3-ol (44 %)	
cis-4-Methylpent-2-ene	4-Methylpentan-2-ol (91 %)	97
	4-Methylpentan-3-ol (9 %)	
trans-4-Methylpent-2-ene	4-Methylpentan-2-ol (82 %)	99
	4-Methylpentan-3-ol (18 %)	
cis-4,4-Dimethylpent-2-ene	4,4-Dimethylpenten-2-ol (98 %)	85
	4,4-Dimethylpentan-3-ol (2 %)	
trans-4,4-Dimethylpent-2-ene	4,4-Dimethylpentan-2-ol (95 %)	91
	4,4-Dimethylpentan-3-ol (5 %)	
Trisubstituted internal		
2-Methylbut-2-ene	2-Methylbutan-2-ol	94
2,4,4-Trimethylpent-2-ene	2,4,4-Trimethylpentan-2-ol	86
Tetrasubstituted internal		
2,3-Dimethylbut-2-ene	2,3-Dimethylbutan-2-ol	85

alkenes. An equally convenient method of Markovnikov addition is by oxymercuration followed by demercuration with $NaBH_4$ (equation (1.47))[221].

$$RCH{:}CH_2 + Hg(OAc)_2 \xrightarrow{\text{1:1 THF } H_2O, \text{ 25 C } 10 \text{ min}} RCH(OH){\cdot}CH_2HgOAc$$
$$\xrightarrow{\text{NaOH} + NaBH_4, \text{ 25°C}} RCH(OH)Me + Hg \tag{1.47}$$

Mono-, di-, tri-, and tetra-alkylethylenes give high yields of the 'Markovnikov' alcohols (see Tab. 1.7). This procedure also gives chiral alcohols, for example by using $Hg^{II}(+)$-tartrate in $3:2:2$ $MeCN$–THF–H_2O, dec-1-ene, oct-1-ene, and hept-1-ene can be hydrated to the related alkan-2-ols with about 15% excess of one chiral isomer[222]. Hydroxy-, alkoxy-, and acetoxy-mercurations are well known. If alkyl hydroperoxides are used, then peroxymercuration takes place (equation (1.48))[223, 224].

$$RCH:CH_2 + Hg(OAc)_2 + Bu^tO\cdot OH \xrightarrow{CH_2Cl_2} Bu^tO\cdot OCHR\cdot CH_2\cdot HgOAc + HOAc \quad (1.48)$$

The mechanism of transoxymercuration (equation (1.49)) is now thought to involve a bisalkene–mercury cation (57), rather than a mercurinium ion (58)[225].

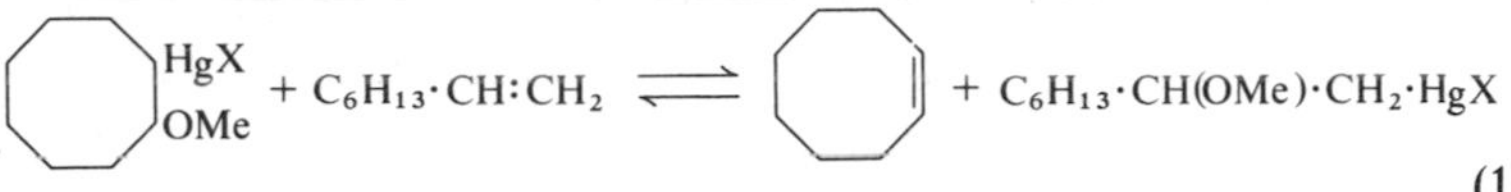

$$(1.49)$$

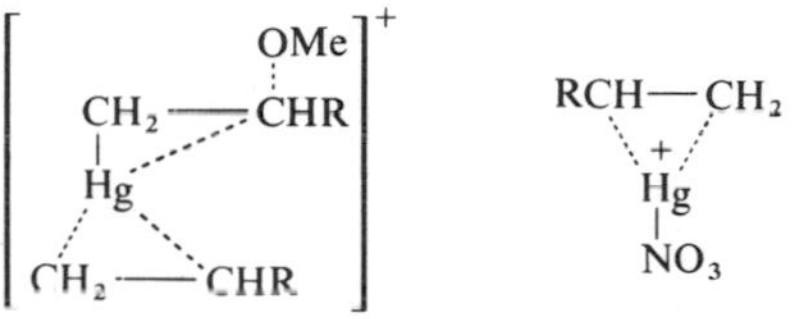

(57)

(58)

Alkenes can be oxidised by mercuric salts both by addition and by redox reactions. A recent review concentrates on the latter, in which the oxygen adds to the alkene and the mercury salt is reduced[226]. A typical reaction is given in equation (1.50).

$$CH_3\cdot CH:CH_2 \xrightarrow{HgSO_4-H_2SO_4} CH_2:CH\cdot CHO \quad (1.50)$$

Unsaturated alcohols (59) are previously unreported products of oxidation of alkenes by aqueous $HgSO_4$ (equation (1.51))[227]. These alcohols also appear

$$CH_3:CH\cdot CH_2R + HgSO_4 + H_2O \rightarrow CH_3:CH\cdot CH(OH)R + Hg + H_2SO_4 \quad (1.51)$$
$$(59)$$

$$(59) + HgSO_4 \rightarrow CH_2:CH\cdot COR + Hg + H_2SO_4 \quad (1.52)$$

Table 1.8 Products from mercuric sulphate oxidation of alkenes.
(From Tinker[227], by courtesy of the Elsevier Publishing Company)

Alkene	Primary product	Yield* (mole %)	Secondary product	Yield† (mole %)
Propene	Allyl alcohol	51	Acrolein	87
But-1-ene	But-1-en-3-ol	41	But-1-en-3-one	100
trans-But-2-ene	Butan-2-one	81	—	—
Pent-1-ene	Pent-1-en-3-ol	33	Pent-1-en-3-one	92
	Pent-2-en-1-ol	8	Pent-1-en-3-one	86
cis-Pent-2-ene	Pentan-2-one	53	—	—
	Pentan-3-one	27	—	—
Hex-1-ene	Hex-2-en-3-ol	35	Hex-1-en-3-one	91
Dodec-1-ene	Dodec-1-en-3-ol	6	Not examined	
Cyclohexene	Cyclopentenecarboxaldehyde	74	—	

*Based on reduced mercury. †Based on reacted alcohol.

to be intermediates in the oxidation of alkenes to unsaturated ketones (equation (1.52)). Typical alcohols and ketones are given in Table 1.8.

Facile conversion of alkenes into ketones is achieved by using $Hg(OAc)_2$ with $PdCl_2$ [228]. This is because the palladium compounds formed are unstable in solution and give rapid reductive elimination to the ketone and palladium metal (e.g. equation (1.53)). The reaction is carried out by adding the organo-

$$Me(CH_2)_3 \cdot CH(OH) \cdot CH_2PdCl_2^- \rightarrow Me(CH_2)_3 \cdot COMe + Pd + HCl + Cl^- \quad (1.53)$$

mercurial, formed from the alkene and $Hg(OAc)_2$ in $THF–H_2O$ or MeOH, to Li_2PdCl_4 in THF. Palladium is precipitated and a quantitative yield of the ketone is produced in about 2 h. Typically, hex-1-ene is converted into hexan-2-one, and 3,3-dimethylbut-1-ene into 3,3-dimethylbutan-2-one, both in 100 % yield. The reaction becomes catalytic with respect to $PdCl_2$ if $CuCl_2$ is added to reoxidise the palladium (see Ref. 196 for a similar catalysis of alkene arylation).

The mechanism of hydration of simple alkenes has been firmly established as $A–S_E2$ (equation (1.54))[229]. 2,3-Dimethylbut-2-ene and *trans*-cyclo-octene

$$\diagdown C:C \diagup + H^+ \xrightarrow{\text{slow}} H\overset{|}{C}-\overset{|}{C}+ \xrightarrow[-H^+]{H_2O} H\overset{|}{C}-\overset{|}{C}-OH \quad (1.54)$$

both show general acid catalysis in $H_3PO_4–H_2PO_4^-$ and $HSO_4^-–SO_4^{2-}$ buffers. Rates decrease with decreasing buffer concentration in contrast to the results of other recent work[230], but here the acid concentration was high and specific ionic interactions would completely mask general acid catalysis.

1.4.2.9　Oxidation

Alkenes may be oxidised both by oxygen gas and by organic or inorganic oxidising agents, sometimes with a catalyst sometimes without. An interesting oxidation by O_2 in the absence of a catalyst is that of n-butenes to give acetic acid, a reaction used commercially by Farbenfabriken Bayer AG[231]. A mixture of butenes is treated with acetic acid at 100–120 °C at 15–25 atm. when s-butyl acetate is formed. This may then be oxidised by air at c. 200 °C and 60 atm. to give acetic acid (3 mol).

The rate of oxidation of propene by oxygen to give epoxy-propane is increased by using dimethyl phthalate as the solvent[232]. If the solvent is PhCl, $PhNO_2$, PhEt, silicone oil, or tetradecane, then peroxidation predominates. Peroxidation of $Me_2C:CHEt$ to $Me_2C:CH \cdot CH(O \cdot OH)Me$ occurs at 50–70 °C in the absence of solvent with air as the oxidising agent[233].

The use of homogeneous catalysts for alkene oxidations has been reviewed[234]. A very interesting new oxidising agent is bistriphenylsilyl chromate (60)[235].

$$(Ph_3SiO)_2CrO_2 + R^1R^2C:CR^3R^4 \longrightarrow \begin{array}{c} Ph_3SiO \\ \\ Ph_3SiO \end{array} \hspace{-0.5em} Cr \hspace{-0.5em} \begin{array}{c} O-CR^1R^2 \\ | \\ O-CR^3R^4 \end{array}$$
$$(60) \qquad \qquad \qquad \qquad \qquad \qquad \qquad (1.55)$$

$$\longrightarrow \begin{array}{c} Ph_3SiO \\ \\ Ph_3SiO \end{array} \hspace{-0.5em} Cr + \begin{array}{c} R^1 \\ \\ R^2 \end{array} \hspace{-0.5em} C=O + \begin{array}{c} R^3 \\ \\ R^4 \end{array} \hspace{-0.5em} C=O$$
$$(61)$$

This cleaves alkenes (equation (1.55)) to aldehydes and ketones and produces reduced organochromium species. It is also a catalyst for ethylene polymerisation, the active species being the divalent complex (61), since it is well known that low-valent transition metal centres are the active sites in many polymerisation catalysts (equation (1.56)).

$$(61) + CH_2:CH_2 \rightarrow Ph_3SiO \cdot Cr \cdot OSiPh_3 \rightarrow Ph_3SiO \cdot Cr \cdot CH_2 \cdot CH_2 \cdot OSiPh_3$$
$$\overset{\uparrow}{CH_2 = CH_2}$$

$$\xrightarrow{CH_2:CH_2} Ph_3SiO \cdot Cr(CH_2 \cdot CH_2)_n \cdot OSiPh_3 \xrightarrow{CH_2:CH_2} Ph_3SiO \cdot Cr \cdot CH_2 \cdot CH_3 +$$
$$CH_2:CH(CH_2 \cdot CH_2)_{n-1}OSiPh_3 \qquad (1.56)$$

The mechanism of alkene oxidation by chromyl chloride has been the subject of dispute. It is now thought to be as given in equation (1.57), involving the intermediate adduct (62)[236]. A simple one-step synthesis of aldehydes and

$$R_2C:CR_2 + CrO_2Cl_2 \longrightarrow R-\overset{\overset{\displaystyle R}{|}}{\underset{\underset{\displaystyle CrOCl_2}{\overset{\displaystyle |}{O}}}{C}}=\overset{\overset{\displaystyle R}{|}}{\underset{}{C}}-R \xrightarrow{\text{hydrolysis, 0 5 C}} R_3C \cdot COR \qquad (1.57)$$
$$(62)$$

ketones is the reaction of alkenes with $Tl(NO_3)_3$ in methanol at room temperature[237]. The reaction products, $TlNO_3$ and a ketal or acetal are separated by filtration and the ketal or acetal hydrolysed to give the ketone or aldehyde. Selenium dioxide may be used to oxidise alkenes[238, 239]; usually allylic products are formed by oxidation at the position α to the most substituted end of the double bond[238]. However, in acetic acid in the presence of sulphuric acid diacetoxy-adducts are formed; hex-1-ene, for example, gives 1,2-diacetoxyhexane[239].

Epoxidation of propene occurs when it reacts with peroxyacetic acid at $< 50°C$[240]. H_2O_2 also epoxidises alkenes in neutral solution if an isocyanate is added as a co-reagent (equation (1.58))[241] and peroxybenzimidic acid (63), formed in situ from benzonitrile and alkaline H_2O_2, has also been used[242].

$$2RN:CO + H_2O_2 + R_2C:CR_2 \xrightarrow[\text{CHCl}_3]{\text{Ether or}} R_2C \overset{\displaystyle O}{\overset{\diagup \diagdown}{-}} CR_2 + (RNH)_2CO + CO_2 \quad (1.58)$$

A system, closely similar to the enzyme, oxidases, which converts alkenes into epoxides, contains chelates of transition metals. Soluble Cu,Mo,Cr or

$$\underset{HO \cdot O}{\overset{Ph}{\diagdown}} C : NH$$
$$(63)$$

W azo-complexes in benzene solution with molecular oxygen under pressure convert propene into epoxypropane[243]. It is not known whether the reaction involves radicals or ionic intermediates.

Oxymercuration products from alkenes can be oxidised electrolytically to give a variety of carboxylic acids[244]. These are formed by carbenium-ion mechanisms, and C—C bond cleavage can occur; the acids produced, therefore, are those with the same number of carbon atoms as the alkene, and lower acids. For example, pent-1-ene gives four acids; n-pentanoic (26%), n-butanoic (14%), acetic (36%), and formic (14%) (percentages are moles % of total organic products).

A new photochemical process for alkenes produces ethers[245]. The reaction, which takes place in aqueous or alcoholic solution and involves a cation radical, is illustrated in equation (1.59) with 2,3-dimethylbut-2-ene. The hydrogen atoms produced can react with the alkene as shown in equation (1.60).

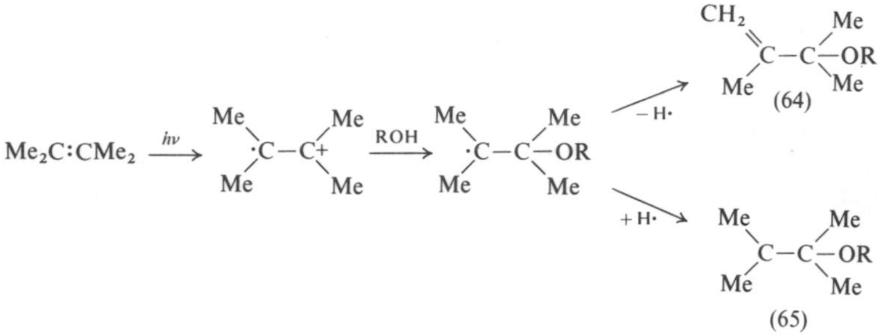

$$
\text{Me}_2\text{C:CMe}_2 \xrightarrow{h\nu} \underset{\text{Me}}{\overset{\text{Me}}{\text{·C}}}-\underset{\text{Me}}{\overset{\text{Me}}{\text{C}^+}} \xrightarrow{\text{ROH}} \underset{\text{Me}}{\overset{\text{Me}}{\text{·C}}}-\underset{\text{Me}}{\overset{\text{Me}}{\text{C}}}-\text{OR}
$$

(64)

(65)

(1.59)

$$
\text{Me}_2\text{C:CMe}_2 + \text{H·} \longrightarrow \underset{\text{Me}}{\overset{\text{Me}}{\text{·C}}}-\underset{\text{Me}}{\overset{\text{Me}}{\text{C}}}{\overset{}{\leftarrow}}\text{H}
$$

(66)

(67)

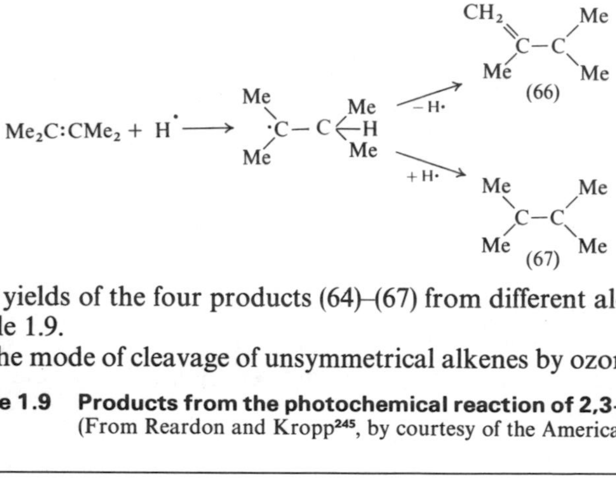

(1.60)

The yields of the four products (64)–(67) from different alcohols are given in Table 1.9.

The mode of cleavage of unsymmetrical alkenes by ozone can be predicted

Table 1.9 **Products from the photochemical reaction of 2,3-dimethylbut-2-ene.**
(From Reardon and Kropp[245], by courtesy of the American Chemical Society)

Solvent	Reactant	Yield(%)			
		(64)	(65)	(66)	(67)
30% aq. MeCN	3	16	23	1.5	1
MeOH	42	2	2.5	1	2
EtOH	3	23	16	6	7
Bu^{n}OH	5	23	10	7	7 (others 8)

by consideration of the stabilisation of the carbenium ions produced and alkene geometry[246].

1.4.2.10 Reactions with sulphur compounds

Alkenes can also be oxidised to ketones in a one-step process by using an aqueous suspension of sulphur as the oxidising agent[247]. Sulphonation products are obtained if sulphuric acid[248] or SO_3 in an air-stream[249] is used. In the last case a complex mixture of sulphonated products is formed; the alkenes also polymerise.

The photochemical addition of thiocyanogen to alkenes is analogous to the photochemical addition of halogens and of thiols[250]. A radical chain mechanism (equation (1.61)) will explain the experimental results.

$$(SCN)_2 \xrightarrow{h\nu} 2 \cdot SCN$$

$$\text{C:C} + \cdot SCN \rightleftharpoons -\overset{SCN}{\underset{|}{C}}-\overset{.}{\underset{|}{C}}- \xrightarrow{(SCN)_2} -\overset{SCN}{\underset{|}{C}}-\overset{SCN}{\underset{|}{C}}- + \cdot SCN \quad (1.61)$$

1.4.2.11 Halogenation

By using Bu_3SnH to remove the bromine reductively (equation (1.62)) it is possible to add HF indirectly to alkenes by using the well-known addition of BrF, formed from N-bromoacetamide in anhydrous hydrogen fluoride (equation (1.62))[251].

$$\text{C:C} + BrF \xrightarrow{-80\ ^\circ C} \overset{F}{\underset{Br}{C-C}} \xrightarrow{Bu_3SnH} \overset{F}{\underset{H}{C-C}} \quad (1.62)$$

Photodifluoroamination of alkenes not only substitutes NF_2 for H (equations (1.13–1.17) and (1.63)) but also results in the addition of F and NF_2 to the double bond (equation (1.63)); the fluorine is mainly on the terminal carbon atom (68)[64].

$$Me \cdot CH:CH_2 + N_2F_4 \rightarrow NF_2 \cdot CH_2 \cdot CH:CH_2 + CH_2F \cdot CH(NF_2) \cdot Me + CH_2(NF_2) \cdot CHF \cdot Me$$

$$(1.63)$$

Similarly, a mixture of products is obtained from the gas-phase reaction of chloramine with alkenes (equations (1.9–1.12) and (1.64))[62]. In non-polar

$$Me_2C:CH_2 \rightarrow Me_2CCl \cdot CH_2Cl + Me_3CCl + MeC(CH_2Cl):CH_2 \quad (1.64)$$

solutions trichloramine generates N_2 quantitatively and vicinal dichlorides are formed[252], most probably by a radical mechanism (equation (1.65)).

$$3 \text{ C:C} + 2NCl_3 \longrightarrow 3 \overset{Cl\ \ Cl}{C-C} + N_2 \quad (1.65)$$

Chloroaminations (e.g. equation (1.66)) of alkenes have been reviewed[253].

$$Me \cdot CH:CH_2 + Et_2NCl \rightarrow Me \cdot CHCl \cdot CH_2 \cdot NEt_2 \quad (1.66)$$

Chlorination of 2-methylpentenes with Cl_2 in $CHCl_3$ at $c.\ 0\ ^\circ C$ gives, for the pent-1-ene, substitution in the 2-methyl group (69), and for the pent-2-ene, isomerisation to the terminal alkene occurs with allylic substitution (70)[254].

$$\begin{array}{cc} CH_2:CPr^n \cdot CH_2Cl & CH_2:CMe \cdot CHEtCl \\ (69) & (70) \end{array}$$

Allylic substitution also occurs when alk-1-enes are chlorinated with N-t-butyl-N-chloro- or N-chloro-N-cyclohexyl-ethanesulphonamide[255].

The reaction of t-butyl hypochlorites with alkenes in nitromethane gives,

surprisingly, products that are not formed from the alkene[256]. Chloronitromethane (71) is the product, whatever alkene is used, but in the absence of an
alkene, the hypochlorite and nitromethane do not react. Studies with different alkenes show that an alkene forms a π-complex with the hypochlorite
and that this then reacts with the solvent (equation (1.67)).

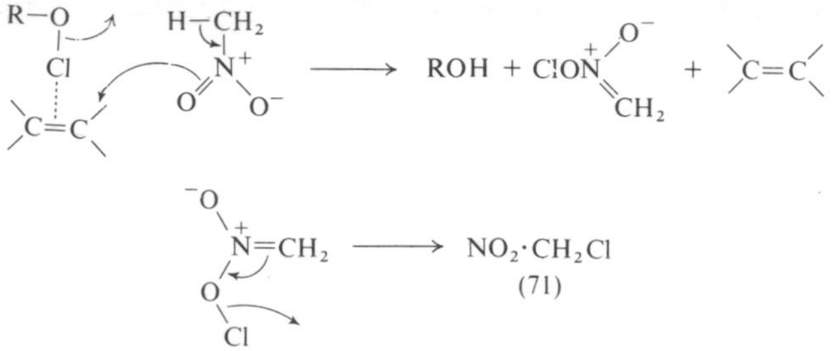

$$\text{R–O} \quad H\text{–CH}_2 \longrightarrow \text{ROH} + \text{ClON} \begin{smallmatrix} O^- \\ \| \\ \end{smallmatrix}_{CH_2} + \quad \text{C=C}$$

$$\underset{(71)}{\overset{O}{\underset{O}{N}}}\!=\!CH_2 \longrightarrow NO_2\cdot CH_2Cl \tag{1.67}$$

In contrast, polyfluoroalkyl hypochlorites add quantitatively to the double
bond (equation (1.68))[257].

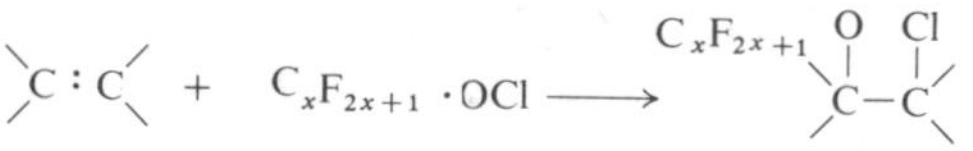

$$\text{C:C} + C_xF_{2x+1}\cdot OCl \longrightarrow \underset{}{\overset{C_xF_{2x+1}}{\underset{}{\overset{O}{\|}}}\underset{}{\overset{Cl}{\underset{}{|}}} C\!-\!C \tag{1.68}$$

Ring-halogenated alkylcyclopropanes can be formed by heating alkenes
with $PhHg\cdot CHCl_2$ to c. 140 °C. The mercury compound is a source of
chlorocarbene, which adds to the double bond (equation (1.69))[258].

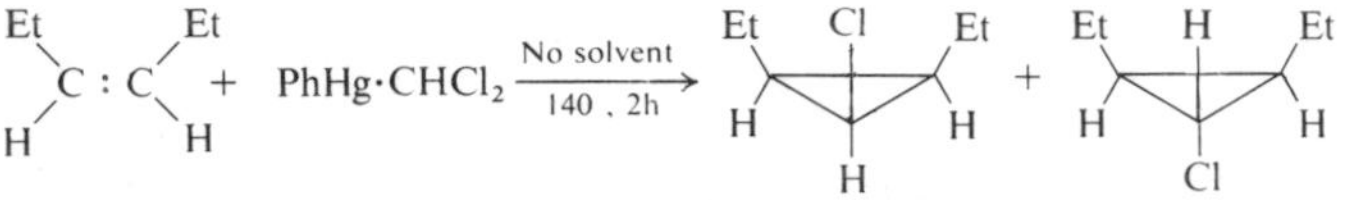

$$\underset{H}{\overset{Et}{\underset{}{}}}C : C\underset{H}{\overset{Et}{\underset{}{}}} + PhHg\cdot CHCl_2 \xrightarrow[140 \,.\, 2h]{\text{No solvent}} \text{(products)} \tag{1.69}$$

With $SbCl_5$ in CCl_4 as the source of chlorine, selective cis-chlorination occurs[259]. A mixture of products, formed by radical mechanisms, is obtained if
C_2H_4 and CCl_4 react at 160 °C, and 60–70 atm in the presence of Et_3PO_3
and a metal salt: $FeCl_2$ or $FeCl_3$ is the most effective metal salt, and then
over 90% of the product is 1,1,1,3-tetrachloropropane[260]. Alkenes react with
$TeCl_4$ to form addition compounds[261, 262], which on thermolysis give allylic
chlorides (equation (1.70))[262]. Under appropriate conditions allyl chloride is

$$TeCl_4 + CH_2\text{:}CH\cdot CH_3 \rightarrow Cl_2Te(C_3H_6Cl)_2 \xrightarrow{\Delta} CH_2\text{:}CH\cdot CH_2Cl$$
$$+ CH_2Cl\cdot CHCl\cdot CH_3 + HCl \tag{1.70}$$

formed directly[263, 264]. Cupric chloride is used to add chlorine to an alkene in
a liquid phase reaction[265–268], which, in contrast to the radical chlorination in
the vapour phase, is thought to be ionic. Methoxy-products are also formed
when methanol is used as the solvent[266]. The proposed mechanism is given
in equation (1.71).

$$C_2H_4 + CuCl_2 \rightleftharpoons C_2H_4Cl\cdots CuCl \xrightarrow{CuCl_2} ClCH_2\cdot CH_2Cl + 2CuCl$$
$$\xrightarrow{MeOH} ClCH_2\cdot CH_2OMe + HCuCl$$
$$HCuCl + CuCl_2 \rightarrow 2CuCl + HCl \tag{1.71}$$

Support for this mechanism is provided by the observation that reactivity decreases with increase in size and complexity of the alkene. This corresponds to the amount of complex formation of alkenes with Ag^+. The radical-initiated addition of polyhalogenoalkanes to alkenes is very effectively catalysed by a 1:1 mixture of $CuCl_2$ and ethanolamine (equation (1.72))[269]. This redox system offers several advantages over other radical initiation methods, in particular yields of 1:1 adducts are maximised, a large excess of alkyl polyhalide is unnecessary and vigorous reaction conditions requiring special apparatus are not needed.

$$Cu^+ + ammine + CCl_4 \rightarrow Cu^+-ammine\ Cl + CCl_3 \cdot \qquad (1.72)$$

A simple efficient synthesis of chloroiodoalkanes also uses Cu^{II} halides (e.g. equation (1.73))[270]. A catalyst of $CuCl_2$–$PdCl_2$–$CaCl_2$ is used for the

$$Me{\cdot}CH{:}CH_2 + CuCl_2 + I_2 \rightarrow Me{\cdot}CHCl{\cdot}CH_2I\ (73\%) + MeCHI{\cdot}CH_2Cl\ (27\%) \quad (1.73)$$

oxychlorination of C_2H_4 to give $HOCH_2{\cdot}CH_2Cl$. Insertion mechanisms are suggested, but the role of the $CaCl_2$ is not clear: it may be a stabiliser of intermediate complexes[271].

As for the chlorinations above, bromination by $CuBr_2$ in methanol gives dibromo- and methoxybromo-alkanes[272]. Methyl and acetyl hypobromite react with alk-1-enes by both an ionic and a radical mechanism[273]. The reaction products from the two mechanisms are different, the 1-bromo-compound (e.g. (72)) being formed under polar conditions (low alkene concentration, radical inhibitors present) and the 2-bromo-compound (e.g. (73)) being formed under conditions favouring radicals (high alkene concentration).

$$RCH(OMe){\cdot}CH_2Br \qquad\qquad RCHBr{\cdot}CH_2OMe$$
$$(72) \qquad\qquad\qquad (73)$$

The conjugate bromination of alkenes and oxiranes gives di-β-bromo-ethers (equation (1.74))[274].

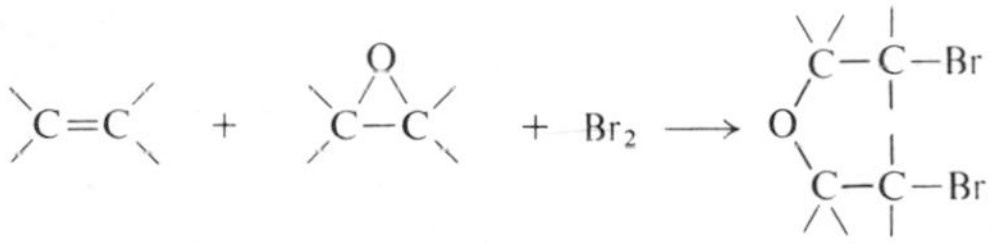

$$(1.74)$$

An efficient synthesis of primary alkyl bromides is provided by a hydro-boration-transmetallation-bromination reaction sequence (equation (1.75))[275].

$$R_3^1B + R_2^2C{:}CR_2^2 \rightarrow R_2^1B{\cdot}CR_2^2{\cdot}CR_2^2R^1 \xrightarrow[OH^-]{HgO-H_2O} (R^1CR_2^2{\cdot}CR_2^2)_2Hg \xrightarrow{Br_2} R^1CR_2^2{\cdot}CR_2^2Br$$
$$(1.75)$$

Aliphatic iodonitrates are formed when alkenes react with iodinium nitrate[276, 277]. The reagent is prepared in chloroform–pyridine solution, by treating ICl with $AgNO_3$, AgCl being precipitated. Immediate addition of the alkene gives the iodonitrate. For example, 2,3-dimethylhex-1-ene reacts as in equation (1.76). When the carbenium ion intermediate formed by I^+

$$CH_2{:}CMe{\cdot}CHMePr^n + INO_3 \rightarrow CH_2I{\cdot}CMe(ONO_2){\cdot}CHMePr^n \ (40\%)$$
$$+ CH_2(ONO_2){\cdot}CMeI{\cdot}CHMePr^n \ (60\%) \qquad (1.76)$$

attack is tertiary, then the nucleophilic pyridine competes with the nitrate
and other products are formed (equation (1.77)).

$$Me_2C{:}CHMe \longrightarrow \quad (1.77)$$

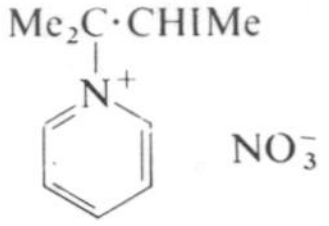

Iodine isocyanate adds to alkenes via a three-membered iodonium ion
intermediate[278]. The iodo-isocyanates may be treated with methanol to give
iodocarbamates, and these can be converted into carbamates with zinc
(equation (1.78)).

$$CH_2{:}CHBu^n + INCO \rightarrow Bu^nCHI{\cdot}CH_2{\cdot}NCO \xrightarrow{MeOH} Bu^nCHI{\cdot}CH_2{\cdot}NH{\cdot}CO_2Me \quad (1.78)$$
$$\xrightarrow{Zn} Bu^nCH_2{\cdot}CH_2{\cdot}NH{\cdot}CO_2Me$$

1.4.2.12 Organometallic complexes

Intense activity concerning the preparation and reactions of complexes of
alkenes with transition metals has necessitated a complete review elsewhere[1].
Selected recent work of special interest is reviewed here. Although many
thousand alkene–transition-metal complexes are known, new ones are being
continually prepared[280-283].

$$Pd(\text{acetylacetonate}) + Et_2AlOEt + C_2H_4 + PPh_3 \ (\text{or } (\text{cyclohexyl})_3P \text{ or } (o\text{-tolyl})_3\,P{:}O)$$
$$\longrightarrow (Ph_3P)_2PdC_2H_4 \qquad (1.79)$$
$$(74)$$

$$[IrCl(\text{cyclo-octene})_2]'_2 + C_2H_4 \xrightarrow[-50\ C]{\text{heptane}} IrCl(C_2H_4)_4 \qquad (1.80)$$
$$(75)$$

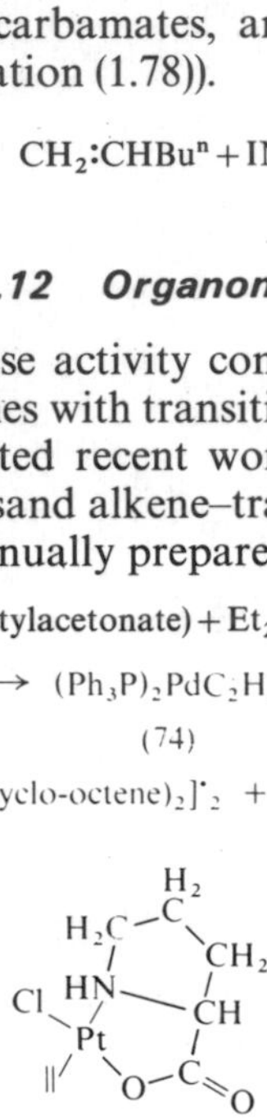

(alkene = C_2H_4, C_3H_6, *cis*- or *trans*-but-2-ene, isobutene,
or 2-methyl-but-2-ene)

$$(76) \qquad\qquad\qquad\qquad (1.81)$$

$$Ph_3SnFe(CO)_2C_5H_5 + C_2H_4 \xrightarrow[20\ ^\circ C,\ 2h]{C_6H_6,\ h\nu} Ph_3SnFe \leftarrow$$
$$(77)$$

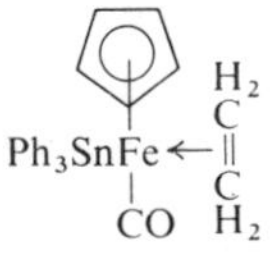

The first zerovalent palladium–ethylene complex ((74), equation (1.79))[280],
ethylene iridium(I) complex ((75), equation (1.80))[281], platinum(II)–alkene
complex containing an amino-acid(76)[282], and an ethylene–iron–tin complex
((77), equation (1.81))[283], have been reported. The rate of insertion of ethylene
into *trans*-PtHBr(PMePh$_2$)$_2$ is found to be much increased if the Br ligand
is replaced by acetone and the complex *trans*-PtH(PMePh$_2$)$_2$(Me$_2$CO)$^+$PF$_6^-$
is used[284]. This is evidence that the insertion reaction has a cationic mechanism. The rotation of the alkene in platinum(II)–alkenes has been further

studied, both experimentally[285] and theoretically[286]. For PtX–acetylacetone–alkene complexes (X = Cl or Br, alkene = ethylene, propene, cis- or trans-but-2-ene, or 2,3-dimethylbut-2-ene), rates of rotation have been estimated and barriers to rotation obtained[285]. As might be expected, the rotation barrier increases with the complexity of the alkene; for ethylene, propene, cis- and trans-but-2-ene, $\Delta G^{\ddagger}$ is 12.7, 12.8, 13.0 and 16.1 kcal mol^{-1} respectively.

The ability of alkenes to complex with a metal has been used to prepare g.l.c. columns for separating alkenes; silver nitrate–benzyl cyanide is a well-known and much used system. A number of rhodium(I) complexes, in particular dicarbonyl-3-trifluoroacetylcamphoratorhodium(I) (78) have now been studied[287], although their value compared with known column packings is not clear.

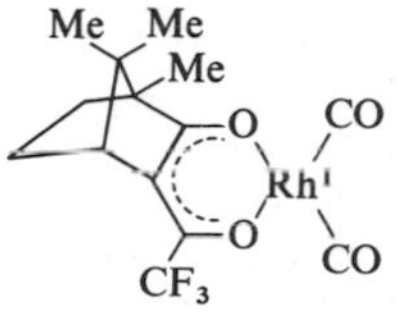

(78)

A number of reviews have recently appeared. These include the following subjects: alkene and acetylene complexes of Ni, Pd and Pt [288], alkene complexes of transition metals as reactive intermediates[289], reactions of homogeneous catalysts in alkene chemistry[290] and palladium complexes in preparative organic chemistry[291].

1.4.2.13 Miscellaneous

Carbocations are of interest, because of their involvement in chemical reactions of many types. Recent work on electrophilic substitution of alkanes (Section 1.3.2.2) is concerned with carbenium and carbonium ions. The preparation of stable carbenium (trivalent) ions from alkenes has been achieved[292]. The necessary conditions require the use of solutions in 'super-acids' and specific conditions vary from one alkene to another. If, for example, a solution of 2-methylbut-2-ene in SO_2 or SO_2ClF is mixed with HSO_3F–SbF_5 or HF–SbF_5 in SO_2 or SO_2ClF at low temperatures then Me_2EtC^+, $Me_2Pr^iC^+$, and Me_3C^+ are all formed. The properties of alkene carbocations and carbanions have been reviewed[293].

Ethylene and propene in an electric discharge give carbenium ions and these react to give numerous products[67].

1.5 ALKADIENES

1.5.1 Alka-1,2-dienes

1.5.1.1 Preparation

Treatment of phenyl-substituted-gem-dibromocyclopropanes with methyl-lithium gives allenes (equation (1.82)) or bicyclobutanes (equation (1.83)) depending on the structure of the dibromocyclopropane[294]. The course of the reaction depends on structural factors. An extensive review of the synthesis of allene derivatives has been published[295].

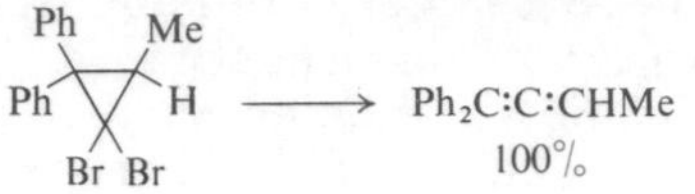

$$\text{(1.82)}$$

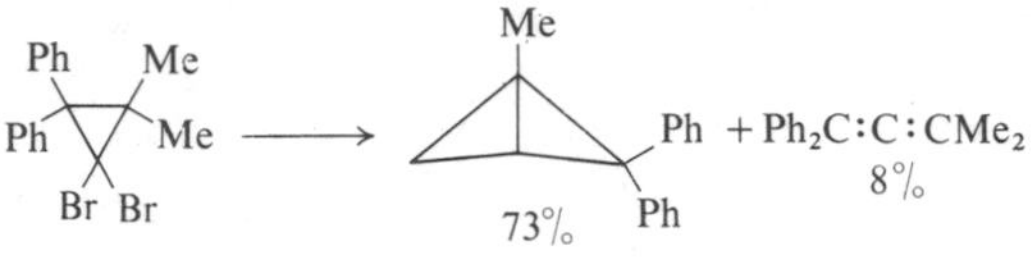

$$\text{(1.83)}$$

1.5.1.2 Reactions

Allene–alkene and allene–allene cyclo-additions have been reviewed[296]. The thermal dimerisation of chiral penta-2,3-diene gives six cyclobutane-type dimers (79)[297]. The products obtained support a tetrahedral (41) or a diradical

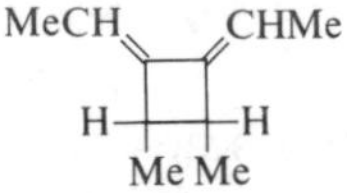

(79)

intermediate for the reaction. Dihalogenocarbenes and carbethoxycarbene add to alkylallenes to give methylene- and alkylidene-cyclopropanes (equation (1.84))[298]. With monoalkylallenes, carbethoxycarbene also gives spiropentanes, which are also the main product from the reaction of mono-, di-,

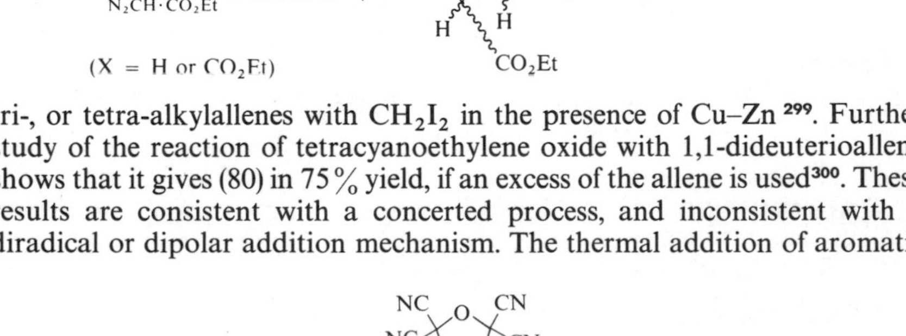

$$\text{(1.84)}$$

(X = H or CO_2Et)

tri-, or tetra-alkylallenes with CH_2I_2 in the presence of Cu–Zn [299]. Further study of the reaction of tetracyanoethylene oxide with 1,1-dideuterioallene shows that it gives (80) in 75% yield, if an excess of the allene is used[300]. These results are consistent with a concerted process, and inconsistent with a diradical or dipolar addition mechanism. The thermal addition of aromatic

(80)

thiones to allenes gives the product (81) (equation (1.85))[301]. In addition to this product the heterocyclic molecule (82) is obtained in 64% yield if the reaction is initiated photochemically[302].

$$R^2C{:}S \ + \ Me_2C{:}C{:}CMe_2 \longrightarrow CH_2{:}CMe{\cdot}C(SCHR^2){:}CMe_2$$
$$(R = Ph \ or \ p\text{-}MeC_6H_4) \qquad\qquad (81)$$

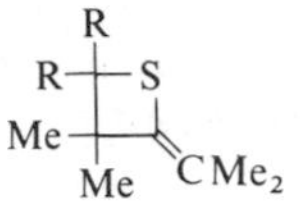

(82)

$$(1.85)$$

The reactions of tetramethylallene with activated isocyanates, electron-deficient alkenes and acetylenes have also been studied[303]. Toluene-p-sulphonyl isocyanate readily gives 3-isopropylidene-4,4-dimethyl-1-(toluene-p-sulphonyl)azetidin-2-one (equation (1.86)), and trichloroacetyl isocyanate reacts over 18 h to give the cyclic product (83) which during 9 days rearranges

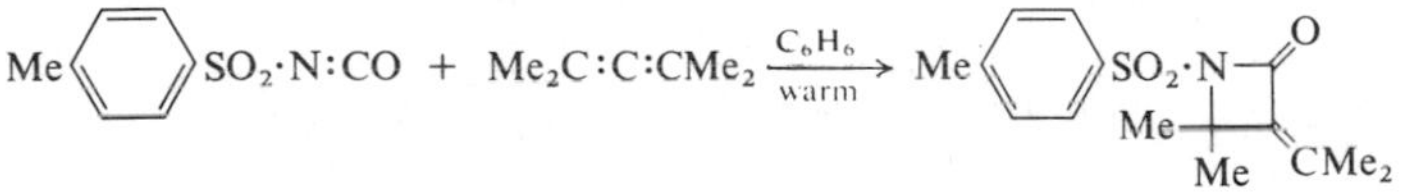

$$(1.86)$$

completely to (84). Acrylonitrile reacts at 150 °C to give a 2:3 mixture of

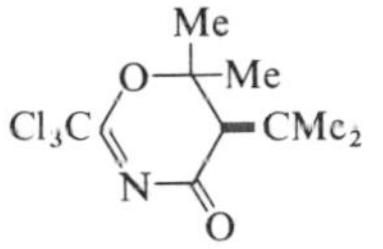

(83)

$$Cl_3C{\cdot}CO{\cdot}NH{\cdot}CO{\cdot}C(CMe_2){\cdot}CMe{:}CH_2$$
(84)

3-isopropylidene-2,2-dimethylcyclobutanecarbonitrile (85) and 3-(2-cyano-ethyl)-2,4-dimethylpenta-2,4-diene (86). A convenient synthesis of tetrakis-

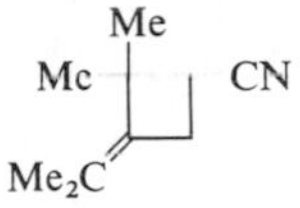

(85)

$$Me_2C{:}C(CH_2{\cdot}CH_2CN){\cdot}CMe{:}CH_2$$
(86)

(trimethylsilyl)allene is the metallation of allene with Bu^{n}Li and its reaction with Me$_3$SiCl [304]. The stereochemistry of addition reactions of allenes has been extensively studied[305–309], including oxymercuration[305], free-radical reactions with hypochlorous acid[306] and with t-butyl hypochlorite[307], reactions with I$_2$, Br$_2$, ICl, IBr and BrCl [308], and the gas-phase addition of HBr [309]. With hypochlorous acid[306] the chlorine becomes attached to the central carbon atom and the OH to the more substituted carbon atom, as shown in the examples in equations (1.87) and (1.88).

$$CH_2{:}C{:}CHEt + HOCl \xrightarrow{70\%} CH_2{:}CCl{\cdot}CH(OH)Et \ (90\%)$$
$$+ CH_2OH{\cdot}CCl{:}CHEt \ (10\%) \qquad\qquad (1.87)$$
$$MeCH{:}C{:}CMe_2 + HOCl \xrightarrow{72\%} MeCH{:}CCl{\cdot}C(OH)Me_2 \qquad (1.88)$$

The products of reaction of Bu^{t}OCl with penta-2,3-diene and with pent-2-yne[307], are the same because the mesomeric radical shown in equation (1.89) is involved. This reaction has also been studied for allene and propyne

$$-C\equiv C-\overset{\cdot}{\underset{\diagdown}{C}}\diagup \longleftrightarrow -\overset{\cdot}{C}=C=\overset{\diagup}{\underset{\diagdown}{C}}\tag{1.89}$$

by using e.s.r. spectroscopy[310]. Thus the radicals obtained by hydrogen abstraction with Bu^{t}O radicals from each hydrocarbon are identical, and the radical obtained by hydrogen abstraction from methyl-substituted allenes has the alkylene-allyl structure (87) rather than the localised dienyl structure (88)[310].

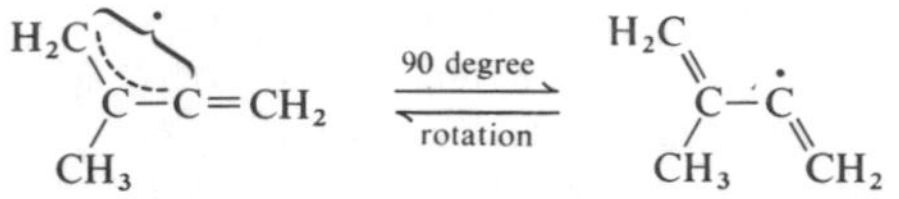

(87)   (88)

A number of reactions are catalysed by Group VIII-metal complexes and allene complexes of these metals have been reported[311-317]. Cyclo-oligomerisation of allene catalysed by (Ph$_3$P)$_2$Ni^0 gives a mixture of trimers, tetramers, pentamers, hexamers, waxes and polymers[311]. The complex (Ph$_3$P)$_2$NiC$_3$H$_4$ (89) is first formed, and has been isolated. The reaction of allene with C$_2$H$_4$ or CH$_2$:CH·CO$_2$Me catalysed by this complex is as given in equation (1.90).

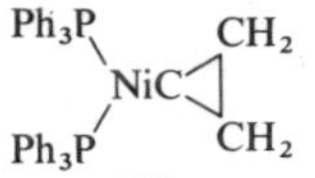

(89)

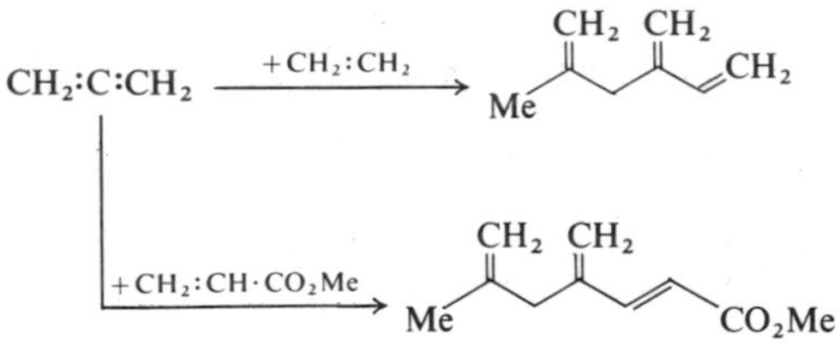

(1.90)

Complexes of iridium and rhodium[312, 313] (e.g. (90)[312] and (91)[313]), palladium (92)[314, 315], and platinum (93)[313, 316] have been reported. The palladium complex (92) has a tri-allene ligand. In the 2,4-dimethylpenta-2,3-diene complexes

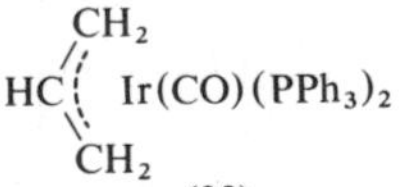

(90)

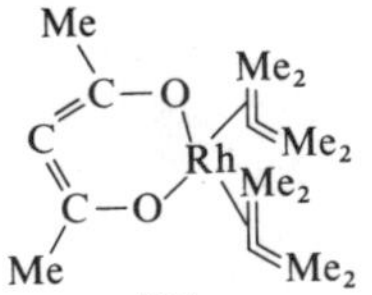

(91)

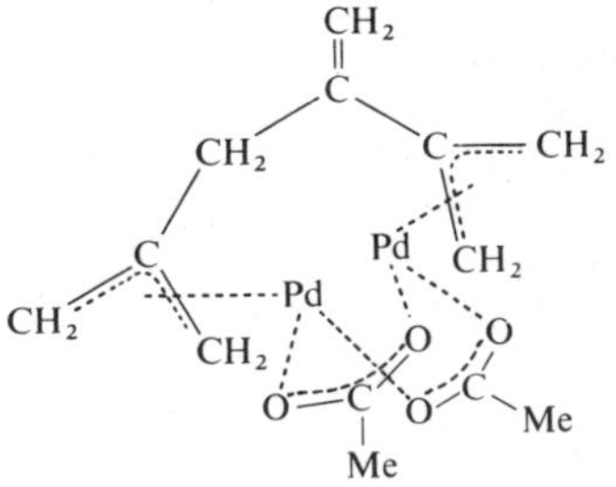

[(Me$_2$C:C:CMe$_2$)PtCl$_2$]$_2$

(93)

(92)

(91) and (93) n.m.r. confirms that the diene is linked through one of its double bonds[313]. That these ligands are perpendicular to the co-ordination plane as in (91) and not in the plane has been shown by a crystal structure determination[317]. A stereoscopic diagram of the acetylacetonatobis(tetramethylallene)-rhodium(I) (91) is given in Figure 1.8.

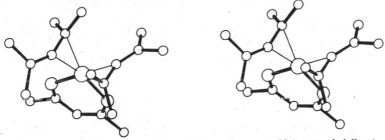

Figure 1.8 Stereoscopic picture of acetylacetonatobis(tetramethylallene)-rhodium(I)[317]. (To view, place a postcard with its short edge between the pictures and view one picture with each eye. Allow the eyes to relax until the images fuse together)

1.5.2 Alka-1,3-dienes

1.5.2.1 Preparation

It has been reported[318] that alkynes react with di-isobutylaluminium hydride to give alka-1,3-dienes (equation (1.91)). Alk-1-ynes could not be used because

$$EtC\vdots CEt + Bu_2^i AlH \longrightarrow EtCH\vdots CEt(AlBu_2^i) \xrightarrow[70\,°C,\,24h]{EtC\vdots CEt,} \begin{array}{c} Et \quad\quad Et \\ \diagdown \quad / \\ C=C \quad AlBu_2^i \\ / \quad\quad \diagdown \\ H \quad\quad C=C \\ \quad\quad / \quad \diagdown \\ \quad\quad Et \quad\quad Et \end{array}$$

$$(1.91)$$

$$\xrightarrow{H_3O^+} \begin{array}{c} Et \quad\quad Et \\ \diagdown \quad / \\ C=C \quad H \\ / \quad\quad \diagdown \\ H \quad\quad C=C \\ \quad\quad / \quad \diagdown \\ \quad\quad Et \quad\quad Et \end{array}$$

$$RCH\vdots CH(AlBu_2^i) \xrightarrow{RC\vdots CH} RCH\vdots CH_2 + RC\vdots CAlBu_2^i \qquad (1.92)$$
$$\quad(94)$$

of the competing cross-metallation (equation (1.92)). However, if the vinylalane (94) is treated with cuprous chloride in tetrahydrofuran the isomerically pure *trans-trans*-1,3-diene is obtained (equation (1.93))[319], presumably by the formation and destruction of vinyl copper complex.

$$\begin{array}{c} R \quad\quad H \\ \diagdown \quad / \\ C\vdots C \\ / \quad\quad \diagdown \\ H \quad\quad AlBu_2^i \end{array} \xrightarrow{CuCl} \begin{array}{c} R \quad\quad H \\ \diagdown \quad / \\ C\vdots C \quad H \\ / \quad\quad \diagdown \\ H \quad\quad C\vdots C \\ \quad\quad / \quad \diagdown \\ \quad\quad H \quad\quad R \end{array} \qquad (1.93)$$

1.5.2.2 *Reactions*

Oligomerisation of buta-1,3-diene is still a much studied topic, and argument continues as to whether or not such reactions are concerted. The recent isolation of bis(buta-1,3-diene)tricyclohexylphosphinenickel(0) (95) and tricyclohexylphosphine(divinylcyclobutane)nickel(0) (96)[320], has allowed the products of further reaction of these two model intermediates to be examined[321]. They both give the same products indicating that the dimerisation is not concerted but is a multistep process involving, among other intermediates, an α,ω-bisallyl-C_8-chain. Cyclodimerisation occurs with a $[\mathrm{Nihal}_2(\mathrm{PBu}_3^n)_2 +$

$$
\begin{array}{cc}
\mathrm{CH_2{=}\!\!{=}CH_2} & \\
\mathrm{(cyclohexyl)_3P-Ni} & \\
\mathrm{CH_2{=}\!\!{=}CH_2} & \\
(95) & (96)
\end{array}
$$

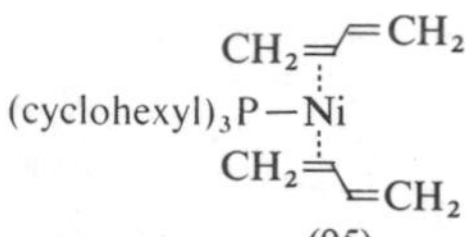

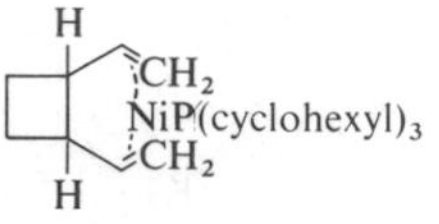

RLi] catalyst[322], or with a $[\textit{trans}\text{-NiCl}(o\text{-tolyl})(\mathrm{PEt}_3)_2 + \text{alcohol}]$ catalyst[323] to give 2-methylenevinylcyclopentane(97). In the absence of an alcohol the latter catalyst gives cyclo-octa-1,5-diene and 4-vinylcyclohexene in low yield.

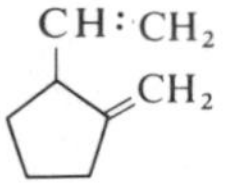

$$
\begin{array}{c}
\mathrm{CH{:}CH_2} \\
\text{⟨structure⟩} \quad \mathrm{CH_2} \\
(97)
\end{array}
$$

By using palladium complexes as catalysts octa-1,3,7-triene (98) is obtained as the major product[324], but in acetone solution octa-1,4,7-triene (99) and octa-2,4,6-triene (100) are also produced. A palladium complex (101) of the type that is an intermediate in such oligomerisations can be isolated from

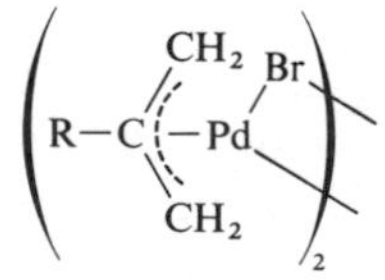

$$
\begin{array}{ccc}
(98) & (99) & (100)
\end{array}
$$

the reaction of $[(\pi\text{-allyl})\mathrm{PdBr}]_2$ with buta-1,3-diene, when the buta-1,3-diene is inserted into the π-allyl ligand (equation (1.94))[325].

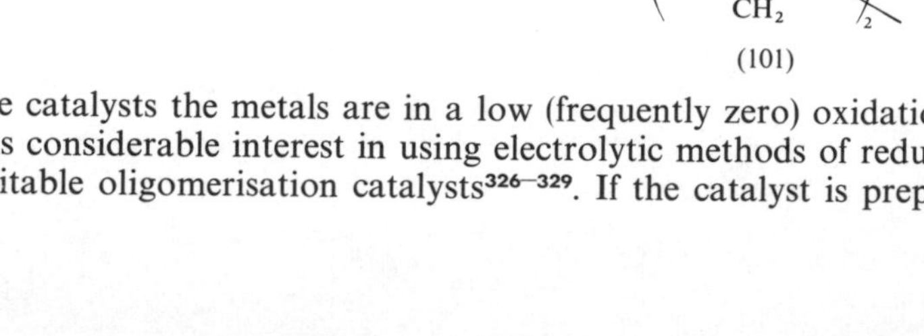

$$
\left(\begin{array}{c} \mathrm{CH_2} \; \mathrm{Br} \\ \mathrm{R-C{-}Pd} \\ \mathrm{CH_2} \end{array} \right)_2 + 2\mathrm{CH_2{:}CH{\cdot}CH{:}CH_2} \longrightarrow \left(\begin{array}{c} \mathrm{CH_2} \\ \mathrm{R-C} \\ \mathrm{CHY} \\ \mathrm{CH_2} \\ \mathrm{CH} \quad \mathrm{Br} \\ \mathrm{CH{-}Pd} \\ \mathrm{CH_2} \end{array} \right)_2 \tag{1.94}
$$

$$
(101)
$$

In these catalysts the metals are in a low (frequently zero) oxidation state. There is considerable interest in using electrolytic methods of reduction to give suitable oligomerisation catalysts[326-329]. If the catalyst is prepared by reducing (acetylacetonato)$_2$Ni^{II} in the presence of a tetra-alkylammonium bromide the oligomerisation product is cyclododeca-1,5,9-triene(102)[326]. If triphenylphosphine is also present then cyclo-octa-1,5,-diene(103) and

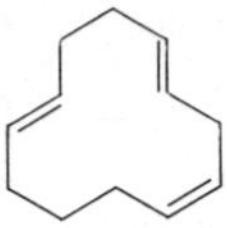

(102) 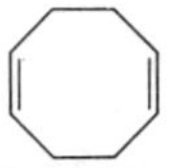   (103) 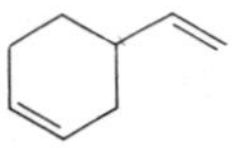   (104) 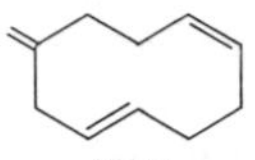   (105) 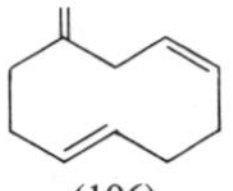   (106)

4-vinylcyclohexene (104) are formed instead[326], and these are also the two products if the catalyst is NiCl$_2$(CH$_2$OMe)$_2$ and triphenylphosphine, in acetonitrile or NN-dimethylformamide[327]. Another similar system, NiCl$_2$ with pyridine, gives linear octatrienes (98–100) and no cyclic oligomers[328], as do the systems CoBr$_2$ or Co(acetylacetone)$_3$ in the presence of Ph$_3$P, (PhO)$_3$P, Ph$_3$As, α,α'-bipyridyl, or salicylaldehyde[329].

Mixed oligomerisations, or telomerisations (mixed polymerisations of two or more compounds) have also been much studied. The co-oligomerisation of buta-1,3-diene and allene (1:10) with a nickel complex in an autoclave gives 8- and 9-methylene-cis, $trans$-cyclodeca-1,5-diene (105 and 106)[330]. The yield is maximised (to $c.$ 70%) by passing the allene continuously into the mixture and so keeping its concentration low. Alkyl octa-2,7-dienyl ethers are the major products when buta-1,3-diene and alcohols react together in the presence of a reducing agent and a nickel-complex catalyst (e.g. equation (1.95))[331]. Again the active catalyst is Ni^0 formed by reduction of the added complex.

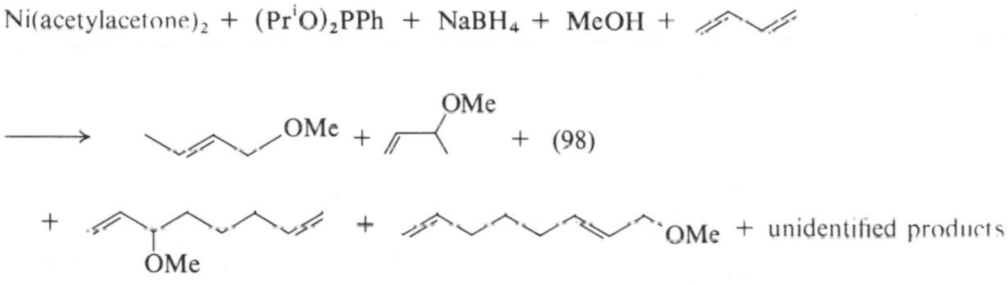

Ni(acetylacetone)$_2$ + (Pr^{i}O)$_2$PPh + NaBH$_4$ + MeOH +

(1.95)

+ (98)

+ unidentified products

Carbonylation of buta-1,3-diene or isoprene, in ethanolic solution gives esters of the 'monomeric' acid when PdCl$_2$ is used as the catalyst (equation (1.96))[332] and esters of the 'dimeric' acid when Pd(acetylacetone)$_2$ with PPh$_3$ is used (equation (1.97))[333].

CH$_2$:CMe·CH:CH$_2$ + CO + EtOH $\xrightarrow[\text{100 C}]{\text{PdCl}_2}$ Me$_2$C:CH·CH$_2$·CO$_2$Et

+ CH$_2$:CMe·CH$_2$·CH$_2$·CO$_2$Et

+ Me$_2$C(O Et)·CH$_2$·CH$_2$·CO$_2$Et

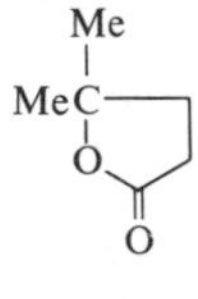

(1.96)

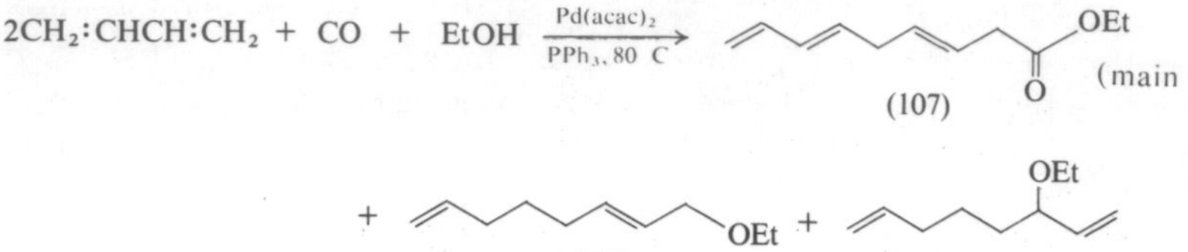

$$2CH_2\!:\!CHCH\!:\!CH_2 + CO + EtOH \xrightarrow[\text{PPh}_3, 80\ C]{\text{Pd(acac)}_2}$$

(107) (main product)

(1.97)

\+ (108) + (109)

The ethers (108) and (109) are not converted into the ester (107)[333], which must, therefore come from buta-1,3-diene as shown in equation (1.98).

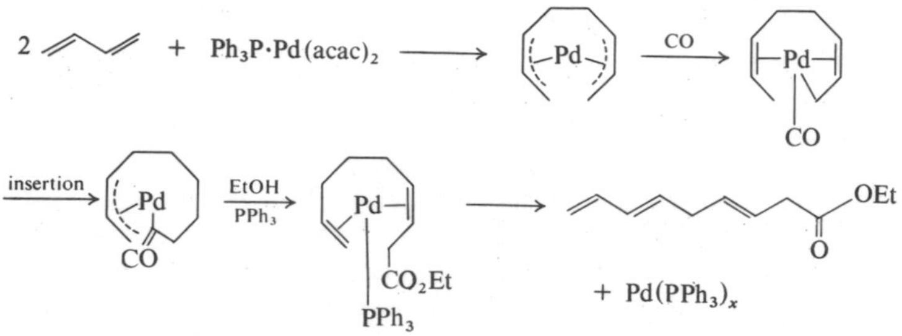

$$2 \quad + \quad Ph_3P\!\cdot\!Pd\,(acac)_2 \longrightarrow \xrightarrow{CO}$$

(1.98)

$$\xrightarrow{\text{insertion}} \xrightarrow[\text{PPh}_3]{\text{EtOH}} \longrightarrow$$

$$+ \; Pd(PPh_3)_x$$

The reaction of buta-1,3-diene with aldehydes with palladium phosphine catalysts gives, by an insertion mechanism similar to that in equation (1.98), three products[334–336]: two isomeric divinyltetrahydropyrans, (110)[334–336] and (111)[334] and the alcohol (112)[335, 336].

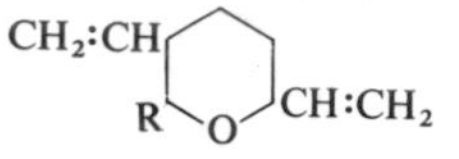

(110) 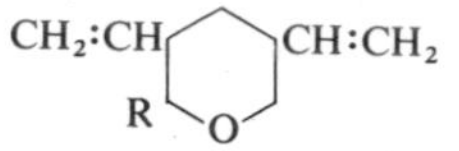   (111) 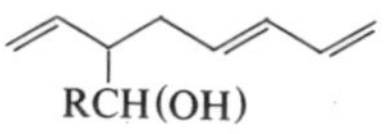   (112)

(R = H,[334, 335] Me[335] or Ph[336])

The same catalyst gives octadienyl acetates in the reaction of buta-1,3-diene with acetic acid[337]. Addition of tertiary amines causes a marked increase in reaction rate. Primary and secondary amines cannot be used as these react with the acetates formed to give octadienylamines. Catalytic codimerisation of buta-1,3-diene with acrylic[338] or glyoxylic[339] esters, in the presence of $Fe(CHAc_2)_3$–Ph_3P–Et_3Al or $SnCl_4$ respectively, produces esters such as (113) and (114)[338].

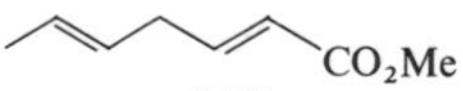

(113) 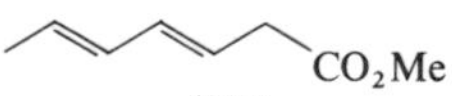   (114)

By using palladium–phosphine-complex catalysts, addition products (115) and (116) (equation (1.99)) are formed when buta-1,3-diene reacts with active methylene compounds such as β-keto-esters, β-diketones, dialkyl malonates,

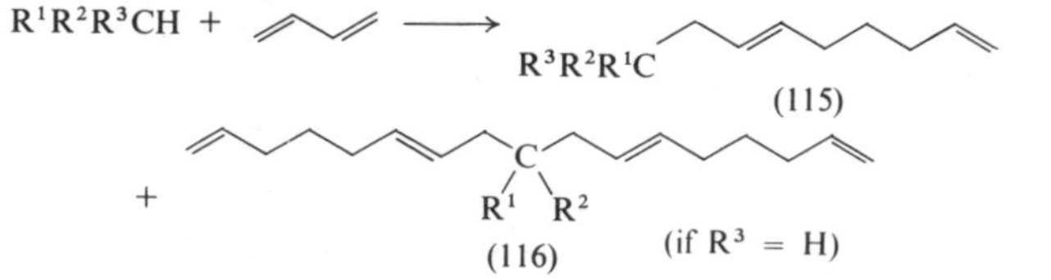

$$R^1R^2R^3CH + \longrightarrow$$

(115)

(1.99)

\+

(116)   (if R^3 = H)

α-formylketones, α-formyl-, α-cyano- and α-nitro-esters, cyanoacetamide and ethyl phenylsulphonylacetate[340]. If a chelate phosphine complex such as $PdBr_2(Ph_2PCH_2 \cdot CH_2PPh_2)_2$, is used then dimerisation of the buta-1,3-diene does not occur[341], as only one buta-1,3-diene molecule at a time can coordinate to the palladium.

The products are the 1:1 adducts $R^1R^2CH \cdot CHMe \cdot CH:CH_2$ and $R^1R^2CH \cdot CH_2 \cdot CH:CHMe$. Norbornadiene and buta-1,3-diene react to give five different possible products (117)–(121), depending on the catalyst used[342-344]. Thus Fe(cyclo-octatetraene)$_2$ or $FeCl_3$–Pr^iMgCl gives (117), Fe(acetylacetone)$_3$–$AlEt_2Cl$–$(Ph_3PCH_2)_2$ gives (119) and (120)[342] Co(acetylacetone)$_3$–Et$_3$Al gives (121)[343, 344], but with PPh$_3$ added, CoCl$_2$ gives (119)[344] and NiCl$_2$–AlEt$_2$Cl–PPh$_3$ gives (118) as the main product with a little (117)[344].

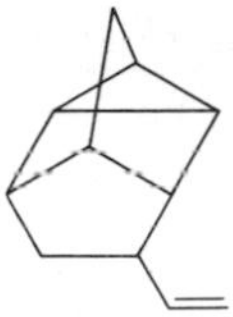

(117) 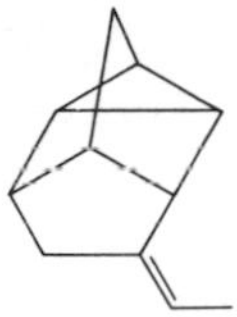   (118) 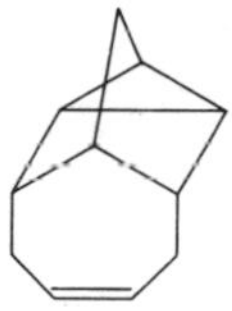   (119)

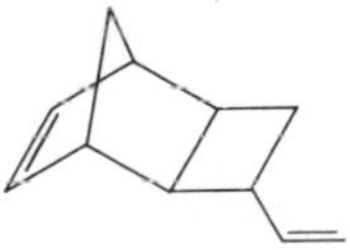

(120) 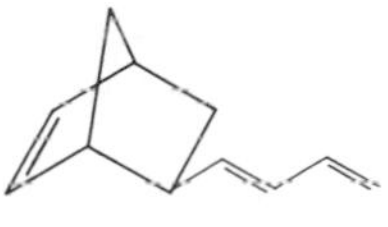   (121)

Isomeric *syn*- and *anti*-1-methyl-(π-allyl)Co(CO)$_3$, (122) and (123), have been prepared from buta-1,3-diene and HCo(CO)$_4$ [345]. These, together with the *cis*-σ-complex (124), are intermediates in the hydrogenation of buta-1,3-diene with HCo(CO)$_4$.

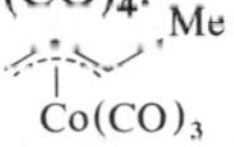

(122)    (123) 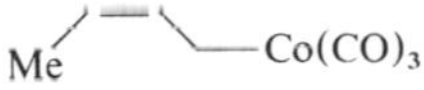   (124)

Palladium–triphenylphosphine complexes catalyse the reaction of buta-1,3-diene or isoprene with phenyl isocyanate to give the piperidones (125)–(128) (equations (1.100) and (1.101)[346].

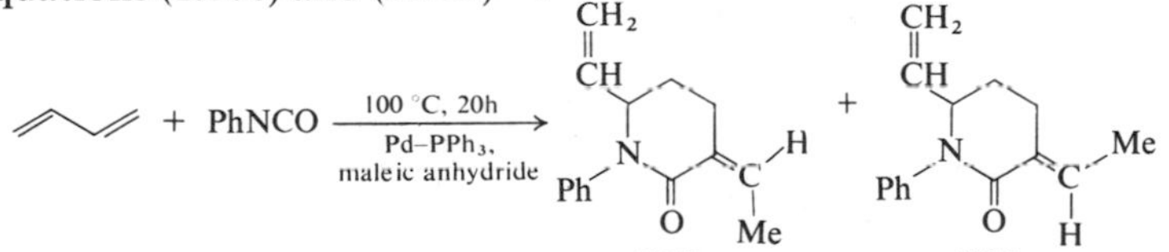

(125)   (126)   (1.100)

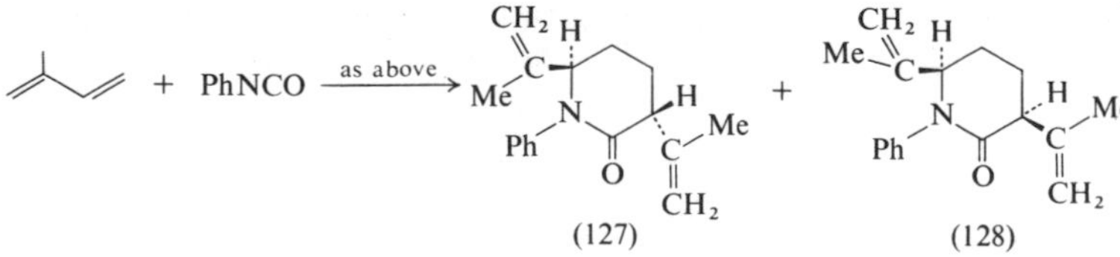

(127)   (128)   (1.101)

The electrolysis of a system containing buta-1,3-diene and styrene gives 1,4-dimethoxy-1-phenylhex-5-ene (129), and 1,6-dimethoxy-1-phenylhex-4-

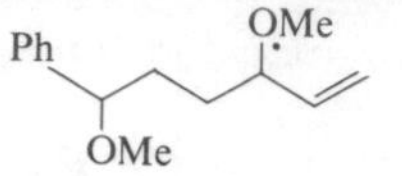

(129) 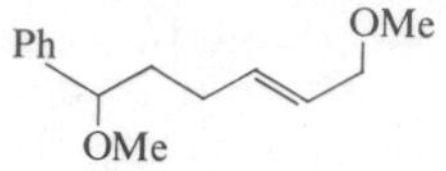   (130)

ene (130)[347], whereas the co-electrolysis of buta-1,3-diene with Grignard reagents gives a variety of products by a radical mechanism[348]. Hydrosilylation catalysed by palladium and nickel complexes[349] and hydrostannylation[350] of 1,3-dienes, involves addition of R_3SiH to the buta-1,3-diene dimer[349] to give (131), and radical addition of R_3SnH to give both

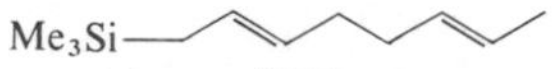

(131)

the 1:4- and 1:2-adduct of penta-1,3-diene respectively[350]. A stereospecific addition of secondary amines to buta-1,3-diene in the presence of Bu^nLi in cyclohexane gives, for example, 1-diethylamino-cis-but-2-ene from diethylamine (equation (1.102))[351]. The amine is essential for the coupling reaction, although too much decreases the stereospecificity, and it is thought that the reactant is the complex, $Et_2NLi\cdot2Et_2NH$. Tetrafluoro-

$$Et_2NH + Bu^nLi \rightarrow Et_2NLi + Bu^nH$$

$$Et_2NLi + CH_2{:}CH{\cdot}CH{:}CH_2 \xrightarrow{Et_2NH} Et_2NC_4H_6Li \xrightarrow{Et_2NH} Et_2NCH_2{\cdot}CH{:}CHMe$$
$$+ Et_2NLi \qquad (1.102)$$
$$(recycled)$$

hydrazine reacts with both isolated and conjugated double bonds[352]. Conjugated double bonds react with one N_2F_4 molecule to give 1:4 addition, and if forcing conditions are used an explosion occurs.

Buta-1,3-diene reacts with water in the presence of a palladium–phosphine catalyst and CO_2 to give octadienols in high yield (equation (1.103))[353]. The bromination of buta-1,3-diene in methanol gives the bromoether (132), with

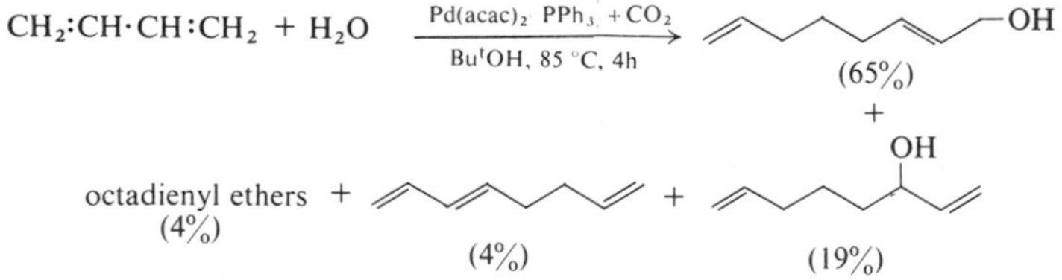

$$CH_2{:}CH{\cdot}CH{:}CH_2 + H_2O \xrightarrow[Bu^tOH,\ 85\ °C,\ 4h]{Pd(acac)_2\ PPh_3\ +CO_2}$$

only a small amount of the isomer (133) and none of the isomer (134). This is evidence that the intermediate in bromination is best described by the structure (135) with, perhaps slight delocalisation across the allylic system[354].

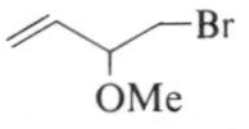

(132) 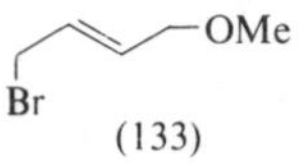   (133) 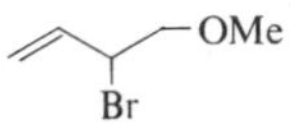   (134) 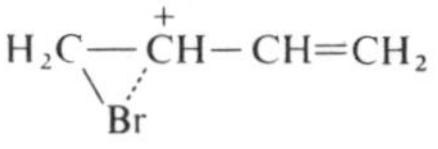   (135)

The excited state of *cis*- and that of *trans*-penta-1,3-diene are of different reactivity, because when they are irradiated with u.v. light the two reaction products, 1,3-dimethylcyclopropene and 3-methylcyclobutene are formed at different rates from the two isomers[355]. The intermediate in the photoisomerisation is the diradical (136) proposed earlier[356]. The direct *cis-trans*

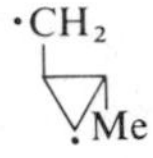

(136)

photo-isomerisation has also been studied[357]. The photosensitised addition of buta-1,3-diene, isoprene, or cyclopentadiene to α-acetoxyacrylonitrile gives, for example, three adducts (137) to (139), as well as the three buta-1,3-diene dimers (140), (141) and (104) (equation (1.104))[358]. The reaction takes place by formation of the diradical (142). This then undergoes 1,4- or 1,6-ring closure. The photosensitised addition of acrylonitrile to buta-1,3-diene also

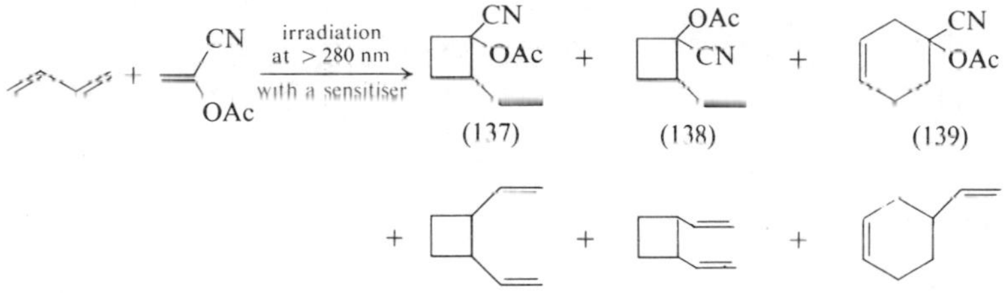

(137)   (138)   (139)

(140)   (141)   (104)

(1.104)

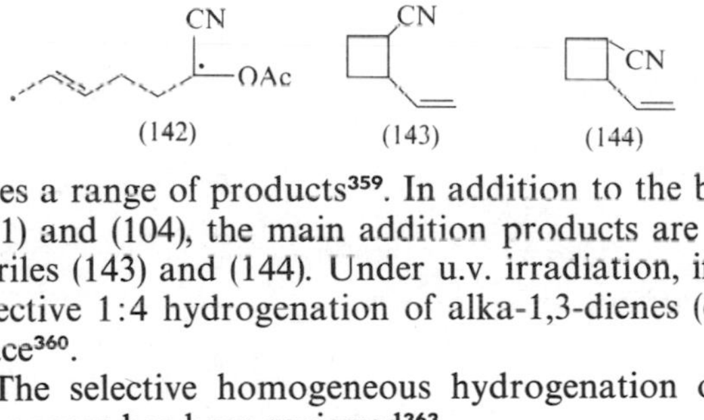

(142)   (143)   (144)

gives a range of products[359]. In addition to the buta-1,3-diene dimers (140), (141) and (104), the main addition products are the vinylcyclobutyl carbonitriles (143) and (144). Under u.v. irradiation, in the presence of $Cr(CO)_6$, selective 1:4 hydrogenation of alka-1,3-dienes (e.g. equation (1.105)) takes place[360].

The selective homogeneous hydrogenation of dienes and polyenes to monoenes has been reviewed[363].

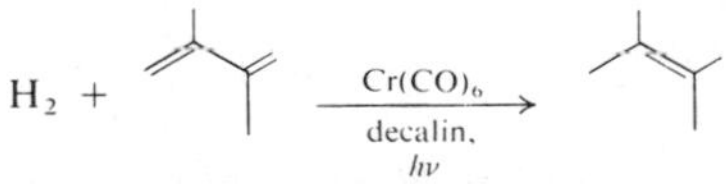

(1.105)

Oxetanes are formed when benzophenone reacts with conjugated dienes under u.v. irradiation (e.g. equation (1.106)[361]. The mechanism involves attack of the triplet benzophenone on the ground-state diene molecule.

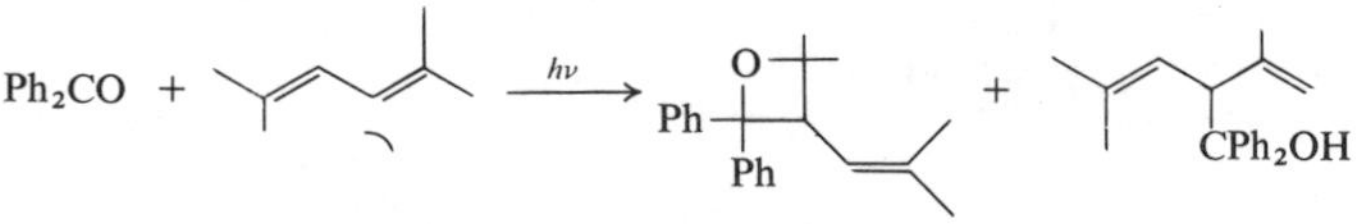

(1.106)

Tetramethyldiphosphine adds to buta-1,3-diene by a $Me_2P\cdot$-radical attack, to give the expected 1:4 addition products (equation (1.107))[362].

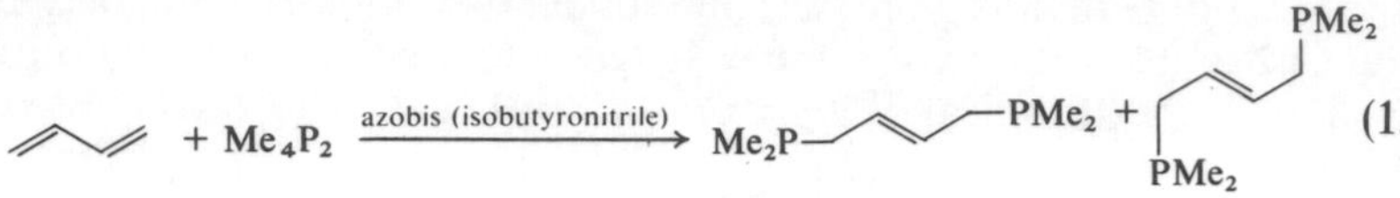

$$\text{(1.107)}$$

A number of complexes involving buta-1,3-diene have been prepared. These include a complex (145) formed from diborane and buta-1,3-diene in tetrahydrofuran[364], a but-2-enyl-buta-1,3-diene complex of cobalt (146) formed from buta-1,3-diene in the presence of Zn, EtOH and PPh_3 [365], (π-allyl)nickel phosphite complexes formed from the reaction of $HNi\,P(OR)_3^+$ (R = Me or Et) with conjugated dienes[366] and buta-1,3-diene iron carbonyl complexes formed from buta-1,3-diene, $FeCl_3$ and Pr^i—MgCl in ethereal solution[367].

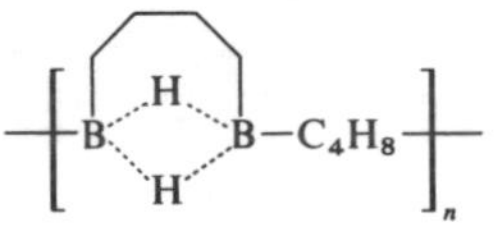

(145)

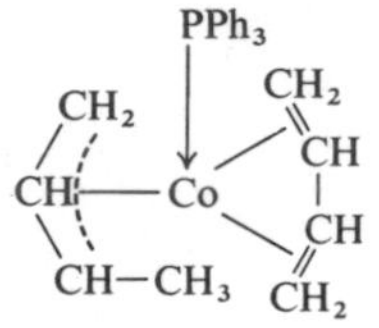

(146)

1.5.3 Alka-1,4-dienes

1.5.3.1 Preparation

Allylboranes react with vinyl ethers to form pure 1,4-dienes and their derivatives (e.g. equation (1.108))[368].

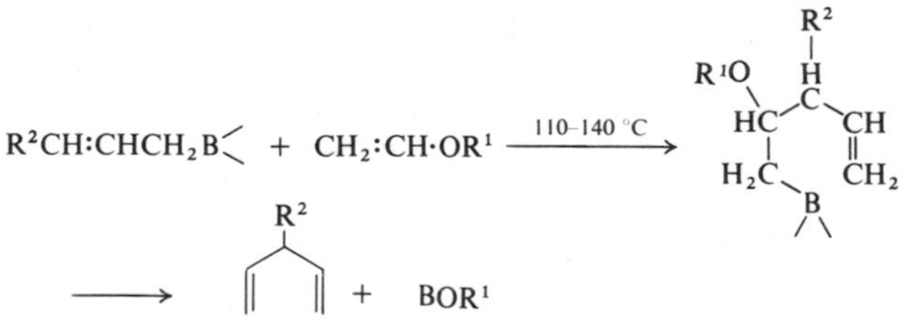

$$R^2CH{:}CHCH_2B{<} \;+\; CH_2{:}CH{\cdot}OR^1 \xrightarrow{110-140\,°C} \qquad \text{(1.108)}$$

$$\longrightarrow \quad + \quad BOR^1$$

[R^1 = Et, Bu^n; R^2 = H (80% yield) or Me (50% yield)]

2-Methylpenta-1,4-diene is formed from n-butyl vinyl ether and trismethyl-allylborane, and from isopropenyl ethyl ether and triallylborane (equation (1.109)). Allylboranes react with monosubstituted alkynes to give vinyl-

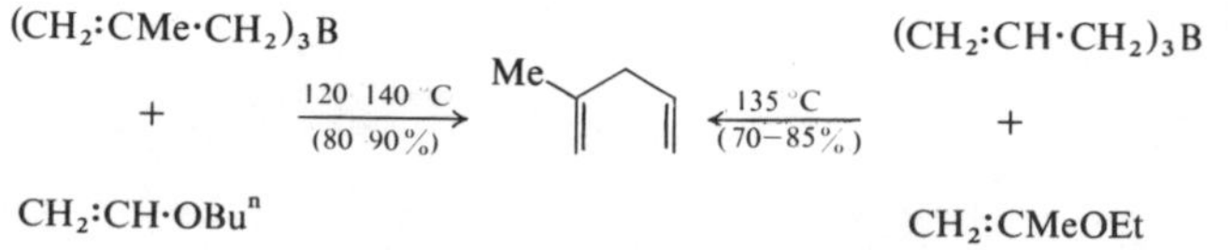

$$(CH_2{:}CMe{\cdot}CH_2)_3B \qquad\qquad (CH_2{:}CH{\cdot}CH_2)_3B$$

$$+ \quad \xrightarrow[\,(80-90\%)\,]{120-140\,°C} \quad \xleftarrow[\,(70-85\%)\,]{135\,°C} \quad + \qquad \text{(1.109)}$$

$$CH_2{:}CH{\cdot}OBu^n \qquad\qquad CH_2{:}CMeOEt$$

borane derivatives, and subsequently substituted penta-1,4-dienes (equation (1.110))[369].

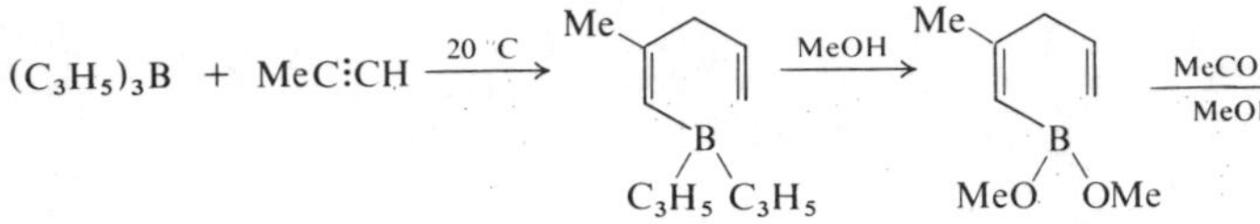

$$(C_3H_5)_3B \; + \; MeC\dot{:}CH \xrightarrow{20\ C} \quad \xrightarrow{MeOH} \quad \xrightarrow[MeOH]{MeCO_2H} \qquad (1.110)$$

1.5.3.2 Reactions

$(C_2H_4)(PAr_3)_2Ni^0 + HCl$ (Ar = o-tolyl or phenyl) gives a very active catalyst for skeletal isomerisation of 1,4-dienes (equation (1.111))[370]. The active catalyst is a nickel hydride. By studying a diene labelled with deuterium,

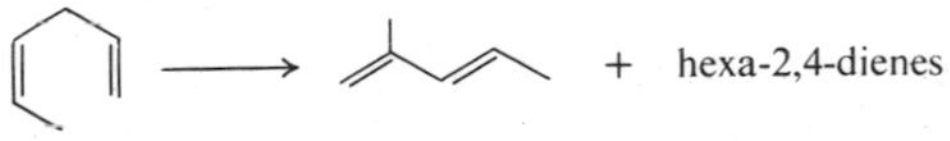

$$\bigcirc\!\!\!\!\parallel \longrightarrow \bigwedge\!\!\!\bigvee \quad + \quad hexa\text{-}2,4\text{-dienes} \qquad (1.111)$$

the position of particular carbon atoms in the reactant can be identified in the product (equation (1.112))[371].

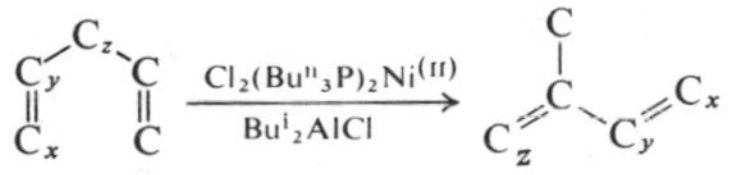

$$\xrightarrow[Bu^i_2AlCl]{Cl_2(Bu^n_3P)_2Ni^{(II)}} \qquad (1.112)$$

1.5.4 Alka-1,5-dienes

1.5.4.1 Preparation

The widespread natural occurrence of acyclic 1,5-dienes has stimulated

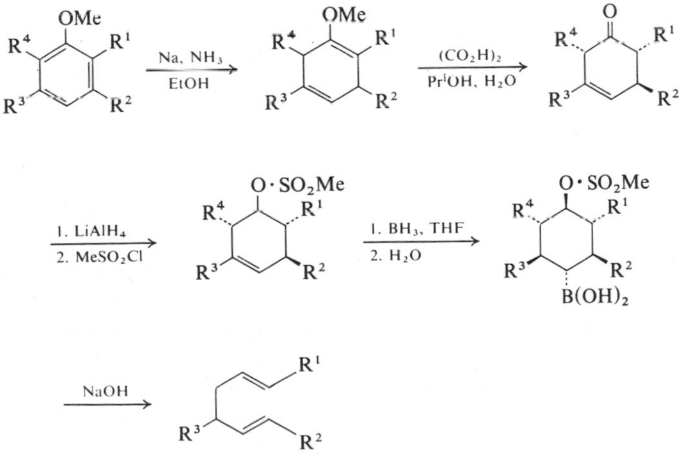

$$\xrightarrow[EtOH]{Na,\ NH_3} \qquad \xrightarrow[Pr^iOH,\ H_2O]{(CO_2H)_2} \qquad (1.113)$$

$$\xrightarrow[2.\ MeSO_2Cl]{1.\ LiAlH_4} \qquad \xrightarrow[2.\ H_2O]{1.\ BH_3,\ THF}$$

$$\xrightarrow{NaOH}$$

($R^1 = R^2 = R^4 = H$, $R^3 = Et$, or $R^1 = R^2 = H$, $R^3 = R^4 = Me$, or $R^1 = R^2 = R^3 = Me$, $R^4 = H$)

interest in the design of efficient synthetic routes to these compounds. One such route is given in equation (1.113)[372].

1.5.4.2   Reaction

Hexa-1,5-diene reacts with two molecules of N_2F_4 to give the fully saturated product[352].

1.5.5   Alkatrienes

1.5.5.1   Reactions

$Me_2C:C:C:CMe_2$ reacts with iodine in $CHCl_3$ to give the 2,3-di-iodo-adduct, but no 1,4-di-iodo-adduct. In methanol the 1,4-adduct predominates and in $CHCl_3$ bromine gives the 1,4-dibromo-adduct[373].

1.6   ALKYNES

1.6.1   Preparation and isolation

Terminal alkynes can be prepared from terminal alkenes by bromination followed by dehydrobromination at room temperature with a solution of $NaNH_2$ or NaH in Me_2SO[374]. The yields of the higher alkynes are much higher than in their usual preparation from sodium acetylide and an alkyl halide in liquid NH_3. Heating them with $NaNH_2$ converted the alk-1-ynes into alk-2-ynes[374]. Long-chain alkynes are also formed by the usual reaction mentioned above ($NaC_2H + RhaI$), with Me_2SO as the solvent[375].

The polyynes (147) and (148) have been isolated from the plant *Psilothonna tagetes*[376].

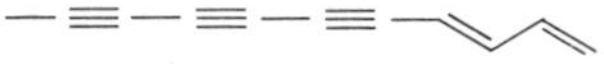

(147)    (148)

1.6.2   Reactions

1.6.2.1   Hydroboration

Monohydroboration of terminal alkynes with (cyclohexyl)$_2$BH or $(Me_2Pr^iC)_2BH$ puts the boron almost exclusively on the terminal carbon atom[377]. If unsymmetrical disubstituted acetylenes are used, the direction of addition depends on the size of the substituents. Alkenic side-chains can be introduced stereoselectively into cyclic systems by the reaction given in equation (1.114)[378]. The second stage, the treatment of the vinylborane with 6 M-NaOH and then I_2 in THF, results in migration of the 2-methylcyclohexyl moiety with retention of configuration, giving an 85% yield of *trans*-1-methyl-2-(*cis*-hex-1-enyl)cyclohexane.

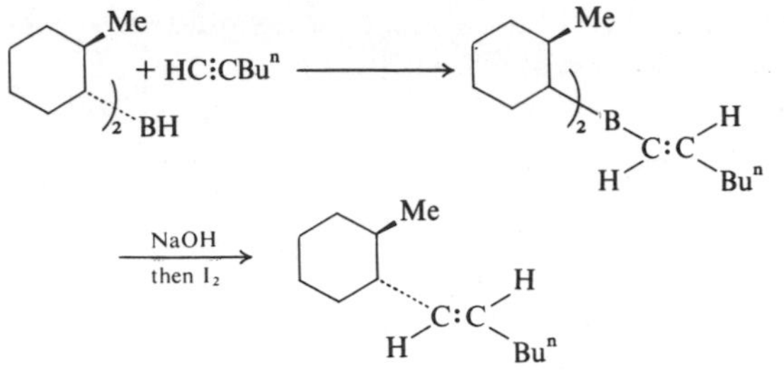

$$(1.114)$$

Conjugated diynes on hydroboration give alka-1,3-dienes containing two boron atoms[379], and these may be converted into the dienes by protonolysis.

1.6.2.2 Reactions with organic compounds

Alkinylation of pyridine by using the coupling reaction (equation (1.115)) has been studied with a wide range of alkynes and alkadiynes[380]. Carbo-

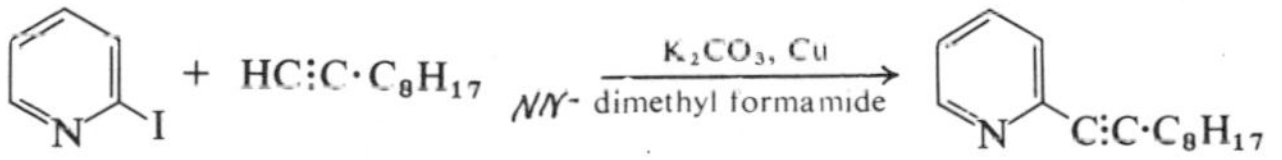

$$(1.115)$$

ethoxycarbene, prepared by the copper-catalysed decomposition of ethyl diazoacetate, with $MeC{:}C{\cdot}CH{:}CHMe$ gives three adducts[381], ethyl trans-trans-2-methyl-3-(prop-1-ynyl)cyclopropane-1-carboxylate(149) and cis-cis-2-methyl-3-(prop-1-ynyl)cyclopropane-1-carboxylate(150) by addition to the double bond, and ethyl 2-methyl-3-(prop-1-enyl)-cycloprop-2-ene-1-carboxylate(151) by addition to the triple bond. The yields are 22, 5 and 42% respectively and this shows that the triple bond is the more reactive. The photoaddition of vinylene carbonate(152) to an alkyne gives the cyclobutene (153) (equation (1.116))[382, 383].

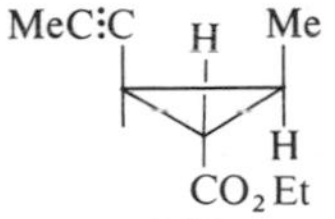

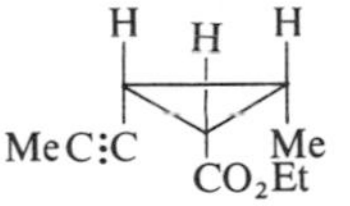

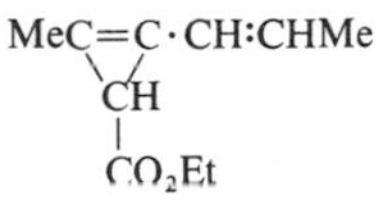

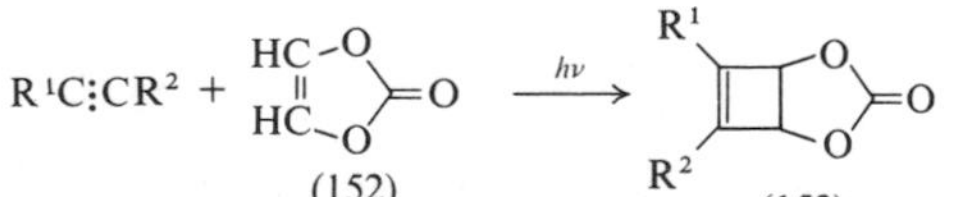

$$(1.116)$$

The formation of benzene by the mercury-photosensitised trimerisation of acetylene has been studied by using $C_2H_2 + C_2D_2$. The results can be explained as well by a modified excited-state mechanism as by the accepted free-radical mechanism[384]. Conjugated alkadiynes also trimerise in the presence of a catalyst: $Ni(CO)_2PPh_3$ gives the compound (154) (equation

(1.117)) as the only product, while (cyclopentadienyl) $Co(CO)_2$ gives, in addition, the compound (155)[385].

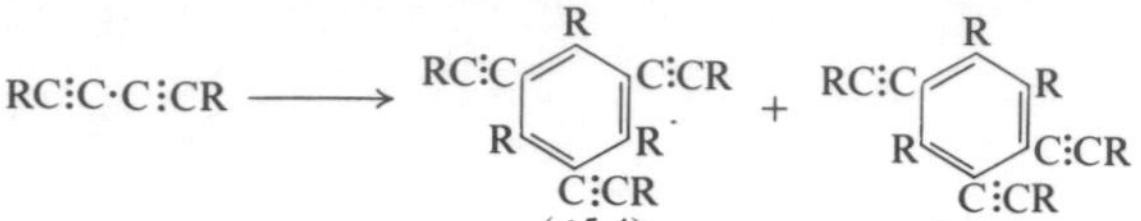

$$RC\text{:}C\cdot C\text{:}CR \longrightarrow (154) + (155) \qquad (1.117)$$

The photocyclo-addition of acetylene to benzene gives cyclo-octatetraene, by insertion of the acetylene into the benzene ring[386].

1.6.2.3   Oxidation and reactions with sulphur compounds

Nitric oxide reacts with pent-1-yne by a free radical mechanism (equation (1.118)) to give a mixture of pent-1-yn-3-ol, pent-1-yn-3-one and pent-1-yn-3-yl nitrate[387]. The nitrate is formed by nitration of the alcohol. Acetylene is

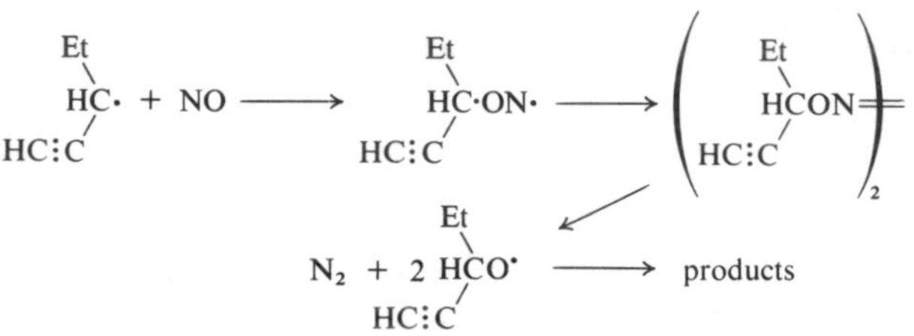

$$(1.118)$$

oxidised to oxalic acid by 48 % nitric acid at 24 °C in the presence of palladium[388]. α-Diketones are formed from alkynes by oxidation with RuO_4[389]. The catalyst is generated *in situ* from catalytic amounts of RuO_2 and an oxygen donor such as sodium metaperiodate or sodium hypochlorite. Acids are also formed, and these are the only products from terminal alkenes (equation (1.119)).

$$R^1C\text{:}CR^2 \xrightarrow[\text{CCl}_4-\text{H}_2\text{O, 0°C}]{\text{RuO}_2-\text{NaOCl}} R^1CO\cdot COR^2 + R^1CO_2H \qquad (1.119)$$

$(R^1 = R^2 = Bu^n$, ketone, 70 %; acid, 20 %; $= R^2 = Pr^n$, ketone, 58.5 %; acid, 40 %; $R^1 = Bu^t, R^2 = H$, acid, 60 %)

Acetylene reacts with organic disulphides to give a mixture of products

$$HC\text{:}CH + Bu^iS\cdot SBu^i \xrightarrow[\text{12–19 atm.}]{\text{170 180°C}} H_2C\text{:}CH\cdot SBu^i + Bu^iS\cdot CH\text{:}CH\cdot SBu^i$$
$$\qquad\qquad\qquad\qquad\qquad (38\%)\qquad\qquad\qquad (15\%)$$
$$+ MeCH(SBu^i)_2 + Bu^iS\cdot CH_2\cdot CH_2\cdot SBu^i \qquad (1.120)$$
$$(11\%)\qquad\qquad (2\%)$$

(equation (1.120))[390]. SCl_2 reacts with alkynes to give the corresponding divinyl sulphide (e.g. (156)) in quantitative yield[391].

$$EtCCl\text{:}CEt\cdot S\cdot CEt\text{:}CClEt$$
$$(156)$$

1.6.2.4   Halogenation

Arenesulphenyl chlorides and areneselenyl halides add to alkynes to give aryl halogenoalkenyl sulphides and selenides in high yield (equation (1.121))[392].

$$R^1C_6H_4AX + R^2C\vdots CR^2 \rightarrow R^1C_6H_4A\cdot CR^2\vdots CR^2X \qquad (1.121)$$

$(R^1 = H, \textit{o}\text{-}NO_2, \text{ or } \textit{p}\text{-}NO_2; A = S, Se; X = Cl, Br; R^2 = H, Pr^n Bu^n \text{ or } Ph)$

Chlorosulphonyl isocyanate adds to alkynes[393]. The adduct with hex-3-yne (157) gives hexan-3-one on hydrolysis and methyl 2-ethyl-3-oxopentanoate (equation (1.122)) on methanolysis.

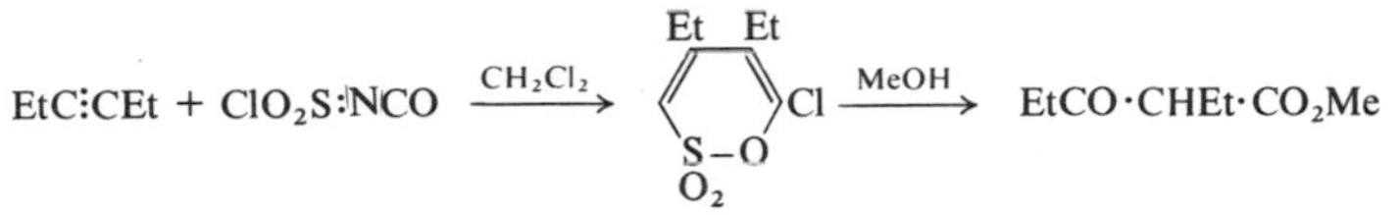

$$EtC\vdots CEt + ClO_2S\vdots NCO \xrightarrow{CH_2Cl_2} \text{(157)} \xrightarrow{MeOH} EtCO\cdot CHEt\cdot CO_2Me \quad (1.122)$$

But-2-yne, chlorine, and benzonitrile react to give the compound (158)[394].

$$PhCCl\vdots N\cdot C(\vdots CR_2)\cdot CCl_2Me$$
(158)

1.6.2.5   Organometallic complexes

PdCl$_2$ reacts with but-2-yne to give the complex (159)[395]. H(CO)(Ph$_3$P)$_2$Ir^1 reacts with alk-1-ynes by loss of hydrogen to form novel acetylides (160)[396]. The complex trans-PtClMeQ$_2$ (Q = PMe$_2$Ph or AsMe$_3$) reacts with an alkyne and silver hexafluorophosphate in methanol to form cationic methoxy-carbene complexes (161)[397]. Reactions of alkynes that involve organopalladium complexes[291] and stereoselective reactions of alkynes that involve organonickel complexes[398], have been reviewed.

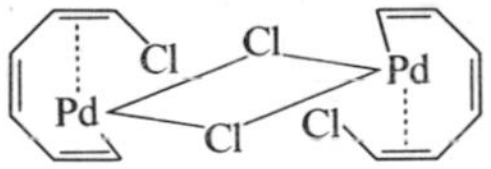

$$Ir(C\vdots CR)(CO)(PPh_3)_{2 \text{ or } 3}$$
(160)

(R = Me, Et, or Bu^i)

$$\left[\begin{array}{c} Q \\ \downarrow \\ Me-Pt \leftarrow C \\ \uparrow \\ Q \end{array} \begin{array}{c} OMe \\ \diagup \\ \diagdown \\ CHR^1R^2 \end{array} \right]^{+} PF_6^{-}$$
(161)

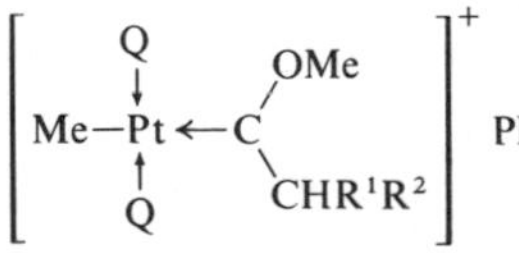

1.6.2.6   Miscellaneous

Base-catalysed isomerisations[399] and nucleophilic additions[400] of alkynes have been reviewed.

Acetylene and propyne react at the surface of liquid sodium in different ways[401]. Acetylene gives disodium acetylide, hydrogen, and ethylene, and propyne forms sodium propynide and but-1-ene.

Acetylene is a competitor inhibitor of nitrogen for the nitrogenase of *Azotobacter vinelandii*, and it is believed to be bound to the same active sites as nitrogen. This enzyme contains molybdenum, iron, cysteine and 'labile sulphur'. Acetylene is also reduced to ethylene at room temperature in a mixture of a metal chloride, 1-thioglycerol, $Na_2S_2O_4$, NaOH, and water, and, significantly, molybdenum chloride is by far the most active metal chloride[402]. Presumably the catalysts are Mo^{IV}–thiol complexes, and, as with the enzyme, the stereochemistry of the reduction is predominantly *cis*[403]. It is also significantly accelerated by the addition of biological phosphorylating agents, such as ATP, to the system. The observed kinetic

$$(Mo_{red}) + C_2H_2 \rightleftharpoons (Mo{\cdot}C_2H_2)$$
$$(Mo{\cdot}C_2H_2) + H_2O \rightarrow (Mo_{ox}) + C_2H_4$$
$$(Mo_{ox}) + 2e \rightarrow (Mo_{red}) \tag{1.123}$$

behaviour is consistent with the reaction of acetylene with a catalytically active reduced-molybdenum complex (Mo_{red}) that is continuously regenerated (equation (1.123)). The iron in the enzyme may be an electron-transfer catalyst.

References

1. Davidson, J. M. (1972). *M.T.P. International Review of Science (Inorganic Chemistry Series 1)*, **6**, 347
2. Dewar, M. J. S. and de Llano, C. (1969). *J. Amer. Chem. Soc.*, **91**, 789
3. Pople, J. A. (1970). *Accounts Chem. Res.*, **3**, 217
4. Pople, J. A., Lathan, W. A. and Hehre, W. J. (1971). *J. Amer. Chem. Soc.*, **93**, 808
5. Random, L., Lathan, W. A., Hehre, W. J. and Pople, J. A. (1971). *J. Amer. Chem. Soc.*, **93**, 5339
6. Hehre, W. J., Stewart, R. F. and Pople, J. A. (1969). *J. Chem. Phys.*, **51**, 2657
7. Ditchfield, R., Hehre, W. J. and Pople, J. A. (1971). *J. Chem. Phys.*, **54**, 724
8. Lathan, W. A., Hehre, W. J. and Pople, J. A. (1970). *Tetrahedron Letters*, 2699
9. Clark, D. T. and Lilley, D. M. J. (1970). *Chem. Commun.*, 549
10. Salahub, D. R. and Sandorfy, C. (1971). *Theoret. Chim. Acta*, **20**, 227
11. Hess, B. A. and Schaad, L. J. (1971). *J. Amer. Chem. Soc.*, **93**, 305
12. Maksić, Z. B. and Randić, M. (1970). *J. Amer. Chem. Soc.*, **92**, 424
13. Miyazaki, T. (1970). *Tetrahedron Letters*, 1363
14. Underwood, G. R. and Iorio, J. M. (1971). *J. Org. Chem.*, **36**, 3987
15. Streitwieser, A. (1961). *Molecular Orbital Theory for Organic Chemists*. (New York: John Wiley)
16. Lowe, J. P. (1970). *J. Amer. Chem. Soc.*, **92**, 3799
17. Pitzer, R. M. and Lipscomb, W. N. (1963). *J. Chem. Phys.*, **39**, 1995
18. Stevens, R. M. (1970). *J. Chem. Phys.*, **52**, 1397
19. Epstein, I. R. and Lipscomb, W. N. (1970). *J. Amer. Chem. Soc.*, **92**, 6094
20. Allen, L. C. and Basch, H. (1971). *J. Amer. Chem. Soc.*, **93**, 6373
21. Veillard, A. (1970). *Theoret. Chim. Acta*, **18**, 21
22. Levy, B. and Moireau, M. C. (1971). *J. Chem. Phys.*, **54**, 3316
23. Random, L. and Pople, J. A. (1970). *J. Amer. Chem. Soc.*, **92**, 4786
24. Jorgensen, W. L. and Allen, L. C. (1971). *J. Amer. Chem. Soc.*, **93**, 567
25. Weiss, S. and Leroi, G. E. (1968). *J. Chem. Phys.*, **48**, 962
26. Allen, L. C. (1968). *Chem. Phys. Letters*, **2**, 597
27. Allen, L. C. and Scarzafava, E. (1971). *J. Amer. Chem. Soc.*, **93**, 311

28. Schaad, L. J. (1970). *Tetrahedron*, **26**, 4115
29. Parczewshi, A. (1970). *Tetrahedron*, **26**, 3539
30. Lehn, J. M. (1971). *Conformational Analysis*. (New York: Academic Press)
31. Brundle, C. R., Robin, M. B., Basch, H., Pinsky, M. and Bond, A. (1970). *J. Amer. Chem. Soc.*, **92**, 3863
32. Pitzer, K. S. (1945). *J. Amer. Chem. Soc.*, **67**, 1126
33. Turner, D. W., Baker, C., Baker, A. D. and Brundle, C. R. (1970). *Molecular Photoelectron Spectroscopy*. (London: Wiley)
34. Beckey, H. D. and Comes, F. J. (1970). *Advan. Analyt. Chem. Instrument.*, **8**, 1
35. Baker, A. D., Betteridge, D., Kemp, N. R. and Kirby, R. E. (1971). *J. Molec. Structure*, **8**, 75
36. Worley, S. D. (1971). *Chem. Rev.*, **71**, 295
37. Politzer, P. and Harris, R. R. (1971). *Tetrahedron*, **27**, 1567
38. Bykov, G. V. (1964). *Electronic Charges of Bonds in Organic Molecules*. (New York: Macmillan)
39. Bach, R. D. and Henneike, H. F. (1970). *J. Amer. Chem. Soc.*, **92**, 5589
40. Herndon, W. C. and Giles, W. B. (1970). *Molec. Photochem.*, **2**, 277
41. Woodward, R. B. and Hoffmann, R. (1969). *Angew. Chem. Int. Ed. Engl.*, **8**, 781
42. Cassar, L., Eaton, P. E. and Halpern, J. (1970). *J. Amer. Chem. Soc.*, **92**, 3519
43. Katz, T. J. and Cerefice, S. A. (1969). *J. Amer. Chem. Soc.*, **91**, 6519
44. Heimbach, P. and Hey, H. (1970). *Angew. Chem. Int. Ed. Engl.*, **9**, 528
45. Mol, J. C., Moulijn, J. A. and Boelhouwer, C. (1968). *J. Catalysis*, **11**, 87
46. Mol, J. C., Visser, F. R. and Boelhouwer, C. (1970). *J. Catalysis*, **17**, 114
47. Mango, F. D. and Schachtschneider, J. H. (1971). *J. Amer. Chem. Soc.*, **93**, 1123
48. Bailey, G. C. (1969). *Catalysis Rev.*, **3**, 37
49. Muetterties, E. L. (1970). *Accounts Chem. Res.*, **3**, 266
50. Caldow, G. L. and MacGregor, R. A. (1970). *Inorg. Nucl. Chem. Letters*, **6**, 645
51. Caldow, G. L. and MacGregor, R. A. (1971). *J. Chem. Soc. A*, 1654
52. van der Lugt, W.Th. (1970). *Tetrahedron Letters*, 2281
53. Mango, F. D. (1971). *Tetrahedron Letters*, 505
54. Lewandos, G. S. and Pettit, R. (1971). *Tetrahedron Letters*, 789
55. Mango, F. D. (1971). *Chem. Tech.*, **1**, 758
56. Warkentin, J. and Hine, K. E. (1970). *Can. J. Chem.*, **48**, 3545
57. Pucci, S., Aglietto, M. and Luisi, P. L. (1970). *Gazz. Chim. Ital.*, **100**, 159
58. Han, J. and Calvin, M. (1970). *Chem. Commun.*, 1490
59. Mauger, J. and Maurin, J. (1970). *Bull. Soc. Chim. Fr.*, 2329
60. Smit, P. and Den Hertog, H. J. (1971). *Tetrahedron Letters*, 595
61. Kelly, C. C., Yu, W. H. S. and Wijnen, M. H. J. (1970). *Can. J. Chem.*, **48**, 603
62. Prakash, H. and Sisler, H. H. (1970). *J. Org. Chem.*, **35**, 3111
63. Kostyuchenko, V. M., Gershenovich, A. I. and Panov, E. P. (1971). *Zh. Prikl. Khim. (Leningrad)*, **44**, 1368
64. Bumgardner, C. L., Lawton, E. L., McDaniel, K. G. and Carmichael, H. (1970). *J. Amer. Chem. Soc.*, **92**, 1311
65. Müller, E. and Böttcher, A. E. (1970). *Tetrahedron Letters*, 3083
66. Falconer, W. E., Sunder, W. A. and Walker, L. G. (1971). *Can. J. Chem.*, **49**, 3892
67. Paciorek, K. L. and Kratzer, R. H. (1970). *Can. J. Chem.*, **48**, 1777
68. Olah, G. A., Halpern, Y., Shen, J. and Mo, Y. K. (1971). *J. Amer. Chem. Soc.*, **93**, 1251
69. Olah, G. A. and Olah, J. A. (1971). *J. Amer. Chem. Soc.*, **93**, 1256
70. Olah, G. A. and Lin, H. C. A. (1971). *J. Amer. Chem. Soc.*, **93**, 1259
71. Olah, G. A. (1971). *Chem. Tech.*, **1**, 566
72. Gol'dshleger, N. F., Tyabin, M. B., Shilov, A. E. and Shteinman, A. A. (1969). *Zhur. Fiz. Khim.*, **43**, 2174
73. Hodges, R. J., Webster, D. E. and Wells, P. B. (1971). *Chem. Commun.*, 462
74. Hodges, R. J., Webster, D. E. and Wells, P. B. (1971). *J. Chem. Soc. A*, 3230
75. Tyabin, M. B., Shilov, A. E. and Shteinman, A. A. (1971). *Dokl. Akad. Nauk. SSSR*, **198**, 380
76. Hodges, R. J., Webster, D. E. and Wells, P. B. (1972). *J. J. C. S. Dalton Trans.*, in press
77. Hodges, R. J., Webster, D. E. and Wells, P. B. (1972). *J. J. C. S. Dalton Trans.*, in press
78. Olah, G. A. (1970). *Science*, **168**, 1298
79. Olah, G. A. and Halpern, Y. (1971). *J. Org. Chem.*, **36**, 2354

80. Olah, G. A., Shen, J. and Schlosberg, R. H. (1970). *J. Amer. Chem. Soc.*, **92**, 3831
81. Valibekov, Y. V. and Bolotov, G. M. (1970). *Gas (Rome)*, **20**, 156
82. Szymanski, A. and Podgorski, A. (1970). *Nukleonika*, **15**, 417
83. Kobzev, Y. N., Kozlov, G. I. and Khudyakov, G. N. (1970). *Khim. Vys. Energ.*, **4**, 519
84. Bashleev, V. B., Lapushonok, L. Y. and Solov'ev, R. A. (1970). *Izv. Sib. Otd. Akad. Nauk. SSSR, Ser. Tekh. Nauk.*, 93
85. Mogel, G. and Eremin, E. N. (1970). *Zh. Fiz. Khim.*, **44**, 1383
86. Takahashi, T., Tsutsui, S., Nakashio, F., Takeshita, K. and Sakai, W. (1970). *Kogyo Kagaku Zasshi*, **73**, 161
87. Valibekov, Y. V. (1970). *Dokl. Akad. Nauk. Tadzh. SSR*, **13**, 33
88. Makino, M. and Kawana, Y. (1970). *Kogyo Kagaku Zasshi*, **73**, 306
89. Klug, M. J. and Markovetz, A. J. (1971). *Advan. in Microbiol. Physiol.* (London: Academic Press), **5**, 1
90. Killinger, A. (1970). *Arch. Mikrobiol.*, **73**, 153
91. Durnjak, Z., Roche, B. and Azoulay, E. (1970). *Arch. Mikrobiol.*, **72**, 135
92. Merdinger, E. and Merdinger, R. P. (1970). *App. Microbiol.*, **20**, 651
93. Barua, P. K., Bhagat, S. D., Pillai, K. R., Singh, H. D., Baruah, J. N. and Iyengar, M. S. (1970). *App. Microbiol.*, **20**, 657
94. Cardini, G. and Jurtshuk, P. (1970). *J. Biol. Chem.*, **245**, 2789
95. Lebeault, J. M., Roche, B., Duvnjak, Z. and Azoulay, E. (1970). *Arch. Mikrobiol.*, **72**, 140
96. Liu, C. and Johnson, M. J. (1971). *J. Bacteriol.*, **106**, 830
97. Ratledge, C. (1970). *Chem. and Ind.*, 843
98. Fujii, K. and Fukui, S. (1970). *Eur. J. Biochem.*, **17**, 552
99. Thorpe, R. F. and Ratledge, C. (1972). *J. General Microbiol.*, **72**, 15
100. Vestal, J. R. and Perry, J. J. (1971). *Can. J. Microbiol.*, **17**, 445
101. Blevins, W. T. and Perry, J. J. (1971). *Z. für Allgem. Mikrobiol.*, **11**, 181
102. Kornblum, N., Boyd, S. D., Pinnick, H. W. and Smith, R. G. (1971). *J. Amer. Chem. Soc.*, **93**, 4316
103. Monson, R. J. (1971). *Tetrahedron Letters*, 567
104. Hall, D. W., Dormish, F. L. and Hurley, E. (1970). *Ind. Eng. Chem. Prod. Res. Develop.*, **9**, 234
105. Abruscato, G. J. and Tidwell, T. T. (1970). *J. Amer. Chem. Soc.*, **92**, 4125
106. Bertini, F., Grasselli, P., Zubiani, G. and Cainelli, G. (1970). *Chem. Commun.*, 144
107. Trost, B. M. and Chen, F. (1971). *Tetrahedron Letters*, 2603
108. Reucroft, J. and Sammes, P. G. (1971). *Quart. Rev. Chem. Soc.*, **25**, 135
109. Hubert, A. J. and Reimlinger, H. (1969). *Synthesis*, 97
110. Hubert, A. J. and Reimlinger, H. (1970). *Synthesis*, 405
111. Mackenzie, K. (1970). *The Chemistry of Alkenes, Vol. 2*. Ed. Zabicky, J. (London: Interscience), 115
112. Sprung, J. L., Akimoto, H. and Pitts, J. N. (1971). *J. Amer. Chem. Soc.*, **93**, 4358
113. Vollershtein, E. L., Beilin, S. I. and Dolgoplosk, B. A. (1970). *Dokl. Akad. Nauk. SSSR*, **193**, 1335
114. Cerceau, C., Laroche, M., Pazdzerski, A. and Blouri, B. (1970). *Bull. Soc. Chim. Fr.*, **88**, 2323
115. Kloosterziel, H. and Van Drunen, J. A. A. (1970). *Rec. Trav. Chim.*, **89**, 37
116. Hirai, H., Sawai, H., Ochiai, E. and Makishima, S. (1970). *J. Catal.*, **17**, 119
117. Yagupsky, M. and Wilkinson, G. (1970). *J. Chem. Soc. A*, 941
118. Strohmeier, W. and Rehder-Stirnweiss, W. (1970). *J. Organometallic Chem.*, **22**, C27
119. Kovács, J., Speier, G. and Markó, L. (1970). *Inorg. Chim. Acta.*, **4**, 412
120. Hudson, B., Webster, D. E. and Well, P. B. (1972). *J. C. S. Dalton Trans.* 1204
121. Augustine, R. L. and Van Peppen, J. F. (1970). *Chem. Commun.*, 495
122. Lyons, J. E. (1971). *Chem. Commun.*, 562
123. Ewing, D. F., Hudson, B., Webster, D. E. and Wells, P. B. (1972). *J. C. S. Dalton Trans.*, 1287
124. Bingham, D., Webster, D. E. and Wells, P. B. (1972). *J. C. S. Dalton Trans.*, 1928
125. Suzuki, T. and Takegami, Y. (1971). *Kogyo Kagaku Zasshi*, **74**, 1371
126. Wang, J. L. and Menapace, H. R. (1971). *J. Catalysis*, **23**, 144
127. Corain, B. (1971). *Chem. and Ind.*, 1465
128. Hudson, B., Webster, D. E. and Wells, P. B. Unpublished observations
129. Warwel, S., Hemmerich, H. and Asinger, F. (1971). *Angew. Chem. Int. Ed. Engl.*, **10**, 281
130. Lewandos, G. S. and Pettit, R. (1971). *J. Amer. Chem. Soc.*, **93**, 7087

131. Zuech, E. A., Hughes, W. B., Kubicek, D. H. and Kittleman, E. T. (1970). *J. Amer. Chem. Soc.*, **92,** 528·
132. Hughes, W. B. (1970). *J. Amer. Chem. Soc.*, **92,** 532
133. Moulijn, J. A. and Boelhouwer, C. (1971). *Chem. Commun.*, 1170
134. Bencze, L. and Marko, L. (1971). *J. Organometallic Chem.*, **28,** 271
135. Whitesides, G. M. and Ehmann, W. J. (1970). *J. Org. Chem.*, **35,** 3565
136. Mitsui, S. and Kasahara, A. (1970). *The Chemistry of Alkenes, Vol. 2*. Ed. Zabicky, J. (London: Interscience), 175
137. Coffey, R. S. (1970). *Aspects of Homogeneous Catalysis, Vol. 1*. Ed. Ugo, R. (Milan: Carlo Manfredi), 5
138. Yagupsky, M., Brown, C. K., Yagupsky, G. and Wilkinson, G. (1970). *J. Chem. Soc. A,* 937
139. Hidai, M., Kusa, T., Hikita, T., Uchida, Y. and Misono, A. (1970). *Tetrahedron Letters,* 1715
140. Ogata, I., Iwata, R. and Ikeda, Y. (1970). *Tetrahedron Letters,* 3011
141. Van Gaal, H., Cuppers, H. G. A. M. and Van der Ent, A. (1970). *Chem. Commun.*, 1694
142. Legzdins, P., Mitchell, R. W., Rempel, G. L., Ruddock, J. D. and Wilkinson, G. (1970). *J. Chem. Soc. A*, 3322
143. Hui, B. C. and Rempel, G. L. (1970). *Chem. Commun.*, 1195
144. Martin, H. A. and De Jongh, R. O. (1971). *Rec. Trav. Chim.*, **90,** 713
145. Grubb, R. H. and Kroll, L. C. (1971). *J. Amer. Chem. Soc.*, **93,** 3062
146. Lazcano, R. L. and Germain, J. E. (1971). *Bull. Soc. Chim. Fr.*, 1869
147. Vol'pin, M. E., Kukolev, V. P., Chernyshev, V. O. and Kolomnikov, I. S. (1971). *Tetrahedron Letters*, 4435
148. Lefebvre, G. and Chauvin, Y. (1970). *Aspects of Homogeneous Catalysis, Vol. 1*. (R. Ugo, editor). (Milan: Carlo Manfredi), 108
149. Barlow, M. G., Bryant, M. J., Haszeldine, R. N. and Mackie, A. G. (1970). *J. Organometallic Chem.*, **21,** 215
150. Fel'dblyum, V. S., Leshcheva, A. I. and Petrushanskaya, N. V. (1970). *J. Org. Chem. USSR*, **6,** 2419
151. Jones, J. R. and Symes, T. J. (1971). *J. Chem. Soc. C*, 1124
152. Fel'dblyum, V. S., Leshcheva, A. I. and Obeshchalova, N. V. (1970). *J. Org. Chem. USSR*, **6,** 205
153. Fel'dblyum, V. S., Leshcheva, A. I. and Petrushanskaya, N. V. (1970). *Zh. Org. Chem.*, **6,** 1113
154. Maruya, K., Mizoroki, T. and Ozaki, A. (1970). *Bull. Chem. Soc. Jap.*, **43,** 3630
155. Eberhardt, G. G. and Griffin, W. P. (1970). *J. Catal.*, **16,** 245
156. Jones, J. R. (1971). *J. Chem. Soc. C*, 1117
157. Watanabe, S., Suga, K. and Kikuchi, H. (1970). *Aust. J. Chem.*, **23,** 385
158. Henrici-Olivé, G. and Olivé, S. (1970). *Angew. Chem. Int. Ed. Engl.*, **9,** 243
159. Hummel, K. and Ast, W. (1970). *Naturwissenschaften*, **57,** 245
160. Sakuki, T. and Takegami, Y. (1970). *Bull. Chem. Soc. Jap.*, **43,** 1484
161. Takegami, K., Suzuki, T. and Shirai, T. (1970). *Bull. Chem. Soc. Jap.*, **43,** 1478
162. Iwashita, Y. and Sakuraba, M. (1971). *Tetrahedron Letters*, 2409
163. Golub, M. A. (1970). *The Chemistry of Alkenes, Vol. 2*, 411. (J. Zabicky, editor). (London: Interscience)
164. Satlick, W. M. and Pines, H. (1970). *J. Org. Chem.*, **35,** 415
165. Kannan, S. V. and Pines, H. (1971). *J. Org. Chem.*, **36,** 2304
166. Boone, D. E., Eisenbroun, E. J., Flanagan, P. W. and Grigsby, R. D. (1971). *J. Org. Chem.*, **36,** 2042
167. Burger, U. and Huisgen, R. (1970). *Tetrahedron Letters*, 3049
168. Vysochin, V. I., Mikhailova, I. F., Semenova, L. D., Sycheva, T. N. and Barkhash, V. A. (1970). *J. Org. Chem. USSR.*, **6,** 1352
169. Stapp, P. R. (1970). *J. Org. Chem.*, **35,** 2419
170. Stapp, P. R. (1971). *J. Org. Chem.*, **36,** 2505
171. Dalgleish, D. T., Nonhebel, D. C. and Pauson, P. L. (1971). *J. Chem. Soc. C*, 1174
172. Terent'ev, A. B., Chizhov, Y. P. and Brakhme, P. (1970). *Izv. Akad. Nauk. SSSR, Ser. Khim.*, 176
173. Freidlina, R. K., Zagorets, P. A., Bryantsev, I. N. and Terent'ev, A. B. (1971). *Dokl. Akad. Nauk. SSSR*, **197,** 105

174. Isaacs, N. S. and Stanbury, P. F. (1970). *Chem. Commun.*, 1061
175. DoMinh, T. and Strausz, O. P. (1970). *J. Org. Chem.*, **35**, 1766
176. D'yakonov, I. A., Stroiman, I. M. and Vitenberg, A. G. (1970). *J. Org. Chem. USSR*, **6**, 41
177. Jones, M., Baron, W. J. and Shen, Y. H. (1970). *J. Amer. Chem. Soc.*, **92**, 4745
178. Buevich, V. A., Altukhov, K. V. and Perekalin, V. V. (1970). *J. Org. Chem. USSR*, **6**, 661
179. Scheinbaum, M. L. and Dines, M. B. (1971). *Tetrahedron Letters*, 2205
180. Taki, K. (1970). *Bull. Chem. Soc. Jap.*, **43**, 2118
181. Crowley, K. J. and Mazzocchi, P. H. (1970). *The Chemistry of Alkenes, Vol. 2.* (J. Zabicky, editor). (London: Interscience), 267
182. Bauslaugh, P. G. (1970). *Synthesis*, 287
183. Evanega, G. R. and Fabiny, D. L. (1970). *J. Org. Chem.*, **35**, 1757
184. Evanega, G. R. and Fabiny, D. L. (1971). *Tetrahedron Letters*, 1749
185. Ryangi, H., Shima, K. and Sakurai, H. (1970). *Tetrahedron Letters*, 1091
186. Charlton, J. L., Liao, C. C. and de Mayo, P. (1971). *J. Amer. Chem. Soc.*, **93**, 2463
187. Okazaki, R., Okawa, K. and Inamoto, N. (1971). *Chem. Commun.*, 843
188. Morikawa, A., Brownstein, S. and Cvetanoavić, R. J. (1970). *J. Amer. Chem. Soc.*, **92**, 1471
189. Bryce-Smith, D., Foulger, B. E., Gilbert, A. and Twitchett, P. J. (1971). *Chem. Commun.*, 794
190. Heck, R. F. (1969). *J. Amer. Chem. Soc.*, **91**, 6707 and Refs. therein
191. Fujiwara, Y., Moritani, I., Danno, S., Asano, R. and Teranishi, S. (1969). *J. Amer. Chem. Soc.*, **91**, 7166 and references therein
192. Danno, S., Moritani, I., Fujiwara, Y. and Teranishi, S. (1970). *Bull. Chem. Soc. Jap.*, **43**, 3966
193. Moritani, I., Fujiwara, Y. and Danno, S. (1971). *J. Organometallic Chem.*, **27**, 279
194. Fujiwara, Y., Moritani, I., Ikegami, K., Tanaka, R. and Teranishi, S. (1970). *Bull. Chem. Soc. Jap.*, **43**, 863
195. Asano, R., Moritani, I., Sonoda, A., Fujiwara, Y. and Teranishi, S. (1971). *J. Chem. Soc. C*, 3691
196. Shue, R. S. (1971). *Chem. Commun.*, 1510
197. Mizoroki, T., Mori, K. and Ozaki, A. (1971). *Bull. Chem. Soc. Jap.*, **44**, 581
198. Vol'pin, M. E., Volkova, L. G., Levitin, I. Y., Boronina, N. N. and Yurkevich, A. M. (1971). *Chem. Commun.*, 849
199. Hasegawa, E., Yoneda, N., Aomura, K. and Ohtsuka, H. (1971). *Kogyo Kagaku Zasshi*, **74**, 903
200. Eisch, J. J. and Liu, S. J. Y. (1970). *J. Organometallic Chem.*, **21**, 285
201. Coulson, D. R. (1971). *Tetrahedron Letters*, 429
202. Oka, T. and Sato, S. (1970). *Bull. Chem. Soc. Jap.*, **43**, 3711
203. Caldwell, R. A. (1970). *J. Amer. Chem. Soc.*, **92**, 1439
204. Meisels, G. G. (1970). *The Chemistry of Alkenes, Vol. 2.* (J. Zabicky, editor). (London: Interscience), 359
205. Ryang, M. and Tsutsumi, S. (1971). *Synthesis*, 55
206. Yagupsky, G., Brown, C. K. and Wilkinson, G. (1970). *J. Chem. Soc. A*, 1392
207. Wilkinson, G. and Brown, C. K. (1970). *J. Chem. Soc. A*, 2753
208. Yamaguchi, M. (1969). *Shokubai*, **11**, 179
209. Taylor, P. and Orchin, M. (1971). *J. Amer. Chem. Soc.*, **93**, 6504
210. Matsubara, M. (1968–published 1970). *Hokkaido-Ritsu Kogyo Shikenjo Hakoku*, 92
211. Matsubara, M. (1968–published 1970). *Hokkaido-Ritsu Kogyo Shikenjo Hokoku*, 102
212. Yang, Y., Puzitskii, K. V. and Eidus, Y. T. (1970). *Izv. Akad. Nauk SSSR, Ser. Khim.*, 424
213. Matsubara, M. (1968–published 1970). *Hokkaido-Ritsu Kogyo Shikenjo Hokoku*, 111
214. Eidus, Y. T., Puzitskii, K. V. and Yang, Y. (1970). *Izv. Akad. Nauk SSSR, Ser. Khim.*, 1673
215. Ordyan, M. B., Grigoryan, V. S., Avetisyan, R. V. and Eidus, Y. T. (1971). *Izv. Akad. Nauk SSSR, Ser. Khim.*, 116
216. Chalk, A. J. (1970). *Trans. N. Y. Acad. Sci.*, **32**, 481
217. Chalk, A. J. (1970). *J. Organometal. Chem.*, **21**, 207
218. Kumada, M., Kiso, Y. and Umeno, M. (1970). *Chem. Commun.*, 611
219. Hara, M., Ohno, K. and Tsuji, J. (1971). *Chem. Commun.*, 247
220. Dennis, W. E. and Speier, J. L. (1970). *J. Org. Chem.*, **35**, 3879

221. Brown, H. C. and Geoghegan, P. J. (1970). *J. Org. Chem.*, **35**, 1844
222. Carlson, R. M. and Funk, A. H. (1971). *Tetrahedron Letters*, 3661
223. Schmitz, E., Rieche, A. and Brede, O. (1970). *J. Prakt. Chem.*, **312**, 30
224. Bloodworth, A. J. and Ballard, D. H. (1971). *J. Chem. Soc. C*, 945
225. Bach, R. D., Brummel, R. N. and Richter, R. F. (1971). *Tetrahedron Letters*, 2879
226. Arzoumanian, H. and Metzger, J. (1971). *Synthesis*, 527
227. Tinker, H. B. (1971). *J. Organometallic Chem.*, **32**, C25
228. Hunt, D. F. and Rodeheaver, G. T. (1971). *Chem. Commun.*, 818
229. Kresge, A. J., Chiang, Y., Fitgerald, R. H., McDonald, R. S. and Schmid, G. H. (1971). *J. Amer. Chem. Soc.*, **93**, 4907
230. Jensen, J. L. (1971). *Tetrahedron Letters*, 7
231. Schwerdtel, W. (1970). *Hydrocarbon Process*, **49**, 117
232. Andrianov, A. A. and Chernyak, B. I. (1970). *Khim. Prom. (Moscow)*, **46**, 175
233. Kryukov, S. I., Yablonskii, O. P. and Simanov, N. A. (1971). *Zh. Org. Khim.*, **7**, 629
234. Hatch, L. F. (1970). *Hydrocarbon Process*, **49**, 101
235. Baker, L. M. and Carrick, W. L. (1970). *J. Org. Chem.*, **35**, 774
236. Freeman, F., McCart, P. D. and Yamachika, N. J. (1970). *J. Amer. Chem. Soc.*, **92**, 4621
237. McKillop, A., Hunt, J. D., Taylor, E. C. and Kienzle, F. (1970). *Tetrahedron Letters*, 5275
238. Trachtenberg, E. N., Nelson, C. H. and Carver, J. R. (1970). *J. Org. Chem.*, **35**, 1653
239. Javaid, K. A., Sonoda, N. and Tsutsumi, S. (1970). *Bull. Chem. Soc. Jap.*, **43**, 3475
240. Imamura, J., Ohta, N., Ishioka, R. and Sato, S. (1970). *Tokyo Kogyo Shikensho Hokoku*, **65**, 102
241. Matsumura, N., Sonoda, N. and Tsutsumi, S. (1970). *Tetrahedron Letters*, 2029
242. Carlson, R. G., Behn, N. S. and Cowles, C. (1971). *J. Org. Chem.*, **36**, 3832
243. Rouchard, J. and Mawaka, J. (1970). *J. Catal.*, **19**, 172
244. Fleischmann, M., Pletcher, D. and Race, D. M. (1970). *J. Chem. Soc. B*, 1746
245. Reardon, E. J. and Kropp, P. J. (1971). *J. Amer. Chem. Soc.*, **93**, 5593
246. Fliszar, S. and Renard, J. (1970). *Can. J. Chem.*, **48**, 3002
247. Sukuki, S. and Ransley, D. L. (1971). *Ind. Eng. Chem., Prod. Res. Develop.*, **10**, 179
248. Czichocki, G. (1971). *Tenside*, **8**, 117
249. Nakanishi, S. and Yashimura, F. (1970). *Kagyo Kagaku Zasshi*, **73**, 2658
250. Guy, R. G. and Thompson, J. J. (1970). *Chem. Ind. (London)*, 1499
251. Grady, G. L. (1971). *Synthesis*, 255
252. Field, K. W. and Kovacic, P. (1971). *J. Org. Chem.*, **36**, 3566
253. Kovacic, P., Lowery, M. K. and Field, K. W. (1970). *Chem. Rev.*, **70**, 639
254. Kryukov, S. I., Osokin, Y. G., Simanov, N. A. and Yablonskii, O. P. (1970). *Zh. Org. Khim.*, **6**, 1386
255. Ohashi, T., Matsunaga, K., Kurata, Y., Okahara, M. and Komori, S. (1971). *Yukagaku*, **20**, 229
256. Heasley, V. L., Heasley, G. E., McConnell, M. R., Martin, K. A., Ingle, D. M. and Davis, P. D. (1971). *Tetrahedron Letters*, 4819
257. Anderson, L. R., Young, D. E., Gould, D. E., Juurik-Hogan, R., Neuchterlein, D. and Fox, W. B. (1970). *J. Org. Chem.*, **35**, 3730
258. Seyferth, D., Simmons, H. D. and Shih, H. (1971). *J. Organometallic Chem.*, **29**, 359
259. Uemura, S., Sasaki, O. and Okano, M. (1971). *Chem. Commun.*, 1064
260. Asahara, T. and Sato, T. (1971). *Kogyo Kagaku Zasshi*, **74**, 703
261. Arpe, H. and Kucketz, H. (1971). *Angew. Chem. Int. Ed. Engl.*, **10**, 73
262. Ogawa, M. and Ishioko, R. (1970). *Bull. Chem. Soc. Jap.*, **43**, 496
263. Hörnig, L., Grosspietsch, H. and Kuckertz, H. (1970). *Erdol. Kohle. Erdgas. Petrochemie*, **23**, 152
264. Ogawa, M., Inoue, C. and Ishioka, R. (1970). *Kogyo Kagaku Zasshi*, **73**, 1987
265. Ichikawa, K., Uemura, S., Takagaki, Y. and Hiramoto, T. (1970). *Bull. Jap. Petrol. Inst.*, **12**, 77
266. Koyano, T. (1970). *Bull. Chem. Soc. Jap.*, **43**, 1439
267. Koyano, T. (1970). *Bull. Chem. Soc. Jap.*, **43**, 3501
268. Koyano, T. and Watanabe, O. (1971). *Bull. Chem. Soc. Jap.*, **44**, 1378
269. Burton, D. J. and Kehoe, L. J. (1970). *J. Org. Chem.*, **35**, 1339
270. Baird, W. C., Surridge, J. H. and Buza, M. (1971). *J. Org. Chem.* **36**, 2088
271. Stangl, H. and Jira, R. (1970). *Tetrahedron Letters*, 3589
272. Koyano, T. (1971). *Bull. Chem. Soc. Jap.*, **44**, 1158

273. Heasley, V. L., Frye, C. L., Heasley, G. E., Martin, K. A., Redfield, D. A. and Wilday, P. S. (1970). *Tetrahedron Letters*, 1573
274. Movsumzade, M. M., Shabanov, A. L., Movsumzade, S. M. and Gurbanov, P. A. (1971). *J. Org. Chem. USSR*, 410
275. Tufariello, J. J. and Hovey, M. M. (1970). *Chem. Commun.*, 372
276. Diner, U. E. and Lown, J. W. (1970). *Chem. Commun.*, 333
277. Diner, U. E. and Lown, J. W. (1971). *Can. J. Chem.*, **49**, 403
278. Hassner, A., Hoblitt, R. P., Heathcock, C., Kropp, J. E. and Lorber, M. (1970). *J. Amer. Chem. Soc.*, **92**, 1326
279. Susuki, T. and Tsuji, J. (1970). *J. Org. Chem.*, **35**, 2982
280. van der Linde, R. and de Jongh, R. O. (1971). *Chem. Commun.*, 563
281. van der Ent, A. and van Soest, T. C. (1970). *Chem. Commun.*, 225
282. Konya, K., Fujita, J. and Nakamoto, K. (1971). *Inorg. Chem.*, **10**, 1699
283. Kolobova, N. E., Skripkin, V. V. and Anisimov, K. N. (1970). *Izv. Akad. Nauk SSSR, Ser. Khim.*, **10**, 2225
284. Clark, H. C. and Kurosawa, H. (1971). *Chem. Commun.*, 957
285. Holloway, C. E., Hulley, G., Johnson, B. F. G. and Lewis, J. (1970). *J. Chem. Soc. A*, 1653
286. Wheelock, K. S., Nelson, J. H., Cusachs, L. C. and Jonassen, H. B. (1970). *J. Amer. Chem. Soc.*, **92**, 5110
287. Schurig, V. and Gil-av, E. (1971). *Chem. Commun.*, 650
288. Nelson, J. H. and Jonassen, H. B. (1971). *Coord. Chem. Rev.*, **6**, 27
289. Biellmann, J. F., Hemmer, H. and Levisalles, J. (1970). *The Chemistry of Alkenes, Vol. 2.* 215. (J. Zabicky, editor). (London: Interscience)
290. Levisalles, J. (1970). *Chim. Ind. Genie Chim.*, **103**, 525
291. Hüttel, R. (1970). *Synthesis*, 225
292. Olah, G. A. and Halpern, Y. (1971). *J. Org. Chem.*, **36**, 2354
293. Rickey, H. G. (1970). *The Chemistry of Alkenes, Vol. 2.*, 39. (J. Zabicky, editor). (London: Interscience)
294. Moore, W. R. and Hill, J. B. (1970). *Tetrahedron Letters*, 4553
295. Mavrov, M. V. and Kucherov, V. F. (1970). *Reakts. Metody. Issled. Org. Soedin.*, **21**, 90
296. Baldwin, J. E. and Fleming, R. H. (1970). *Fortschr. Chem. Forsch.*, **15**, 281
297. Gajewski, J. J. and Black, W. A. (1970). *Tetrahedron Letters*, 899
298. Battioni, P., Vo-Quang, L. and Vo-Quang, Y. (1970). *Bull. Soc. Chim. Fr.*, 3938
299. Battioni, P., Vo-Quang, L. and Vo-Quang, Y. (1970). *Bull. Soc. Chim. Fr.*, 3942
300. Dolbier, W. R. and Dai, S. H. (1970). *Tetrahedron Letters*, 4645
301. Gotthardt, H. (1971). *Tetrahedron Letters*, 2343
302. Gotthardt, H. (1971). *Tetrahedron Letters*, 2345
303. Martin, J. C., Carter P. L. and Chitwood, J. L. (1971). *J. Org. Chem.*, **36**, 2225
304. Jaffe, F. (1970). *J. Organometallic Chem.*, **23**, 53
305. Linn, W. S., Weters, W. L. and Caserio, M. C. (1970). *J. Amer. Chem. Soc.*, **92**, 4018
306. Bianchini, J. P. and Cocordano, M. (1970). *Tetrahedron*, **26**, 3401
307. Byrd, L. R. and Caserio, M. C. (1970). *J. Amer. Chem. Soc.*, **92**, 5422
308. Caserio, M. C., Findley, M. C. and Waters, W. L. (1971). *J. Org. Chem.*, **36**, 275
309. Tien, R. Y. and Abell, P. I. (1970). *J. Org. Chem.*, 956
310. Kochi, J. K. and Krusic, P. J. (1970). *J. Amer. Chem. Soc.*, **92**, 4110
311. De Pasquale, R. J. (1971). *J. Organometallic Chem.*, **32**, 381
312. Brown, C. K., Mowat, W., Yagupsky, G. and Wilkinson, G. (1971). *J. Chem. Soc. A*, 850
313. Vrieze, K., Volger, H. C. and Praat, A. P. (1970). *J. Organometallic Chem.*, **21**, 467
314. Okamoto, T., Sakakibara, Y. and Kunichika, S. (1970). *Bull. Chem. Soc. Jap.*, 2658
315. Okamoto, T. (1971). *Bull. Chem. Soc. Jap.*, 1353
316. Deeming, A. J., Johnson, B. F. G. and Lewis, J. (1970). *Chem. Commun.*, 598
317. Hewitt, J. G. and De Boer, J. J. (1971). *J. Chem. Soc. A*, 817
318. Zweifel, G., Polston, N. L. and Whitney, C. C. (1968). *J. Amer. Chem. Soc.*, **90**, 6243
319. Zweifel, G. and Miller, R. L. (1970). *J. Amer. Chem. Soc.*, **92**, 6678
320. Jolly, P. W., Tkatchenko, I. and Wilke, G. (1971). *Angew. Chem. Int. Ed. Engl.*, **10**, 328
321. Jolly, P. W., Tkatchenko, I. and Wilke, G. (1971). *Angew. Chem. Int. Ed. Engl.*, **10**, 329
322. Kiji, J., Masui, K. and Furukawa, J. (1970). *Tetrahedron Letters*, 2561
323. Kiji, J., Masui, K. and Furukawa, J. (1970). *Chem. Commun.*, 1310
324. Arakawa, T. and Miyake, H. (1971). *Kogyo Kagaku Zasshi*, **74**, 1143
325. Medema, D. and Van Helden, R. (1971). *Rec. Trav. Chem.*, **90**, 304

326. Lehmkuhl, H. and Leuchte, W. (1970). *J. Organometallic Chem.*, **23**, C30
327. Hughes, W. B. (1971). *J. Org. Chem.*, **36**, 4073
328. Ohta, T., Ebina, K. and Yamazaki, N. (1971). *Bull. Chem. Soc. Jap.*, 1321
329. Matschiner, H., Kerrinnes, H. J. and Issleib, K. (1971). *Z. Anorg. Allgem. Chem.*, **380**, 1
330. Heimbach, P., Selbeck, H. and Troxler, E. (1971). *Angew. Chem. Int. Ed. Engl.*, 659
331. Shields, T. C. and Walker, W. E. (1971). *Chem. Commun.*, 193
332. Hosaka, S. and Tsuji, T. (1971). *Tetrahedron*, **27**, 3821
333. Billups, W. E., Walker, W. E. and Shields, T. C. (1971). *Chem. Commun.*, 1067
334. Haynes, P. (1970). *Tetrahedron Letters*, 3687
335. Manyik, R. M., Walker, W. E., Atkins, K. E. and Hammack, E. S. (1970). *Tetrahedron Letters*, 3813
336. Ohno, K., Mitsuyasu, T. and Tsuji, J. (1971). *Tetrahedron Letters*, 67
337. Walker, W. E., Manyik, R. M., Atkins, K. E. and Farmer, M. L. (1970). *Tetrahedron Letters*, 3817
338. Singer, H., Umbach, W. and Dohr, M. (1971). *Synthesis*, 265
339. Klimova, E. I., Treshchova, E. G. and Arbuzov, Y. A. (1970). *Zh. Org. Khim.*, **6**, 413
340. Hata, G., Takahashi, K. and Miyake, A. (1971). *J. Org. Chem.*, **36**, 2116
341. Takahashi, K., Miyake, A. and Hata, G. (1971). *Chem. and Ind.*, 488
342. Greco, A., Carbonaro, A. and Dall'asta, G. (1970). *J. Org. Chem.*, **35**, 271
343. Takahashi, A. and Inukai, T. (1970). *Chem. Commun.*, 1473
344. Carbonaro, A., Cambisi, F. and Dell'asta, G. (1971). *J. Org. Chem.*, 1443
345. Rupilius, W. and Orchin, M. (1971). *J. Org. Chem.*, **36**, 3604
346. Ohno, K. and Tsuji, J. (1971). *Chem. Commun.*, 247
347. Schäfer, H. and Steckhan, E. (1970). *Tetrahedron Letters*, 3835
348. Schäfer, H. and Kuntzel, H. (1970). *Tetrahedron Letters*, 3333
349. Takahashi, S., Shibano, T., Kojima, H. and Hagihara, N. (1971). *Organometallic Chem. Syn.*, **1**, 193
350. Albert, H. J., Neumann, W. P., Kaiser, W. and Ritter, H. P. (1970). *Chem. Ber.*, **103**, 1372
351. Imai, N., Narita, T. and Tsuruta, T. (1971). *Tetrahedron Letters*, 3517
352. Fokin, A. K., Maklyacv, F. L. and Kutepov, A. P. (1970). *Zh. Org. Khim.*, **6**, 1771
353. Manyik, R. M., Atkins, K. E. and Walker, W. E. (1971). *Chem. Commun.*, 330
354. Heasley, V. L. and Chamberlain, P. H. (1970). *J. Org. Chem.*, **35**, 539
355. Boué, S. and Srinivasan, R. (1970). *J. Amer. Chem. Soc.*, **92**, 3226
356. Srinivasan, R. and Boué, S. (1970). *Tetrahedron Letters*, 203
357. Salticl, J., Metts, L. and Wrighton, M. (1970). *J. Amer. Chem. Soc.*, **92**, 3227
358. Dilling, W. L., Kroening, R. D. and Little, J. C. (1970). *J. Amer. Chem. Soc.*, **92**, 928
359. Dilling, W. L. and Kroening, R. D. (1970). *Tetrahedron Letters*, 695
360. Nasielski, J., Kirsch, P. and Wilputte-Steinert, L. (1971). *J. Organometallic Chem.*, **27**, C13
361. Barltrop, J. A. and Carless, H. A. J. (1971). *J. Amer. Chem. Soc.*, **93**, 4794
362. Hewertson, W. and Taylor, I. C. (1970). *J. Chem. Soc. C*, 1990
363. Andreeta, A., Conti, F. and Ferrari, G. F. (1970). *Aspects of Homogeneous Catalysis, Vol. 1*, 204. (R. Ugo, editor) (Carlo Manfredi: Milan)
364. Brown, H. C., Negishi, E. and Burke, P. L. (1971). *J. Amer. Chem. Soc.*, **93**, 3400
365. Vitulli, G., Porri, L. and Segre, A. L. (1971). *J. Chem. Soc. A*, 3246
366. Tolman, C. A. (1970). *J. Amer. Chem. Soc.*, **92**, 6785
367. Carbonaro, A. and Greco, A. (1970). *J. Organometallic Chem.*, **25**, 477
368. Mikhailov, B. M. and Bubnov, Y. N. (1971). *Tetrahedron Letters*, 2127
369. Mikhailov, B. M., Bubnov, Y. N., Korobeinikova, S. A. and Frolov, S. I. (1971). *J. Organometallic Chem.*, **27**, 165
370. Miller, R. G., Pinke, P. A., Stauffer, R. D. and Golden, H. J. (1971). *J. Organometallic Chem.*, **29**, C42
371. Miller, R. G., Pinke, P. A. and Baker, D. J. (1970). *J. Amer. Chem. Soc.*, **92**, 4490
372. Marshall, J. A. and Babler, J. H. (1970). *Tetrahedron Letters*, 3861
373. Sharma, R. K. and Malval, P. (1970). *N. Mex. Acad. Sci. Bull.*, **11**, 20
374. Klein, J. and Gurfinkel, E. (1970). *Tetrahedron*, **26**, 2127
375. Castek, A., Gospocic, L., Kljaic, K. and Prostenik, M. (1970). *Bull. Sci. Cons. Acad. Sci. Arts RSF Yougoslavie, Sect. A.*, **15**, 157
376. Bohlmann, F. and Zdero, C. (1971). *Chem. Ber.*, **104**, 954

377. Zweifel, G., Clark, G. M. and Polston, N. L. (1971). *J. Amer. Chem. Soc.*, **93**, 3395
378. Zweifel, G., Fisher, R. P., Snow, J. T. and Whitney, C. C. (1971). *J. Amer. Chem. Soc.*, **93**, 6309
379. Zweifel, G. and Polston, N. L. (1970). *J. Amer. Chem. Soc.*, **92**, 4068
380. Shvartsberg, M. S., Kotlyarevskii, I. L., Kozhevnikova, A. N. and Andrievskii, V. N. (1970). *Izv. Akad. Nauk. SSSR, Ser. Khim.*, **5**, 1144
381. Gmyzina, R. N., D'yakonov, I. A. and Danilkina, L. P. (1970). *Zh. Org. Khim.*, **6**, 2168
382. Grubbs, R. H. (1970). *J. Amer. Chem. Soc.*, **92**, 6693
383. Tancrede, J. and Rosenblum, M. (1971). *Synthesis*, 219
384. Shida, S. and Tsukada, M. (1970). *Bull. Chem. Soc. Jap.*, **43**, 2740
385. Chalk, A. J. and Jerussi, R. A. (1972). *Tetrahedron Letters.*, 61
386. Bryce-Smith, D., Gilbert, A. and Grzonka, J. (1970). *Chem. Commun.*, 498
387. Reed, S. F. (1970). *J. Org. Chem.*, **35**, 3961
388. Kukushkin, Y. N., Kobzev, V. V. and Morozova, L. P. (1970). *Zh. Prikl. Khim. (Leningrad)*, **43**, 2759
389. Gopal, H. and Gordon, A. J. (1971). *Tetrahedron Letters*, 2941
390. Atavin, A. S., Gusarova, N. K., Amosova, S. V., Trofimov, B. A. and Kalabin, G. A. (1970). *Zh. Org. Khim.*, **6**, 2386
391. Barton, T. J. and Zika, R. G. (1970). *J. Org. Chem.*, **35**, 1729
392. Kataev, E. G. and Mannafov, T. G. (1970). *Zh. Org. Khim.*, **6**, 1959
393. Moriconi, E. J., White, J. G., Franck, R. W., Jansing, J., Kelly, J. F., Salamone, R. A. and Shimakawa, Y. (1970). *Tetrahedron Letters*, 27
394. Theilacker, W. and Thiem, K. W. (1970). *Chem. Ber.*, **103**, 670
395. Dietl, H., Reinheimer, H., Moffat, J. and Maitlis, P. M. (1970). *J. Amer. Chem. Soc.*, **92**, 2276
396. Brown, C. K. and Wilkinson, G. (1971). *Chem. Commun.*, 70
397. Chisholm, M. H. and Clark, H. C. (1970). *Chem. Commun.*, 763
398. Chiusoli, G. P. (1970). *Aspects of Homogeneous Catalysis, Vol. 1*, (R. Ugo, editor). (Carlo Manfredi: Milan), 77
399. Bushby, R. J. (1970). *Quart. Rev., Chem. Soc.*, **24**, 585
400. Miller, S. I. and Tanaka, R. (1970). *Selec. Org. Transform.*, **1**, 143
401. Addison, C. C., Hobdell, M. R. and Pulham, R. J. (1971). *J. Chem. Soc. A*, 1704
402. Schrauzer, G. N. and Schlesinger, G. (1970). *J. Amer. Chem. Soc.*, **92**, 1808
403. Schrauzer, G. N. and Doemeny, P. A. (1971). *J. Amer. Chem. Soc.*, **93**, 1608
404. Zimmerman, H. E. (1969). *Angew. Chem. Int. Ed. Engl.*, **8**, 1
405. Littler, J. S. (1971). *Tetrahedron*, **27**, 81
406. Olah, G. A. (1972). *J. Amer. Chem. Soc.*, **94**, 808
407. Bertram, J., Fleischmann, M. and Pletcher, D. (1971). *Tetrahedron Letters*, 349

2
Halogeno Compounds

G. M. BROOKE
Durham University

2.1   INTRODUCTION

This review is concerned with the formation and cleavage of carbon–halogen bonds. The reactions are classified in terms of known or expected mechanisms at sp^3-, sp^2- (vinylic) and sp^1-hybridised carbon. The effect of a halogen atom on the properties of other functional groups in the same molecule, unless direct interaction is involved, is not included in this chapter.

2.2   PREPARATION OF ALKYL HALIDES

2.2.1   Electrophilic additions to alkenes

The selective introduction of small numbers of fluorine atoms into organic molecules has always been a problem. Two significant advances have been made in this field by using fluoro-oxytrifluoromethane, CF_3OF, which reacts with alkenes[1,2] by an ionic mechanism and with saturated hydrocarbons[3] by a radical mechanism (Section 2.2.5). In the former reactions, the products can be rationalised in terms of an intermediate β-fluorocarbonium ion which may interact with the counter ions CF_3O^- or F^- (from $CF_3O^- \rightleftharpoons CF_2O + F^-$) or with the solvent during work-up. Testosterone acetate, which is deactivated towards electrophilic attack, also reacted (though

$$(R = F \text{ or } CF_3O)$$

slowly) with CF_3OF to give, after an alkaline work-up, 4-fluorotestosterone. The stereospecific formation of *cis* addition products (above) may be explained in terms of the rapid collapse of an ion pair (Section 2.5.1.2). The stereospecificity of addition of CF_3OF to *cis*- and *trans*-but-2-ene, however, merits special investigation in view of the current interest in 1,2-fluorinium ions.

The overall addition of BrF (from $BrF_3 + Br_2$) to perfluorohept-1-ene gives specifically perfluoro-2-bromoheptane, but the mechanism of the reaction has not been established[4]. The internal alkene perfluorohept-2-ene showed no selectivity, equal amounts of the 2- and the 3-bromo-compound being formed.

Chloroiodoalkanes are conveniently prepared by the reaction of alkenes with iodine and cupric chloride[5]. Unsymmetrical alkenes give mixtures of products (e.g. $Me \cdot CHCl \cdot CH_2I$ (73%) and $Me \cdot CHI \cdot CH_2Cl$ (27%) from $Me \cdot CH{:}CH_2$), the ratios of which resemble those from ICl additions.

The reactions of alkenes with iodine and water to form 1,2-iodohydrins have unfavourable equilibrium constants[6]. However, conversion of the hydrogen iodide formed into iodine by the addition of iodic acid enables high yields of the adducts to be obtained.

Russian workers have reviewed electrophilic additions to olefins in liquid hydrogen fluoride, where the initially formed carbonium ion is captured by the fluoride ion[7]. Mixtures of Markovnikov and anti-Markovnikov products are formed by the addition of sulphenyl chlorides to alkenes[8] (see Section 2.3.1).

2.2.2 Nucleophilic additions to alkenes and to carbonyl compounds

The nucleophilic addition of fluoride ion to polyfluoroalkenes has received considerable attention. The resulting carbanions displace fluoride from polyfluoro-aromatic[9] and -heterocyclic compounds[10, 11]. Small amounts

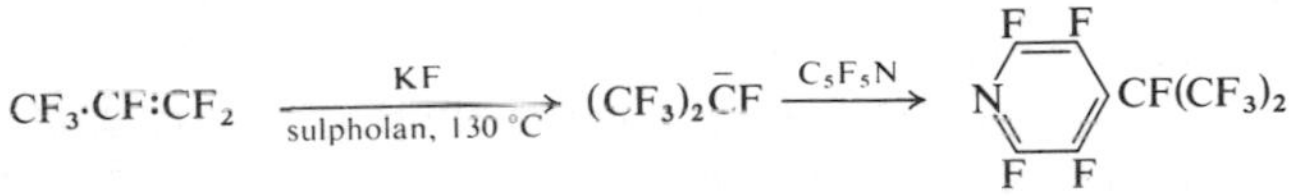

$$CF_3 \cdot CF{:}CF_2 \xrightarrow[\text{sulpholan, 130 °C}]{KF} (CF_3)_2\bar{C}F \xrightarrow{C_5F_5N} \underset{F\ \ F}{\overset{F\ \ F}{N}} CF(CF_3)_2$$

of trimethylamine (2% w/w) promote the anionic $[(CF_3)_2\bar{C}F]$ oligomerisation of perfluoropropene in dipolar aprotic solvents to give isomeric dimers and trimers, depending on the reaction conditions[12].

Silver fluoride reacts with perfluoro-propene[13] and -allene[14] in acetonitrile at 25 °C to form stable silver compounds, $(CF_3)_2CFAg$ and $CF_3 \cdot C Ag{:}CF_2$, respectively. Perfluoroalkylmercury compounds have been prepared by analogous reactions[15], e.g. $(CF_3)_3C \cdot Hg \cdot C(CF_3)_3$ from $(CF_3)_2 C{:}CF_2$ and HgF_2.

Apart from perfluorocyclobutanol, α-fluoroalcohols have eluded isolation and characterisation. The simplest member of the series, fluoromethanol is stabilised by protonation in strongly acidic media[16].

$$CH_2O + HF \longrightarrow [FCH_2OH] \xrightarrow[SbF_5, -78 °C]{HF} FCH_2\overset{+}{O}H_2\ SbF_6^-$$

The selective conversion of carbonyl groups in aldehydes or ketones into CF_2 groups may be conveniently performed at room temperature and atmospheric pressure in glass apparatus by using $MoF_6 + BF_3$ in CH_2Cl_2 [17]. The usual reagent for this reaction, SF_4, requires special handling techniques.

2.2.3 Radical additions to alkenes

Various xenon fluorides react with olefins to give complex mixtures of products via radical intermediates[18]. Ethylene and XeF_4 gave $CH_2F \cdot CH_2F$ (45%), $Me \cdot CHF_2$ (35%), and $CH_2F \cdot CHF_2$ (20%).

Good yields of 1:1 addition products are obtained from polyhalogenoalkanes and alkenes by using a redox system of radical initiation[19, 20].

$$CCl_3Br + CH_2{:}CH \cdot (CH_2)_6H \xrightarrow[\text{ethanolamine}]{CuCl_2} CCl_3 \cdot CH_2 \cdot CHBr \cdot (CH_2)_6H\ (96\%)$$

Sulphonyl chlorides ($RSO_2 \cdot Cl$) add to styrenes under the influence of $CuCl_2$—$Et_3 \overset{+}{N}HCl^-$, giving β-chlorosulphones ($ArCHCl \cdot CH_2 \cdot SO_2 R$) [21]. Trichloramine in CH_2Cl_2 at 25 °C is a highly effective reagent for converting alkenes into vic-dichlorides, and where comparisons are possible, it appears to be superior to other chlorinating agents used for this purpose (e.g. Cl_2, PCl_5, SO_2Cl_2) [22]. In contrast, $(CF_3)_2NCl$ forms mixtures of 1:1 adducts with unsymmetrical alkenes under both radical and ionic conditions [23].

2.2.4 Nucleophilic substitution at carbon

Cyanuric chloride is an effective reagent for converting alcohols into alkyl chlorides by an $S_N i$ process [24]. With added sodium iodide, alkyl iodides are formed in good yields, presumably via cyanuric iodide, but with added sodium bromide both the alkyl bromide and chloride are formed [25].

Extensive rearrangement of alkyl groups, especially primary, occurs during the thermolysis of alkyl chloroformates to alkyl chlorides [26] (e.g. the n-butyl derivative gives 78% of 2-chlorobutane). This is in contrast to the analogous reaction with chlorosulphites, and indicates the intermediacy of ion pairs, $R^+ ClCO_2^-$.

The oxidation of alkyl carbazates ($RO \cdot CO \cdot NH \cdot NH_2$) by N-halogeno-succinimides (NHalS) gives alkyl halides, CO_2, and N_2 [27]. Apart from the

$$ROH \xrightarrow[\text{(ii) } NH_2 \cdot NH_2]{\text{(i) } COCl_2} RO \cdot CO \cdot NH \cdot NH_2 \xrightarrow{Hal^+} \underset{\underset{Hal}{R}}{\overset{O}{\underset{}{\parallel}}} \xrightarrow[-N_2]{-CO_2} R\,Hal$$

observation that the method succeeds best when alkene formation is difficult (e.g. 1-adamantyl carbazate gave 1-iodo- and 1-bromo-adamantane in 63% and 49% yield respectively but no 1-chloro-compound), little else is known about the reaction mechanism.

In contrast to the relative unreactivity of neopentyl compounds towards nucleophilic attack at carbon in both the $S_N 1$ and the $S_N 2$ mechanism, neopentyl alcohol is readily converted into the chloro-compound with $Ph_3 P$ and CCl_4. Stereochemical studies have shown that the reaction proceeds by an intramolecular process with inversion at carbon, on the basis of which the decomposition of a halogenoalkoxyphosphorane through a four-centre transition state was proposed [28]. The cleavage of ethers by triphenylphosphonium dihalides to alkyl halides has been extended to oxiranes to give vicinal dichlorides and dibromides [29].

2.2.5 Radical substitution at carbon

Hydrogen attached to saturated carbon atoms is replaced photochemically by fluorine by using $CF_3 OF$ in inert solvents [3]. Cyclohexane was converted into fluorocyclohexane (44%, in $CFCl_3$ at -78 °C) and isobutyric acid gave the 2- and the 3-fluoroacid (31 and 39% respectively), but particularly important was the fluorination of bioactive molecules, e.g. L-isoleucine.

$$Me{\cdot}CH_2{\cdot}CHMe{\cdot}CH(\overset{+}{N}H_3){\cdot}CO_2H \xrightarrow[hv]{CF_3OF,\ HF}$$

$$FCH_2{\cdot}CH_2{\cdot}CHMe{\cdot}CH(\overset{+}{N}H_3){\cdot}CO_2H$$

2.2.6 Miscellaneous methods

Potassium tetrafluorocobaltate(III) is a relatively mild fluorinating agent, which converts tetrachloroethylene in the vapour phase at 180 °C into $CFCl_2{\cdot}CFCl_2$ (41%) and $CFCl_2{\cdot}CF_2Cl$ (8%)[30]. The suggestion has been made that fluorinations with high valency metallic fluorides are basically oxidation reactions[31].

Treatment of phenylcyclopropane with $PbF_2(OAc)_2$ gives among the products, $PhCHF{\cdot}CH_2{\cdot}CH_2OAc$ (38%) and $PhCHF{\cdot}CH_2{\cdot}CH_2F$ (10%), which are presumed to arise from an initial ring opening with $\overset{+}{P}bF(OAc)_2^-$[32].

Brown and his co-workers have developed methods for the anti-Markovnikov hydroiodination and hydrobromination of terminal alkenes by a hydroboration–halogenation sequence[33, 34].

$$(RCH_2{\cdot}CH_2)_3B \xrightarrow[\text{(ii) 2NaOH in MeOH, 25 °C}]{\text{(i) 2I}_2} 2RCH_2{\cdot}CH_2I + RCH_2{\cdot}CH_2{\cdot}B(OH)_2$$

Only two of the three alkyl groups are utilised in the iodination, but hydroboration of the alkene with di-isoamylborane circumvented this difficulty since the primary group migrates in preference to the isoamyl group. Complete conversion of $(RCH_2{\cdot}CH_2)_3B$ into $3RCH_2{\cdot}CH_2Br$ was effected by bromine followed by NaOMe MeOH at 0 °C. An *exo*: *endo* product ratio of 1:3 for the bromination of tri-*exo*-2-norbornylborane provided a clue to the mechanism of the reaction[35]. Although only one norbornyl group migrated under the reaction conditions, the stereochemistry of the major product was unexpected and was rationalised on the basis of an initial coordination of the boron with methoxide, followed by rear-side attack by bromine on the carbon. Bromine in CH_2Cl_2 in the dark at 25 °C will convert one alkyl group in a trialkylborane into an alkyl bromide by a free-radical mechanism[36].

$$(R_2CH)_3B + Br_2 \longrightarrow (R_2CH)_2BBr + R_2CHBr$$

Hydroboration by 9-borabicyclo[3,3,1]nonane, even of internal alkenes, and bromination of the 9-B-alkyl compound gives alkyl bromides in high yields[37], e.g. $Me{\cdot}CH{:}C(CH_3)_2 \rightarrow Me{\cdot}CHBr{\cdot}CH(CH_3)_2$ (87%).

Sulphuryl chloride, which chlorinates dialkyl sulphides and sulphoxides exclusively α to the sulphur, tends not to attack this position in sulphones[38]. With unsymmetrical compounds, substitution occurs in the longer chain, e.g. n-butyl ethyl sulphone gives only the β- (12%), γ- (41%), δ- (31%), and γδ-substitution product (10%) in the n-butyl group. A radical mechanism for this curious reaction has been discounted.

α-Bromination of acyl chlorides is readily effected by N-bromosuccinimide in the presence of acid catalysts, but only with difficulty by bromine as in the Hell–Volhard–Zelinsky reaction even though acyl halides are believed to be intermediates[39]. Selective α-bromination and α-iodination of esters under very mild conditions is achieved through the halogenation of α-lithio derivatives, prepared from the esters and lithium N-isopropylcyclohexyl-amide in tetrahydrofuran at $-78\,°C$[40] [$CH_2{:}CH{\cdot}(CH_2)_7{\cdot}CH_2{\cdot}CO_2Et \rightarrow CH_2{:}CH{\cdot}(CH_2)_7{\cdot}CHI{\cdot}CO_2Et$ (80%)].

The lability of the C—Cl bond in the CCl_3 group of polychlorohydro-carbons has been shown in their reactions with alkenes, with iron penta-carbonyl as catalyst[41].

$$CCl_3{\cdot}CH_2{\cdot}CH_2Cl + CH_2{:}CH_2 \longrightarrow ClCH_2{\cdot}CH_2{\cdot}CCl_2{\cdot}CH_2{\cdot}CH_2Cl \ (56\%)$$

2.3   PREPARATION OF VINYLIC HALIDES

2.3.1   Electrophilic additions to alkynes

The overall Markovnikov addition of BrF to unsymmetrical alkynes is effected by N-bromoacetamide in anhydrous hydrogen fluoride[42] (e.g. $Bu^n{\cdot}C{:}CH \rightarrow trans\text{-}Bu^n{\cdot}CF{:}CHBr$, 95%). Tolan reacts with iodine in an acetic acid–peroxyacetic acid mixture along similar lines to give the *trans*-iodoacetoxylated product[43]. The stereochemical courses of these reactions may be accounted for by postulating bridged halogenonium ion inter-mediates (see p. 73).

In contrast, reactions with acyl chlorides $RCO{\cdot}Cl$ give mixtures of *cis* and *trans* adducts, $-C(COR){:}C(Cl)-$. Acetylenic thioethers and phosgene yield predominantly *cis*-adducts (60–87%) in the products[44].

$$R^1C{:}C{\cdot}SR^2 \xrightarrow{COCl_2} R^1C{:}\overset{+}{C}{\cdot}SR^2 \longrightarrow R^1C(CO{\cdot}Cl){:}CCl{\cdot}SR^2$$
$$\underset{^-OCCl_2}{|}$$

The stereochemistry of addition of the 1:1 benzoyl chloride—aluminium trichloride complex to hex-3-yne is influenced by the polarity of the solvent[45], which controls the position of the following equilibrium:

$$PhCO{\cdot}Cl{\cdot}AlCl_3 \rightleftharpoons Ph\overset{+}{C}O + AlCl_4^-$$

It is proposed that formation of the *cis*-adduct occurs through the undis-sociated complex by cycloaddition of C—Cl across the triple bond, whereas with the ionised material, cycloaddition of $-\overset{+}{C}{=}O$ across the triple bond and ring opening by chloride ion gives the *trans*-adduct. The ratio of *cis*- to *trans*-adducts in pure CH_2Cl_2 at $-40\,°C$ is 6.7:1, whereas in 1:1 CH_2Cl_2–$PhCO{\cdot}Cl$, it is 2:1, which is a reflection of the higher ionising power of this solvent.

Toluene-p-sulphenyl chloride adds to terminal alkynes to give exclusively anti-Markovnikov (A.M.) products (*trans-p*-Tol $S{\cdot}CR{:}CHCl$) in all solvents for R = alkyl, but for R = phenyl, while exclusive A.M. products are formed in ethyl acetate, in the more polar acetic acid 71% of the adduct consisted of trans-p-Tol S·CH:CPhCl[46]. Sulphur dichloride reacts with phenylacetylene to give only trans-PhC(SCl):CHCl. Bridged covalent or ion-pair intermediates ((1) or (2)), which are attacked at the less hindered carbon atom, account for trans A.M. adducts in solvents of low polarity, whereas competing electronic control through (3) occurs in ionising solvents[47].

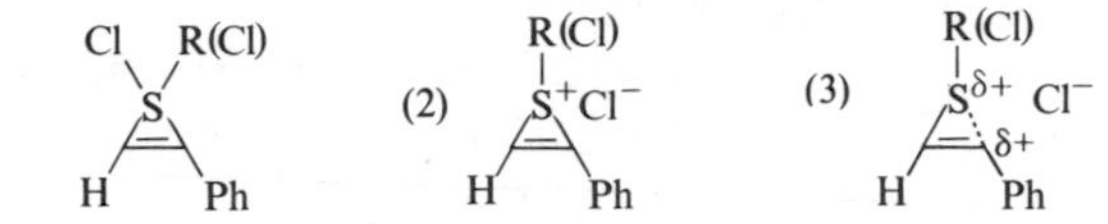

2.3.2 Nucleophilic additions to alkynes

Fluoride ions react with alkynes in dipolar aprotic solvents to give vinylic carbanions, which may be trapped by polyfluoro-aromatic[48] or -heterocyclic[49] compounds. Diethyl acetylenedicarboxylate and caesium fluoride in sulpholan react with pentafluoropyridine at 100 °C to form adducts, mainly trans-EtO$_2$C·CF:CX·CO$_2$Et, where X = 4-tetrafluoropyridyl or X = H, (from the solvent). In the analogous reaction with perfluorobut-2-yne, the initially formed vinylic carbanion may be trapped as before, or undergo further reaction with the butyne to give perfluoro-dienyl and -trienyl carbanions, which may also be trapped. Under vigorous conditions, fluoride ion promotes polymerisation of the butyne to F[C(CF$_3$):C(CF$_3$)]$_n$. The first stable perfluorovinylsilver compound (trans-CF$_3$·CF:CAg·CF$_3$) was prepared from perfluorobut-2-yne and silver fluoride in acetonitrile at 25 °C[50].

Propiolic acid[51] and 1-alkynylphosphines[52] (e.g. Bu$_2^i$PC:CH) react with hydrogen halides to give predominantly trans-adducts by way of O- and P-protonated intermediates respectively.

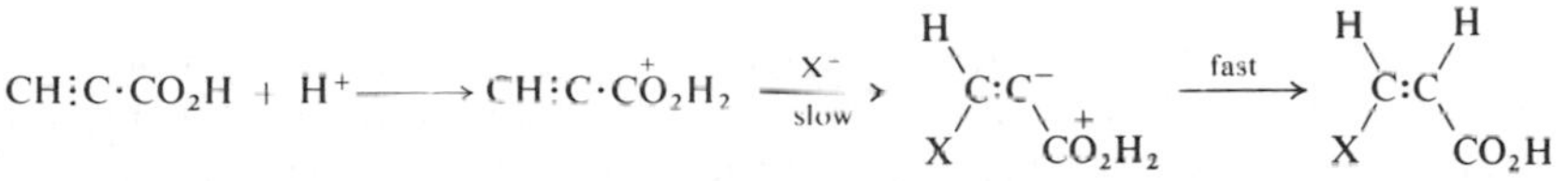

Halogenoallenes (R^{1}R^{2}C:C:CXY; X = I, Y = H; X = Br, Y = Br; X = Br; Y = Cl) are prepared from acetylenic alcohols R^{1}R^{2}C(OH)·C:CY and a hydrogen halide, HX, in the presence of Cu^{I}X[53].

2.3.3 Radical additions to alkynes

Sulphonyl halides, RSO$_2$·X, add to alkynes under the influence of light or in the presence of a copper catalyst. Terminal alkynes, R^{1}C:CH, give alkyl 2-halogenoalkenyl sulphones, R^{1}CX:CH·SO$_2$R^2, which have the trans configuration, from the photochemical reaction with sulphonyl iodides[54]. However, both cis and trans products are formed from sulphonyl chlorides with Cu^I or Cu^{II} chloride or cyanide as catalyst; solvents of high dielectric constant and added quaternary ammonium chlorides strongly favour the formation of the trans-adduct[55].

2.3.4 Miscellaneous methods

The halogenolysis (I_2, Br_2, or Cl_2) of *cis-* and *trans-β*-styrylpyridinecobal-oxime (which are readily prepared in pure form) in acetic acid, give the corresponding *cis-* and *trans*-2-halogenostyrene stereospecifically[56]. Electro-philic attack by halogen followed by rapid carbon–cobalt bond cleavage accounts for these observations.

Corey has examined the reactions of *β*-oxidophosphonium ylids with halogenating agents[57]. The betaine from ethylidenetriphenylphosphorane and heptanal was treated with n-butyl lithium to give the required ylid, which with N-chlorosuccinimide (NClS) or with iodobenzene dichloride gave geometrically isomeric chloroalkenes.

$$
\begin{array}{ccc}
\underset{H}{\overset{n\text{-}C_6H_{13}}{\diagdown}}C=C\underset{Me}{\overset{Cl}{\diagup}} & \xleftarrow{\;\text{PhICl}_2\;}\; n\text{-}C_6H_{13}\cdot C\cdot CHMe\overset{+}{:}\overset{|}{P}Ph_3 \;\xrightarrow{\;\text{NClS}\;} & \underset{H}{\overset{n\text{-}C_6H_{13}}{\diagdown}}C=C\underset{Cl}{\overset{Me}{\diagup}}
\end{array}
$$

Chlorofluoromethylenetriphenylphosphorane generated *in situ* from sodium chlorofluoroacetate and triphenylphosphine gives 2-substituted 1-chloroperfluoroalkenes with a wide variety of polyfluorinated ketones[58]. Evidence was presented for the formation of the ylid by the following mechanism rather than by prior decomposition of the carboxylate to the carbene, $\ddot{C}FCl$.

$$
Ph_3P + CFCl_2\cdot CO_2^- \xrightarrow{-Cl^-} Ph_3\overset{+}{P}CFCl\cdot \overset{-}{C}O_2 \xrightarrow{-CO_2} Ph_3P:CFCl
$$

$$
\downarrow{\scriptstyle RCOR_f}
$$

$$
R_fCR:CFCl
$$
$$
(cis \text{ and } trans)
$$

However no ylid-type intermediate is involved in the conversion of alde-hydes and ketones into 1,1-dichloro- or 1,1-dibromo-alkenes by using an excess of CCl_4 or CBr_4 respectively with tris(dimethylamino)phosphine[59].

Carbonium ions are believed to be involved in the formation of vinyl halides in high yields from 5,5-dialkyl-N-nitroso-oxazolidones in 2-meth-oxyethanol saturated with a lithium halide (chloride, bromide or iodide) on treatment with 2-methoxyethoxide[60]. The oxidation of p-substituted propio-phenone hydrazones with iodine and base (Et_3N) gives 1-aryl-1-iodopropenes, an α-iodocarbonium ion probably being an intermediate[61].

2.4 PREPARATION OF 1-HALOGENOALKYNES

The six dihalogenoacetylenes containing chlorine, bromine, or iodine have been resynthesised or made for the first time in pure form from lithium chloro-or bromo-acetylide or dilithium acetylide and a halogen and their general properties have been summarised[62].

$$
ClCH:CHCl \xrightarrow{\;2\text{PhLi}\;} [ClC:CLi] \xrightarrow{\;X_2\;} ClC:CX \qquad (X = Cl, Br, \text{ or } I)
$$

Dichloroacetylene may also be prepared in an easily handled ethereal solution by dehydrochlorination of trichloroethylene with potassium hydroxide in ethylene glycol at 140 °C[63].

Difluoroacetylene is known and chlorofluoroacetylene is produced by dehydrochlorination of 1,1-dichloro-2-fluoroethylene at 120 °C under reduced pressure[64], but all attempts to make lithium fluoroacetylide for reaction with bromine or iodine were unsuccessful[62].

2.5   NEIGHBOURING GROUP INTERACTIONS OF HALOGENS

2.5.1   Interactions with carbonium ions

Intense interest has been shown over the past 5 years in reactions which involve intermediate halogen-containing carbonium ions, $X(C)_nC^+$. The formation of stable ions in strongly acidic media (SbF_5–SO_2 or FSO_3H–SbF_5–SO_2) and their direct observation by n.m.r. spectroscopy, and the study of reactions carried out in the weakly nucleophilic solvent trifluoroacetic acid, have been especially responsible for the developments in this area. The formation of the ions and the consequences of halogen interaction will be discussed in terms of increasing distance of the halogen from the cationic centre.

2.5.1.1   Halogenocarbonium ions (n = 0)

Following the preparation of the first stable halogenocarbonium ions, difluorophenylcarbonium ion and fluorodiphenylcarbonium ion[65], Olah and his co-workers prepared the 2-fluoropropylcarbonium ion by two routes[66].

$$Me \cdot CF_2 \cdot Me \xrightarrow[60\,°C]{SbF_5–SO_2} Me \cdot \overset{+}{C}F \cdot Me \longleftrightarrow Me \cdot C(:F^+) \cdot Me$$

$$\underset{\uparrow}{\big|} \xrightarrow{FSO_3H–SbF_5–SO_2} Me \cdot CF{:}CH_2$$

Evidence for appreciable resonance interaction in the ion was obtained from the large downfield shift of the ^{19}F n.m.r. resonance relative to those of its precursor. Attempts to prepare the simpler ions, $Me \cdot \overset{+}{C}HF$ and $Me \cdot \overset{+}{C}F_2$, were unsuccessful.

2.5.1.2   1,2-Halogenonium ions (n = 1)

A cyclic three-membered brominium ion formed by direct bonding between the β-bromine and the carbonium ion was originally postulated as an intermediate to account for the stereospecific trans-addition of bromine to cis- and trans-but-2-ene[67]. Participation by bromine also accounted for the rate constant of acetolysis of trans-2-bromocyclohexyl p-bromobenzenesulphonate[68] being appreciably higher than that expected for its unassisted solvolysis, as estimated from the rate for the cis-isomer[69]. Participation by iodine in the reaction of the analogous iodo-ester was even more effective [68], but the chloro-compound was solvolysed without significant interaction[69].

$$Me_2CX \cdot CFMe_2 \xrightarrow[-60\ C]{SbF_5-SO_2} Me_2C-CMe_2 \quad SbF_6^-$$
$$\underset{X}{\overset{\backslash+/}{}}$$

$$(X = Cl, Br, or I)$$

Olah has formed stable tetramethylethylene halogenonium ions in strongly acidic media[70]. The n.m.r. data for the product of ionisation of the 1,2-difluoro-compound ($X = F$) was interpreted in terms of a pair of rapidly equilibrating classical ions. Clark, however, has argued that the chemical shift and coupling constant results are not incompatible with a fluorinium ion, and has produced theoretical evidence which, by analogy definitely favours the bridged ion[71]. Calculations show that the ethylene fluorinium ion is not only more stable than the classical 2-fluoroethyl cation (by 3.58 kcal mol^{-1}) but that there is also a high energy barrier (24.65 kcal mol^{-1}) for conversion of the 2-fluoroethyl cation into the fluorinium ion, so that the equilibration of classical 2-fluoroethyl cations will be slow. Furthermore, it is not clear that the necessary suprafacial 1,2-fluorine shift, unlike the 1,2-hydrogen shift, is a symmetry-allowed process.

The simplest stable 1,2-halogenonium ions, the ethylene-iodinium and -brominium ions have been prepared[72], but the trimethylethylenechlorinium ion is the least substituted 1,2-chlorinium ion to have been formed[73]. Acyclic dimethylhalogenonium hexafluoroantimonates are crystalline solids, stable at room temperature under nitrogen[74].

$$MeX + MeF \rightarrow SbF_5 \xrightarrow[-78°C]{SO_2} Me \cdot \overset{+}{X} \cdot Me\ SbF_6^- \ (or\ Sb_2F_{11}^-)$$

$$(X = Cl, Br, or I)$$

These compounds are powerful alkylating agents[75], e.g. $Et_2O \longrightarrow Et_2\overset{+}{O}Me$. The stereospecificity of addition of halogens and pseudohalogens to alkenes gives useful information concerning the magnitude of the interactions between a carbonium ion and a β-halogen atom. The additions of IN_3[76] $INCO$[77], Br_2[78], BrN_3[79], $BrN(CF_3)_2$[23], and Cl_2[80] to cis- and to trans-but-2-ene by an ionic mechanism are stereospecific, proceeding via tightly bound 1,2-halogenonium ions. The 1,2-iodinium ion is the most stable 1,2-halogenium ion, since even when the carbonium ion could be stabilised by conjugation with an aromatic ring, as in the Markovnikov additions of IN_3[81] and $INCO$[77] to isomeric 2-deuteriostyrenes, stereospecific reactions are observed. In contrast, BrN_3 reacts with cis-2-deuteriostyrene to give a 1:1 mixture of the threo- and the erythro-adduct[81]. However, that a β-bromine can cause an electrostatic interaction with a benzylic cationic centre and prevent completely free rotation about the C—C bond before reaction with other anions, is shown by the addition of bromine to cis- and trans-1-phenylpropene in solvents of different polarity[82]. The products from the addition to the cis-isomer are much more solvent-dependent than for the trans-isomer, since with the cis-isomer, conformational interactions after initial attack by Br^+ are at a maximum, but identical product distributions are never attained, even in the most polar solvents used. Weak electrostatic interaction between the bromine and the benzylic carbonium ionic centre is sufficient to prevent completely free rotation about the carbon–carbon bond (cf.(4)).

(4)

It is of interest to note that BrN_3 adds non-stereospecifically to *cis*-2-deuteriostyrene under photolytic conditions to give a 1:1 mixture of *threo*- and *erythro*-PhCHBr·CHDN$_3$, by an initial attack by N_3 [81], while the ease of addition of IN_3 to CF_2:CFX (X = Cl or CF_3) to give N_3CF_2CFXI indicates an initial nucleophilic attack by N_3^- [83].

The electrophilic addition of chlorine[80] and of fluorine[84] to *cis*- and *trans*-1-phenylpropene are also non-stereospecific and intermediate open carbonium ions have been proposed. The high tendency towards *cis*-addition in both cases (e.g. 78 and 68% from the *cis*- and the *trans*-hydrocarbon respectively, with F_2 in CCl_3F at $-78\,^{\circ}$C) is the result of ion-pairing phenomena in non-polar solvents.

Methyl migrations to open the 1,2-halogenonium ion ring do not occur in the reactions of 3,3-dimethylbut-1-ene with Cl_2 [85], Br_2 [86], Br_2 in methanol[86], IN_3 [87], INCO [77], or INO_3 [88] by an ionic mechanism. Simple adducts are formed which have the anti-Markovnikov orientation (where this is observable). Steric factors which overwhelm electronic effects are invoked to explain the opening of the halogenonium ring since the isomeric hex-1-ene forms the Markovnikov adduct exclusively with IN_3 in acetonitrile [$CH_3\cdot(CH_2)_3\cdot CHN_3\cdot CH_2I$] [76] and INCO in ether gives a 7:3 ratio of the Markovnikov to the anti-Markovnikov adduct[77]. Phenyl migration does take place and opens 1,2-halogenonium ions during the reactions of Br_2 [89], BrN_3 [87], IN_3 [90] and INCO [90] with 3,3,3-triphenylpropene by an ionic mechanism.

$$Ph_3C\cdot CH:CH_2 \xrightarrow{IN_3} [Ph_3C\cdot \underset{+}{\overset{\quad}{CH}}-CH_2] \longrightarrow [Ph_2\overset{+}{C}\cdot CHPh\cdot CH_2I]$$
$$\downarrow N_3$$
$$Ph_2CN_3\cdot CHPh\cdot CH_2I$$
$$(99\%)$$

The addition of HCl[91] and BrCl[92] to allene gives products [CH_2:CCl·CH$_3$ and CH_2:CBr·CH$_2$Cl respectively] which suggest different positions for the initial electrophilic attack by H^+ and Br^+. The formation of a bridged 1,2-brominium ion which opened at the sterically less hindered terminal carbon atom was proposed to unify these results and has received support by the direct observation of the ion in a strongly acidic medium by ^1H n.m.r. spectroscopy[93]. In order to account for the stereospecific *trans*-addition of bromine to hex-3- and hex-1-yne, an unsaturated bridged brominium ion was suggested[94].

A bridged brominium ion in equilibrium with an open β-bromocarbonium ion is believed to be involved in the debromination of ($\pm$)-stilbene dibromide by two-electron reductants such as I^- and Br^- [95].

1,2-Halogen migrations have been observed in a number of reactions

which involve intermediate carbonium ions. Ring opening of stable tri-
methylethylenehalogenonium ions (from 2-fluoro-3-halogeno-2-methyl-
butane with Cl, Br or I as substituents) with methanol occurs at the less
hindered carbon atom to give 2-halogeno-3-methoxy-2-methylbutanes,
presumably by a bimolecular process[73]. In contrast the solvolysis of 2-
chloropropyl *p*-nitrobenzoate[96] and the reaction of trimethoxonium tetrafluoroborate with 2-bromo- and 2-iodopropyl methyl ether[94] in trifluoroacetic acid all give rearranged trifluoroacetates as the sole product, the
intermediate halogenonium ion opening to a secondary rather than the less
stable primary carbonium ion, before final reaction with the weakly nucleophilic solvent. Only 8% of chlorine migration is observed in the *acetolysis* of

$$
\text{Me CHCl·CH}_2\text{O·COAr} \xrightarrow[\text{CF}_3\cdot\text{CO}_2\text{H}]{} \quad \underset{\substack{\diagdown \overset{+}{} \diagup \\ \text{Cl}}}{\text{MeCH}-\text{CH}_2} \longrightarrow \quad \text{MeCH(CH}_2\text{Cl)O·CO·CF}_3
$$

$$
\text{Ar} = p\text{-NO}_2\text{C}_6\text{H}_4
$$

1-[^{14}C-2]-chloroethyl *p*-nitrobenzoate[98]. Deamination of 2-halogeno-1-
propylamines with nitrous acid gives mixtures of alcohols containing the
rearranged products, 1-halogenopropan-2-ols (76% and 53% for the Br
and the Cl compound respectively)[99].

2.5.1.3   1,3-Halogenonium ions (n = 2)

The deamination of a deuterium-labelled derivative of 3-bromo-2,2-bis
(bromomethyl)propylamine in acetic acid gives products which show the
presence of rapidly equilibrating 1,3-brominium ions[100]. The completely

$$
(\text{BrCH}_2)_3\text{C·CD}_2\overset{+}{\text{N}}\text{H}_3\text{Cl}\bar{\text{O}}_4 \xrightarrow[\text{HOAc}]{\text{HONO}} \quad \underset{\substack{| \\ \text{CH}_2\text{Br}}}{\overset{\substack{\text{Br}\searrow \; \text{CH}_2\overset{\curvearrowright}{\text{Br}^+} \\ | \qquad |}}{\text{CH}_2-\text{C}-\!\!-\text{CD}_2}} \longleftrightarrow \underset{\substack{| \\ \text{CH}_2\text{Br}}}{\overset{\substack{\overset{+}{\text{Br}}-\!\!\text{CH}_2 \\ | \qquad |}}{\text{CH}_2-\text{C}-\text{CD}_2\text{Br}}} \longleftrightarrow \text{etc.}
$$

statistical distribution of deuterium in these ions is never attained and is
explained on the basis of solvent attack on the unbridged ion R$\overset{+}{\text{C}}$D$_2$ formed
initially. In trifluoroacetic acid 80% of the trifluoroacetate produced arises
from equilibrated ions. 1,3-Bromine migration is observed to the extent of
18% in the products from the deamination of 3-bromopropylamine, but no
such migration occurs with the chloro-analogue[101].
   Halogen migrations in solvolyses have been observed only in systems
which contain the best leaving group known, the trifluoromethylsulphonate
group. Trifluoroacetolysis of such 3-halogenobutyl esters gives rearranged
3-halogeno-1-methylpropyl trifluoroacetates in the following molar percentages: 58, 55 and 5 for the I, Br and Cl compounds respectively[102].

2.5.1.4   1,4-Halogenonium ions (n = 3)

1,4-Halogen participation was originally postulated to occur in the acetolysis of 4-iodo- and 4-bromobutyl tosylate[103]. The rate enhancements were
not very pronounced but recent work shows that participation is much more

important in trifluoroacetic acid as solvent. The configuration of the *erythro*-
and of the *threo*-4-chloro-1-methylpentyl tosylate are largely retained
(c. 90%) in the products, and after making allowance for the inductive effect
of chlorine the accelerator effect of 1,4-chlorine participation is 99 and 65
for the *erythro*- and the *threo*-isomer respectively[104]. The first kinetic evidence
for 1,4-fluorine participation is suggested in the trifluoroacetolysis of 4-
fluoro-1-methylbutyl tosylate where a modest acceleratory effect of participation of 2.4 is observed.

Stable 1,4-halogenonium ions are formed by the ionisation of 1,4-dihalo-
genobutanes (with Cl, Br or I, but *not* F as substituents) in SbF_5-SO_2 at
$-60\,^{\circ}C$[105]. The formation of the 1,4-chlorinium ion but *not* the 1,2-chlori-
nium ion[77] from unsubstituted butane and ethane derivatives respectively,
reflects the expected greater stability of the five-membered over the three-
membered ring.

1,4-Halogen migration products are formed in the solvolysis of 4-chloro-
pentyl p-nitrobenzoate[96, 106], and in the reactions of 4-chloro- and 4-bromo-
pentyl methyl ether with trimethyloxonium tetrafluoroborate[97] in trifluoroacetic acid: 99.5%, 100% and 89.6% of the corresponding 4-halo-
geno 1 methylbutyl trifluoroacetates respectively. In the more nucleo-

$$\text{MeCHBr·(CH}_2)_3\text{·OMe} \xrightarrow[\text{CF}_3\text{·CO}_2\text{H}]{\text{Me}_3\overset{+}{\text{O}}\text{BF}_4^{-}} \cdot\text{CH}\overset{\text{CH}_2\text{·CH}_2}{\underset{\text{Br}^+}{\diagup\diagdown}}\text{CH}_2 \longrightarrow \text{CF}_3\text{·CO·O·CHMe·(CH}_2)_3\text{Br}$$

philic solvent acetic acid, 4-chloropentyl p nitrobenzoate gives only 5% of
rearranged ester[106], while 22% of 4-bromo-3-d$_2$-butyl acetate is formed
from 4-bromo-2-d$_2$-butyl p-nitrobenzoate[107]. No fluorine migration occurs
in the trifluoroacetolysis of 4-fluoro-1-d$_2$-butyl trifluoromethyl sulphonate[108].
The trifluoroacetolysis of the ester of a primary alcohol containing a secondary fluorine at position 4, the system most likely to reveal a 1,4-fluorine
shift, has not been reported.

Since the earlier demonstration of a 1,4-chlorine shift in the addition of
trifluoroacetic acid to 5-chlorohex-1-ene[109], halogen migrations in general
have now been found in addition to acetylenic compounds[108].

$$\text{HC:C·CH}_2\text{·CH}_2\text{·CH}_2\text{X} \xrightarrow{\text{CF}_3\text{·CO}_2\text{H}} \text{CH}_2\text{:C}\overset{\text{CH}_2-\text{CH}_2}{\underset{\text{X}^+}{\diagup\diagdown}}\text{CH}_2 \longrightarrow \text{CH}_2\text{:CX·(CH}_2)_3\text{·O·CO·CF}_3$$

$$\text{CF}_3\text{·CO·O·CF Me·(CH}_2)_3\text{·O·CO·CF}_3$$

(5)

The reaction with 5-fluoropent-1-yne is of particular interest since this
appears to be analogous to those of the other halogeno-compounds to the
extent of 15%, followed by addition of trifluoroacetic acid to the initially
formed vinylic fluoride[110]. The formation of the adduct (5) was unexpected
until subsequent kinetic studies on the addition of trifluoroacetic acid to
various 2-halogenopropenes showed that the 2-fluoro-compound was at
least 200 times more reactive than the chloro-compound, the most reactive
of the other halogeno-alkenes.

2.5.1.5 1,5- and 1,6-Halogenonium ions (n = 4 and 5)

Trifluoroacetolysis of 5-bromo- and 5-chloro-hexyl p-nitrobenzoate gives
high yields of products resulting from 1,5-halogen shifts[106]. Attempts to
prepare the stable 1,5-brominium ion from 1,5-dibromopentane in SbF_5–SO_2
at −60 °C were unsuccessful: the 4-methyl-1,4-brominium ion was formed[105].

The first example of a 1,6-halogen shift has been reported recently[96]:
17% of primary chlorides are formed in the trifluoroacetolysis of 6-chloro-
heptyl p-nitrobenzoate.

2.5.2 Interactions with radicals

The interaction of neighbouring halogen with radical centres has been
proposed to explain a number of observations which have their counterparts in the carbonium ion reactions described above.

The chlorination of t-butyl bromide gives one product only, by a 1,2-
bromine migration[111].

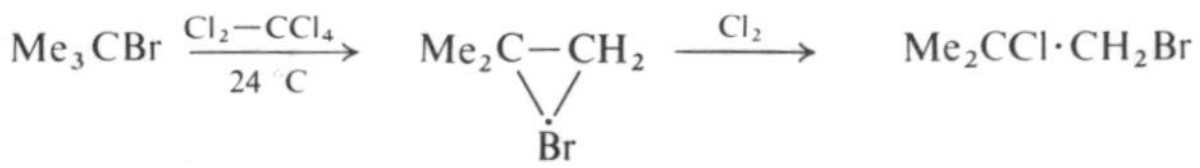

$$Me_3CBr \xrightarrow[\text{24 C}]{Cl_2-CCl_4} Me_2C-CH_2 \xrightarrow{Cl_2} Me_2CCl \cdot CH_2Br$$
$$\underset{Br}{\diagdown/}$$

1,2- and 1,3-Bridging in the transition state is proposed to explain the enhanced rates of abstraction of iodine by phenyl radicals from 2-bromo-1-
iodoethane and 1,3-di-iodopropane, by 90 and 35% respectively over and
above that anticipated from inductive effects alone[112]. There was no participation by a β-chlorine atom.

The debromination of meso- and (±)-2,3-dibromobutane by Bu^tSnH
proceeds via bridged species in equilibrium with open radicals to give
mixtures of cis- and trans-but-2-ene in the ratios 10:90 and 66:34 respectively[113]. For the corresponding dichloro-compounds, the ratios of cis-
to trans-alkene were almost identical (c. 20:80), again suggesting essentially
no 1,2-chlorine participation.

2.5.3 Sigmatropic halogen migrations

The rearrangement of 1,1,1-trichloroalk-2-en-4-ones to 1,1,5-trichloroalk-1-
en-4-ones is a novel example of a 1,5-sigmatropic migration of chlorine[114].

$$R^1CH_2 \cdot CO \cdot CR^2 {:} CH \cdot CCl_3 \xrightarrow{H^+} R^1CHCl \cdot CO \cdot CHR^2 \cdot CH{:}CCl_2$$

2.6 REACTIONS OF ALKYL HALIDES

2.6.1 Reactions with nucleophiles

Sodium borohydride in aprotic solvents (dimethyl sulphoxide, sulpholan,
diglyme, or dimethyl formamide) reduces a wide range of alkyl halides to

hydrocarbons in the presence of other reducible groups ($-CO_2H$, $-CO_2R$, $-NO_2$)[115, 116], while sodium cyanoborohydride in N^1,N^2,N^3-hexamethyl phosphoric triamide will reduce alkyl bromides and iodides selectively even with an oxirane ring, or a ketonic or aldehydic carbonyl group in the same molecule[117]. There is evidence for the operation of a simple S_N2 mechanism with $NaBH_4$[118] but the reduction of $(-)$-(R)-3-chloro-3,7-dimethyl-octane in dimethyl sulphoxide did show the intermediacy of an organo-borane, formed by an elimination–hydroboration sequence, the protonolysis of which gave a racemic hydrocarbon[119].

The direct attack of a nucleophile on the halogen atom of an alkyl halide in the presence of a proton donor in effect causes the overall reduction of the compound. Lithium iodide and boron trifluoride in ether or tetrahydrofuran, followed by aqueous $Na_2S_2O_3$ to remove I_3^-, brings about the quantitative replacement of halogen by hydrogen in a number of cyclic α-halogenoketones at room temperature[120]. Sodium diphenylphosphide in liquid ammonia

$$IBr + 2I^- \longrightarrow I_3^- + Br^-$$

reacts with 3-bromopropyne to form propyne, but with the chloro-compound, nucleophilic displacement of halogen occurs[121]. Terminal perfluoro-alkenes are formed from primary perfluoroalkyl iodides and methylmag-nesium chloride at temperatures above 25 °C, by an initial attack on iodine followed by β-elimination of fluorine[122] (e.g. n-$C_8F_{17}I \rightarrow n$-C_6F_{13}·CF:CF_2).

There are few recorded examples of S_N2 displacements of tertiary halides, the carbonium ion mechanism being preferred. Kornblum has shown that p-nitrocumyl chloride reacts with sodium thiophenoxide by a bimolecular process[123], and the amines[124] (e.g. quinuclidine), under mild conditions to give tertiary alkyl substitution products by a radical-anion chain mechanism. As a consequence of the mechanism which was proposed, once the forma-tion of chain-carrying p-nitrocumyl radicals had been initiated, a nucleo-phile, unreactive in the sense of not being able to initiate the radical reaction, could bring about chlorine replacement. p-Nitrocumyl chloride failed to react with sodium nitrite during 90 min in the dark, but the addition of 5 mol % of the lithium salt of 2-nitropropane gave a 93 % yield of α,p-dinitrocumene after 90 min[125].

Ion cyclotron resonance spectroscopy has revealed some novel nucleo-philic displacement reactions in the gas phase[126]. Examination of a mixture of HCl and MeF shows the formation of $Me\overset{+}{C}lH$ in a bimolecular process.

$$HCl + Me\overset{+}{F}H \longrightarrow Me\overset{+}{C}lH + HF$$

In mixtures of water with MeCl and EtCl, reactions of both protonated halides with water were observed, but they followed different paths : $Me\overset{+}{C}lH$ was deprotonated to give MeCl and H_3O^+, while $Et\overset{+}{C}lH$ underwent nucleophilic substitution to give $Et\overset{+}{O}H_2$ and HCl. The order of proton affinity, $MeCl < H_2O < EtCl$, afforded a rationalisation of these results.

Displacement of HF from $CH_3\overset{+}{F}H$ by such unusual nucleophiles as N_2 and CO, forming MeN_2^+ and $Me\overset{+}{C}O$ respectively, confirmed predictions for such reactions based on thermochemical data.

The replacement of the halogen in α-halogeno-esters[127] and -ketones[128] by an alkyl group from trialkylboranes is promoted by potassium t-butoxide in t-butyl alcohol.

$$R_3^1B + BrCHXY \longrightarrow R^1CHXY \ (X = CO_2Et \ or \ COR^2; \ Y = H, R, Cl \ or$$
$$Br)$$

The utilisation of only one of the three alkyl groups is overcome by the use of B-alkyl-9-borabicyclo[3,3,1]nonane compounds[129, 130]. The use of the highly hindered base, potassium 2,6-di-t-butylphenoxide has extended the scope of these reactions to include the alkylation of the highly reactive α-bromoacetone[131], and mono-[132] and di-chloroacetonitrile[133]. All three halogen atoms in $CHClF_2$ can be replaced by alkyl groups from trialkylboranes by using lithium triethylcarboxide, the conditions being sufficiently mild so as not to isomerise certain organoboranes[134]. Subsequent oxidation with aqueous alkaline hydrogen peroxide gives trialkyl carbinols.

Alkyl chlorides can readily be converted into the corresponding bromides by halogen exchange. The rare reverse process may now be carried out by using silver dichlorofluoroacetate in a suitable solvent at reflux temperature[135] (e.g. n-$C_8H_{17}Br \longrightarrow$ n-$C_8H_{17}Cl$ in 89% yield in monoglyme).

Primary alkyl bromides (RBr) react with disodium tetracarbonylferrate(II) in the presence of triphenylphosphine in tetrahydrofuran at 25 °C to give high yields of the aldehyde, RCHO[136]. Dibenzyl ketone (90%) is formed from benzyl bromide and carbon monoxide with aqueous potassium hexacyanodinickelate[137].

2.6.2 Reactions via carbonium ions

The anti-Markovnikov addition of hydrogen bromide to 3,3,3-trifluoro-propene with aluminium tribromide as catalyst[138] is a reaction widely quoted to illustrate the reversal of the usual order of carbonium ion stability, i.e. that a secondary is more stable than a primary carbonium ion. However, the dimerisation of the alkene to trans-$CF_3 \cdot CH{:}CH \cdot CHMe \cdot CF_3$ with DSO_3F is not compatible with the initial formation of a saturated carbonium ion by D^+ addition[139]. Moreover, although recent work has confirmed the exclusive formation of the anti-Markovnikov adduct with HBr, two products are formed with HCl–$AlCl_3$ ($CF_3 \cdot CH_2 \cdot CH_2Cl$ and $ClCF_2 \cdot CH_2 \cdot CH_2Cl$, the proportions being dependent on the concentration of $AlCl_3$), and with HSO_3Cl the major product was $ClSO_2 \cdot CF_2 \cdot CH_2 \cdot CH_2Cl$[140]. It was proposed that in all these acidic systems, ionisation of a C—F bond of the alkene occurs to give the 1,1-difluoroallyl cation. This in turn reacts with the solvent anion A^-, and the resulting allylic compound then either reacts with more of the added acid HA, or with the HF initially generated to give the overall anti-Markovnikov addition product.

$$[CF_2 \cdots CH \cdots CH_2]^+ \xrightarrow{A^-} CF_2{:}CH{\cdot}CH_2A \xrightarrow{HA \,(or\, HF)} ACF_2{\cdot}CH_2{\cdot}CH_2A$$

$$(or\ CF_3{\cdot}CH_2{\cdot}CH_2A)$$

The absence of products resulting from a 1,2-hydrogen shift in the solvolysis of 1,1,1-trifluoro-2-propyl tosylate[139] ($CF_3{\cdot}\overset{+}{C}H{\cdot}Me \nleftrightarrow CF_3{\cdot}CH_2\overset{+}{C}H_2$) underlined the need for the alternative to the previously accepted mechanism of additions to the alkene.

The formation of 2-octyl fluoride from the corresponding bromide with a mixture of HNO_3 and HF occurs via a carbonium ion, formed by attack of the nitronium ion on the halogen[141].

Tertiary alkyl halides are rapidly converted into quaternary hydrocarbons with trialkylaluminium compounds at $-78\,°C$ [142]. Neopentane is formed quantitatively in 10 min from t-butyl chloride and trimethylaluminium, while the same hydrocarbon is formed only slowly from the isomeric isobutyl chloride.

2.0.3 Metallation reactions of alkyl and vinyl halides

Ashby has described the preparation of the first alkylmagnesium fluorides from alkyl fluorides (methyl[143], ethyl[143] and n-hexyl fluoride[144]) and magnesium. The rates of reaction are significantly affected by the solvent, the reaction temperature and the catalyst (which is essential to achieve any reaction). n-Hexylmagnesium fluoride is formed in 92% yield in 4 h in 1,2-dimethoxyethane at reflux temperature with iodine as catalyst and gives the expected products in typical Grignard reactions, with yields comparable with those obtained from the corresponding bromo-compound. Unlike the other Grignard reagents, the fluoro-compounds are dimeric[145].

Lithium dialkylcopper reagents [LiR_2^1Cu] react with alkyl and vinyl bromides or iodides [R^2X] under very mild conditions either by cross-coupling (to give $R^1{-}R^2$) or by halogen–copper exchange (to give R^2Cu)[146]. The former reaction is of wide applicability[147, 148].

$$2LiMe + Cu_2I_2 \xrightarrow[0\,°C]{ether} Li^+[CuMe_2]^- \xrightarrow[6\,h,\,0\,°C]{n-C_{10}H_{21}I} n{\text{-}}C_{11}H_{24}\ (90\%)$$

The extent of the exchange reaction depends strongly upon the reaction conditions (solvent, temperature), and is particularly favoured by copper reagents which have extended alkyl groups[147]. Corey has shown that lithium dialkylcopper reagents react very much faster with bromides and iodides than with carbonyl compounds, and has found conditions for effecting an initial halogen–metal exchange in a δ-halogenoketone followed by intramolecular carbanionic addition to the carbonyl group[149].

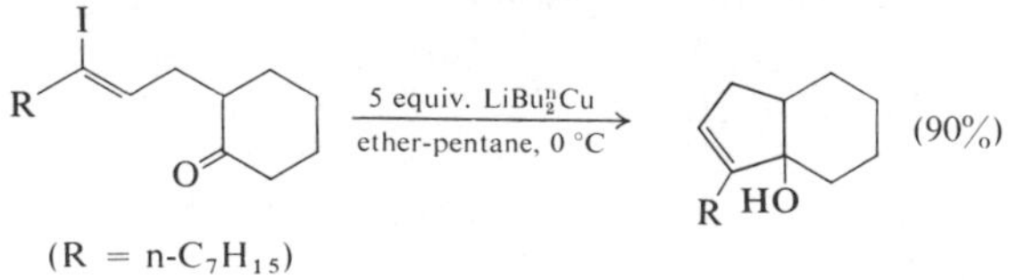

$(R = n{\text{-}}C_7H_{15})$

Lithium dimethylcopper was ineffective in promoting this cyclisation and in order to cyclise halogenoketones which contained the halogen attached to an sp^3-hybridised carbon atom, the anion prepared by the reaction of nickel tetraphenylporphine with lithium naphthalene (2 mol equiv.) in tetrahydrofuran was necessary[149]. 7-Iodoheptan-2-one gave 1-methylcyclohexanol (88%) by reaction with 3 mol equiv. of the reagent at 0 °C for 20 h.

Methylene dibromide and di-iodide react with magnesium to form stable solutions of methylene magnesium bromide and iodide respectively which readily react in ether with aldehydes and ketones to form terminal alkenes[150]. Extension of the reaction to other alkylidene dihalides was less successful, considerable reduction of the carbonyl group occurring with ethylidene dibromide. The Reformatsky procedure for the synthesis of β-hydroxy esters has been dramatically improved by the inclusion of trimethyl borate in the reactants and by conducting the reaction at room temperature[151]. The essentially quantitative yields of products are believed to be due to the neutralisation of the basic products by the mildly acidic borate.

Ethyl trichloroacetate forms the stable alkoxide $CCl_2{:}C(OZnCl){\cdot}OEt$ with zinc in tetrahydrofuran at -15 °C, which in turn reacts with aldehydes and ketones to give α,α-dichloro-β-hydroxy esters[152].

2.7   REACTIONS OF VINYL HALIDES

2.7.1   Reactions with nucleophiles

A novel allylic displacement of halogen, involving an intermediate carbanion, has been proposed for the reaction of ethanolic sodium ethoxide with certain fluoroalkenes $PhCR{:}CF_2$ (6; $R = CF_2Cl$ and 7; $R = CF_2{\cdot}CF_3$)[153].

$$(6) + EtO^- \xrightarrow[-77°C]{slow} [Ph\bar{C}(CF_2Cl){\cdot}CF_2{\cdot}OEt] \xrightarrow{-Cl^-} F_2C{:}CPh{\cdot}CF_2{\cdot}OEt$$

The rates of disappearance of the three alkenes, (6), (7) and (8) ($R = CF_3$) under identical reaction conditions were very similar. Following the initial formation of the ether linkage, subsequent reaction in each case required the formation and breaking of bonds of vastly different strength: (6) and (7) gave products due to the allylic displacement of chlorine (100%) and fluorine (60%) respectively, while overall addition of ethanol to the double bond and replacement of vinylic fluorine occurred with (8). It was concluded that the rate determining step in each case is the attack of ethoxide on $PhCR{:}CF_2$ to form a carbanion.

The displacement of chloride by methanolic methoxide in a series of cis- and trans-β-substituted-1-chloropolyfluoroalkenes occurs stereospecifically with retention of the olefin geometry[154]. The results are rationalised by postulating an irreversible trans-addition of the incoming nucleophile across the double bond to form an unstable carbanion followed by rapid cis-elimination of the chloride.

2.7.2   Reactions via vinyl cations

The relative inertness of vinyl halides towards heterolytic cleavage has been attributed to the low stability of vinyl cations and/or to an unusually strong carbon–halogen bond due to its partial double bond character or to the increased σ character of the bond[155]. While these carbonium ions have not been prepared in strongly acidic media as have their trisubstituted counterparts, evidence for their existence in solvolyses is now well established, following the work of Grob in 1964, who studied substituent effects in the solvolysis of substituted 1-bromostyrenes in 80% ethanol[156]. More recent work with triarylvinyl halides has consolidated the general understanding of the mechanism[157]. The stereochemistry of the solvolysis products has shown the intermediacy of a linear sp^1-hybridised cation. Acetolysis of either cis- or trans-1,2-dianisyl-2-phenylvinyl bromide or chloride gives 1:1 mixtures of isomeric acetates after 10% product formation, and this ratio remains constant throughout the reaction[158]. The possible isomerisation of the starting materials or products under the conditions before or after reaction respectively was ruled out by control experiments.

$$R^1R^2C:CR^1Hal \longrightarrow R^1R^2C:\overset{+}{C}R^1 \xrightarrow{HOAc} R^1R^2C:C(OAc)R^1 \quad (cis:trans = 1:1)$$

Allylic double bonds also stabilise vinylic cations generated from acyclic 2-bromo-1,3-dienes[159]. That the reactivity of these halides is due to the delocalisation of the positive charge on a linear cation ($>C=\overset{+}{C}-\overset{+}{C}=C<$ $\longleftrightarrow$ $>\overset{+}{C}-C=C=C<$) is elegantly demonstrated by the inertness of 2-bromocyclohexa-1,3-diene where geometrical considerations preclude such stabilisation[160].

Allenyl bromide, unlike simple vinyl halides is solvolysed in 50% aqueous ethanol[161], and in the solvolysis of 1-chloro-1-p-substituted phenyl -3,3-diphenylallene in 80% aqueous acetone, correlation of the rate constants with σ^+ constants gives the relatively low ρ value of -2.0, indicating substantial localisation of charge at the 3-position[162].

$$Ph_2C:C:\overset{+}{C}\cdot C_6H_4X\text{-}p \longleftrightarrow Ph_2\overset{+}{C}\cdot C:C\cdot C_6H_4X\text{-}p$$

The mesomerism of allenyl and propargylic cations has been shown by n.m.r. studies of stable alkynyl carbonium ions in strongly acidic media[163].

2.8   REACTIONS OF 1-HALOGENOALKYNES

Three mechanisms for the nucleophilic replacement of halogen at sp^1-hybridised carbon have been recognised.

Viehe has shown that sodium thiophenoxide reacts with 1-chloro-t-butylacetylene in NN-dimethylformamide to give $Bu^tC:C\cdot SPh$ but in ethanol β-addition of thiophenol occurs to form cis- and trans-$Bu^tC(SPh):CHCl$ in the ratio 7:93[164]. Significantly, the reaction of either of these alkenes with lithium diethylamide gives $Bu^tC\!:\!C\cdot SPh$. It was proposed that the same intermediate vinyl carbanion is formed initially in all these reactions, and in the absence of a proton donor it loses chloride and rearranges through a bridged sulphonium ion to the overall product.

$$Bu^tC\!:\!CCl + PhS^- \longrightarrow PhS\cdot CBu^t\!:\!\bar{\bar{C}}Cl \longrightarrow \underset{\substack{| \\ Ph}}{\overset{Bu^t}{\underset{S}{\overset{\diagdown}{C}}}}\!\!\!\!\!\overset{=}{\underset{+}{C}}\!\!\!\bar{\bar{C}} \longrightarrow Bu^tC\!:\!C\cdot SPh$$

The formation of N-substituted *trans*-3-halogenoacrylamides from isocyanides with 1-chloro- or 1-iodo-acetylenic compounds in boiling aqueous methanol could also arise by β-addition of the carbon of the isocyanide to the halogenoacetylene[165].

Two other mechanisms operate in the reactions of phosphorus-containing nucleophiles with 1-halogenoacetylenes.

$$RC\!:\!CX + Nu^- \xrightarrow{\text{slow}} RC\!:\!C^- + NuX \xrightarrow{\text{fast}} RC\!:\!CNu + X^- \tag{2.1}$$

$$RC\!:\!CX + Nu^- \xrightarrow{\text{slow}} [R\bar{C}\!:\!CXNu] \xrightarrow{\text{fast}} RC\!:\!CNu + X^- \tag{2.2}$$

Tri-n-butylphosphine reacts with 1-bromo-2-phenylacetylene to give $[Bu_3^n\overset{+}{P}\cdot C\!:\!CPh]Br^-$ by mechanism (2.1) since the ratio of the rate constants for the reactions of the bromo- and chloro-compound in NN-dimethylformamide, $k(Br)/k(Cl) = 14$, reflects the ease of rupture of the weaker C—Br bond in the rate determining step, and addition of methanol traps the acetylide to give phenylacetylene exclusively[166]. From the formation of both phenylacetylene and $[Bu_3^nP^+\cdot C\!:\!CPh]Cl^-$ from the chloroacetylene in DMF–MeOH, it is inferred that mechanisms (2.1) and (2.2) operate simultaneously. Similar studies with triethylphosphite and 1-bromo-2-phenylacetylene indicated reaction by mechanisms (2.1) and (2.2), while the chloro-compound reacts either by mechanism (2.2) alone, or by mechanisms (2.1) and (2.2) simultaneously, depending on the conditions[166, 167]. No products resulting from the trapping of vinyl carbanions (proposed as intermediates in (2.2)) were found in any of these reactions, but the expected one was identified among the products from tri-isopropyl phosphite and 1-bromoethynyl-cyclohexan-1-ol[167].

A kinetic study of the reaction between bromoacetylene and triethylamine eliminated all possible mechanisms except two: initial α-addition and initial β-addition, but could not distinguish between them[168].

Nucleophilic substitution of chlorine in 1-chloro-2-phenylacetylene by alkoxides in aprotic media gives acetylenic ethers which previously were accessible only by multi-stage reactions[169]. However, alkoxides react with 1-bromo-n-pentylacetylene in toluene to give products arising from a vinylidene carbene[170].

$$Bu^n\cdot CH_2\cdot C\!:\!CBR \xrightarrow{-HBr} Bu^n\cdot CH\!:\!C\!=\!\ddot{C}$$

The Chodkiewicz-Cadiot coupling of terminal acetylenes with 1-iodo- and 1-bromo-acetylenes has been extended to include 1-chloro-compounds[171].

2.9 MISCELLANEOUS REACTIONS INVOLVING FREE RADICALS AND CARBENES

Homolytic fission of the C—Cl bond in perchloro-9-phenylfluorene takes place at 230–240 °C with the formation of the remarkably stable perchloro-9-phenylfluorenyl radical which is completely dissociated[172]. Treatment of 3,3,3-trichloropropene with copper powder in pyridine at 25 °C gives a mixture of all three possible dienes from the dimerisation of the radical $CH_2{:}CH\dot{C}Cl_2 \longleftrightarrow \dot{C}H_2CH{:}CCl_2$ [173].

The thermal and sodium iodide-induced decomposition of phenyl halogenoalkylmercury compounds (PhHgR) is an efficient route to many carbenes: $\ddot{C}F_2(R = CF_3)$[174]; $\ddot{C}FCl(R = CCl_2F)$[175]; $CF_3\ddot{C}Cl(R = CCl_2{\cdot}CF_3$ or $CClBr{\cdot}CF_3)$[176]; $Me\ddot{C}Cl(R = CCl_2{\cdot}Me)$[177]. Carbenes prepared by elimination of lithium halide from a lithium compound (RLi) include $\ddot{C}Cl_2(R = CCl_3)$[178]; $\ddot{C}HF(R = CHBrF)$[179]; $\ddot{C}F_2$ and $\ddot{C}FCl(R = CF_2Cl)$[180]. Fluoro-olefins are formed via carbene intermediates in the thermolysis of polyfluoroalkyltrifluorosilanes.

$$CHF_2{\cdot}CF_2{\cdot}SiF_3 \xrightarrow{-SiF_4} CHF_2{\cdot}\ddot{C}F \longrightarrow CHF{:}CF_2$$

However, with polyfluoroalkyltrichlorosilanes, some fluorochloro-alkenes are also obtained by prior halogen exchange between the α-fluorine atoms and the chlorine atoms on silicon.

$$(CH_2F{\cdot}CF_2{\cdot}SiCl_3 \longrightarrow CH_2F{\cdot}CCl_2{\cdot}SiClF_2 \longrightarrow CHF{:}CHCl, 7\%)[181].$$

Fluorine migration is responsible for the formation of trifluoroethylene (10–32%) from $CF_3{\cdot}\ddot{C}H$ generated from the diazoalkane precursor[182].

References

1. Barton, D. H. R., Godinho, L. S., Hesse, R. H. and Pechet, M. M. (1968). *Chem. Commun.*, 804
2. Barton, D. H. R., Danks, L. J., Ganguly, A. K., Hesse, R. H., Tarzia, G. and Pechet, M. M. (1969). *Chem. Commun.*, 227
3. Kollonitsch, J., Barash, L. and Doldouras, G. A. (1970). *J. Amer. Chem. Soc.*, **92**, 7494
4. Lo, E. S., Readio, J. D. and Iserson, H. (1970). *J. Org. Chem.*, **35**, 2051
5. Baird, W. C., Surridge, J. H. and Buza, M. (1971). *J. Org. Chem.*, **36**, 2088
6. Cornforth, J. W. and Green, D. T. (1970). *J. Chem. Soc. C*, 846
7. German, L. S. and Knunyantz, I. L. (1969). *Angew. Chem. Internat. Edn.*, **8**, 349
8. Mueller, W. H. and Butler, P. B. (1968). *J. Amer. Chem. Soc.*, **90**, 2075
9. Chambers, R. D., Jackson, J. A., Musgrave, W. K. R. and Story, R. A. (1968). *J. Chem. Soc. C*, 2222
10. Chambers, R. D., Cheburkov, Yu. A., MacBride, J. A. H. and Musgrave, W. K. R. (1971). *J. Chem. Soc. C*, 532
11. Drayton, C. J., Flowers, W. T. and Haszeldine, R. N. (1971). *J. Chem. Soc. C*, 2750
12. Brunskill, W., Flowers, W. T., Gregory, R. and Haszeldine, R. N. (1970). *Chem. Commun.*, 1444
13. Miller, W. T. and Burnard, R. J. (1968). *J. Amer. Chem. Soc.*, **90**, 7367
14. Banks, R. E., Haszeldine, R. N., Taylor, D. R. and Webb, G. (1970). *Tetrahedron Lett.*, 5215

15. Dyatkin, B. L., Sterlin, S. R., Martynov, B. I. and Knunyants, I. L. (1970). *Tetrahedron Lett.,* 1387
16. Olah, G. A. and Mateescu, G. D. (1971). *J. Amer. Chem. Soc.,* **93,** 782
17. Mathey, F. and Bensoam, J. (1971). *Tetrahedron,* **27,** 3965
18. Shieh, T. C., Feit, E. D., Chernick, C. L. and Yang, N. C. (1970). *J. Org. Chem.,* **35,** 4020
19. Burton, D. J. and Kehoe, L. J. (1970). *J. Org. Chem.,* **35,** 1339
20. Burton, D. J. and Kehoe, L. J. (1971). *J. Org. Chem.,* **36,** 2596
21. Truce, W. E. and Goralski, C. T. (1971). *J. Org. Chem.,* **36,** 2536
22. Field, K. W. and Kovacic, P. (1971). *J. Org. Chem.,* **36,** 3566
23. Fleming, G. L., Haszeldine, R. N. and Tipping, A. E. (1971). *J. Chem. Soc. C,* 3829
24. Sandler, S. R. (1970). *J. Org. Chem.,* **35,** 3967
25. Sandler, S. R. (1971). *Chem. and Ind.,* **49,** 1416
26. Clinch, P. W. and Hudson, H. R. (1971). *J. Chem. Soc. B,* 747
27. Clive, D. L. J. and Denyer, C. V. (1971). *Chem. Commun.,* 1112
28. Weiss, R. G. and Snyder, E. I. (1970). *J. Org. Chem.,* **35,** 1627
29. Thakore, A. N., Pope, P. and Oehlschlager, A. C. (1971). *Tetrahedron,* **27,** 2617
30. Tatlow, J. C., Burdon, J. and Coe, P. L. (1970). Ger. offen 1, 925, 836 *(Chem. Abs.,* **72,** 78 104t)
31. Burdon, J. (1971). *6th International Symposium on Fluorine Chemistry, Durham,* July 18–23. Conference Abstract A13.
32. Bornstein, J. and Skarlos, L. (1971). *Chem. Commun.,* 796
33. Brown, H. C., Rathke, M. W. and Rogic, M. M. (1968). *J. Amer. Chem. Soc.,* **90,** 5038
34. Brown, H. C. and Lane, C. F. (1970). *J. Amer. Chem. Soc.,* **92,** 6660
35. Brown, H. C. and Lane, C. F. (1971). *Chem. Commun.,* 521
36. Brown, H. C. and Lane, C. F. (1970). *J. Amer. Chem. Soc.,* **92,** 7212
37. Lane, C. F. and Brown, H. C. (1971). *J. Organometal. Chem.,* **26,** C51
38. Tabushi, I., Tamura, Y. and Yoshida, Z. (1971). *Tetrahedron Lett.,* 3893
39. Gleason, J. G. and Harpp, D. N. (1970). *Tetrahedron Lett.,* 3431
40. Rathke, M. W. and Lindert, A. (1971). *Tetrahedron Lett.,* 3995
41. Chekovakaya, E. Ts., Kuz'mina, N. A. and Freidlina, R. Kh. (1970). *Izv. Akad. Nauk SSSR, Ser. Khim.,* **10,** 2343
42. Dear, R. E. A. (1970). *J. Org. Chem.,* **35,** 1703
43. Ogata, Y. and Urasaki, I. (1971). *J. Org. Chem.,* **36,** 2164
44. Van Den Bosch, G., Bos, H. J. T. and Arens, J. F. (1970). *Rec. Trav. Chim. Pays-Bas,* **89,** 133
45. Martens, H. and Hoornaert, G. (1970). *Tetrahedron Lett.,* 1821
46. Calo, V., Modena, G. and Scorrano, G. (1968). *J. Chem. Soc. C,* 1339
47. Barton, T. J. and Zika, R. G. (1970). *J. Org. Chem.,* **35,** 1729
48. Flowers, W. T., Haszeldine, R. N. and Marshall, P. G. (1970). *Chem. Commun.,* 371
49. Chambers, R. D., Musgrave, W. K. R. and Partington, S. (1970). *Chem. Commun.,* 1050
50. Miller, W. T., Snider, R. H. and Hummel, R. J. (1969). *J. Amer. Chem. Soc.,* **91,** 6532
51. Bowden, K. and Price, M. J. (1970). *J. Chem. Soc. B,* 1466, 1472
52. Borkent, G. and Drenth, W. (1970). *Rec. Trav. Chim. Pays-Bas,* **89,** 1057
53. Greaves, P. M., Kalli, M., Landor, P. D. and Landor, S. R. (1971). *J. Chem. Soc. C,* 667
54. Truce, W. E. and Wolf, G. C. (1971). *J. Org. Chem.,* **36,** 1727
55. Amiel, Y. (1971). *J. Org. Chem.,* **36,** 3691, 3697
56. Johnson, M. D. and Meeks, B. S. (1970). *Chem. Commun.,* 1027
57. Corey, E. J., Shulman, J. I. and Yamamoto, H. (1970). *Tetrahedron Lett.,* 447
58. Burton, D. J. and Krutzsch, H. C. (1970). *J. Org. Chem.,* **35,** 2125
59. Combret, J. C., Villieras, J. and Lavielle, G. (1971). *Tetrahedron Lett.,* 1035
60. Newman, M. S. and Beard, C. D. (1970). *J. Amer. Chem. Soc.,* **92,** 4309
61. Campbell, J. R., Pross, A. and Sternhell, S. (1971). *Aust. J. Chem.,* **24,** 1425
62. Kloster-Jenson, E. (1971). *Tetrahedron,* **27,** 33
63. Siegel, J., Jones, R. A. and Kurlansik, L. (1970). *J. Org. Chem.,* **35,** 3199
64. Delavarenne, S. Y. and Viehe, H. G. (1970). *Chem. Ber.,* **103,** 1198
65. Olah, G. A., Cupas, C. A. and Comisarow, M. B. (1966). *J. Amer. Chem. Soc.,* **88,** 362
66. Olah, G. A., Chambers, R. D. and Comisarow, M. B. (1967). *J. Amer. Chem. Soc.,* **89,** 1268
67. Roberts, I. and Kimball, G. E. (1937). *J. Amer. Chem. Soc.,* **59,** 947
68. Winstein, S., Grunwald, E. and Ingraham, L. L. (1948). *J. Amer. Chem. Soc.,* **70,** 821

69. Grunwald, E. (1951). *J. Amer. Chem. Soc.*, **73**, 5458
70. Olah, G. A. and Bollinger, J. M. (1967). *J. Amer. Chem. Soc.*, **89**, 4744
71. Clark, D. T. (1971). *23rd International Congress of Pure and Applied Chemistry*, 31. (London: Butterworths)
72. Olah, G. A., Bollinger, J. M. and Brinich, J. (1968). *J. Amer. Chem. Soc.*, **90**, 2587
73. Olah, G. A. and Bollinger, J. M. (1968). *J. Amer. Chem. Soc.*, **90**, 947
74. Olah, G. A. and DeMember, J. R. (1970). *J. Amer. Chem. Soc.*, **92**, 718
75. Olah, G. A. and DeMember, J. R. (1970). *J. Amer. Chem. Soc.*, **92**, 2562
76. Fowler, F. W., Hassner, A. and Levy, L. A. (1967). *J. Amer. Chem. Soc.*, **89**, 2077
77. Hassner, A., Hoblitt, R. P. and Heathcock, C. (1970). *J. Amer. Chem. Soc.*, **92**, 1326
78. Rolston, J. H. and Yates, K. (1969). *J. Amer. Chem. Soc.*, **91**, 1469
79. Hassner, A. and Boerwinkle, F. (1968). *J. Amer. Chem. Soc.*, **90**, 216
80. Fahey, R. C. and Schubert, C. (1965). *J. Amer. Chem. Soc.*, **87**, 5172
81. Hassner, A., Boerwinkle, F. P. and Levy, A. B. (1970). *J. Amer. Chem. Soc.*, **92**, 4879
82. Rolston, J. H. and Yates, K. (1969). *J. Amer. Chem. Soc.*, **91**, 1477
83. Banks, R. E. and McGlinchey, M. J. (1971). *J. Chem. Soc. C*, 3971
84. Merritt, R. F. (1967). *J. Amer. Chem. Soc.*, **89**, 609
85. Ecke, G. C., Cook, N. E. and Whitmore, F. C. (1950). *J. Amer. Chem. Soc.*, **72**, 1511
86. Puterbaugh, W. H. and Newman, M. S. (1957). *J. Amer. Chem. Soc.*, **79**, 3469
87. Hassner, A. and Fowler, F. W. (1968). *J. Org. Chem.*, **33**, 2686
88. Diner, U. E. and Lown, J. W. (1971). *Can. J. Chem.*, **49**, 403
89. Hassner, A. and Teeter, J. S. (1971). *J. Org. Chem.*, **36**, 2176
90. Hassner, A. and Teeter, J. S. (1970). *J. Org. Chem.*, **35**, 3397
91. Taylor, D. R. (1967). *Chem. Rev.*, **67**, 317
92. Peer, H. G. (1962). *Rec. Trav. Chim. Pays-Bas*, **81**, 113
93. Bollinger, J. M., Brinich, J. M. and Olah, G. A. (1970). *J. Amer. Chem. Soc.*, **92**, 4025
94. Pincock, J. A. and Yates, K. (1970). *Can. J. Chem.*, **48**, 3332
95. Mathai, I. M., Schug, K. and Miller, S. I. (1970). *J. Org. Chem.*, **35**, 1733
96. Peterson, P. E. and Coffey, J. E. (1971). *J. Amer. Chem. Soc.*, **93**, 5208
97. Peterson, P. E. and Slama, F. J. (1968). *J. Amer. Chem. Soc.*, **90**, 6516
98. Reutov, O. A., Smolina, T. A., Groplus, E. D. and Bekh, A. K. (1971). *Dokl. Akad. Nauk SSSR*, **197**, 604
99. Reutov, O. A., Gudkova, A. S., Ostapchuk, G. M. and Akimova, V. G. (1968). *Izv. Akad. Nauk SSSR Ser. Khim.*, **8**, 1922
100. Reinecke, C. E. and McCarthy, J. R. (1970). *J. Amer. Chem. Soc.*, **92**, 6376
101. Reutov, O. A., Smolina, T. A., Polevaya, O. Yu., Gopius, E. D. and Betaneli, L. V. (1971). *Dokl. Akad. Nauk. SSSR*, **197**, 848
102. Peterson, P. E. and Boron, W. F. (1971). *J. Amer. Chem. Soc.*, **93**, 4076
103. Glick, R. E. (1954). *Ph.D. Thesis*, University of California at Los Angeles, California
104. Peterson, P. E., Bopp, R. J., Chevli, D. M. Curran, E. L., Dillard, D. E. and Kamat, R. J. (1967). *J. Amer. Chem. Soc.*, **89**, 5902
105. Olah, G. A. and Peterson, P. E. (1968). *J. Amer. Chem. Soc.*, **90**, 4675
106. Peterson, P. E. and Coffey, J. F. (1968). *Tetrahedron Lett.*, 3131
107. Trahanovsky, W. S., Smyser, G. L. and Doyle, M. P. (1969). *Tetrahedron Lett.*, 3127
108. Peterson, P. E., Bopp, R. J. and Ajo, M. M. (1970). *J. Amer. Chem. Soc.*, **92**, 2834
109. Peterson, P. E. and Tao, E. V. P. (1964). *J. Amer. Chem. Soc.*, **86**, 4503
110. Peterson, P. E. and Bopp, R. J. (1967). *J. Amer. Chem. Soc.*, **89**, 1283
111. Juneja, P. S. and Hodnett, E. M. (1967). *J. Amer. Chem. Soc.*, **89**, 5685
112. Danen, W. C. and Winter, R. L. (1971). *J. Amer. Chem. Soc.*, **93**, 716
113. Stunk, R. J., DiGiacomo, P. M., Aso, K. and Kuivila, H. G. (1970). *J. Amer. Chem. Soc.*, **92**, 2849
114. Kiehlmann, E., Loo, P-W., Menon, B. C. and McGillivray, N. (1971). *Can. J. Chem.*, **49**, 2964
115. Hitchins, R. O., Hoke, D., Keogh, J. and Koharski, D. (1969). *Tetrahedron Lett.*, 3495
116. Bell, H. M., Vanderslice, C. W. and Spehar, A. (1969). *J. Org. Chem.*, **34**, 3923
117. Hutchins, R. O., Maryanoff, B. E. and Milewski, C. A. (1971). *Chem. Commun.*, 1097
118. Volpin, M., Dvolaitzky, M. and Levitin, I. (1970). *Bull. Soc. Chim. Fr.*, **4**, 1526
119. Jacobus, J. (1970). *Chem. Commun.*, 338
120. Townsend, J. M. and Spencer, T. A. (1971). *Tetrahedron Lett.*, 137
121. Hewertson, W. and Taylor, I. C. (1970). *Chem. Commun.*, 119

122. Lo, E. S. (1971). *J. Org. Chem.*, **36**, 364
123. Kornblum, N., Davies, T. M., Earl, G. W., Holy, N. L., Kerber, R. C., Musser, M. T. and Snow, D. H. (1967). *J. Amer. Chem. Soc.*, **89**, 725
124. Kornblum, N. and Stuchal, F. W. (1970). *J. Amer. Chem. Soc.*, **92**, 1804
125. Kornblum, N., Swiger, R. T., Earl, G. W., Pinnick, H. W. and Stuchal, F. W. (1970). *J. Amer. Chem. Soc.*, **92**, 5513
126. Holz, D., Beauchamp, J. L. and Woodgate, S. D. (1970). *J. Amer. Chem. Soc.*, **92**, 7484
127. Brown, H. C., Rogic, M. M., Rathke, M. W. and Kabalka, G. W. (1968). *J. Amer. Chem. Soc.*, **90**, 818; 1911
128. Brown, H. C., Rogic, M. M. and Rathke, M. W. (1968). *J. Amer. Chem. Soc.*, **90**, 6218
129. Brown, H. C. and Rogic, M. M. (1969). *J. Amer. Chem. Soc.*, **91**, 2146
130. Brown, H. C., Rogic, M. M., Nambu, H. and Rathke, M. W. (1969). *J. Amer. Chem. Soc.*, **91**, 2147
131. Brown, H. C., Nambu, H. and Rogic, M. M. (1969). *J. Amer. Chem. Soc.*, **91**, 6852
132. Brown, H. C., Nambu, H. and Rogic, M. M. (1969). *J. Amer. Chem. Soc.*, **91**, 6854
133. Nambu, H. and Brown, H. C. (1970). *J. Amer. Chem. Soc.*, **92**, 5790
134. Brown, H. C., Carlson, B. A. and Prager, R. H. (1971). *J. Amer. Chem. Soc.*, **93**, 2070
135. Vida, J. A. (1970). *Tetrahedron Letters*, 3447
136. Cooke, M. P. (1970). *J. Amer. Chem. Soc.*, **92**, 6081
137. Hashimoto, I., Tsuruta, N., Ryang, M. and Tsutsumi, S. (1970). *J. Org. Chem.*, **35**, 3748
138. Henne, A. L. and Kaye, S. (1950). *J. Amer. Chem. Soc.*, **72**, 3369
139. Myhre, P. C. and Andrews, G. D. (1970). *J. Amer. Chem. Soc.*, **92**, 7595
140. Myhre, P. C. and Andrews, G. D. (1970). *J. Amer. Chem. Soc.*, **92**, 7596
141. Svetlakov, N. V., Moisak, I. E. and Averko-Antonovich, I. G. (1969). *Zh. Org. Khim.*, **5**, 2105
142. Kennedy, J. P. (1970). *J. Org. Chem.*, **35**, 532
143. Ashby, E. C. and Yu, S. H. (1971). *J. Org. Chem.*, **36**, 2123
144. Ashby, E. C., Yu, S. H. and Beach, R. G. (1970). *J. Amer. Chem. Soc.*, **92**, 433
145. Ashby, E. C. and Yu, S. H. (1971). *J. Organometal. Chem.*, **29**, 339
146. Corey, E. J. and Posner, G. H. (1967). *J. Amer. Chem. Soc.*, **89**, 3911
147. Corey, E. J. and Posner, G. H. (1968). *J. Amer. Chem. Soc.*, **90**, 5615
148. Whitesides, G. M., Fischer, W. F., Filippo, J. S., Bashe, R. W. and House, H. O. (1969). *J. Amer. Chem. Soc.*, **91**, 4870
149. Corey, E. J. and Kuwajima, I. (1970). *J. Amer. Chem. Soc.*, **92**, 395
150. Bertini, F., Grasselli, P., Zubiani, G. and Cainelli, G. G. (1970). *Tetrahedron*, **26**, 1281
151. Rathke, M. W. and Lindert, A. (1970). *J. Org. Chem.*, **35**, 3966
152. Castro, B., Villieras, J. and Ferracutti, N. (1969). *Bull. Soc. Chim. Fr.*, **10**, 3521
153. Koch, H. F. and Kielbania, A. J. (1970). *J. Amer. Chem. Soc.*, **92**, 729
154. Burton, D. J. and Krutzsch, H. C. (1971). *J. Org. Chem.*, **36**, 2351
155. Hanack, M. (1970). *Accounts Chem. Res.*, **3**, 209
156. Grob, C. A. and Cseh, G. (1964). *Helv. Chim. Acta.*, **47**, 194
157. Miller, L. L. and Kaufman, .D. A. (1968). *J. Amer. Chem. Soc.*, **90**, 7282
158. Rappoport, Z. and Apeliog, Y. (1969). *J. Amer. Chem. Soc.*, **91**, 6734
159. Grob, C. A. and Spaar, R. (1969). *Tetrahedron Lett.*, 1439
160. Grob, C. A. and Pfaendler, H. R. (1970). *Helv. Chim. Acta.*, **53**, 2119; 2130
161. Lee, C. V., Hargrove, R. J., Dueber, T. F. and Stang, P. J. (1971). *Tetrahedron Lett.*, 2519
162. Schiavelli, M. A., Hixon, S. C. and Moran, H. W. (1970). *J. Amer. Chem. Soc.*, **92**, 1082
163. Richey, H. G. and Richey, J. M. (1970). *Carbonium Ions*, Vol. 2. (G. A. Olah and P. v R. Schleyer, editors), 931. (New York: Wiley–Interscience)
164. Delavarenne, S. Y. and Viehe, H. G. (1970). *Chem. Ber.*, **103**, 1216
165. Johnson, F., Gulbenkian, A. H. and Nasutavicus, W. A. (1970). *Chem. Commun.*, 608
166. Fujii, A., Dickstein, J. I. and Miller, S. I. (1970). *Tetrahedron Lett.*, 3435
167. Burt, D. W. and Simpson, P. (1971). *J. Chem. Soc. C,* 2872
168. Tanaka, R. and Miller, S. I. (1971). *J. Org. Chem.*, **36**, 3856
169. Tanaka, R. and Miller, S. I. (1971). *Tetrahedron Lett.*, 1753
170. Cymerman Craig, J. and Baird, C. D. (1971). *Chem. Commun.*, 691
171. Philippe, J-L., Chodkiewicz, W. and Cadiot, P. (1970). *Tetrahedron Lett.*, 1795
172. Ballester, M., Castaner, J. and Pujadas, J. (1971). *Tetrahedron Lett.*, 1699
173. Dolbier, W. R. and Harman, C. A. (1971). *Chem. Commun.*, 150

174. Seyferth, D. and Hopper, S. P. (1971). *J. Organometal. Chem.*, **2**, C62
175. Seyferth, D. and Darragh, K. V. (1970). *J. Org. Chem.*, **35**, 1297
176. Seyferth, D. and Mueller, D. C. (1971). *J. Amer. Chem. Soc.*, **93**, 3714
177. Seyferth, D. and Mueller, D. C. (1971). *J. Organometal. Chem.*, **28**, 325
178. Köbrich, G., Büttner, H. and Wagner, E. (1970). *Angew. Chem. Internat. Edn.*, **9**, 169
179. Schlosser, M. and Heinz, G. (1971). *Chem. Ber.*, **104**, 1934
180. Schlosser, M. and Le Van Chau (1971). *Angew. Chem. Internat. Edn.*, **10**, 138
181. Bevan, W. I., Haszeldine, R. N. and Middleton, J. (1970). *J. Organometal. Chem.*, **23**, C17
182. Atherton, J. H., Fields, R. and Haszeldine, R. N. (1971). *J. Chem. Soc. C*, 366

3
Alcohols, Ethers and Related Compounds

S. G. WILKINSON
University of Hull

3.1   INTRODUCTION

As the list of contents shows, this chapter is concerned with progress in the chemistry of some of the major classes of compounds which contain singly-bound oxygen. Compounds which additionally contain doubly-bound oxygen (e.g. carboxylic acids and esters, peroxy-acids and diacyl peroxides) are not included. For practical reasons, it has also been necessary to omit certain classes of compounds (e.g. acetals, halogenohydrins and hypohalites) which potentially fell within the scope of this chapter.

The very recent appearance of major works on the chemistry of most of the functional groups to be discussed has provided an excellent base for the construction of this first biennial review. Thus it has been possible to write this supplementary review of the more recent literature (essentially that published in the period 1970–1971) without obsessive concern for balance within each subject. So that each account should be reasonably coherent, the topics selected have generally been those of which several reports have appeared during the period covered by the review. Clearly, therefore, this chapter may reflect current interests and activities more accurately than progress and attainment, and the importance of relatively isolated pieces of work may not have been recognised or adequately represented. As the headings indicate, the topics have been organised mainly according to an operational rather than a mechanistic scheme, although much of the subject matter is mechanistic in nature.

3.2   MONOHYDRIC ALCOHOLS

With the notable omission of a chapter on oxidation and reduction, a comprehensive account of the chemistry of alcohols is contained in a two-part volume[1] of the series edited by Patai. Fortunately, the oxidation of alcohols has been reviewed frequently and the major methods currently in use are described in other recent books[2-5]. Other topics which have recently been reviewed include the following: the dehydration of alcohols with aluminium oxide[6], the oxidative cyclisation of alcohols with lead tetra-acetate[7], and the chemistry of the higher (fatty) alcohols[8]. Other reviews of asymmetric synthesis[9] and of optical resolution[10] have included their application to alcohols. Although much work has been done on the radiolysis of alcohols, this topic has not been included in the present chapter because of lack of space.

3.2.1 Preparation

With the availability of a wide variety of excellent methods for the reduction of carbonyl compounds to alcohols, the emphasis in recent publications has been on selectivity. In this connection, detailed studies of the reducing characteristics of diborane[11] and disiamylborane[12] $[(Mc_2CH \cdot CHMe)_2BH]$ have been undertaken. Both reagents readily reduce the carbonyl group of aldehydes and ketones but show useful differences in reactivity towards other functional groups. For example, a ketonic carbonyl group may be selectively reduced in the presence of a carboxy group with disiamylborane but not with diborane. The generally lower activity of disiamylborane is probably due to steric factors[12]. Sodium cyanotrihydridoborate ($NaBH_3CN$) is another selective reducing agent[13]; aldehydes and ketones are reduced to alcohols at pH 3–4. Other hydride reducing agents described are sulphurated sodium borohydride[14] ($NaBH_2S_3$) and magnesium aluminium hydride[15]; the former reagent is intermediate in activity between lithium aluminium hydride and sodium borohydride. Suitable oximes are also reduced to alcohols by heating them under reflux with alkaline sodium borohydride[16]. The potential value of this reaction is based mainly on the accessibility of oximes from compounds other than aldehydes and ketones. The use of catalytic hydrogenation for the reduction of carbonyl compounds is well known; recent examples include the use of rhodium complexes for the homogeneous catalytic hydrogenation of ketones[17] and the use of an iridium heterogeneous catalyst for the preparation of unsaturated alcohols from $\alpha\beta$-unsaturated aldehydes[18]. The reactions between carbonyl compounds and alkyl halides in the presence of lithium to give alcohols may be a convenient, one-step alternative to the Grignard method[19]; reactions were complete in c. 2 h and yields were usually in the range 50–90%. The reduction of carboxylic esters of phenols and acidic alcohols (notably 2,2,2-trifluoroethanol) with sodium borohydride provides a route for the preparation of alcohols from acids under unusually mild conditions[20]. The choice of trifluoroethyl esters as intermediates was influenced by their relative accessibility, for example, through the use of diazo-1,1,1-trifluoroethane.

The introduction of the oxymercuration–demercuration procedure for the preparation of alcohols from alkenes by Markovnikov hydration was a major advance. It provided a mild alternative to acid-catalysed hydration, and complemented the hydroboration method for anti-Markovnikov hydration. A detailed account of Brown's convenient and efficient procedure has now been published[21]. Oxymercuration is achieved rapidly at room temperature in aqueous tetrahydrofuran and the organomercurial product is then reduced by the addition of alkaline sodium borohydride as indicated in the following scheme. The method has been applied with partial success

$$RCH{:}CH_2 \xrightarrow[\substack{Aq.\ tetrahydrofuran\\25\,°C}]{Hg(OAc)_2} RCH(OH){\cdot}CH_2HgOAc \xrightarrow[\substack{NaOH\\25\,°C}]{NaBH_4} RCH(OH){\cdot}CH_3 + Hg$$

to the asymmetric synthesis of alcohols by the use of optically active mercury(II) carboxylates for oxymercuration[22].

Of other preparative methods, the more recent applications of organoboranes, to which H. C. Brown has again made the outstanding contributions, are reviewed in Chapter 9 of this volume. The preparation of alcohols from alkenes by ozonolysis in methanol or butan-1-ol, followed by two-stage catalytic hydrogenation, has been much improved[23], but this method seems to be of limited value in synthesis. Further studies on allylic oxidation of alkenes to alcohols by mercury(II) salts[24] and by selenium dioxide[25, 26] have also been reported.

The industrial conversion of alkenes into primary alcohols by catalytic hydroformylation (the 'OXO' process) has been much studied. Reports on this work are largely confined to the patent literature and rather inaccessible reviews, and are too numerous to be considered in detail. Among developments of interest[27] have been the use of metals other than cobalt (e.g. iron, rhodium) and the extensive use of mixed catalysts containing phosphorus (e.g. Ref. 28) or tertiary amines (e.g. Ref. 29). From a perusal of the patent literature, it appears that industrial interest in the production of alcohols by autoxidation of aluminium and other metal alkyls[30] and by autoxidation of alkanes in the presence of boric acid[31] has been maintained.

Inexpensive methods using deuterium oxide for the preparation of α-deuterated alcohols have been described[32, 33]. Improved procedures for the perdeuteration[34] and O-deuteration[35] of t-butyl alcohol, and a synthesis of 4-deuteriobutan-1-ol[36] have also been reported. Tritium-labelled alcohols may be prepared conveniently and economically by reduction of an aldehyde or ketone with lithium borohydride which has previously undergone partial hydrolysis with tritiated water, in order to effect isotope exchange[37]. Similar work has been done with sodium cyanotrihydridoborate[13].

3.2.2 Analytical methods

In the last 2 years there has been a considerable burst of interest in the use of paramagnetic shift reagents in n.m.r. spectroscopy. Such reagents seem certain to find widespread application in the study of complex alcohols by ^1H n.m.r. techniques. The chemical shifts induced by association of the reagent with the hydroxy group (probably by pseudo-contact interaction) decrease as the distance of the protons from the hydroxy group increases. As a result, otherwise complex spectra may readily be resolved and interpreted. Complexes of dipivaloylmethane ($Bu^tCO{\cdot}CH_2{\cdot}COBu^t$) with certain lanthanides are particularly effective shift reagents. Thus, the europium complex causes large shifts of proton resonances to lower field[38, 39]; a simple example of the effect observed is given in Figure 3.1. By contrast, the praseodymium complex causes even larger shifts to higher field[40]. The europium complex has recently been used successfully in the quantitative analysis of multicomponent mixtures of alcohols by n.m.r.[41] spectroscopy. The spectral changes induced by complexes of β-dicarbonyl compounds with cobalt(II), nickel(II), and copper(II) have also been reported[42] but are less striking.

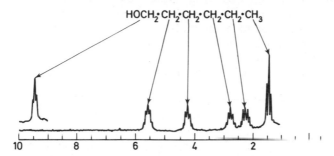

$HOCH_2 \cdot CH_2 \cdot CH_2 \cdot CH_2 \cdot CH_2 \cdot CH_3$

Figure 3.1   100 MHz ^{1}H n.m.r. spectrum of hexan-1-ol in carbon tetrachloride in the presence of the europium complex of dipivaloyl-methane. Upper left-hand trace offset 1 p.p.m. Chemical shifts in δ units (p.p.m.) relative to internal tetramethylsilane
(From Sanders and Williams[38], by courtesy of the Chemical Society)

The ready formation of adducts with hexafluoroacetone provides a simple and sensitive, but not absolutely reliable, means of classifying alcohols[43]. Thus, the chemical shift of the ^{19}F resonance singlet for the adduct (measured downfield from that observed for hexafluoroacetone hydrate) was characteristically 2.45–2.70 p.p.m. for primary alcohols, 1.75–1.95 p.p.m. for secondary alcohols, or 1.2–1.4 p.p.m. for tertiary alcohols. The ^{1}H and ^{13}C n.m.r. spectra of propan-2-ol and its derivatives in which the six methyl protons are successively replaced by methyl groups have been studied[44]. The increasing downfield chemical shift for the carbon attached to the hydroxy group with increasing methyl substitution was interpreted in terms of an electron-*withdrawing* inductive effect of methyl. Inevitably the validity of this interpretation was soon challenged[45]. Although the n.m.r. results could not be explained, the idea that chemical shifts could always be simply and unambiguously correlated with inductive effects was disputed[45]. The ^{13}C chemical shifts for a more extensive range of alcohols have also been determined[46] and linear relationships for the prediction of shifts for other alcohols have been derived.

The major features of the mass spectrometric decomposition of alcohols have been known for some time. Recent studies have concerned isotope effects on the electron-impact mass spectra of ethanol and its deuterated derivatives[47] and the mechanism of formation of the ion at m/e 31 in the mass spectrum of t-butyl alcohol[48, 49]. Other investigations have been directed to the problem of identifying alcohols which do not give a positive molecular ion by the electron-impact method. Negative-ion mass spectrometry has been used successfully for this purpose with long-chain alcohols[50]. The relatively new technique of chemical-ionisation mass spectrometry should also be particularly useful in this respect. When isobutane is used as the reactant, spectra with dominant $(M+1)^+$ ions (from alcohols up to C_3) or $(M-17)^+$ ions (from higher alcohols) are obtained[51]. Useful spectra are also obtained by using a mixture of methane with either acetaldehyde or acetone

as the reactant gas[52]. The combination of chemical-ionisation mass spectrometry with gas–liquid chromatography, in which methane was used as the
carrier-reactant gas, has been described briefly[53]. Another new technique,
Plasma Chromatography™–mass spectrometry[54], may be used for the
detection of trace amounts of alcohols.

Among other analytical procedures, sensitive methods for the determination[55] and identification[56, 57] of alcohols are based on the preparation of
highly coloured esters. All classes of aliphatic alcohols react rapidly and
quantitatively with 2-ketopropanoyl chloride 2,6-dinitrophenylhydrazone,
on the addition of triethylenediamine (1,4-diazabicyclo[2.2.2]octane) to the
dried reactants[55]. The similar reaction between 4-chlorocarbonyl-4'-N,N-
dimethylaminoazobenzene (a stable acid chloride) and primary or secondary
alcohols gives highly coloured esters with characteristic melting points[56].
Procedures for the identification of the esters of lower alcohols (C_1–C_{10}) by
chromatography (paper and thin-layer) have been devised[57].

Gas–liquid chromatography (g.l.c.) is an established technique for the
determination of optical purity of dissymmetric alcohols by the separation
of diastereoisomeric esters. In recent studies, the chromatographic separations achieved[58] for several diastereoisomeric pairs of esters of 3β-acetoxy-
androst-5-ene-17β-carboxylic acid were similar to those for the corresponding
(−)-(R)-menthyloxycarbonyl derivatives[59]. Other similar separations have
been reported[60, 61] for the diastereoisomeric urethanes from the reaction of
(+)-(R)-1-phenylethyl isocyanate with asymmetric secondary alcohols;
thin-layer chromatography has also been used for the differentiation of these
derivatives[62]. The resolution of the N-trifluoroacetyl-L-alanyl esters of 3,3-
dimethylbutan-2-ol on a long column was adequate for preparative g.l.c. to
be successful[63].

3.2.3 Acidity and basicity

For various reasons, interest in the acidities of alcohols has recently been
heightened. Because of the paucity of data and the difficulties associated with
the direct determination of pK_a values, further studies on the calculation of
such values have been carried out[64]. Published data were used to derive
linear free energy relationships by means of which pK_a values for primary
alcohols could be calculated from values of Taft's polar substituent constants
(σ*). Equation (3.1) applies to alcohols in which C-2 is sp^2- or sp^1-hybridised,
and equation (3.2) to alcohols in which C-2 is sp^2-hybridised.

$$pK_a(RCH_2OH) = -1.316\sigma^*(R) + 15.74 \qquad (3.1)$$

$$[n = 6, r = 0.999, s = 0.0273]$$

$$pK_a(RCH_2OH) = -1.316\sigma^*(R) + 16.23 \qquad (3.2)$$

It was also shown that values of σ* (and hence pK_a) could be calculated from
values for $v_{C=O}$ in the infrared spectra of the 3-phenylpropanoates of the
alcohols, for which linear correlations were obtained[64].

As determined by ion cyclotron resonance spectroscopy in the gas phase,
the order of acidities of alcohols is the opposite of that found for the liquid

phase[65]. This remarkable finding has been supported by the results of other experiments[66, 67], and by theoretical calculations[68-72], and the first steps leading to a standard acidity scale for alcohols in the gas phase have been taken[67]. The stabilising influence of alkyl groups on alkoxide ions in the gas phase conflicts with their usual description as electron-releasing substituents. However, the gas-phase basicities of alcohols are in the order predicted on the basis of the usual inductive effects of alkyl groups. These findings are most easily reconciled if it is assumed that alkyl groups have a greater ability (relative to hydrogen) to stabilise both positive and negative centres as a result of their greater polarisability[65]. The inverse order of acidities for alcohols in the liquid phase might then be explained in terms of a steric effect of alkyl groups on the extent of solvation of alkoxide ions, rather than in terms of an electron-withdrawing inductive effect of alkyl groups[44].

The usual inductive effect of alkyl groups should facilitate the removal of a lone-pair electron from oxygen, and the expected correlation between first ionisation potentials and values of σ^* and σ_I has been demonstrated[73, 74]. The correlation can be extended to include the base-strengthening effect of alkyl substitution[75, 76] and thus provides a means of calculating basic ionisation constants (pK_{BH+}). By taking the value of pK_{BH+} for H_3O^+ as -3.43, various relationships were derived: those for σ^* and ionisation potential (E_I) are given in equations (3.3) and (3.4) respectively, and calculated values of pK_{BH+} are given in Table 3.1. The value of the constant calculated[76] for

$$pK_{BH+} = -2.18 - 2.36\sigma^* \tag{3.3}$$

$$pK_{BH+} = +5.73 - 0.727E_I \tag{3.4}$$

Table 3.1 Basic ionisation constants of alcohols
(From Levitt and Levitt[76], by courtesy of Pergamon Press)

| Alcohol | Calculated pK_{BH+} values | |
	From σ^*	From E_I
MeOH	-2.18	-2.17
EtOH	-1.94	-1.92
Pr^nOH	-1.90	-1.70
Bu^nOH	-1.87	-1.68
Pr^iOH	-1.73	-1.67
Bu^tOH	-1.47	-1.49

ethanol agrees well with that (-1.94) determined from n.m.r. chemical shift data for sulphuric acid solutions[77]. The enthalpy of protonation in sulphuric acid or fluorosulphonic acid has been proposed[78] as a criterion of base strength.

3.2.4 Oxidation to carbonyl compounds

Numerous studies of the oxidation of alcohols have been described during the period covered by this review. Among new oxidants are potassium ferrate(VI)[79] and various 'positive-halogen' compounds: iodinium nitrate[80],

1-chlorobenzotriazole[81], N-bromodiphenylmethanimine[82] (Ph$_2$C:NBr), and monochloramine[83]. Potassium ferrate(VI) was used in aqueous alkaline solution; alkenes and aldehydes were not oxidised under these conditions. On oxidation with iodinium nitrate in chloroform–pyridine, secondary alcohols gave ketones in fair yields, but primary alcohols gave low conversions[80] and alkenic alcohols were not oxidised[84]. A mechanism involving abstraction of the α-hydrogen as a hydride ion was suggested, as for the oxidation of alcohols by N-bromosuccinimide[85, 86] in polar media. On the other hand, the reaction conditions for and characteristics of the oxidations with 1-chlorobenzotriazole, N-bromodiphenylmethanimine, and mono-chloramine indicate that these reactions proceed by radical chain mechanisms. The compounds in this last group are efficient oxidants and should find considerable use in synthesis.

The importance of oxidations by compounds of chromium(VI) continues to stimulate investigations on both their practical and theoretical aspects. Secondary alcohols dissolved in ether are conveniently and efficiently oxidised to ketones[87] by reaction with the stoichiometric amount of sodium dichromate in aqueous sulphuric acid at 25–30 °C. Attempts to improve the yields of aldehydes from primary alcohols by oxidation with neutral sodium dichromate were successful only for benzyl alcohols[88]. The acceptability of the chromium trioxide–pyridine complex as an oxidant should be enhanced by further improvements in the preparation and use of the reagent[89, 90], and an alternative reagent (pyridine dichromate) has been described[91] and made commercially available.

The mechanism of the oxidation of alcohols by chromic acid has been studied for many years, and the initial steps leading to the formation of chromium(IV) seem to be firmly established (equations (3.5) and (3.6)). The second step ((3.6), decomposition of the intermediate chromate ester) is rate-limiting and involves abstraction of the secondary hydrogen as a hydride ion. The correlation between rate constants and σ* values, and the value (-1.60) of ρ* for the overall reaction support this mechanism[92].

$$R_2CHOH + HCrO_4^- + H^+ \longrightarrow R_2CHO \cdot CrO_3H + H_2O \qquad (3.5)$$

$$R_2CHO \cdot CrO_3H \longrightarrow R_2CO + Cr^{IV} \qquad (3.6)$$

Speculation about the subsequent steps of the reaction, which lead to chromium(III) and account for two-thirds of the total carbonyl product, has led to a number of studies on the role of chromium(IV). It has been confirmed that chromium(IV) is an active oxidant[93–95] and that a free radical mechanism (equation (3.7), (3.8) and (3.9)) probably applies to oxidations done in the presence of water[95, 96].

$$R_2CHOH + Cr^{IV} \longrightarrow R_2\dot{C}OH + Cr^{III} \qquad (3.7)$$

$$R_2\dot{C}OH + Cr^{VI} \longrightarrow R_2CO + Cr^V \qquad (3.8)$$

$$R_2CHOH + Cr^V \longrightarrow R_2CO + Cr^{III} \qquad (3.9)$$

By carrying out the oxidation in the presence of cerium(III) or cerium(IV), which catalyse the disproportionation of chromium(IV) and chromium(V), these subsequent reactions can be prevented and the oxidation confined to the two-electron process with chromium(VI) as the sole oxidant[97]. The oxidation of alcohols by chromium(VI) has been discussed by Littler in terms of the Zimmerman treatment of electrocyclic reactions[98]. On the basis of a five-membered cyclic transition state leading to hydride abstraction in step (3.6), the acyclic mechanisms for oxidation of alcohols by manganese(VII) and lead(IV) and the inactivity of iodine(VII) as an oxidant could be rationalised[98].

Further studies on the kinetics of oxidation of simple alcohols by cerium(IV) salts in aqueous perchlorate media[99, 100], in dilute nitric acid[101, 102] and in sulphuric acid or sulphuric acid–acetic acid mixtures[103] have been made. The relaxation kinetics of the intermediate cerium(IV)–alcohol complex formed during the oxidation have also been investigated[104]. Oxidative cleavage (equation (3.10)) is the major reaction which occurs on treatment of alkylphenylmethanols with cerium(IV) ammonium nitrate in 50% aqueous acetonitrile at 90 °C, when the alkyl group (R) is isopropyl or t-butyl[105].

$$PhCH(OH)R + Ce^{IV} \longrightarrow PhCHO + R^{\cdot} + Ce^{III} \qquad (3.10)$$

Ruthenium tetroxide is another powerful reagent for the oxidation of secondary alcohols to ketones, although its major applications so far seem to have been in alicyclic chemistry and in situations where other oxidants were ineffective[106, 107]. Further studies of the oxidations and decarbonylations which occur when alcohols are treated with ruthenium trichloride in the presence of triphenylphosphine have been reported[108].

Among other investigations, kinetic studies of the oxidation of alcohols by vanadium(V)[109], thallium(III)[110], manganese(III)[111], bromine[112, 113], N-chlorosuccinimide[114], Fenton's reagent[115], peroxydisulphates[116, 117] and nitric acid[118] may be noted. The abstraction of an α-hydrogen atom has been favoured as the primary process in the oxidation of alcohols by peroxydisulphates[117] and in their photo-oxidation by uranium(VI) salts[119–121]. However, this appears less certain in view of the detection by means of e.s.r. spectroscopy of methoxy radicals as intermediates in these and other photochemical oxidations of methanol, and the use of N-benzylidene-t-butylamine N-oxide in spin-trapping experiments[122, 123].

3.2.5 Dehydration to alkenes

A general review of the dehydration of alcohols, with emphasis on reactions in the gas phase, has been written by Knözinger[124]. Only methods of potential importance for laboratory syntheses will be considered here.

$N^1N^2N^3$-Hexamethyl phosphoric triamide is a novel reagent for the dehydration of alcohols[125–127]. Good yields (70–98%) of alkenes were obtained from secondary alcohols in reactions done at 220–240 °C without added catalyst. The initial suggestion[125] that the reaction proceeds via an alcohol–reagent complex was later discarded[126] in favour of an E2 mechanism and the formation of an intermediate ester (equation (3.11)). Although El and Ei mechanisms for the formation of the normal (Saytzeff) alkene were

$$ROH + (Me_2N)_3PO \longrightarrow (Me_2N)_2PO(OR) + Me_2NH \qquad (3.11)$$

apparently ruled out by Monson[125, 126], a carbenium ion (or ion-pair)

mechanism was proposed by other workers[127] to explain, for example, the
dehydration of the alcohol (1) and the rearranged and fragmented products
from the alcohol (2). The products from alcohols such as (2) closely paralleled

$$\text{Bu}_2^t\text{CHOH} \qquad\qquad \text{Bu}_2^t\text{C(OH)·CH}_2\text{Bu}^t$$

$$(1) \qquad\qquad\qquad (2)$$

those obtained by thermolysis of the corresponding p-nitrobenzoates[127].
The dehydration of (2) by $N^1N^2N^3$-hexamethyl phosphoric triamide is a
useful additional method for the preparation of tri-t-butylethylene (inaccessible until recently[128, 129]), and the general reaction should be useful as a
route to other hindered alkenes.

The potassium xanthates derived from tertiary alcohols are readily isolable
and on thermolysis give significantly better yields of alkenes than do the
corresponding S-methyl xanthates (in the Chugaev reaction)[130]. In other
studies[131, 132] of the Chugaev reaction a competing decomposition pathway
leading to unexpected products (including the original alcohol) was detected.
This pathway was considered to involve homolytic fission of the S—alkyl
bond, and its competitive importance was directly related to the stability of
the alkyl radical produced[132]. Preliminary studies of two other new dehydration reactions have been reported. The N-carboalkoxysulphamate
esters derived from secondary and tertiary alcohols decompose readily to
give alkenes (equation (3.12))[133]. The reaction apparently involves the

$$\text{R}_2^1\text{CH·CHR}^2\text{·OSO}_2\text{·}\bar{\text{N}}\text{·CO}_2\text{R}^3 \text{ Et}_3\overset{+}{\text{N}}\text{H} \longrightarrow \text{R}_2^1\text{C:CHR}^2 + \text{Et}_3\overset{+}{\text{N}}\text{H } \bar{\text{O}}\text{SO}_2\text{·NH·CO}_2\text{R}^3 \quad (3.12)$$

rate-determining formation of a tight ion-pair, followed by a fast stereospecific loss of a cis-β-proton. The sulphurane (3) is an extremely powerful
reagent for the direct dehydration of secondary and tertiary alcohols[134].
Thus, t-butyl alcohol in chloroform was converted quantitatively into

$$\text{Ph}_2\text{S(OR}_F)_2 \qquad (\text{R}_F = \text{PhC(CF}_3)_2)$$

$$(3)$$

isobutene in seconds at $-50\,°\text{C}$; in general, primary alcohols gave mixed
ethers (R·O·R_F) in place of alkenes.

3.2.6   Replacement of the hydroxy group by halogen

The range of methods available for the conversion of alcohols into alkyl
halides has recently been reviewed[135]. Two additional methods[136, 137] for the
preparation of chlorides have been described briefly; in both cases an $S_N i$
mechanism was thought to operate. The chlorination of alcohols in alkaline
micelles by dichlorocarbene proceeded rapidly at room temperature[137].
However, the reaction was not attempted with aliphatic alcohols, and
restrictions might be placed on its applicability by the nature of the reactive
species and by the strongly alkaline conditions used to generate the diradical.
No base was required for the conversion of anhydrous alcohols into chlorides
by the use of cyanuric chloride[136]. Excellent conversions of allylic alcohols
into the corresponding chlorides, without rearrangement, were obtained[138]

by the reaction with methanesulphonyl chloride and a mixture of lithium chloride, N,N-dimethylformamide, and 2,4,6-trimethylpyridine at 0 °C.

Other investigations have been concerned with the applications and mechanisms of established reactions. As part of a continuing study in this area, the mechanistic features of the thermal decomposition of alkyl chlorocarbonates in the liquid phase have been examined[139]. Although the decomposition to alkyl chlorides has been considered to follow an S_Ni mechanism, the extensive rearrangements found for primary and secondary alkyl groups[139] suggest the formation of ion-pair intermediates. The rate of decomposition was increased, and the formation of alkenes and rearranged chlorides prevented or much reduced, by the addition of pyridinium chloride. Rearrangements similar to those observed for the chlorocarbonates also occurred in the reactions of branched-chain alcohols with boron trichloride or tribromide[140]. Although the major products (e.g. (4)) were consistent with the formation of a carbonium ion intermediate, some minor products (e.g. (5), (6)) were apparently formed by other mechanisms (product isomerisation, radical addition of hydrogen bromide to alkene by-products) (equation (3.13)).

$$EtMe_2COH \xrightarrow{BBr_3} EtMe_2CBr + Me_2CH \cdot CHBrMe + EtMeCH \cdot CH_2Br \qquad (3.13)$$

	(4)	(5)	(6)
Product distribution	86%	12%	2%

In another study[141], optically active (5) was prepared directly from (+)-3-methylbutan-2-ol by reaction with triphenylphosphine dibromide in the presence of pyridine; most other reagents tried gave the rearranged bromide (4) as the major product. Studies on the stereochemical course of halogenation with triphenylphosphine and a tetrahalogenomethane have been extended to neopentyl alcohol (as the 1-deuterio-derivative)[142]. The extent of inversion of configuration was greater with carbon tetrachloride than with carbon tetrabromide. By means of deuterium labelling, it has been shown[143] that a protonated cyclopropane pathway is involved to a small extent in the conversion of propan-1-ol into 1-chloropropane by treatment with zinc chloride and hydrochloric acid (but not with thionyl chloride). Anomalous reactions of polyhalogenoalcohols with various halogenating agents have been described[144].

3.2.7   Miscellaneous reactions

The oxidative cyclisation of alcohols to tetrahydrofurans is an important synthetic process[7]. Compared with reactions in which lead tetra-acetate is used as the oxidant, better yields of tetrahydrofuran derivatives from acyclic primary and secondary alcohols are obtained by using silver(I) oxide and bromine (or mercury(II) oxide and iodine)[145]. Further improvements, which suppress the formation of carbonyl by-products, involve the use of silver carbonate or silver nitrate in place of silver oxide[146] or the addition of an ether such as tetrahydrofuran itself to the reaction mixture[147]. The mechanism of formation of the ketonic by-products from secondary alcohols with lead tetra-acetate has been studied[148].

The susceptibility of alkoxyphosphonium cations to nucleophilic attack at carbon with inversion of configuration has been utilised for the conversion of alcohols into carboxylic and phosphoric esters[149] and into primary amines[150]. The presumed intermediate alkoxyphosphonium salt (7) was generated by reaction of the alcohol with triphenylphosphine and diethyl azodicarboxylate (equations (3.14) and (3.15)).

$$EtO_2C \cdot N\text{:}N \cdot CO_2Et + Ph_3P + R^1OH + HZ \rightarrow [Ph_3\overset{+}{P}OR^1\overset{-}{Z}] + EtO_2C \cdot NH \cdot NH \cdot CO_2Et \quad (3.14)$$

$$(7)$$

$$(7) \rightarrow R^1Z + Ph_3PO \quad\quad\quad\quad (3.15)$$

$$Z = R^2CO_2, \ (R^2O)_2 \ PO \cdot O \quad \text{or}$$

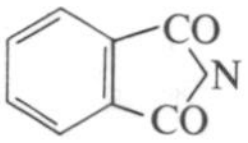

An alternative method for the replacement of hydroxy by amino, which has been found[151] to work well for the synthesis of polyamines, is indicated by the following reaction sequence.

$$ROH \xrightarrow{PhSO_2Cl} RO \cdot SO_2Ph \xrightarrow{NaN_3} RN_3 \xrightarrow{LiAlH_4} RNH_2$$

Acid-labile alcohols have been converted into esters in good yields by reaction of their lithium salts with acid chlorides[152]; the method was only tried with acid chlorides which lack α-hydrogens. The use of molecular sieves in the alkoxide-catalysed alcoholysis of lower esters, with adsorption of the displaced alcohol, provides an efficient method for the preparation of esters from tertiary alcohols in particular[153, 154]. A facile method for the preparation of esters from alkyl t-butyl ethers is given in Section 3.4.3. Techniques for the preparation of pure, alcohol-free alkoxides have been described[155]. The alkoxides from branched-chain alcohols and alkali metals formed adducts with tetrahydrofuran which could be crystallised from this solvent; bound tetrahydrofuran was readily removed by drying the adduct. A different method was necessary for the sodium alkoxides from straight-chain alcohols.

Two new methods for the selective protection of hydroxy groups have been introduced by Barnett and Needham. Tritylone ethers (8) were prepared by acid-catalysed reaction of the two hydroxy-compounds with removal of water as the benzene azeotrope[156, 157].

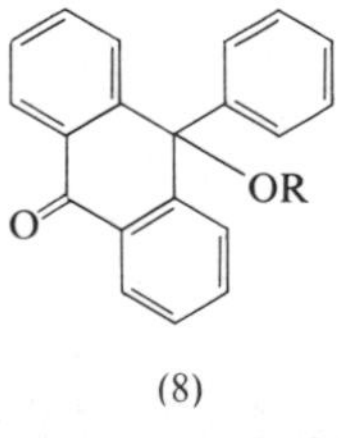

(8)

As with the trityl group, reaction occurs preferentially with primary hydroxy groups. Tritylone ethers are more acid-stable than trityl ethers[157] and, unlike trityl ethers, the ether linkage may be cleaved specifically by means of the Wolff–Kishner reduction, with the formation of 9-phenylanthracene (equation (3.16)). The second method involves the use of the 9-anthryl group for the protection of the hydroxy group[158, 159]. The regeneration of hydroxy-

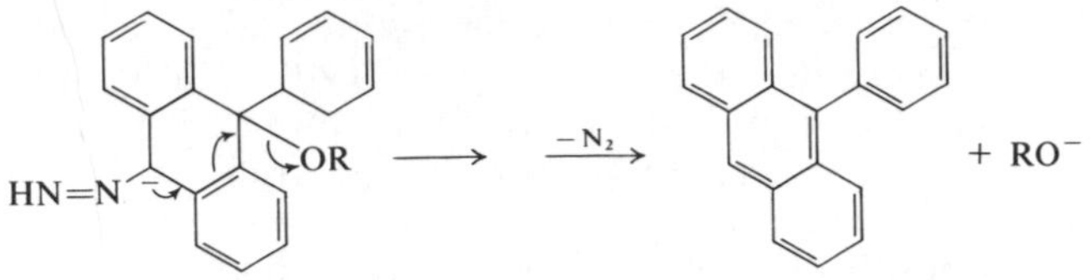

$$+ RO^- \qquad (3.16)$$

compounds from alkyl 9-anthryl ethers (9) can be achieved by the mild and novel method of singlet-oxygen oxidation followed by catalytic reduction (equation (3.17)). The initial ethers (9) were prepared either by the

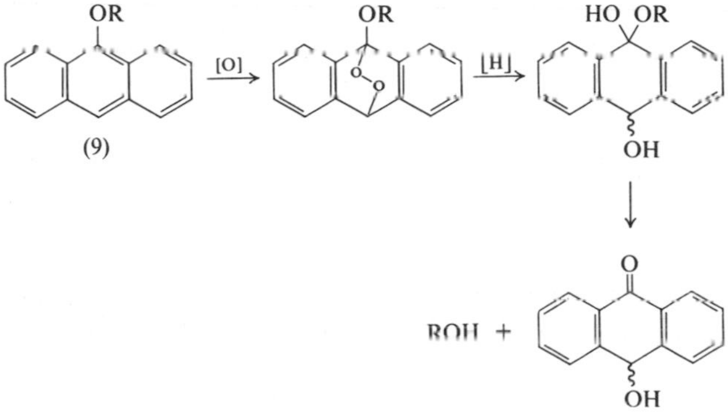

$$(3.17)$$

Williamson synthesis, i.e. reaction between the tosylate of the alcohol and the phenolate ion from anthrone, or by the acid-catalysed alcoholysis of 9-methoxyanthracene[159].

3.3. DI- AND POLY-HYDRIC ALCOHOLS

Although a good deal of additional work on the chemistry of di- and poly-hydric alcohols has been published, it covers many different classes of compounds, and many of the reports concern compounds which contain additional functional groups (e.g. unsaturated and halogenated diols). Rather than give a systematic but fragmented account of the whole subject, discussion is now limited to a few special topics, in the expectation that important omissions will be made good in later issues of this series. Sugar alcohols are excluded from the review.

3.3.1 Preparation

A thorough study of the oxymercuration–demercuration products from dienes and unsaturated alcohols has been carried out[160]. From certain alkenols to which the Markovnikov rule could be applied, the sole or major product was that predicted by the rule (equations (3.18) and (3.19)). From

$$CH_2{:}CH{\cdot}CH_2{\cdot}CH_2OH \longrightarrow Me{\cdot}CH(OH){\cdot}CH_2{\cdot}CH_2OH \qquad (3.18)$$

$$CH_2{:}CH{\cdot}CH(OH){\cdot}Me \longrightarrow Me{\cdot}CH(OH){\cdot}CH(OH){\cdot}Me + HOCH_2{\cdot}CH_2{\cdot}CH(OH){\cdot}Me$$

	(10)	(11)	
Product distribution	95%	5%	(3.19)

but-2-en-1-ol, the proportions of the diols produced, (10) and (11), were the inverse of those from but-3-en-2-ol (equation (3.19)), i.e. the original hydroxy group directs the incoming hydroxy group to the more remote position. In situations where a five- or six-membered ring could be formed readily, cyclic ethers were obtained quantitatively as a result of intramolecular nucleophilic attack by the hydroxy group (equation (3.20)). This result could be avoided and the appropriate diol could be obtained ultimately, by

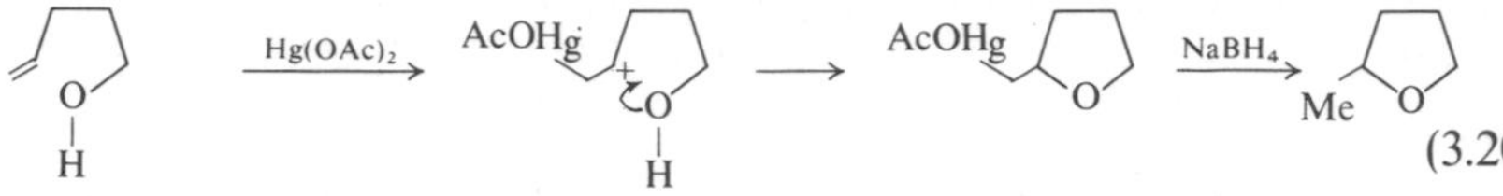

$$\underset{\overset{|}{H}}{O} \xrightarrow{\text{Hg(OAc)}_2} \overset{\text{AcOHg}}{\underset{\overset{|}{H}}{O}} \longrightarrow \overset{\text{AcOHg}}{O} \xrightarrow{\text{NaBH}_4} \underset{\text{Me} \quad O}{} \tag{3.20}$$

acetylating the alcohol before oxymercuration. The products from both conjugated and non-conjugated acyclic dienes were those expected on the basis of stepwise oxymercuration and the behaviour established for the alkenols. Given the availability of the appropriate starting compounds, this reaction clearly may be applied to the synthesis of a wide variety of dihydric alcohols.

For the *trans*-hydroxylation of alkenes, *o*-sulphoperoxybenzoic acid is claimed to have the advantage over other peroxy acids in common use that esterification of the diol does not occur[161]. It has been confirmed[162] that stereospecific *trans*-hydroxylation is the result of the following reaction sequence: epoxidation of an alkene, acid-catalysed ring-opening of the epoxide in dimethyl sulphoxide (equation (3.21)), hydrolysis of the hydroxy-alkoxydimethylsulphonium salt (12).

$$\overset{O}{\overset{/\backslash}{R_2C-CR_2}} + Me_2SO + HX \longrightarrow \overset{OH}{\underset{\overset{|}{OSMe_2} \ X^-}{\overset{|}{R_2C-CR_2}}} \tag{3.21}$$

$$(12)$$

$$(HX = 2,4,6\text{-trinitrobenzenesulphonic acid})$$

The stereospecificity found in the hydroxylation of various *trans*-alk-4-en-1-ols with peroxyformic acid was not absolute; about 5% of the 1,4,5-triol formed was the *threo*-isomer corresponding to *cis*-addition[163]. The classical method for *cis*-hydroxylation of alkenes by permanganate has been simplified and improved[164]; powdered potassium permanganate was added to the alkene dissolved in acetone–water (7:1 v./v.).

The synthesis of ditertiary vicinal glycols containing at least three identical radicals is possible by the reaction of diethyl oxalate with Grignard reagents in the presence of water[165]. The intermediate halogenomagnesium alkoxide (13) is precipitated from ethereal solution as formed, and is subsequently treated with additional Grignard reagent to obtain the required product in moderate yield (c. 20%) (equation (3.22)). The traditional route to

$$EtO_2C \cdot CO_2Et \xrightarrow{R^1MgX} EtO \cdot CR^1(OMgX) \cdot CO_2Et \xrightarrow{R^2MgX} R^1R^2C(OMgX) \cdot CR^2(OMgX) \xrightarrow{H_2O}$$

$$(13)$$

$$R^1R^2C(OH) \cdot C(OH)R_2^2 \tag{3.22}$$

symmetrical ditertiary glycols is the pinacolic reduction of ketones. Merely by using dichloromethane or tetrahydrofuran as solvent, in which aluminium pinacolates are soluble, the mechanics of the reaction have been simplified and the yields significantly improved[166]. In another study[167] of the reaction it was shown that only surface amalgamation of magnesium was necessary, and that the reduction of acetone had a stoichiometry of 1:1 (ketone: magnesium). A new reaction which may compete with pinacolic reduction is that between carboxylic acids and alkyl-lithium compounds in the presence of titanium(III) chloride[168]. Although a rather complex mixture of products was obtained, the reaction of pentanoic acid with n-butyl-lithium gave 5,6-dibutyldecane-5,6-diol (22%).

3.3.2   Oxidative cleavage of 1,2-diols

Methods for the specific cleavage of vicinal glycols have recently been reviewed[169], and the use of periodate has been the subject of a monograph[170]. Lead tetra-acetate and sodium metaperiodate, the established reagents for such oxidations, are in general so satisfactory that other oxidants such as sodium bismuthate, phenyliodosodiacetate, nickel peroxide and peroxydisulphates in the presence of silver(I) have found only occasional use. The mechanism of periodate oxidation has been studied extensively[171] and recent papers have extended earlier kinetic studies of the oxidation of propane-1,2-diol[172] and pinacol[173]. Despite kinetic differences, it was concluded[173] that the oxidation of both compounds proceeds by a mechanism which involves the initial formation of a periodate monoester, followed by ring closure to a cyclic diester, and finally the decomposition of a monoanion of the diester to the reaction products. The mechanism was used to explain the different types of pH dependence observed for different 1,2-diols. The cleavage of 1,2-diols with nickel peroxide probably does not involve a cyclic intermediate[174], nor does it involve abstraction of an α-hydrogen as occurs in the oxidation of monohydric alcohols with this reagent[175]. Instead, a concerted mechanism was proposed[174], in which the diol is bound to nickel through both hydroxy groups in an intermediate complex of the type indicated (14).

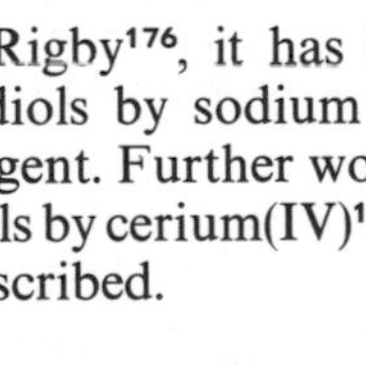

(14)

Contrary to the findings of Rigby[176], it has been stated[177] that aldehydes formed on oxidation of 1,2-diols by sodium bismuthate may be oxidised further by an excess of the reagent. Further work on the kinetics and mechanisms of the cleavage of 1,2-diols by cerium(IV)[178–180], acid permanganate[181,182] and cobalt(III)[183] has been described.

3.3.3   Acid-catalysed dehydration and rearrangement

As found for other molecular rearrangements which may involve bridged-ion intermediates, the pinacol-pinacolone rearrangement of (−)-(4S)-2,3,4-trimethylhexane-2,3-diol (15) to (+)-3,3,4-trimethylhexan-2-one (16) proceeds with retention of configuration by the asymmetric migrating group[184]. The effects of the cyclohexyl group on the product distribution from rearrange-

$$EtCHMe \cdot CMe(OH) \cdot CMe_2(OH) \qquad MeCO \cdot CMe_2 \cdot CHMeEt$$

$$(15) \qquad\qquad\qquad (16)$$

ments of pinacols in concentrated sulphuric acid at $0\,°C$ have been determined[185]. As found with other pinacols[186], the products depended on the balance of importance between relative migratory aptitudes and abilities to stabilise an initial carbonium ion for the different groups present. Thus, the migratory aptitude of cyclohexyl was shown to be greater than that of methyl in the rearrangement of 2,3-dicyclohexylbutane-2,3-diol (equation (3.23)).

$$RMeC(OH) \cdot C(OH)RMe \longrightarrow MeCO \cdot CR_2Me + RCO \cdot CRMe_2$$

$$(18) \qquad\qquad\qquad (3.23)$$

$$R = cyclohexyl \qquad 70\% \qquad\qquad 30\%$$

However, in the rearrangement of other ditertiary glycols such as ((17); R = cyclohexyl), the preferential formation (70%) of (18) showed that the

$$Me_2C(OH) \cdot C(OH)R_2$$

$$(17)$$

cyclohexyl group favoured loss of the adjacent hydroxy group and migration of a methyl group. This type of result can be accommodated by a mechanism which involves the reversible formation of cationic intermediates by loss of a tertiary hydroxy group, and a consideration of the relative kinetics of the alternative steps[186]. The rearrangement of acid-labile pinacols containing a furyl group has been carried out in the vapour phase over alumina; the relative migratory aptitudes were the same as those determined in acidic solutions[187].

The complex reactions which accompany or follow the dehydration of 1,3-diols over acid catalysts (aluminium oxide, calcium phosphate) have been studied[188, 189]. The intermediate formation of oxetanes was considered to be a reaction of major importance. The similarly complex reactions which occur on heating 2-substituted 1,3-diols with aqueous sulphuric acid have been studied extensively by Mazet and his colleagues[190-194]. In earlier work they had confirmed that the major initial products from 2,2-dimethylpropane-1,3-diol and related diols are carbonyl compounds (chiefly aldehydes) formed by dehydration–rearrangement. The kinetics and mechanistic details of the intermediate alkyl migrations (equation (3.24)) have now been investigated more fully[190-192]. The relative migratory aptitude of R^1 and the assistance

$$R^1 \overset{+}{C}H_2 \qquad\qquad CH_2R^1$$
$$\underset{R^2}{\overset{\diagdown}{C}}\diagup\diagdown CH_2OH \longrightarrow R^2-\overset{+}{C}\diagup\diagdown CH_2OH \qquad (3.24)$$

given to the migration of R^1 by the non-migrating group (R^2) were assessed. When both R^1 and R^2 are methyl or primary alkyl groups the alternative aldehydes (19) and (20) are produced in similar amounts, but when R^2 is a

$$R^1CH_2 \cdot CHR^2 \cdot CHO \qquad R^2CH_2 \cdot CHR^1 \cdot CHO$$

$$(19) \qquad\qquad\qquad (20)$$

secondary or tertiary alkyl group (19) can be obtained in major yield[191]. Similar studies have been carried out on 1,3-diols of type (21). Migration of a

$$R^2CH(OH)\cdot CR^1Me\cdot CH_2OH \qquad R^1 = H \text{ or } Me$$

$$(21) \qquad\qquad R^2 = alkyl$$

methyl group from C-2 was detected only when R^1 was methyl[193]. Increasing the size of R^2 reduced the tendency of methyl to migrate from C-2 to C-3, and increased the amount of cyclisation to tetrahydrofurans[194]. The cyclisation of 1,4,5-triols to tetrahydrofurans under acidic conditions has also been reported[163, 195].

3.3.4 Cyclic boronates

The facile reaction of boronic acids $[RB(OH)_2]$ with 1,2- and 1,3-diols to give cyclic boronate esters has been known for some time, and has found use in carbohydrate synthesis. More recently, boronates have assumed importance in the identification of diols and polyols (and compounds such as alkenes which are readily converted into diols) by g.l.c.[196–198]. For this purpose, the more volatile n-butylboronates have been preferred to the phenylboronates. The esters are formed rapidly and simply at room temperature, while further advantages (in speed or degree of separation) accrue during g.l.c. from the use of these derivatives in place of acetates or trimethylsilyl ethers. Boronates are, also amenable to mass spectrometric analysis[196, 197, 199].

3.3.5 Synthesis of simple glycerides

Although good methods for the synthesis of chemically and stereochemically defined glycerides have been available for some time, certain problems have remained. Whereas saturated 1,2-diacyl-sn-glycerols (L-1,2-diglycerides) are readily obtained from D-mannitol via 1,2-O-isopropylidene-sn-glycerol, the corresponding unsaturated lipids and the enantiomeric 2,3-diacyl-sn-glycerols have been less accessible. Solutions to both problems are provided by new procedures[200–202] which use 2,2,2-trichloroethoxycarbonyl as a protective group and 1,2-O-isopropylidene-sn-glycerol as starting material. Cyclic carbonates, formed as intermediates in the above syntheses[200, 202], are also utilised in an alternative route to 1,2-diacyl-sn-glycerols starting from 3,4-O-isopropylidene-D-mannitol[203] (full details of a further paper[204] about this route were not available at the time of writing). Other approaches to the synthesis of saturated 1,2-diacyl-sn-glycerols have been described[205, 206]. Symmetrical 1,3-diacylglycerols may be synthesised simply and efficiently from dihydroxyacetone[207]. A non-stereospecific route to triacylglycerols containing three different fatty acyl residues, from allyl alcohol or epichlorohydrin, has been reported[208].

3.4 ETHERS

The chemistry of ethers was comprehensively reviewed in 1967 (Ref. 209) and relatively few investigations reported during the last 2 years have been concerned directly with aliphatic ethers. The selected studies will therefore be described under three general headings.

3.4.1 Preparation

A preliminary study has been made[210] of a novel method which involves the desulphurisation of sulphenates ($R^1S \cdot OR^2$) with tervalent phosphorus compounds (notably tri-n-butylphosphine) under mild conditions. A quantitative yield of t-butyl methyl ether was obtained from O-methyl t-butylsulphenate, but yields for higher alkyl t-butyl ethers were inferior.

Work on vinyl ethers has included several synthetic studies. A full range of specifically deuterated methyl vinyl ethers has been prepared by reduction of methoxyacetylene or 1-deuterio-2-methoxyacetylene with lithium aluminium hydride or lithium aluminium deuteride[211]. The mechanism previously proposed[212] for the vinyl transetherification reaction catalysed by mercury(II) acetate has been confirmed and catalytically-active organo-mercurial intermediates have been isolated[213]. A general method for the synthesis of monohalogenoalkyl vinyl ethers via an Arbuzov rearrangement (equation (3.25)) has been described[214].

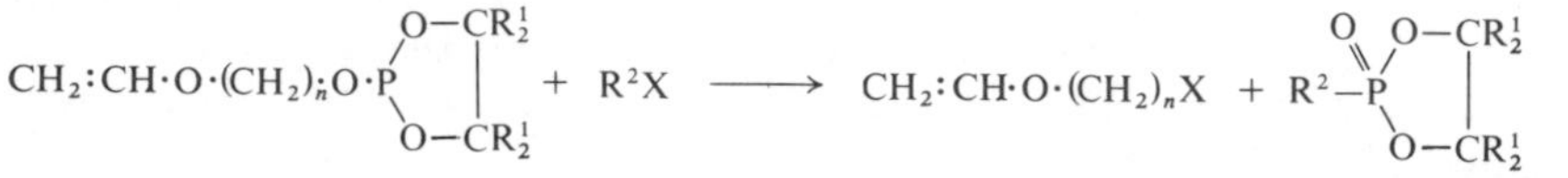

$$CH_2\!:\!CH \cdot O \cdot (CH_2)_n O \cdot P \begin{smallmatrix} O-CR_2^1 \\ | \\ O-CR_2^1 \end{smallmatrix} + R^2X \longrightarrow CH_2\!:\!CH \cdot O \cdot (CH_2)_n X + R^2-\!\!\overset{O}{\underset{|}{P}} \begin{smallmatrix} O-CR_2^1 \\ | \\ O-CR_2^1 \end{smallmatrix}$$

$$(3.25)$$

The required starting compounds were prepared by the reaction of the appropriate vinyloxyalkanols with substituted ethylene phosphorochloridites. A range of β,β-dichlorovinyl ethyl ethers with electron-withdrawing substituents in the β-position of the ethyl group has been synthesised by nucleophilic substitution reactions of β-chloroethyl β,β-dichlorovinyl ether[215].

3.4.2 Physical studies

Detailed analyses of the infrared and Raman spectra of various isotopic species of simple dialkyl ethers have been published[216-221] and an infrared study of conformations of dialkoxyethanes has been made[222]. Mass spectral data for a number of ethers have been recorded in a specialist publication (*Arch. Mass Spectral Data*), and the mechanism of β-cleavage[223] and of secondary reactions which involve four-centre rearrangements of fragment ions from α-cleavage have been studied[224, 225]. The fragmentation patterns of various classes of unsaturated ethers have been reported[226].

3.4.3 Reactions

Several papers have dealt with reactions of ethers with carboxylic acids and their functional derivatives. Carboxylic acids may be rapidly and efficiently esterified by heating them with alkyl t-butyl ethers[227, 228]. Except with the stronger organic acids (e.g. trifluoroacetic acid), catalysis by acids[227, 228] or by salts such as lithium perchlorate[228] is necessary. The acid-catalysed reaction apparently does not involve prior cleavage of the t-butyl–oxygen bond but probably proceeds by nucleophilic attack by the ether on the protonated carboxylic acid, followed by loss of the t-butyl carbonium ion and water from the reaction intermediate (22). The cleavage of ethers by

$$R^1C(OH)_2 \cdot \overset{+}{O}R^2Bu^t \qquad R^1CO \cdot \overset{+}{O}R^2R^3 \quad R^4SO_3^-$$

$$(22) \qquad\qquad\qquad (23)$$

acid halides or anhydrides in the presence of a Lewis acid, with the formation of esters and alkyl halides, is a mechanistically-related established procedure. A recent study[229] has shown that iron pentacarbonyl is a milder though less efficient catalyst for this reaction; the activity of the catalyst is not due to hydrolysis to the Lewis acid, iron(III) chloride. A more interesting development is the use of mixed sulphonic-carboxylic acid anhydrides as powerful reagents for ether cleavage[230]. The expected intermediate complex (23), formed by nucleophilic attack by the ether on the anhydride or on the derived acylium ion, could not be detected by n.m.r. spectroscopy. Because of the low nucleophilicity of sulphonate ions, decomposition of (23) to a mixture of a carboxylic and a sulphonic ester is expected to favour an S_N1 mechanism to a greater degree than does the corresponding reaction involving an acid halide. Consequently, in the cleavage of unsymmetrical ethers, the mixed anhydride method leads more specifically to cleavage of the bond from oxygen to the more highly substituted carbon centre (to give the more stable carbonium ion).

During studies of solvolysis with 2,2,2-trifluoroethanol, a method for the cleavage of 2,2,2-trifluoroethyl ethers was required[231]. Reductive cleavage with sodium naphthalene was found to be a useful albeit slow method for this purpose (equation (3.26)).

$$(3.26)$$

$$RO^- + CH_2{:}CF_2 \longleftarrow R\overset{\frown}{O}{-}CH_2{-}\overset{\frown}{C}F_2$$

Oxidation by selenium dioxide leads to the cleavage of allyl and propargyl ethers[232], presumably as a result of allylic oxidation and the intermediate formation of a hemiacetal.

The direct oxidation of ethers ($R^1CH_2 \cdot OR^2$) to esters ($R^1CO \cdot OR^2$) by trichloroisocyanuric acid in an excess of water[233] apparently does not involve hypochlorous acid as the active oxidant[234]. The preferred site of oxidation of unsymmetrical ethers is determined by steric factors (Newman's 'rule of six'). Thus, ethyl isobutyl ether gave ethyl isobutyrate, while ethyl isopropyl ether gave isopropyl acetate. The similar oxidation of ethers to esters with ruthenium tetroxide may find greater use following the introduction of sodium hypochlorite as a cheap regenerative oxidant[106, 235].

Palladium catalysts such as bis(triphenylphosphine)palladium chloride are effective in causing a reversible exchange of allylic groups of ethers

($R^1O \cdot CH_2 \cdot CH:CHR^2$) with active-hydrogen compounds such as alcohols, acids, primary and secondary amines, and methyl acetoacetate[236].

The products and stereochemistry of Wittig rearrangements of some allyl ethers have been studied[237].

3.5 EPOXIDES

While the review by Rosowsky[238] remains the major work of reference on this subject, several aspects of the chemistry of epoxides are discussed in a book[209] edited by Patai. Other recent reviews have dealt with the determination of epoxides[239], radical reactions[240], cyclisation reactions[241] and miscellaneous new reactions[242] of epoxides. Although the expanding chemistry of α-halogenoepoxides has not been included in this survey, a review of the rearrangement reactions[243] of these compounds has appeared. Much of the discussion in this section is concerned with the mechanistic features of the ring-opening reactions of epoxides. The kaleidoscopic variation of mechanism with epoxide structure and reaction conditions, which is so characteristic of these reactions, is still very evident in recent studies. Here, more than in other sections, it was felt that the discussion would be improved by the inclusion of reactions of some arylalkyl compounds.

3.5.1 Preparation

New reagents for the epoxidation of alkenes are o-sulphoperoxybenzoic acid[161] and thallium(III) acetate[244]. The former is readily obtained as a stable solution in aqueous acetone by treatment of o-sulphobenzoic acid anhydride with hydrogen peroxide. Epoxides may be obtained in reasonable yields under neutral conditions by treatment of alkenes with hydrogen peroxide and an aryl isocyanate[245]. Best results were obtained when a non-polar solvent and a 2:1 molar ratio of isocyanate to peroxide were used, as expected from equation (3.27). Compared with conventional peroxy-acids, peroxy-

$$R^1CH:CHR^1 + 2R^2NCO + H_2O_2 \longrightarrow R^1CH \overset{O}{\overset{\diagup \diagdown}{-\!\!-\!\!-}} CHR^1 + R^2NH \cdot CO \cdot NHR^2 + CO_2 \qquad (3.27)$$

benzimidic acid [$PhC(:NH) \cdot O \cdot OH$], which is formed *in situ* from benzonitrile and alkaline hydrogen peroxide, is a relatively indiscriminate reagent, unsuitable for the selective epoxidation of polyenoic compounds[246]. The influence of solvents on the rate and mechanism of epoxidation with peroxy-acids has been studied further[247, 248]. In particular, a decrease in rate constant for epoxidation in an aprotic solvent could be correlated with an increase in solvent basicity[247, 248]. The effect was attributed to intermolecular hydrogen-bonding between the peroxy-acid and the solvent (the extent of which was determined from the shift in frequency of the O—H stretching vibration in the infrared spectrum), with a consequent decrease in the concentration of intramolecularly hydrogen-bonded peroxy-acid (24), commonly considered to be the effective oxidant. The high rate constants and the activation parameters for epoxidations in 2,2,2-trifluoroethanol suggested a different mechanism in this alcohol at least, with efficient solvation of the transition state and concerted proton transfers (25) between the solvent and the peroxy-

$$
\begin{array}{cc}
\text{RC}\overset{\nearrow \text{O}_{\cdots}\text{H}}{\underset{\text{O}-\text{O}}{}} & \text{R}^1\text{C}\overset{\text{O}}{\underset{\text{O}-\text{O}}{}} \\
\end{array}
$$

(24) Site of nucleophilic attack by the alkene

(25)

acid[247]. The epoxidation of alkenes by hydroperoxides in the presence of metal ions (notably molybdenum(VI)) is discussed in Section 3.6.5. The similar epoxidation with molecular oxygen in the presence of molybdenum chelates has been investigated by Rouchaud and his colleagues[249, 250]. The potential of the route to epoxides from iodohydrins has been increased[251] by new procedures[252] for the preparation of iodohydrins.

3.5.2 Deoxygenation and reduction

Reactions for the conversion of an epoxide into the salt of a halogenohydrin and for the reductive elimination reaction of the salt to give an alkene have been combined in a procedure for the direct conversion of an epoxide into the alkene by treatment with magnesium amalgam and magnesium bromide[253]. Reductive elimination with the zinc–copper couple has also been used for this conversion[254]. However, low yields were often obtained even after prolonged boiling in ethanol, and the reaction was sometimes complicated by saturation or cis–trans isomerisation of the alkene. The reductive elimination with a chromium(II)–ethylenediamine complex[255] was in general more rapid and more efficient, but less stereospecific[254]. The deoxygenation of epoxides by atomic carbon is non-stereospecific[256].

The expectation that the reduction of an epoxide by the borohydride anion would be a stereospecific *trans* process has been verified[251]. Treatment of epoxides with sulphurated sodium borohydride leads to disulphides not alcohols[257]. In a study[258] of the reduction of epoxides with alkali metals in liquid ammonia, it was confirmed that the reaction proceeds via a carbanionic intermediate and, where alternatives were possible, that the final product was that from the more stable carbanion. Further studies on catalyst selectivity in the gas-phase hydrogenation of epoxides have been reported[259, 260].

3.5.3 Rearrangement

The rearrangement of epoxides to carbonyl compounds may be effected by numerous catalysts and has been widely studied[238]. Several aspects of the boron trifluoride-catalysed reaction have recently been investigated by Hartshorn and his colleagues[261-263]. Earlier studies by these workers had indicated that a discrete carbonium ion species was formed in the rate-determining step of some rearrangements (equation (3.28)). The value of c. 2 for the deuterium isotope effect for the subsequent hydride shift, determined by analysis of the products from 1-deuterio-1,2-epoxy-2-methyl-propane, was considered to be in accord with a transition state close to the

$$R_2C\overset{O}{\underset{\diagdown}{\diagup}}CHR \xrightarrow{BF_3} R_2\overset{+}{C}-\overset{\overset{\overline{O}BF_3}{|}}{CHR} \longrightarrow R_2CH\cdot CO\cdot R \qquad (3.28)$$

carbonium-ion species[261]. Similar studies[262, 263] of the rearrangements of cis- and trans-1-deuterio-1,2-epoxy-2,3,3-trimethylbutane (26) revealed a stereoselective preference (c. 1.9:1) for migration of the hydrogen atom cis to the methyl group in the undeuterated epoxide. This result was explained

$$Bu^tMeC\overset{O}{\underset{\diagdown}{\diagup}}CHD$$

(26)

in terms of the conformations of the intermediate carbonium ion. In each of these rearrangements, the major products were 1,3-dioxolans, formed by further reaction of the initial carbonyl products with the parent epoxide. A study[264] of the boron trifluoride-catalysed reaction of ^{18}O-labelled acetone with cis- and trans-2,3-epoxybutane supported the proposed mechanism[265] for the formation of 1,3-dioxolans.

Lithium salts solubilised in benzene are efficient catalysts for the re-arrangement of epoxides to carbonyl compounds under mild conditions[266]. Lithium bromide forms benzene-soluble complexes with phosphine oxides such as $N^1N^2N^3$-hexamethyl phosphoric triamide, and these complexes catalyse the rearrangement, probably via the lithium salt of a bromohydrin intermediate. A different mechanism was proposed for rearrangements catalysed by lithium perchlorate. In an interesting extension of these reactions, gaseous epoxides were rearranged by passage through columns containing molten eutectic mixtures of lithium salts supported on firebrick[267]. However, the results were complicated by uncatalysed thermal rearrangements which occurred under these conditions.

3.5.4   Nucleophilic substitution

3.5.4.1   Acid-catalysed reactions

Much effort has been devoted towards establishing the mechanistic features of the acid-catalysed nucleophilic substitution reactions of epoxides[238, 268–270]. Both 'normal' and 'abnormal' products[268] may be formed, and the balance of evidence has favoured an A2 or borderline A2 mechanism for reactions of simple aliphatic epoxides. In recent studies increasing use has been made of the value of the entropy of activation ($\Delta S^{\ddagger}$) as a criterion of mechanistic type. In the boron trifluoride-catalysed alcoholysis of 1,2-epoxy-2-methylpropane (equation (3.29)), the 'abnormal' product (27) was formed exclusively or in much the larger amount[271]. From the values (-13.5 to -18.7 cal mol^{-1} K^{-1})

$$Me_2C\overset{O}{\underset{\diagdown}{\diagup}}CH_2 + ROH \longrightarrow Me_2C(OH)\cdot CH_2OR + Me_2C(OR)\cdot CH_2OH \qquad (3.29)$$

(27)

of $\Delta S^{\ddagger}$ and from the relationship between rate constants and dielectric constants of the media, it was inferred that the reactions were of the A2 type. A similar conclusion was reached for the corresponding reaction of ethylene glycol with 1-chloro-2,3-epoxypropane[272]. The values of $\Delta S^{\ddagger}$ were also used in a study[273] of the protic acid-catalysed methanolysis of 1,2-epoxypropane and related epoxides. With 1,2-epoxypropane the 'normal' and the 'abnormal' product were formed in similar amounts, and the separate rate constants and activation parameters for the parallel reactions were determined. The value (-14.8 cal mol^{-1} K^{-1}) of $\Delta S^{\ddagger}$ for the 'abnormal' reaction was taken to indicate a borderline A2 mechanism, compared with an A2 mechanism for the 'normal' reaction ($\Delta S^{\ddagger}$ -17.0 cal mol^{-1} K^{-1}). Related studies[274–276] of the alcoholysis of various (1,2-epoxyethyl)benzenes have indicated that the dominant abnormal reaction also follows the borderline A2 mechanism, except for 1,1-diphenylethylene oxide (A1 or borderline A1 mechanism). In the acid-catalysed methanolysis of $(+)$-(1,2-epoxyethyl)benzene, 89% of inversion of configuration occurred. On the other hand, acid-catalysed hydrolysis of this enantiomer gave an optically-inactive diol, which was taken to indicate an S_N1 (A1) mechanism with racemisation via a long-lived carbonium ion[277]. Further reactions in which ring-opening occurs with considerable retention of configuration have been described[278].

In contrast to the results described above for boron trifluoride-catalysed alcoholysis of 1,2-epoxy-2-methylpropane[271], the similarly catalysed reactions with thiols and selenols[279] gave comparable amounts of both the 'normal' and the 'abnormal' product. This difference was attributed to the greater nucleophilicity (lower selectivity) of thiols and selenols. The participation of the halogen in the acid-catalysed ring-opening reactions of 5-chloro-1,2-epoxyalkanes, with the formation of an intermediate chlorinium ion, has been described briefly[280].

3.5.4.2 Reactions with organometallic compounds

Organocopper reagents, such as lithium dimethylcuprate and lithium diphenylcuprate, show considerable promise for the preparation of alcohols by nucleophilic ring-opening of epoxides[281, 282]. Thus, pentan-3-ol was obtained in 88% yield under mild conditions from 1,2-epoxybutane and lithium dimethylcuprate (equation (3.30)), with little formation of by-products.

$$\text{EtCH—CH}_2 \xrightarrow[\text{Et}_2\text{O, 0 °C, 13·5h}]{\text{2LiCuMe}_2} \text{Et}_2\text{CHOH} \qquad (3.30)$$

Dimethylmagnesium and methyl-lithium are also superior to Grignard reagents in this respect[282]. With vinylethylene oxide [3,4-epoxybut-1-ene (28)], mainly conjugate addition of organocopper reagents occurred to give allylic alcohols (29), predominantly as the trans isomers[282, 283].

$$\text{CH}_2\text{:CH·CH—CH}_2 \qquad\qquad \text{RCH}_2\text{·CH:CH·CH}_2\text{OH}$$

(28)   (29)

The mechanistic features of reactions of organoaluminium compounds with epoxides resemble those of the acid-catalysed ring-opening reactions. The reaction of triethylaluminium (in excess) with some simple aliphatic epoxides apparently follows the borderline S_N2 mechanism to give the 'abnormal' alcohol by backside attack at the more hindered carbon of the epoxide (which is already complexed with a second molecule of the reagent acting as a Lewis acid)[284]. However, the major product from 1,2-epoxypropane was pentan-2-ol instead of 2-methylbutan-1-ol (an explanation of this result was offered)[284]. Whereas the ring-opening of 1,2-epoxypropane on treatment with triethylaluminium in tetradecane (80 °C, 48 h) was apparently regiospecific for production of the 'abnormal' alcohol[284], some 'normal' alcohol was also produced in the reaction of 1,2-epoxybutane with trimethyl-aluminium in hexane (0 °C, 4 h)[285]. With 1,2-epoxy-3-phenylpropane and derived epoxides, the abnormal alcohol was the unique product[285, 286]. Because of the differences in degree of regioselectivity and in rate of reaction, it was suggested[285] that the reaction of the aryl substituted epoxides proceeded via a carbonium ion which was stabilised by phenyl participation (30). The retention of configuration in the reaction products[285, 286] and the behaviour of higher homologous epoxides supported this view. Predominant or

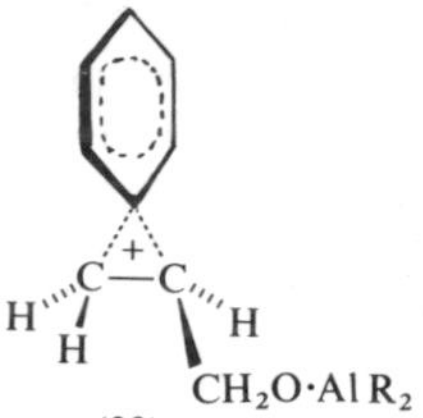

(30)

complete retention of configuration was also observed in the benzylic attack of trialkylaluminiums in hexane on 1-aryl-1,2-epoxy-propanes[286, 287]. By contrast, inversion of configuration occurs during reaction with 2,3-epoxy-butane[284]. Although predominant or complete inversion of configuration has been found in recent studies of reactions of epoxides with dialkyl- and with diallyl-magnesium derivatives, variations in both stereo- and regio-selectivity have been described[282, 287–290]. Reactions of epoxides with organotin compounds[291, 292], with aminoarsines[293] and with trimethylchlorosilane in the presence of magnesium[294] have been reported, and the diverse reactions of epoxynitriles with organometallic compounds have been summarised[295].

3.5.4.3   Reactions with amines

In order to determine the reason for the greater reactivity toward amines of 1,2-epoxy-3-phenoxypropane ((31); Ar = Ph) compared with 1,2-epoxy-3-phenylpropane, the reaction of dibutylamine in t-pentyl alcohol with the ethers (31, 32) of 2,3-epoxypropan-1-ol (glycidol), in which the aromatic ring

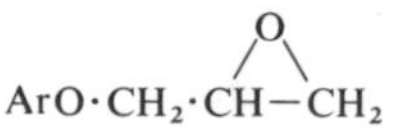

$$ArO \cdot CH_2 \cdot CH—CH_2$$

(31)

$$ArC(CF_3)_2 \cdot O \cdot CH_2 \cdot CH—CH_2$$

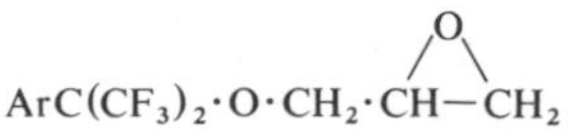

(32)

contained fluoro- or trifluoromethyl-substituents, has been studied[296, 297]. As expected, the reactions followed second-order kinetics and gave solely the 'normal' addition products. Contrary to expectation, the reactivities of these epoxides were less than that of 1,2-epoxy-3-phenoxypropane, except for the compounds (32) which contained *ortho*-fluoro-substituents. The results were taken to indicate that hydrogen-bonding as in (33) was responsible for the enhanced reactivity of 1,2-epoxy-3-phenoxypropane, and that this effect was opposed by the inductive, base-weakening effect of fluoro-substituents[296]. The activating influence of *ortho*- fluoro- (or bromo)-substituents in (32)

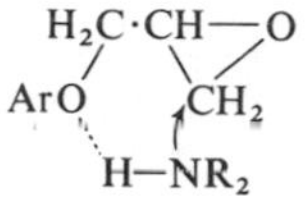

(33)

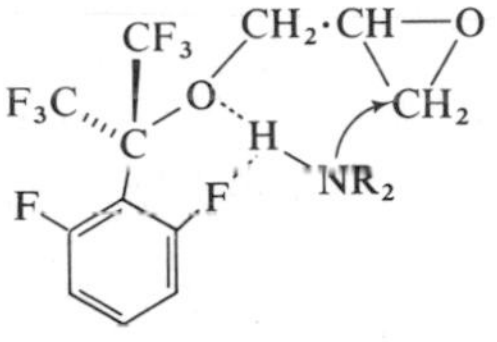

(34)

was attributed to co-operation by a substituent in binding the amine in a manner favourable to reaction, (34).

3.5.5 Miscellaneous reactions

Ring-opening reactions, including a β-elimination process, which occur during fusion with alkali have been studied with long-chain epoxy-acids[298]. Treatment of epoxides with tertiary phosphine dihalides has been used to prepare vicinal dihalides; the hope that only *cis*-dihalides would be obtained from cycloalkene oxides was not realised[299]. The formation of 2,2'-dibromo-derivatives of dialkyl ethers by the bromination of alkene–epoxide mixtures has been studied[300-302]. The homolytic addition of methanethiol to vinylox-iranes such as (35) has provided rare examples of reactions in which fission of the oxirane C—C bond occurs (equation (3.31))[303].

(35)

$$\text{MeSH}$$

$$(3.31)$$

3.6 ALKYL HYDROPEROXIDES

Various topics in hydroperoxide chemistry are reviewed in two excellent recent books about organic peroxy-compounds[304, 305]. In addition to a chapter[306] devoted specifically to hydroperoxides, there are sections in other chapters on particular aspects of hydroperoxide chemistry (e.g. physical properties, methods of determination). A monograph[307] on methods of determination has been published, and further reviews of oxidation reactions in which hydroperoxides are used in the presence of catalytic amounts of metal ions have appeared[308, 309]. Homolytic reactions of hydroperoxides are discussed in a different volume of this series[310].

3.6.1 Preparation

Methods for the synthesis of hydroperoxides have been fully reviewed by Hiatt[306]. Significant improvements to the method of synthesis from alkenes via hydroboration have been described[311]. Controlled autoxidation of the organoborane in tetrahydrofuran at low temperatures gave a diperoxyborate, which was then decomposed by treatment with cold aqueous hydrogen peroxide (equation (3.32)).

$$(RO_2)_2BR + H_2O_2 + H_2O \longrightarrow 2RO_2H + ROH + H_3BO_3 \qquad (3.32)$$

Good yields were obtained, and hydroperoxides corresponding to water-insoluble alcohols could be isolated readily from the reaction mixture. Interest in the chemistry of fluorinated peroxides has led to improved methods for the synthesis of trifluoromethyl hydroperoxide from trifluoromethyl fluoroformyl peroxide[312] or hexafluoroacetone[313]. The formation of inclusion complexes with β-cyclodextrin[314] may be useful both in the isolation and stabilisation of hydroperoxides.

3.6.2 Acidity

The instability towards bases of some hydroperoxides, particularly primary and secondary, has led to attempts to predict pK_a values from linear free energy relationships. For several t-alkyl hydroperoxides in 40% aqueous methanol, a correlation between pK_a values and Taft polar substituent constants was obtained[315]. The value of ρ^* was 0.51 ± 0.07, and the correlation equation (3.33) was derived. Although correlation lines could be drawn for

$$pK_a = -(0.51 \pm 0.07)\sigma^* + 13.13 \pm 0.02 \qquad (3.33)$$

primary and secondary hydroperoxides in water, unusually high values of ρ^* (3.16, 4.1) were obtained therefrom[315, 316]. The predicted pK_a value (13.0–13.2) for 2-chloromethylprop-2-yl hydroperoxide agreed well with that (13.2) obtained from a kinetic study of the alkaline decomposition of this compound[317]. Preparations of the potassium[318] and the quaternary ammonium[319] salt of t-butyl hydroperoxide have been described, and the risk of explosion with the former salt has been noted[320]. The thermal decomposition and other reactions[319, 321] of some quaternary ammonium salts of t-butyl and cumyl hydroperoxide surprisingly indicated that the salts exist in equilibrium with the corresponding ylides and hydroperoxides; however, it should be noted that the salts could not be obtained pure.

3.6.3 Acid-catalysed decomposition

The behaviour of alkyl hydroperoxides in acid solution has been reviewed[306] and clarified by further studies[322, 323]. In concentrated sulphuric acid, the rearrangement to carbonyl compounds and alcohols proceeds via protonation of the hydroxy oxygen and heterolysis of the O—O bond, followed by alkyl migration (equation (3.34)). The relative migratory aptitudes depend on

$$R^1R^2R^3CO \cdot OH \underset{}{\overset{H^+}{\rightleftharpoons}} R^1R^2R^3CO-\overset{+}{O}H_2 \overset{-H_2O}{\rightleftharpoons} R^2R^3\overset{R^1\curvearrowright}{\underset{}{\overset{+}{C}}}-O \longrightarrow \overset{R^2}{\underset{R^3}{C}}\overset{+}{\underset{}{}}-OR^1$$

$$\text{(3.34)}$$

$$\downarrow H_2O$$

$$R^2COR^3 + R^1OH + H^+$$

the class of hydroperoxide[322, 323] and on the strength of the acid used[323], but methyl migrates less readily than other alkyl groups and hydrogen. Whereas t-butyl hydroperoxide in 96% sulphuric acid gave acetone and methanol almost quantitatively[322, 323], in 50% sulphuric acid di-t-butyl peroxide was obtained in 93% yield[323]. In the more dilute acid, the reaction involves protonation of the oxygen attached to carbon, and electrophilic attack by the t-butyl carbonium ion generated on a molecule of unprotonated hydroperoxide (equation (3.35)). Reactions which occur in acidified alcohols have

$$R^1R^2R^3CO \cdot OH \overset{H^+}{\rightleftharpoons} R^1R^2R^3C\overset{H}{\underset{}{\overset{|}{-O^+}}}-OH \overset{-H_2O_2}{\rightleftharpoons} R^1R^2R^3\overset{+}{C} \overset{R^1R^2R^3CO \cdot OH}{\longrightarrow}$$

$$R^1R^2R^3CO \cdot OCR^1R^2R^3$$

$$+ H^+$$

$$\text{(3.35)}$$

also been studied[323].

3.6.4 Base-catalysed decomposition

A summary of base-catalysed decompositions of hydroperoxides has been given by Hiatt[306]. The decomposition of 2-chloromethylprop-2-yl hydroperoxide proceeds by an intramolecular nucleophilic displacement of chlorine and the formation of a 1,2-dioxetane intermediate (36)[317]. Unimolecular

$$\begin{array}{c} O\text{---}O \\ |\quad\; | \\ Me_2C\text{---}CH_2 \end{array}$$

$$(36)$$

molecular decomposition of (36) gives acetone and formaldehyde, while a competitive, base-catalysed decomposition of (36) gives 2-methylpropane-1,2-diol, probably via 2-hydroxybutyraldehyde and a crossed Cannizzaro reaction with formaldehyde[317].

3.6.5 Metal ion-catalysed epoxidation of alkenes and related reactions

The remarkable ability of certain metal ions, particularly those of molybdenum and vanadium, to catalyse heterolytic fission of the O—O bond of hydroperoxides on attack by nucleophiles, has aroused growing interest. The primary reason for this interest has been the industrial potential of such a reaction for the epoxidation of alkenes. Thus, much of the early information was to be found in the patent literature. Although numerous additional patents have been disclosed during the period covered by this survey, the number of relevant papers in scientific journals has grown substantially, and progress in the subject up to 1970 has been reviewed[308]. Although both soluble and insoluble catalysts are effective, the oil-soluble compounds such as metal acetylacetonates, naphthenates and carbonyls have been preferred for kinetic and mechanistic studies. Such studies have recently been reported for the following alkenes: propene[324–326], but-2-ene[327], 2-methylpent-2-ene[328], oct-1-ene[316], styrene and its ring-substituted derivatives[329]. Various conjugated and non-conjugated dienes and allylic compounds have also been studied[330]. Good yields of monoepoxides, based on hydroperoxide conversion, could be obtained for both classes of diene, and with a compound such as 4-vinyl-cyclohexene selective epoxidation of the double bond in the ring could be achieved (reactivity is related to the degree of alkyl substitution of the ethylenic unit). Except with allylic alcohols, molybdenum(VI) compounds are generally the most effective catalysts. The greater activity of vanadyl acetylacetonate, compared with that of molybdenum hexacarbonyl, with allylic alcohols was attributed to its greater ability to form complexes with alcohols and to the formation of a ternary complex (involving the hydroxy group) favourable for epoxidation[330]. This explanation is consistent with the observation[308, 331] that the epoxidation of monoalkenes catalysed by vanadium(V) is the more sensitive to inhibition by alcohols. In contrast to reports for other alkenes[329, 330] the molybdenum(VI)-catalysed epoxidation of but-2-ene[327] was stated to be inhibited by the alcohol formed during the reaction, although it is not clear from the abstract available whether or not the result could be attributed to catalyst degradation[329].

Kinetic studies of the epoxidation of styrene and its derivatives[329] showed that in an inert solvent, such as benzene, the reaction was first order in hydroperoxide, alkene, and catalyst; under other conditions the kinetics were less simple and the behaviour of the catalyst is not fully understood. However, it is generally accepted[308] that the mechanism of epoxidation involves heterolytic fission of the O—O bond of the hydroperoxide during nucleophilic attack by the alkene on an intermediate metal-ion–hydroperoxide complex. In this connection, recent work[332, 333] on the epoxidation of alkenes by co-valent peroxides of molybdenum(VI) is of considerable interest. Indeed, it has been claimed[326, 334] that such compounds (37) are reactive intermediates in

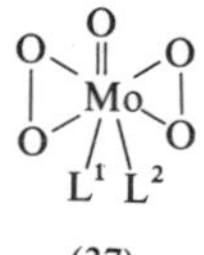

L^1 and L^2 are organic ligands such as aromatic amines, tertiary amides or phosphoramides

(37)

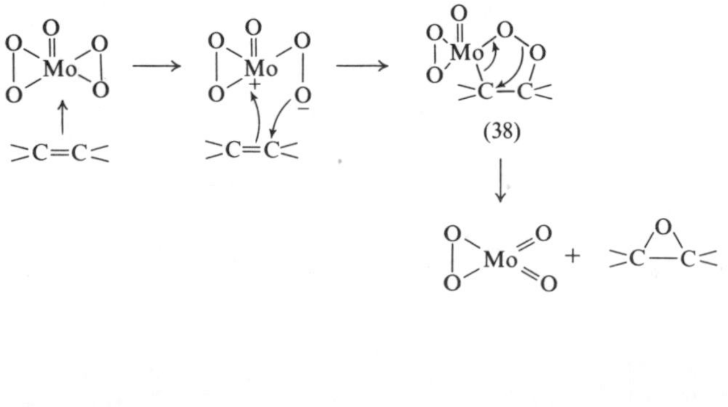

(3.36)

epoxidations by alkyl hydroperoxides in the presence of the metal concerned. The interaction of (37) with the alkene was represented[332] as an aprotic 1,3-dipolar addition reaction (equation (3.36)). The validity of this scheme has been questioned[333]. From a comparison of the relative reactivities of selected pairs of alkenes (norbornene and cyclohexene; cis- and trans-cyclodecene) towards (37) ($L^1 = N^1N^2N^3$-hexamethyl phosphoric triamide, L^2 = no ligand), it was concluded that the reaction pathway did not involve a rate-limiting, five-membered, cyclic transition state as suggested by the above scheme involving structure (38).

In addition to epoxidation, a number of other oxidations can be achieved by the use of alkyl hydroperoxides in the presence of compounds of molybdenum(VI) or vanadium(V). Recent examples include the oxidation of aniline[335], phosphines[320, 336] and sulphides[337]. On the other hand, the selective oxidation of lactams to imides[338] by hydroperoxides occurs only with catalysts such as manganese(III), which normally induce radical reactions, and probably follows a different mechanism.

3.7 DIALKYL PEROXIDES

The chemistry of peroxides has been described in detail in recent books[204, 305] edited by Swern. Several reviews of methods of determination[305, 307, 339] and a review of reactions of peroxides with phosphites, sulphides, and aromatic amines[340] have appeared. Reactions of peroxides are also discussed elsewhere in the present series[310].

3.7.1 Preparation

The general solvomercuration–demercuration reaction described for the preparation of alcohols and polyols has been applied to the synthesis of dialkyl peroxides. Peroxymercuration of alkenes was done with alkyl hydroperoxides in the presence of mercury(II) acetate[341, 342] and the organo-mercurial product was reduced with sodium borohydride to give a dialkyl peroxide in high yield[342]. As expected, the reaction leads to the product of Markovnikov addition and constitutes a particularly useful synthesis of peroxides containing a secondary alkyl group. Several papers have dealt with the synthesis of β-substituted peroxides. β-Bromo- and β-iodo-peroxides may be synthesised[341, 342] by cleavage of the C—Hg bond in the peroxymercuration adduct with the appropriate halogen. An alternative approach—peroxymercuration of alkenes substituted already—has been examined with α,β-unsaturated esters and ketones[343]. The orientation of addition, determined by n.m.r. spectroscopy, was unusually sensitive to structural influences. With compounds alkylated solely at the carbon α to the carbonyl group, the mercury became attached to the β-carbon; with all other compounds the attachment was to the α-carbon. The base-catalysed Michael addition of hydroperoxides to alkenes activated to nucleophilic attack has been used for the preparation of peroxides with β-sulphonate substituents[344], while the base-catalysed addition of hydroperoxides to epoxides has been used in the preparation of unsaturated β-hydroxyalkyl peroxides[345]. The previously unknown bis(perfluoro-t-butyl) peroxide was the unexpected product when perfluoro-t-butyl alcohol was oxidised with chlorine trifluoride[346].

3.7.2 Decomposition reactions

Decomposition reactions of dialkyl peroxides continue to be actively investigated, with emphasis on pathways of induced decomposition. The kinetics of thermal decompositions have been the subject of a recent review[347].

Early studies had suggested that the rate of thermal decomposition of di-t-butyl peroxide was the same in the gas phase as in solution, and that the compound was insensitive to induced decomposition. However, the accelerated decomposition of the pure liquid, with the formation of 1,2-epoxy-2-methylpropane, is attributable[348] to induced decomposition (equations (3.39) and (3.40)) involving the initially-formed t-butoxy radicals (equation (3.37)) or methyl radicals formed by a fragmentation reaction (equation (3.38)).

$$Me_3CO \cdot OCMe_3 \longrightarrow 2Me_3CO^{\cdot} \tag{3.37}$$
$$(39)$$

$$Me_3CO^{\cdot} \longrightarrow Me_2CO + Me^{\cdot} \tag{3.38}$$

$$Me_3CO^{\cdot} \text{ (or } Me^{\cdot}) + (39) \longrightarrow Me_3COH \text{ (or } CH_4) + {\cdot}CH_2 {\cdot} CMe_2 {\cdot} O {\cdot} OCMe_3 \tag{3.39}$$

$$\cdot CH_2 {\cdot} CMe_2 {\cdot} O {\cdot} OCMe_3 \longrightarrow Me_2C\underset{O}{\overset{\displaystyle \diagup\!\!\!\!\diagdown}{}}CH_2 + Me_3CO^{\cdot} \tag{3.40}$$

The similarly enhanced rate of thermal decomposition of di-t-pentyl peroxide[349] cannot be entirely explained by such reactions. Despite the greater reactivity of the secondary hydrogens present in this peroxide, only small amounts of t-pentyl alcohol and epoxyalkenes were formed; acetone was the major product[349]. Thus, the fragmentation (equation (3.41)), which produces the relatively unreactive ethyl radical, apparently occurs more rapidly than attack by the alkoxy radical on the peroxide.

$$EtMe_2CO^{\cdot} \longrightarrow Me_2CO + Et^{\cdot} \tag{3.41}$$

Enhanced rates of thermal decomposition of dialkyl peroxides are also observed in some solvents (e.g. primary and secondary alcohols[350], primary and secondary amines[351], ring-substituted benzyl methyl ethers[352]). The induced decomposition of di-t-butyl peroxide in alcohols[353] proceeds by transfer of a hydrogen atom from an α-hydroxyalkyl radical to peroxide oxygen (equation (3.42)), as indicated by the primary kinetic isotope effect

$$R_2^1\dot{C}OH + R^2O \cdot OR^2 \longrightarrow R_2^1CO + R^2OH + R^2O^{\cdot} \tag{3.42}$$

for the reaction in O-deuteriobutan-2-ol. In amines the reaction probably proceeds by an analogous mechanism. On the other hand, the decomposition of di-t-butyl peroxide induced by α-alkoxybenzyl radicals follows an S_H2 mechanism (equation (3.43)), and the reaction is facilitated by increased

$$ArĊHOR^1 + R^2O \cdot OR^2 \longrightarrow ArCH(OR^1)(OR^2) + R^2O^{\cdot} \qquad (3.43)$$

electrophilicity in the ether radical[352, 354]. Radicals derived from unsubstituted alkyl benzyl ethers are ineffective in inducing decomposition of the peroxide, and the effectiveness of ring-halogenated ethers decreases as the size of the alkyl group increases[355].

The thermal decomposition of di-t-butyl peroxide in various other solvents, which do not necessarily lead to induced decomposition, has been reported[354, 356, 357]. In viscous solvents[357], cage recombination of alkoxy radicals decreases the rate of decomposition. The nature and magnitude of solvent effects on non-induced decomposition have been discussed[357, 358]. The results of studies of the thermal decomposition of di-s-butyl peroxide have been reported in a series of papers by Hiatt and his colleagues. Although induced decomposition appeared to be negligible, no satisfactory explanation for the formation of molecular oxygen during decomposition of the peroxide in solution could be proposed[359]. Attempts to demonstrate the formation of methoxy radicals on low-pressure thermolysis and photolysis of dimethyl peroxide (and its perdeuterio-derivative), by trapping decomposition products in an argon matrix at 8 K and infrared spectroscopic examination thereof, were unsuccessful[360]; the only products detected were methanol and formaldehyde. In connection with this study, the infrared and Raman spectra of dimethyl peroxide have been recorded and bands have been assigned to O—O stretching vibrations[361].

The mass spectra of some simple dialkyl peroxides have been reported[362]. Important fragmentation processes are the elimination of an alkene (equation (3.44)) or an alkyl hydroperoxide (equation (3.45)).

$$R^1O - \overset{+\cdot}{O} - CH_2R^2 \longrightarrow R^1O \cdot \overset{+}{O}H + CH_2 : R^2 \qquad (3.44)$$

$$R^1O - \overset{+\cdot}{O} \qquad H^{\cdot} \longrightarrow R^1O \cdot OH + CH_2 : R^2 \qquad (3.45)$$

3.7.3 Reactions with Grignard reagents

Dialkyl peroxides have been used[363] to induce radical decomposition of aromatic Grignard reagents. The best yields (c. 40%) of biaryls were obtained when the stoichiometric amounts of the peroxides were used. The cleavage of dialkyl peroxides by Grignard reagents also leads to the formation of ethers and alcohols. The reaction is again believed to be a radical process and the influence of structure on reaction rate and product composition has been studied[364]. It has also been shown that the reactivity of the halogen in α-chloroalkyl t-butyl peroxides is great enough for it to be replaced by an alkyl or an aryl group on reaction with a Grignard reagent (1:1 molar ratio), without cleavage of the peroxide bond. Thus the reaction may be used for the preparation of unsymmetrical peroxides[365].

122 ALIPHATIC COMPOUNDS

References

1. (1971). *The Chemistry of the Hydroxyl Group.* (S. Patai, editor). (London: Interscience)
2. Bacon, R. G. R. (1970). *Modern Reactions in Organic Synthesis,* 54. (C. J. Timmons, editor). (London: Van Nostrand Reinhold)
3. Carruthers, W. (1971). *Some Modern Methods of Organic Synthesis,* 244. (Cambridge University Press)
4. Lee, D. G. (1969). *Oxidation: Techniques and Applications in Organic Synthesis,* Vol. 1, 53. (R. L. Augustine, editor). (New York: Marcel Dekker)
5. Moffatt, J. G. (1971). *Oxidation: Techniques and Applications in Organic Synthesis,* Vol. 2, 1. (R. L. Augustine and D. J. Trecker, editors). (New York: Marcel Dekker)
6. Bremer, H., Steinberg, K.-H., Glietsch, J., Lusky, H., Werner, U. and Wendlandt, K.-D. (1970). *Z. Chem.,* **10,** 161
7. Mihailovič, M.Lj. and Čekovič, Ž. (1970). *Synthesis,* 209
8. (1970). *Higher Aliphatic Alcohols (Areas of Application, Methods of Production, Physicochemical Properties).* (S. M. Loktev, editor). See *Chem. Abs.,* **75,** 5183v
9. Inch, T. D. (1970). *Synthesis,* 466
10. Boyle, P. H. (1971). *Quart. Rev. Chem. Soc.,* **25,** 323
11. Brown, H. C., Heim, P. and Yoon, N. M. (1970). *J. Amer. Chem. Soc.,* **92,** 1637
12. Brown, H. C., Bigley, D. B., Arora, S. K. and Yoon, N. M. (1970). *J. Amer. Chem. Soc.,* **92,** 7161
13. Borch, R. F., Bernstein, M. D. and Durst, H. D. (1971). *J. Amer. Chem. Soc.,* **93,** 2897
14. Lalancette, J. M. and Frèche, A. (1970). *Can. J. Chem.,* **48,** 2366
15. James, B. D. (1971). *Chem. Ind. (London),* 227
16. Bell, K. H. (1970). *Aust. J. Chem.,* **23,** 1415
17. Schrock, R. R. and Osborn, J. A. (1970). *Chem. Commun.,* 567
18. Khidekel, M. L., Bakhanova, E. N., Astakhova, A. S., Brikenshtein, Kh.A., Savchenko, V. I., Monakhova, I. S. and Dorokhov, V. G. (1970). *Izv. Akad. Nauk. SSSR, Ser. Khim.,* 499
19. Pearce, P. J., Richards, D. H. and Scilly, N. F. (1970). *Chem. Commun.,* 1160
20. Takahashi, S. and Cohen, L. A. (1970). *J. Org. Chem.,* **35,** 1505
21. Brown, H. C. and Geoghegan, P. J. (1970). *J. Org. Chem.,* **35,** 1844
22. Carlson, R. M. and Funk, A. H. (1971). *Tetrahedron Letters,* 3661
23. White, R. W., King, S. W. and O'Brien, J. L. (1971). *Tetrahedron Letters,* 3587
24. Tinker, H. B. (1971). *J. Organometal. Chem.,* **32,** C25
25. Jerussi, R. A. (1970). *Selective Organic Transformations,* Vol. 1, 301 (B. S. Thyagarajan, editor). (New York: Wiley–Interscience)
26. Bhalerao, U. T. and Rapoport, H. (1971). *J. Amer. Chem. Soc.,* **93,** 4835
27. Eidus, Ya. T., Puzitskii, K. V., Lapidus, A. L. and Nefedov, B. K. (1971). *Usp. Khim.,* **40,** 806
28. Piacenti, F., Bianchi, M., Benedetti, E. and Frediani, P. (1970). *J. Organometal. Chem.,* **23,** 257
29. Fell, B., Shanshool, J. and Asinger, F. (1971). *J. Organometal. Chem.,* **33,** 69
30. Barker, I. R. L. (1971). Ref. 1, 193
31. Ogata, Y. and Kosugi, Y. (1970). *Tetrahedron,* **26,** 2321
32. Zawadzki, R. and Kwart, H. (1969). *Bol. Inst. Quim. Univ. Nac. Auton. Mex.,* **21,** 259
33. Yeo, A. N. H. (1971). *Chem. Commun.,* 609
34. Hocking, M. B. (1971). *Can. J. Chem.,* **49,** 3889
35. Young, A. T. and Guthrie, R. D. (1970). *J. Org. Chem.,* **35,** 853
36. Brown, H. C., Negichi, E. and Gupta, S. K. (1970). *J. Amer. Chem. Soc.,* **92,** 2460
37. Cornforth, R. H. (1970). *Tetrahedron,* **26,** 4635
38. Sanders, J. K. M. and Williams, D. H. (1970). *Chem. Commun.,* 422
39. Sanders, J. K. M. and Williams, D. H. (1971). *J. Amer. Chem. Soc.,* **93,** 641
40. Briggs, J., Frost, G. H., Hart, F. A., Moss, G. P. and Staniforth, M. L. (1970). *Chem. Commun.,* 749
41. Rabenstein, D. L. (1971). *Analyt. Chem.,* **43,** 1599
42. Gillies, E., Szarek, W. A. and Baird, M. C. (1971). *Can. J. Chem.,* **49,** 211
43. Leader, G. R. (1970). *Analyt. Chem.,* **42,** 16
44. Jackman, L. M. and Kelly, D. P. (1970). *J. Chem. Soc. B,* 102
45. Lewis, P. M. E. and Robinson, R. (1970). *Tetrahedron Letters,* 2783

46. Roberts, J. D., Weigert, F. J., Kroschwitz, J. I. and Reich, H. J. (1970). *J. Amer. Chem. Soc.*, **92**, 1338
47. Corval, M. (1970). *Bull. Soc. Chim. Fr.*, 2871
48. Siegel, A. S. (1970). *Org. Mass Spectrom.*, **3**, 1417
49. Tsang, C. W. and Harrison, A. G. (1971). *Org. Mass Spectrom.*, **5**, 877
50. Rankin, P. C. (1970). *Lipids*, **5**, 825
51. Field, F. H. (1970). *J. Amer. Chem. Soc.*, **92**, 2672
52. Hunt, D. F. and Ryan, J. F. (1971). *Tetrahedron Letters*, 4535
53. Schoengold, D. M. and Munson, B. (1970). *Analyt. Chem.*, **42**, 1811
54. Karasek, F. W., Cohen, M. J. and Carroll, D. I. (1971). *J. Chromatogr. Sci.*, **9**, 390
55. Schwartz, D. P. (1970). *Anal. Biochem.*, **38**, 148
56. Churáček, J., Řiha, J. and Jureček, M. (1970). *Fresenius' Z. Anal. Chem.*, **249**, 120
57. Churáček, J., Hušková, M., Pechová, H. and Řiha, J. (1970). *J. Chromatogr.*, **49**, 511
58. Anders, M. W. and Cooper, M. J. (1971). *Analyt. Chem.*, **43**, 1093
59. Westley, J. W. and Halpern, B. (1968). *J. Org. Chem.*, **33**, 3978
60. Pereira, W., Bacon, V. A., Patton, W., Halpern, B. and Pollock, G. E. (1970). *Analyt. Letters*, **3**, 23
61. Hamberg, M. (1971). *Chem. Phys. Lipids*, **6**, 152
62. Freytag, W. and Ney, K. H. (1969). *J. Chromatogr.*, **41**, 473
63. Ayers, G. S., Mossholder, J. H. and Monroe, R. E. (1970). *J. Chromatogr.*, **51**, 407
64. Takahashi, S., Cohen, L. A., Miller, H. K. and Peake, E. G. (1971). *J. Org. Chem.*, **36**, 1205
65. Brauman, J. I. and Blair, L. K. (1970). *J. Amer. Chem. Soc.*, **92**, 5986
66. McAdams, M. J. and Bone, L. I. (1971). *J. Phys. Chem.*, **75**, 2226
67. Bohme, D. K., Lee-Ruff, E. and Young, L. B. (1971). *J. Amer. Chem. Soc.*, **93**, 4608
68. Baird, N. C. (1969). *Can. J. Chem.*, **47**, 2306
69. Lewis, T. P. (1969). *Tetrahedron*, **25**, 4117
70. Hehre, W. J. and Pople, J. A. (1970). *Tetrahedron Letters*, 2959
71. Owens, P. H., Wolf, R. A. and Streitwieser, A. (1970). *Tetrahedron Letters*, 3385
72. Hermann, R. B. (1970). *J. Amer. Chem. Soc.*, **92**, 5298
73. Levitt, L. S. and Levitt, B. W. (1970). *Chem. Ind. (London)*, 990
74. Cocksey, B. J., Eland, J. H. D. and Danby, C. J. (1971). *J. Chem. Soc. B*, 790
75. Levitt, L. S. and Levitt, B. W. (1970). *J. Phys. Chem.*, **74**, 1812
76. Levitt, L. S. and Levitt, B. W. (1971). *Tetrahedron*, **27**, 3777
77. Lee, D. G. and Cameron, R. (1971). *J. Amer. Chem. Soc.*, **93**, 4724
78. Arnett, E. M., Quirk, R. P. and Burke, J. J. (1970). *J. Amer. Chem. Soc.*, **92**, 1260
79. Audette, R. J., Quail, J. W. and Smith, P. J. (1971). *Tetrahedron Letters*, 279
80. Diner, U. E. (1970). *J. Chem. Soc. C*, 676
81. Rees, C. W. and Storr, R. C. (1969). *J. Chem. Soc. C*, 1474
82. McCarty, C. G. and Leeper, C. G. (1970). *J. Org. Chem.*, **35**, 4245
83. Jaffari, G. A. and Nunn, A. J. (1971). *J. Chem. Soc. C*, 823
84. Diner, U. E., Worsley, M. and Lown, J. W. (1971). *J. Chem. Soc. C*, 3131
85. Venkatasubramanian, N. and Thiagarajan, V. (1969). *Can. J. Chem.*, **47**, 694
86. Thiagarajan, V. and Venkatasubramanian, N. (1970). *Indian J. Chem.*, **8**, 809
87. Brown, H. C., Garg, C. P. and Liu, K.-T. (1971). *J. Org. Chem.*, **36**, 387
88. Lee, D. G. and Spitzer, U. A. (1970). *J. Org. Chem.*, **35**, 3589
89. Ratcliffe, R. and Rodehurst, R. (1970). *J. Org. Chem.*, **35**, 4000
90. Stensiö, K.-E. (1971). *Acta Chem. Scand.*, **25**, 1125
91. Coates, W. M. and Corrigan, J. R. (1969). *Chem. Ind. (London)*, 1594
92. Venkatasubramanian, N. and Srinivasan, G. (1970). *Proc. Indian Acad. Sci. A*, **71**, 1
93. Roček, J. and Radkowsky, A. E. (1968). *J. Amer. Chem. Soc.*, **90**, 2986
94. Nave, P. M. and Trahanovsky, W. S. (1970). *J. Amer. Chem. Soc.*, **92**, 1120
95. Rahman, M. and Roček, J. (1971). *J. Amer. Chem. Soc.*, **93**, 5455, 5462
96. Wiberg, K. B. and Mukherjee, S. K. (1971). *J. Amer. Chem. Soc.*, **93**, 2543
97. Doyle, M. P., Swedo, R. J. and Roček, J. (1970). *J. Amer. Chem. Soc.*, **92**, 7599
98. Littler, J. S. (1971). *Tetrahedron*, **27**, 81
99. Wells, C. F. and Husain, M. (1970). *Trans. Faraday Soc.*, **66**, 679
100. Wells, C. F. and Husain, M. (1970). *Trans. Faraday Soc.*, **66**, 2855
101. Mathur, D. L. and Bakore, G. V. (1971). *J. Indian Chem. Soc.*, **48**, 363
102. Mathur, D. L. and Bakore, G. V. (1971). *Bull. Chem. Soc. Jap.*, **44**, 2600

103. Saiprakash, P. K. and Sethuram, B. (1971). *Indian J. Chem.*, **9**, 226
104. Boivin, G. and Zador, M. (1970). *Can. J. Chem.*, **48**, 3053
105. Trahanovsky, W. S. and Cramer, J. (1971). *J. Org. Chem.*, **36**, 1890
106. Wolfe, S., Hasan, S. K. and Campbell, J. R. (1970). *Chem. Commun.*, 1420
107. Moriarty, R. M., Gopal, H. and Adams, T. (1970). *Tetrahedron Letters*, 4003
108. Poddar, R. K. and Agarwala, U. (1971). *Indian J. Chem.*, **9**, 477
109. Saccubai, S. and Santappa, M. (1970). *Indian J. Chem.*, **8**, 533
110. Srinivasan, N. S. and Venkatasubramanian, N. (1970). *Indian J. Chem.*, **8**, 57
111. Wells, C. F. and Barnes, C. (1971). *J. Chem. Soc. A*, 430
112. Thiagarajan, V. and Venkatasubramanian, N. (1970). *Indian J. Chem.*, **8**, 149
113. Kudesia, V. P. (1971). *Bull. Soc. Chim. Belg.*, **80**, 59
114. Srinivasan, N. S. and Venkatasubramanian, N. (1970). *Tetrahedron Letters*, 2039
115. Walling, C. and Kato, S. (1971). *J. Amer. Chem. Soc.*, **93**, 4275
116. Sabesan, A. and Venkatasubramanian, N. (1970). *Indian J. Chem.*, **8**, 251
117. Gallopo, A. R. and Edwards, J. O. (1971). *J. Org. Chem.*, **36**, 4089
118. Strojny, E. J., Iwamasa, R. T. and Frevel, L. K. (1971). *J. Amer. Chem. Soc.*, **93**, 1171
119. Sakuraba, S. and Matsushima, R. (1970). *Bull. Chem. Soc. Jap.*, **43**, 2359
120. Burrows, H. D., Greatorex, D. and Kemp, T. J. (1971). *J. Amer. Chem. Soc.*, **93**, 2539
121. Venkatraman, R. and Rao, S. B. (1971). *Indian J. Chem.*, **9**, 500
122. Hopkins, A. S., Ledwith, A. and Stam, M. F. (1970). *Chem. Commun.*, 494
123. Ledwith, A., Russell, P. J. and Sutcliffe, L. H. (1971). *Chem. Commun.*, 964
124. Knözinger, H. (1971). Ref. 1, p. 641
125. Monson, R. S. (1971). *Tetrahedron Letters*, 567
126. Monson, R. S. and Priest, D. N. (1971). *J. Org. Chem.*, **36**, 3826
127. Lomas, J. S., Sagatys, D. S. and Dubois, J.-E. (1972). *Tetrahedron Letters*, 165
128. Lomas, J. S., Sagatys, D. S. and Dubois, J.-E. (1971). *Tetrahedron Letters*, 599
129. Abroscato, G. J. and Tidwell, T. T. (1970). *J. Amer. Chem. Soc.*, **92**, 4125
130. Rutherford, K. G., Ottenbrite, R. M. and Tang, B. K. (1971). *J. Chem. Soc. C*, 582
131. Gilman, R. E., Henion, J. D., Shakshooki, S., Patterson, J. I. H., Bogdanowicz, M. J., Griffith, R. J., Harrington, D. E., Crandall, R. K. and Finley, K. T. (1970). *Can. J. Chem.*, 48, 970
132. Gilman, R. E. and Bogdanowicz, M. J. (1971). *Can. J. Chem.*, **49**, 3362
133. Burgess, E. M., Penton, H. R. and Taylor, E. A. (1970). *J. Amer. Chem. Soc.*, **92**, 5224
134. Martin, J. C. and Arhart, R. J. (1971). *J. Amer. Chem. Soc.*, **93**, 4327
135. Brown, G. W. (1971). Ref. 1, p. 593
136. Sandler, S. R. (1970). *J. Org. Chem.*, **35**, 3967
137. Tabushi, I., Yoshida, Z. and Takahashi, N. (1971). *J. Amer. Chem. Soc.*, **93**, 1820
138. Collington, E. W. and Meyers, A. I. (1971). *J. Org. Chem.*, **36**, 3044
139. Clinch, P. W. and Hudson, H. R. (1971). *J. Chem. Soc. B*, 747
140. Hudson, H. R., Kinghorn, R. R. F. and Murphy, W. S. (1971). *J. Chem. Soc. C*, 3593
141. Arain, R. A. and Hargreaves, M. K. (1970). *J. Chem. Soc. C*, 67
142. Weiss, R. G. and Snyder, E. I. (1971). *J. Org. Chem.*, **36**, 403
143. Karabatsos, G. J., Zioudrov, C. and Meyerson, S. (1970). *J. Amer. Chem. Soc.*, **92**, 5996
144. Dear, R. E. A., Gilbert, E. E. and Murray, J. J. (1971). *Tetrahedron*, **27**, 3345
145. Mihailović, M. Lj., Čeković, Ž. and Stanković, J. (1969). *Chem. Commun.*, 981
146. Roscher, N. M. (1971). *Chem. Commun.*, 474
147. Deluzarche, A., Maillard, A., Rimmelin, P., Schue, F. and Sommer, J. M. (1970). *Chem. Commun.*, 976
148. Jeremić, D., Milosavljević, S., Andrejević, V., Jakovljević-Marinković, M., Čeković, Ž. and Mihailović. M. Lj. (1971). *Chem. Commun.*, 1612
149. Mitsunobu, O. and Eguchi, M. (1971). *Bull. Chem. Soc. Jap.*, **44**, 3427
150. Mitsunobu, O., Wada, M. and Sano, T. (1972). *J. Amer. Chem. Soc.*, **94**, 679
151. Fleischer, E. B., Gebala, A. E., Levey, A. and Tasker, P. A. (1971). *J. Org. Chem.*, **36**, 3042
152. Kaiser, E. M. and Woodruff, R. A. (1970). *J. Org. Chem.*, **35**, 1198
153. Roelofsen, D. P., De Graaf, J. W. M., Hagendoorn, J. A., Vershoor, H. M. and Van Bekkum, H. (1970). *Recl. Trav. Chim. Pays-Bas*, **89**, 193
154. Wulfman, D. S., McGiboney, B. and Peace, B. W. (1972). *Synthesis*, **49**,
155. Lochmann, L., Čoupek, J. and Lim, D. (1970). *Collection Czech. Chem. Commun.*, **35**, 733
156. Barnett, W. E. and Needham, L. L. (1971). *Chem. Commun.*, 170

157. Barnett, W. E., Needham, L. L. and Powell, R. W. (1972). *Tetrahedron*, **28**, 419
158. Barnett, W. E. and Needham, L. L. (1970). *Chem. Commun.*, 1383
159. Barnett, W. E. and Needham, L. L. (1971). *J. Org. Chem.*, **36**, 4134
160. Brown, H. C., Geoghegan, P. J., Kurek, J. T. and Lynch, G. T. (1970–1). *Organometal. Chem. Syn.*, **1**, 7
161. Bachhawat, J. M. and Mathur, N. K. (1971). *Tetrahedron Letters*, 691
162. Khuddus, M. A. and Swern, D. (1971). *Tetrahedron Letters*, 411
163. Lebouc, A. (1970). *Bull. Soc. Chim. Fr.*, 4099
164. Eremenko, L. T. and Korolev, A. M. (1970). *Izv. Akad. Nauk SSSR, Ser. Khim.*, 147
165. Lapkin, I. I., Svinina, T. A., Kislovets, R. M. and Karavanov, N. A. (1970). *Zh. Org. Khim.*, **6**, 1979
166. Schreibmann, A. A. P. (1970). *Tetrahedron Letters*, 4271
167. Binks, J. and Lloyd, D. (1971). *J. Chem. Soc. C*, 2641
168. Axelrod, E. H. (1970). *Chem. Commun.*, 451
169. Perlin, A. S. (1969). *Oxidation: Techniques and Applications in Organic Synthesis*, Vol. 1, 189. (R. L. Augustine, editor). (New York: Marcel Dekker)
170. Dryhurst, G. (1970). *Periodate Oxidation of Diol and Other Functional Groups: Analytical and Structural Applications*. (New York: Pergamon)
171. Sklarz, B. (1967). *Quart. Rev. Chem. Soc.*, **21**, 3
172. Buist, G. J. and Bunton, C. A. (1971). *J. Chem. Soc. B*, 2117
173. Buist, G. J., Bunton, C. A. and Hipperson, W. C. P. (1971). *J. Chem. Soc. B*, 2128
174. Konaka, R. and Kuruma, K. (1971). *J. Org. Chem.*, **36**, 1703
175. Konaka, R., Terabe, S. and Kuruma, K. (1969). *J. Org. Chem.*, **34**, 1334
176. Rigby, W. (1950). *J. Chem. Soc.*, 1907
177. Berka, A. (1970). *Arch. Pharm. (Weinheim)*, **303**, 233
178. Kurlyankina, V. I., Sarana, N. V. and Koz'mina, O. P. (1970). *Kinet. Katal.*, **11**, 1159
179. Kurlyankina, V. I., Sarana, N. V. and Koz'mina, O. P. (1971). *Zh. Obshch. Khim.*, **41**, 1315
180. Wells, C. F. and Husain, M. (1971). *Trans. Faraday Soc.*, **67**, 1086
181. Nath, P., Banerji, K. K. and Bakore, G. V. (1970). *Indian J. Chem.*, **8**, 1113
182. Nath, P., Banerji, K. K. and Bakore, G. V. (1971). *J. Indian Chem. Soc.*, **48**, 17
183. Meenakshi, A. and Santappa, M. (1970). *J. Catal.*, **19**, 300
184. Beggs, J. J. and Meyers, M. B. (1970). *J. Chem. Soc. B*, 930
185. Gros, G. and Giral, L. (1970). *Bull. Soc. Chim. Fr.*, 1115
186. Collins, C. J. (1960). *Quart. Rev. Chem. Soc.*, **14**, 357
187. Dana, G. and Wiemann, J. (1970). *Bull. Soc. Chim. Fr.*, 3994
188. Freidlin, L. Kh., Sharf, V. Z., Bartok, M. and Nazaryan, A. A. (1970). *Izv. Akad. Nauk SSSR, Ser. Khim.*, 310
189. Sharf, V. Z., Freidlin, L. Kh. and Nazaryan, A. A. (1970). *Izv. Akad. Nauk SSSR, Ser. Khim.*, 597
190. Mazet, M. (1969). *Bull. Soc. Chim. Fr.*, 4309
191. Yvernault, T. and Mazet, M. (1970). *Compt. Rend. Acad. Sci., Ser. C*, **270**, 430
192. Yvernault, T. and Mazet, M. (1971). *Bull. Soc. Chim. Fr.*, 2652
193. Mazet, M. and Brut, M. (1970). *Compt. Rend. Acad. Sci., Ser. C*, **271**, 848
194. Mazet, M. and Desmaison-Brut, M. (1971). *Bull. Soc. Chim. Fr.*, 2656
195. Lebouc, A. (1971). *Bull. Soc. Chim. Fr.*, 3037
196. Brooks, C. J. W. and Maclean, I. (1971). *J. Chromatogr. Sci.*, **9**, 18
197. Eisenberg, F. (1971). *Carbohyd. Res.*, **19**, 135
198. Wood, P. J. and Siddiqui, I. R. (1971). *Carbohyd. Res.*, **19**, 283
199. McKinley, I. R. and Weigel, H. (1970). *Chem. Commun.*, 1002; (1972), 1051
200. Pfeiffer, F. R., Cohen, S. R., Williams, K. R. and Weisbach, J. A. (1968). *Tetrahedron Letters*, 3549
201. Rakhit, S., Bagli, J. F. and Deghenghi, R. (1969). *Can. J. Chem.*, **47**, 2906
202. Pfeiffer, F. R., Miao, C. K. and Weisbach, J. A. (1970). *J. Org. Chem.*, **35**, 221
203. Zhelvakova, É. G., 'Magnashevskii, V. A., Ermakova, L. I., Shvets, V. I. and Preobrazhenskii, N. A. (1970). *Zh. Org. Khim.*, **6**, 1987
204. Shvets, V. I., Kabanova, M. A., Alekseeva, L. D. and Preobrazhenskii, N. A. (1971). *Zh. Org. Khim.*, **7**, 2093
205. Zhelvakova, É. G., Smirnova, G. V., Shvets, V. I. and Preobrazhenskii, N. A. (1970). *Zh. Org. Khim.*, **6**, 1992

206. Kabanova, M. A. and Shvets, V. I. (1971). *Zh. Org. Khim.*, **7**, 1310 —
207. Bentley, P. H. and McCrae, W. (1970). *J. Org. Chem.*, **35**, 2082
208. Carreau, J. P. (1970). *Bull. Soc. Chim. Fr.*, 4104, 4107, 4111
209. (1967). *The Chemistry of the Ether Linkage* (S. Patai, editor). (London: Interscience)
210. Barton, D. H. R., Page, G. and Widdowson, D. A. (1970). *Chem. Commun.*, 1466
211. Dombroski, J. R. and Schuerch, C. (1970). *Macromolecules*, **3**, 257
212. Watanabe, W. H. and Conlon, L. E. (1957). *J. Amer. Chem. Soc.*, **79**, 2828
213. Yuki, H., Hatada, K. and Nagata, K. (1970). *Bull. Chem. Soc. Jap.*, **43**, 1817
214. Shostakovskii, M. F., Atavin, A. S., Trofimov, B. A., Gusarov, A. V., Nikitin, V. M. and Skorobogatova, V. I. (1970). *Zh. Obshch. Khim.*, **40**, 70
215. Atavin, A. S., Zorina, É. F. and Mirskova, A. N. (1971). *Zh. Org. Khim.*, **7**, 229
216. Perchard, J.-P. (1970). *Spectrochim. Acta*, **A26**, 707
217. Perchard, J.-P., Monier, J. C. and Dizabo, P. (1971). *Spectrochim. Acta*, **A27**, 447
218. Allan, A., McKean, D. C., Perchard, J.-P. and Josien, M.-L. (1971). *Spectrochim. Acta*, **A27**, 1409
219. Lasseques, J. C., Grenie, Y. and Forel, M. T. (1970). *Compt. Rend. Acad. Sci., Ser. B*, **271**, 421
220. Perchard, J.-P. (1970). *J. Molec. Struct.*, **6**, 359
221. Perchard, J.-P. (1970). *J. Molec. Struct.*, **6**, 457
222. Iwamoto, R. (1971). *Spectrochim. Acta*, **A27**, 2385
223. Bernasek, S. L. and Cooks, R. G. (1970). *Org. Mass Spectrom.*, **3**, 127
224. Tsang, C. W. and Harrison, A. G. (1970). *Org. Mass Spectrom.*, **3**, 647
225. Lehman, T. A., Elwood, T. A., Bursey, J. T., Bursey, M. M. and Beauchamp, J. L. (1971). *J. Amer. Chem. Soc.*, **93**, 2108
226. Morizur, J.-P. and Djerassi, C. (1971). *Org. Mass Spectrom.*, **5**, 895
227. Derevitskaya, V. A., Klimov, E. M. and Kochetkov, N. K. (1970). *Tetrahedron Letters*, 4269
228. Guenzet, J., El Khatib, M. and Guenzet, N. (1971). *Compt. Rend. Acad. Sci., Ser. C*, **273**, 72
229. Alper, H. and Edward, J. T. (1970). *Can. J. Chem.*, **48**, 1623
230. Karger, M. H. and Mazur, Y. (1971). *J. Org. Chem.*, **36**, 532
231. Sargent, G. D. (1971). *J. Amer. Chem. Soc.*, **93**, 5268
232. Kariyone, K. and Yazawa, H. (1970). *Tetrahedron Letters*, 2885
233. Juenge, E. C. and Beal, D. A. (1968). *Tetrahedron Letters*, 5819
234. Juenge, E. C., Corey, M. D. and Beal, D. A. (1971). *Tetrahedron*, **27**, 2671
235. Berkowitz, L. M. and Rylander, P. N. (1958). *J. Amer. Chem. Soc.*, **80**, 6682
236. Hata, Go, Takahashi, K. and Miyake, A. (1970). *Chem. Commun.*, 1392
237. Rautenstrauch, V. (1970). *Chem. Commun. C*, 4
238. Rosowsky, A. (1964). *Heterocyclic Compounds with Three- and Four-membered Rings*, Part 1, 1 (A. Weissberger, editor). (New York: Interscience Publishers)
239. Hall, R. T. and Mair, R. D. (1971). *Treatise Anal. Chem.*, **14**, 259
240. Meleshevich, A. P. (1970). *Usp. Khim.*, **39**, 444
241. Yandovskii, V. N., Karavan, V. S. and Temnikova, T. I. (1970). *Usp. Khim.*, **39**, 571
242. Swern, D. (1970). *J. Amer. Oil Chem. Soc.*, **47**, 424
243. McDonald, R. N. (1971). *Mechanisms of Molecular Migrations*, Vol. 3, 67. (B. S. Thyagarajan, editor). (New York: Wiley Interscience)
244. Kruse, W. and Bednarski, T. M. (1971). *J. Org. Chem.*, **36**, 1154
245. Matsumura, N., Sonoda, N. and Tsutsumi, S. (1970). *Tetrahedron Letters*, 2029
246. Carlson, R. G., Behn, N. S. and Cowles, C. (1971). *J. Org. Chem.*, **36**, 3832
247. Curci, R., DiPrete, R. A., Edwards, J. O. and Modena, G. (1970). *J. Org. Chem.*, **35**, 740
248. Kavčič, R. and Plesničar, B. (1970). *J. Org. Chem.*, **35**, 2033
249. Rouchaud, J. and De Pauw, M. (1970). *Bull. Soc. Chim. Fr.*, 2905 and following papers
250. Rouchaud, J. (1971). *Bull. Soc. Chim. Fr.*, 1189
251. Cornforth, R. H. (1970). *J. Chem. Soc. C*, 928
252. Cornforth, J. W. and Green, D. T. (1970). *J. Chem. Soc. C*, 846
253. Bertini, F., Grasselli, P., Zubiani, G. and Cainelli, G. (1970). *Chem. Commun.*, 144
254. Kupchan, S. M. and Maruyama, M. (1971). *J. Org. Chem.*, **36**, 1187
255. Kochi, J. K., Singleton, D. M. and Andrews, L. J. (1968). *Tetrahedron*, **24**, 3503
256. Plonka, J. H. and Skell, P. S. (1970). *Chem. Commun.*, 1108

257. Lalancette, J. M. and Frèche, A. (1971). *Can. J. Chem.*, **49**, 4047
258. Kaiser, E. M., Edmonds, C. G., Grubb, S. D., Smith, J. W. and Tramp, D. (1971). *J. Org. Chem.*, **36**, 330
259. Sénéchal, G. and Cornet, D. (1971). *Bull. Soc. Chim. Fr.*, 773
260. Sénéchal, G., Duchet, J.-C. and Cornet, D. (1971). *Bull. Soc. Chim. Fr.*, 783
261. Blackett, B. N., Coxon, J. M., Hartshorn, M. P. and Richards, K. E. (1970). *Aust. J. Chem.*, **23**, 839
262. Blackett, B. N., Coxon, J. M., Hartshorn, M. P. and Richards, K. E. (1970). *J. Amer. Chem. Soc.*, **92**, 2574
263. Blackett, B. N., Coxon, J. M., Hartshorn, M. P. and Richards, K. E. (1970). *Aust. J. Chem.*, **23**, 2077
264. Blackett, B. N., Coxon, J. M., Hartshorn, M. P., Lewis, A. J., Little, G. R. and Wright, G. J. (1970). *Tetrahedron*, **26**, 1311
265. Yandovskii, V. N. and Temnikova, T. I. (1968). *Zh. Org. Khim.*, **4**, 1758
266. Rickborn, B. and Gerkin, R. M. (1971). *J. Amer. Chem. Soc.*, **93**, 1693
267. Kennedy, J. H. and Buse, C. (1971). *J. Org. Chem.*, **36**, 3135
268. Parker, R. E. and Isaacs, N. S. (1959). *Chem. Rev.*, **59**, 737
269. Staude, E. and Patat, F. (1967). Ref. 209, 21
270. Gritter, R. J. (1967). Ref. 209, 373
271. Sekiguchi, S., Matsui, K. and Yasuraoka, Y. (1970). *Bull. Chem. Soc. Jap.*, **43**, 2523
272. Novák, J. and Antošová, J. (1970). *Collection Czech. Chem. Commun.*, **35**, 1096
273. Biggs, J., Chapman, N. B. and Wray, V. (1971). *J. Chem. Soc. B*, 66
274. Biggs, J., Chapman, N. B., Finch, A. F. and Wray, V. (1971). *J. Chem. Soc. B*, 55
275. Biggs, J., Chapman, N. B. and Wray, V. (1971). *J. Chem. Soc. B*, 63
276. Biggs, J., Chapman, N. B. and Wray, V. (1971). *J. Chem. Soc. B*, 71
277. Dupin, C. and Dupin, J.-F. (1970). *Bull. Soc. Chim. Fr.*, 249
278. Robert, A. and Foucaud, A. (1970). *Bull. Soc. Chim. Fr.*, 212
279. Khazemova, L. A. and Al'bitskaya, V. M. (1970). *Zh. Org. Khim.*, **6**, 935
280. Peterson, P. E., Indelicato, J. M. and Bonazza, B. R. (1971). *Tetrahedron Letters*, 13
281. Herr, R. W., Wieland, D. M. and Johnson, C. R. (1970). *J. Amer. Chem. Soc.*, **92**, 3813
282. Herr, R. W. and Johnson, C. R. (1970). *J. Amer. Chem. Soc.*, **92**, 4979
283. Anderson, R. J. (1970). *J. Amer. Chem. Soc.*, **92**, 4978
284. Lundeen, A. J. and Oehlschlager, A. C. (1970). *J. Organometal. Chem.*, **25**, 337
285. Namy, J.-L., Boireau, G. and Abenhaim, D. (1971). *Bull. Soc. Chim. Fr.*, 3191
286. Namy, J.-L. (1971). *Compt. Rend. Acad. Sci., Ser. C*, **272**, 500
287. Namy, J.-L., Abenhaim, D. and Boireau, G. (1971). *Bull. Soc. Chim. Fr.*, 2943
288. Felkin, H., Frajerman, C. and Roussi, G. (1970). *Bull. Soc. Chim. Fr.*, 3704
289. Morrison, J. D., Atkins, R. L. and Tomaszewski, J. E. (1970). *Tetrahedron Letters*, 4635
290. Abenhaim, D., Namy, J.-L. and Boireau, G. (1971). *Bull. Soc. Chim. Fr.*, 3254
291. Iwamoto, N., Ninagawa, A., Matsuda, H. and Matsuda, M. (1971). *Kogyo Kagaku Zasshi*, **74**, 1400
292. Lahournère, J.-C. and Valade, J. (1971). *J. Organometal. Chem.*, **33**, C4
293. Koketsu, J. and Ishii, Y. (1971). *J. Chem. Soc. C*, 2
294. Dunogues, J., Calas, R., Duffaut, N. and Picard, J.-P. (1971). *J. Organometal. Chem.*, **26**, C13
295. Cantacuzène, J. and Normant, J.-M. (1970). *Tetrahedron Letters*, 2947
296. Reines, S. A., Griffith, J. R. and O'Rear, J. G. (1970). *J. Org. Chem.*, **35**, 2772
297. Reines, S. A., Griffith, J. R. and O'Rear, J. G. (1971). *J. Org. Chem.*, **36**, 1209
298. Ansell, M. F., Shepherd, I. S. and Weedon, B. C. L. (1971). *J. Chem. Soc. C*, 1840
299. Thakore, A. N., Pope, P. and Oehlschlager, A. C. (1971). *Tetrahedron*, **27**, 2617
300. Movsumzade, M. M., Shabanov, A. L., Movsumzade, S. M. and Gurbanov, P. A. (1971). *Zh. Org. Khim.*, **7**, 412, 1106
301. Shabanov, A. L., Movsumzade, M. M., Movsumzade, S. M. and Gurbanov, P. A. (1971). *Zh. Org. Khim.*, **7**, 1109
302. Movsumzade, M. M., Shabanov, A. L., Gurbanov, P. A. and Movsumzade, S. M. (1971). *Zh. Org. Khim.*, **7**, 1373
303. Stogryn, E. L. and Gianni, M. H. (1970). *Tetrahedron Letters*, 3025
304. (1970). *Organic Peroxides*, Vol. 1. (D. Swern, editor). (New York: Wiley–Interscience)
305. (1971). *Organic Peroxides*, Vol. 2. (D. Swern, editor). (New York: Wiley–Interscience)
306. Hiatt, R. R. (1971). Ref. 305, 1

307. Johnson, R. M. and Siddiqi, I. W. (1970). *The Determination of Organic Peroxides.* (Oxford: Pergamon Press)
308. Hiatt, R. R. (1971). *Oxidation: Techniques and Applications in Organic Synthesis,* Vol. 2, 113. (R. L. Augustine and D. J. Trecker, editors). (New York: Marcel Dekker)
309. Doumaux, A. R. (1971). *Oxidation: Techniques and Applications in Organic Synthesis,* Vol. 2, 141. (R. L. Augustine and D. J. Trecker, editors). (New York: Marcel Dekker)
310. Hawkins, E. G. E. (1972). *MTP International Review of Science, Organic Chemistry,* Series One, Vol. 10. (W. A. Waters, editor). (London: Butterworths)
311. Brown, H. C. and Midland, M. M. (1971). *J. Amer. Chem. Soc.,* **93,** 4078
312. Bernstein, P. A., Hohorst, F. A. and Des Marteau, D. D. (1971). *J. Amer. Chem. Soc.,* **93,** 3882
313. Ratcliffe, C. T., Hardin, C. V., Anderson, L. R. and Fox, W. B. (1971). *Chem. Commun.,* 784
314. Matsui, Y., Naruse, H., Mochida, K. and Date, Y. (1970). *Bull. Chem. Soc. Jap.,* **43,** 1909, 1910
315. Richardson, W. H. and Hodge, V. F. (1970). *J. Org. Chem.,* **35,** 4012
316. Kato, S., Ishihara, T. and Masuo, F. (1970). *Bull. Jap. Petrol. Inst.,* **12,** 117
317. Richardson, W. H. and Hodge, V. F. (1971). *J. Amer. Chem. Soc.,* **93,** 3996
318. Sokolov, N. A., Usov, L. G. and Shushunov, V. A. (1970). *Zh. Obshch. Khim.,* **40,** 209
319. Lapshin, N. M., Erykalova, M. F., Nikulina, N. P. and Sokolov, N. A. (1970). *Zh. Org. Khim.,* **6,** 2023
320. Hiatt, R. R., Smythe, R. J. and McColeman, C. (1971). *Can. J. Chem.,* **49,** 1707
321. Lapshin, N. M., Erykalova, M. F., Kondrat'eva, K. A. and Aidakin, A. T. (1970). *Zh. Org. Khim.,* **6,** 2026
322. Deno, N. C., Billups, W. E., Kramer, K. E. and Lastomirsky, R. R. (1970). *J. Org. Chem.,* **35,** 3080
323. Turner, J. O. (1971). *Tetrahedron Letters,* 887
324. Farberov, M. I., Stozhkova, G. A., Bondarenko, A. V. and Glusker, A. L. (1970). *Neftekhimiya,* **10,** 218
325. Farberov, M. I., Stozhkova, G. A., Bondarenko, A. V., Kirik, T. M. and Ognevskaya, N. A. (1971). *Neftekhimiya,* **11,** 404
326. Kaloustian, J., Lena, L. and Metzger, J. (1971). *Bull. Soc. Chim. Fr.,* 4415
327. Obukhov, V. M., Farberov, M. I., Bondarenko, A. V. and Lysanov, V. A. (1971). *Neftekhimiya,* **11,** 410
328. Kryukov, S. I., Simanov, N. A. and Farberov, M. I. (1971). *Neftekhimiya,* **11,** 224
329. Howe, G. R. and Hiatt, R. R. (1971). *J. Org. Chem.,* **36,** 2493
330. Sheng, M. N. and Zajacek, J. G. (1970). *J. Org. Chem.,* **35,** 1839
331. Gould, E. S., Hiatt, R. R. and Irwin, K. C. (1968). *J. Amer. Chem. Soc.,* **90,** 4573
332. Mimoun, H., Seree de Roch, I. and Sajus, L. (1970). *Tetrahedron,* **26,** 37
333. Sharpless, K. B., Townsend, J. M. and Williams, D. R. (1972). *J. Amer. Chem. Soc.,* **94,** 295
334. Mimoun, H., Seree de Roch, I., Sajus, L. and Menguy, P. (1968). *Fr. Pat.* 1 549 184
335. Howe, G. R. and Hiatt, R. R. (1970). *J. Org. Chem.,* **35,** 4007
336. Hiatt, R. R. and McColeman, C. (1971). *Can. J. Chem.,* **49,** 1712
337. Tolstikov, G. A., Dzhemilev, U. M., Novitskaya, N. N., Yur'ev, V. P. and Kantynkova, R. G. (1971). *Zh. Obshch. Khim.,* **41,** 1883
338. Doumaux, A. R. and Trecker, D. J. (1970). *J. Org. Chem.,* **35,** 2121
339. Mair, R. D. and Hall, R. T. (1971). *Treatise Analyt. Chem.,* **14,** 295
340. Pobedimskii, D. G. (1971). *Usp. Khim.,* **40,** 254
341. Schmitz, E., Rieche, A. and Brede, O. (1970). *J. Prakt. Chem.,* **312,** 30
342. Ballard, D. H. and Bloodworth, A. J. (1971). *J. Chem. Soc. C,* 945
343. Bloodworth, A. J. and Bunce, R. J. (1971). *J. Chem. Soc. C,* 1453
344. Kropf, H., Bernert, C.-R., Lütjens, J., Pavicic, V. and Weiss, T. (1970). *Tetrahedron,* **26,** 1347
345. Petrovskaya, G. A., Vilenskaya, M. R., Mamchur, L. P. and Yurzhenko, T. I. (1970). *Zh. Org. Khim.,* **6,** 2231
346. Gould, D. E. Ratcliffe, C. T. Anderson, L. R. and Fox, W. B. (1970). *Chem. Commun.,* 216
347. Cubbon, R. C. P. (1970). *Progress in Reaction Kinetics,* Vol. 5, 29. (G. Porter, editor). (Oxford: Pergamon)

348. Bell, E. R., Rust, F. F. and Vaughan, W. E. (1950). *J. Amer. Chem. Soc.*, **72,** 337
349. Huyser, E. S. and Jankauskas, K. J. (1970). *J. Org. Chem.*, **35,** 3196
350. Huyser, E. S. and Bredeweg, C. J. (1964). *J. Amer. Chem. Soc.*, **86,** 2401
351. Huyser, E. S., Bredeweg, C. J. and VanScoy, R. M. (1964). *J. Amer. Chem. Soc.*, **86,** 4148
352. Huang, R. L., Lee, T.-W. and Ong, S. H. (1969). *J. Chem. Soc. C,* 2522
353. Huyser, E. S. and Kahl, A. A. (1970). *J. Org. Chem.*, **35,** 3742
354. Goh, S. H. and Ong, S. H. (1970). *J. Chem. Soc. B,* 870
355. Goh, S. H., Huang, R. L., Ong, S. H. and Sieh, I. (1971). *J. Chem. Soc. C,* 2282
356. Walling, C. and Waits, H. P. (1967). *J. Phys. Chem.,* **71,** 2361
357. Huyser, E. S. and VanScoy, R. M. (1968). *J. Org. Chem.,* **33,** 3524
358. Walling, C. and Bristol, D. (1971). *J. Org. Chem.,* **36,** 733
359. Hiatt, R. R. and Szilagyi, S. (1970). *Can. J. Chem.,* **48,** 615
360. Christe, K. O. and Pilopovich, D. (1971). *J. Amer. Chem. Soc.,* **93,** 51
361. Christe, K. O. (1971). *Spectrochim. Acta,* **A27,** 463
362. Fraser, R. T. M., Paul, N. C. and Phillips, L. (1970). *J. Chem. Soc. B,* 1278
363. Pabiot, J. M. and Pallaud, R. (1970). *Compt. Rend. Acad. Sci., Ser. C,* **270,** 334
364. Jarvie, A. W. P. and Skelton, D. (1971). *J. Organometal. Chem.,* **30,** 145
365. Razuvaev, G. A., Dodonov, V. A. and Zaburdyaeva, S. N. (1970). *Zh. Org. Khim.,* **6,** 657

4
Nitrogen Compounds

PELHAM WILDER, Jr. and J. M. SHEPHERD
Duke University, Durham, N. Carolina

4.1 INTRODUCTION

The purpose of this article is to survey and organise the extensive literature describing the chemistry of aliphatic nitrogen compounds which has appeared in the years 1970 and 1971. For convenience the material has been divided into seven sections: (1) nitriles, (2) isonitriles, (3) nitro compounds, (4) nitroso compounds, (5) oximes, (6) hydrazines and (7) amines. In each section both synthetic methods and reactions are described and, where appropriate, mechanistic and stereochemical details are included. In general an historical development of sections or topics has not been attempted, but references to significant research in the earlier literature are cited occasionally. Also, where necessary for illustration, areas outside the strict realm of aliphatic compounds have been entered.

4.2 CYANIDES

4.2.1 Preparation

4.2.1.1 *Saturated compounds*

A unique preparation of nitriles, mainly aryl but also benzyl, from secondary aroyl amides in yields of up to 90 % by use of chlorotris-(triphenylphosphine)-rhodium, $RhCl(PPh_3)_3$ has been reported[1].

$$Ar CO \cdot NHCH_2 R \longrightarrow Ar CN + RCH_2 OH$$
(Ar = p-tolyl-, m- or p-halogeno phenyl, R = H, Me or Ph)

The reaction may be formulated as a simple elimination of a primary alcohol. N-Benzyl- (or substituted benzyl-)aroylamides give the nitriles derived from both the aryl and the N-benzyl group. Primary aryl amides are also smoothly converted into the corresponding nitrile by catalytic amounts of the rhodium complex; however, negative results were obtained with NN-dialkylamides and N-arylamides.

The use of silicon-containing compounds (silazanes, aminosilanes, alkoxysilanes, and chlorosilanes) in the dehydration of amides provides another convenient method for the preparation of nitriles[2]. Acetonitrile was isolated in excellent yield (85 %) from the reaction of acetamide with hexa-methylcyclotrisilazane, $(HNSiMe_2)_3$. Benzamide and N-methylbenzamide each gave benzonitrile (85 %, 43 %). The nitriles if volatile can easily be distilled from the siloxane polymer that is formed, or if solid they can be filtered off.

Aldehydes are converted into nitriles under extremely mild conditions in a two-step sequence involving (i) an essentially quantitative reaction between the aldehyde and 1,1-diphenylhydrazine to give the corresponding diphenyl-hydrazone, followed by (ii) irradiation under oxygen to yield the nitrile (40–75 %)[3].

The instability and tendency to easy cyclisation of the N-unsubstituted imines (from the reactions of aldehydes with ammonia) has been overcome

$$RCHO + NH_2 \cdot NPh_2 \longrightarrow RCH{:}N \cdot NPh_2 \xrightarrow[O_2]{hv} RCN +$$

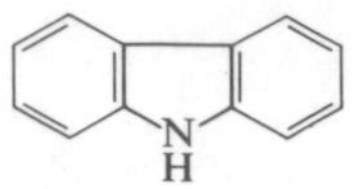

$$(R = PhCH{:}CH\cdot,\ PhCH_2\cdot,\ 1\text{-}C_{10}H_7\cdot,\ p\text{-}MeO\cdot C_6H_4\cdot,\ Ph)$$

by the use of *N*-unsubstituted-imine cobalt complexes[4]. These complexes are prepared by the reaction between the aldehyde and a cobalt hexa-ammine complex. When oxidised with bromine, the imine complex afforded the corresponding nitrile in yields of 50 to 75%.

$$RCHO + Co(NH_3)_6X_2 \rightarrow \left[\begin{array}{c} (RCH{=}NH)_n \\ \downarrow \\ \overset{\cdot}{Co}(NH_3)_{6-n} \end{array} \right] X_2 \xrightarrow{Br_2} RCN$$

$$(R = Ph, PhCH{:}CH\cdot, \text{ or } PhCH_2 \cdot CH_2\cdot)$$

Thermolysis of a series of aldehyde hydrazonium salts at about 200 °C afforded nitriles in high yield[5].

$$CN \cdot (CH_2)_2 \cdot CMe_2 \cdot CHO + H_2N \cdot NZ \rightarrow CN \cdot (CH_2)_2 \cdot CMe_2 \cdot CH{:}N \cdot NZ$$

$$[Z = (CH_2)_{4,\,5,\,6} \text{ or } (CH_2)_2 \cdot O \cdot (CH_2)_2] \downarrow RX \ (R = Me, Et; X = I, BF_4)$$

$$CN \cdot (CH_2)_2 \cdot CMe_2 \cdot CN \xleftarrow{\Delta} CN \cdot (CH_2)_2 \cdot CMe_2 \cdot CH{:}N \cdot \overset{+}{N}ZR \quad \overset{-}{X}$$

With iodides, tar formation was negligible and higher yields of the nitrile were obtained than with tetrafluoroborates.

Occasionally the carboxylic acid–nitrile exchange reaction has been employed in the direct synthesis of nitriles from carboxylic acids by reaction with acetonitrile at high temperatures. The equilibrium proposed is displaced in the direction of the weaker carboxylic acid.

$$R^1CO_2H + R^2CN \rightleftharpoons R^1CO \cdot NH \cdot COR^2 \rightleftharpoons R^1CN + R^2CO_2H$$

Replacement of acetonitrile by short-chain dinitriles has the advantage that once exchanged, the short-chain cyano-acid undergoes cyclisation to the imide and is removed from the equilibrium mixture, thus driving the reaction to completion. Reaction of 1,12-dodecanedioic acid with 2 mol. equivalents of α-methylglutaronitrile afforded a 98% recovery of the cyclic imide and a 97% yield of 1,10-dicyanodecane[6].

$$HO_2C \cdot (CH_2)_{10} \cdot CO_2H + 2CH_2CN \cdot CH_2 \cdot CHMe \cdot CN \xrightarrow{H_3PO_4}$$

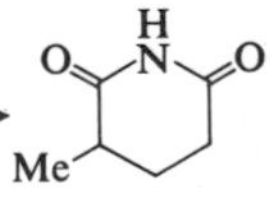

$$+ \ CN \cdot (CH_2)_{10} \cdot CN$$

4.2.1.2 *Unsaturated compounds*

A convenient modification of the Wittig reaction involves the reaction of dibromoacetonitrile with triphenylphosphine to yield the bis-phosphonium salt[7]. When the salt and an aldehyde were heated under reflux with sodium

$$Br_2CH \cdot CN + PPh_3 \xrightarrow{\text{benzene, } \Delta} Ph_3 \overset{+}{P}CH(CN)\overset{+}{P}Ph_3 2\bar{B}r$$

$$\xrightarrow[\substack{\text{ArCHO} \\ \text{or RCHO}}]{\text{NaOH}} \begin{array}{l} ArCH\!:\!CH \cdot CN + 2Ph_3P(\!:\!O) \\ \text{or } RCH\!:\!CH \cdot CN \end{array}$$

hydroxide in a benzene–water solvent, the unsaturated nitrile was obtained in good yield (40–90%), along with triphenylphosphine oxide. The aliphatic aldehyde, 2-ethylhexanal, afforded only the isomerised β,γ-unsaturated nitrile, a well investigated phenomenon in such systems[8]. Another interesting and practical synthesis of unsaturated nitriles is that of substituted α-cyanocinnamides, which are obtained by refluxing in ethanol 1-aryl-2-carbamoyl-2-cyanoethyltriphenylphosphonium chlorides. These salts were formed by reaction of arylmethylenemalononitriles with triphenylphosphine in the presence of hydrochloric acid[9].

$$Ph_3P + XC_6H_4CH\!:\!C(CN)_2 \xrightarrow{HCl/H_2O}$$

$$Ph_3\overset{+}{P}CH(C_6H_4X)\cdot CH(CN)\cdot CO\cdot NH_2 \; Cl^-$$

$$\downarrow$$

$$Ph_3P + HCl + XC_6H_4 \cdot CH\!:\!C(CN)\cdot CO\cdot NH_2$$

Heating N-chlorosulphonyl-β-lactams with NN-dimethylformamide (DMF) afforded unsaturated nitriles in yields of 20–75%[10]. Specifically, it is reported that thermal cleavage of the trisubstituted β-lactam yields a carbonium ion from which a mixture of nitriles was derived. Generally the α,β-

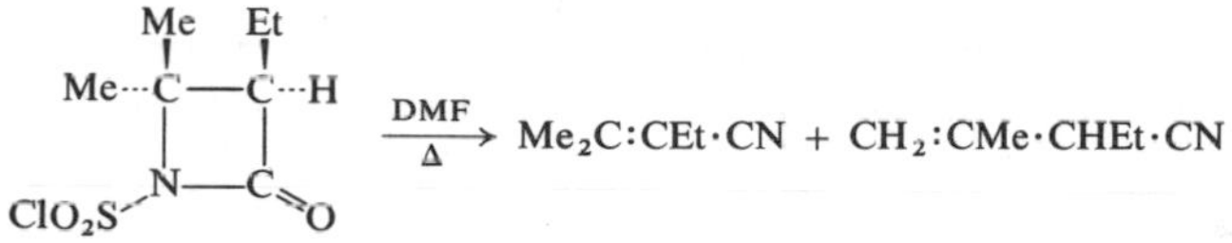

$$\xrightarrow[\Delta]{\text{DMF}} Me_2C\!:\!CEt \cdot CN + CH_2\!:\!CMe \cdot CHEt \cdot CN$$

unsaturated nitrile predominated, especially when the conjugated nitrile was also the more highly substituted alkene. The relatively high acidity of the methine proton may be critical in the deprotonation to the nitrile.

Trialkyl phosphites react with 2,2,3-trihalogenopropionitriles to give the 2-halogenoacrylonitriles in almost quantitative yield[11]. The results have

$$XCH_2 \cdot CX \cdot CN \longrightarrow XCH_2 \cdot C \cdot CN \longrightarrow CH_2\!:\!CX \cdot CN + \bar{X}$$
$$\quad\; | \qquad\qquad\qquad\qquad\quad$$
$$\quad\; X \qquad :P(OR)_3 \qquad\qquad +X\overset{+}{P}(OR)_3$$

$$\bar{X} + X\overset{+}{P}(OR)_3 \longrightarrow RX + (RO)_2P(\!:\!O)\cdot X$$

(R = Et or Me, X = Cl or Br)

been explained by nucleophilic attack by phosphorus on the α-halogen followed by elimination of the β-halogen from the resulting carbanion.

Chlorocyanoacetylene (ClC⋮C·CN, 80%) and cyanoacetylene (H·C⋮C·CN, 40–60%) have been prepared in good yields by thermolysis of trichloropropionitrile and of dichloropropionitrile at temperatures approaching

1000 °C[12]. Malononitrile was prepared by reaction of the chlorocyanoacetylene with ammonia (60–87 %)[12].

4.2.1.3 Cyanation of steroidal compounds

Hydrocyanation of α,β-unsaturated aldehydes by conventional methods leads to 1,2-adducts only. However, treatment of allylidenenanimes with hydrogen cyanide–trialkylaluminium followed by hydrolysis of the resulting 1,3-dicyanopropylamines gave β-cyanoaldehydes in good yields[13]. Another example, that of a Δ^2-2-cyclohexyliminomethyl-17β-hydroxysteroid, gave

$$R^1N{:}CH{\cdot}CR^2{:}CR^3R^4 \xrightarrow{\text{HCN-AlR}_3^5} R^1NH{\cdot}CH(CN){\cdot}CHR^2{\cdot}CR^3R^4{\cdot}CN$$

$$\xrightarrow{\text{(CO}_2\text{H)}_2} CHO{\cdot}CHR^2{\cdot}CR^3R^4{\cdot}CN$$

the dicyanoamine (obtained as its hydrate) in 96 % yield when treated with hydrogen-cyanide–triethylaluminium in tetrahydrofuran. Subsequent heating of the hydrocyanation product with 5 % oxalic acid in tetrahydrofuran–

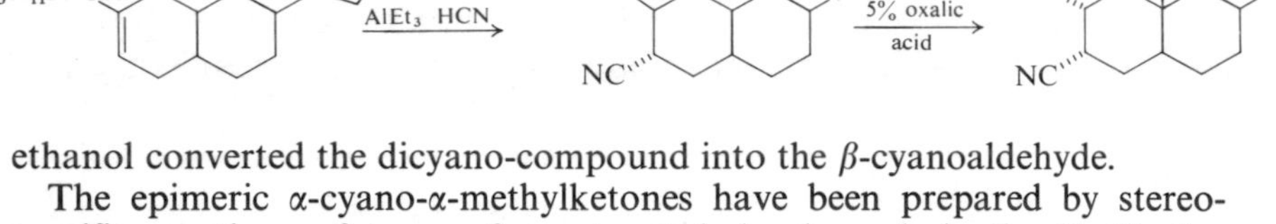

ethanol converted the dicyano-compound into the β-cyanoaldehyde.

The epimeric α-cyano-α-methylketones have been prepared by stereo-specific reactions of trans-α,β-unsaturated decalone derivatives[14]. Similar

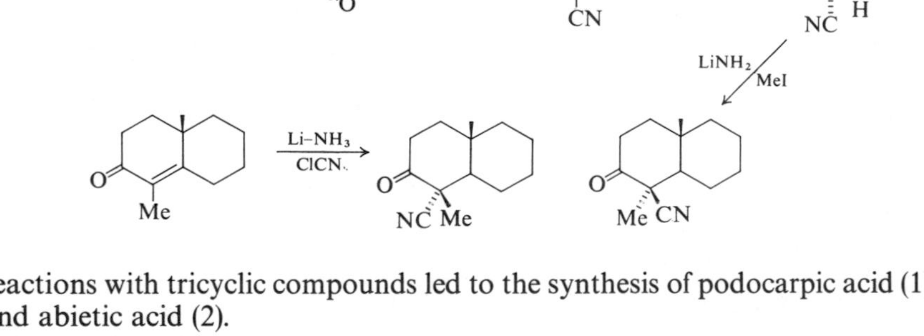

reactions with tricyclic compounds led to the synthesis of podocarpic acid (1) and abietic acid (2).

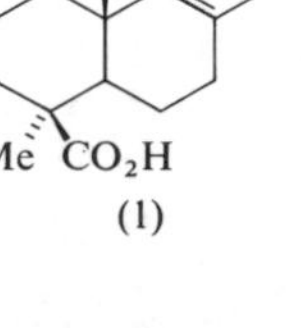

(1) 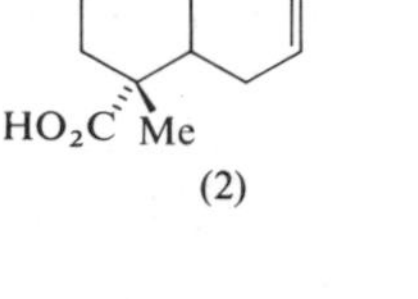　　　　　　　　　　(2)

The methylation of these keto-nitriles to give the equatorial alkylation products is in contrast to the alkylation of the analogous β-keto-esters for which axial alkylation is preferred[15].

Methylation of 2-carbomethoxy- and 2-cyano-4-t-butylcyclohexanone gave predominately axial alkylation products, but for 3-carbomethoxy- and 3-cyano-10-methyl-*trans*-decal-2-one, primarily equatorial alkylation was observed[16].

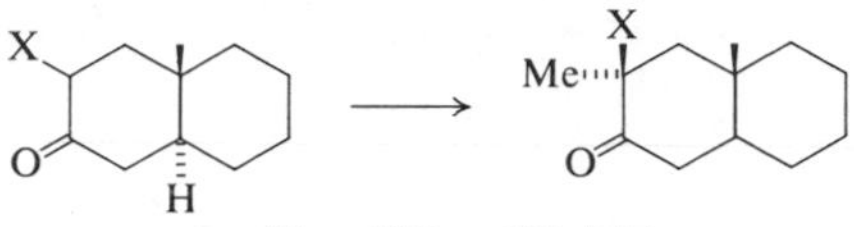

(X = CN or CO$_2$Me)

Mild cleavage of steroidal epoxides with hydrogen cyanide–triethyl-aluminium and with diethylaluminium cyanide in most cases gave *trans*-diaxial β-cyanohydrins[17]. Even a 9α,11α-epoxide, a class of epoxides known to resist cleavage by lithium aluminium hydride or hydrogen halides[18], gave the 11β-cyano-9α-hydroxy-steroid in 50% yield by use of diethylaluminium cyanide.

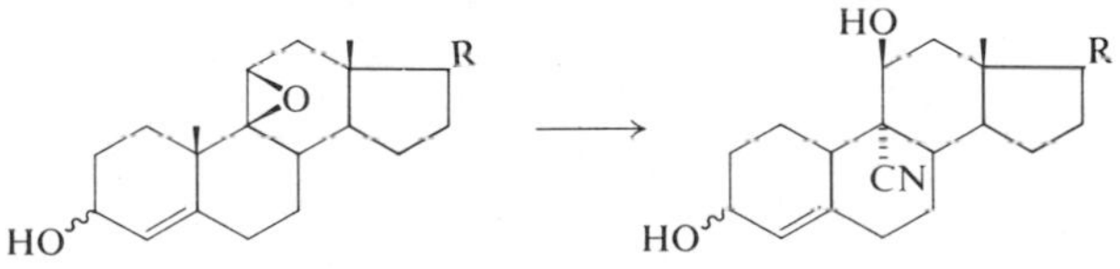

4.2.2 Reactions

4.2.2.1 At an α-carbon atom

Reaction of an alkali metal amide and benzyl cyanide with ethylene di-iodide yields a 'dimer', whereas ethylene dibromide and ethylene dichloride give rise to twofold alkylation, rather than dimerisation[19].

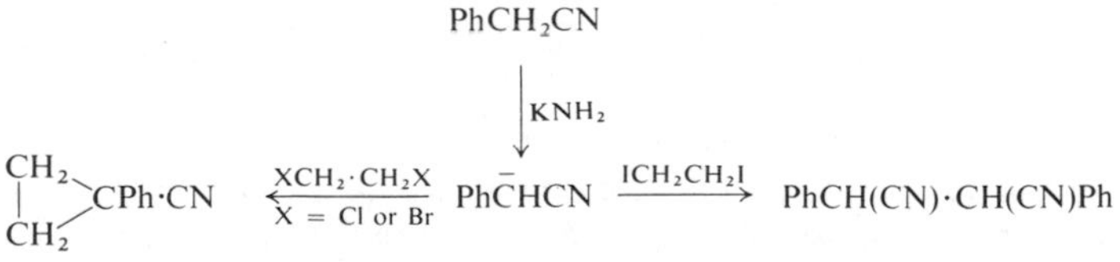

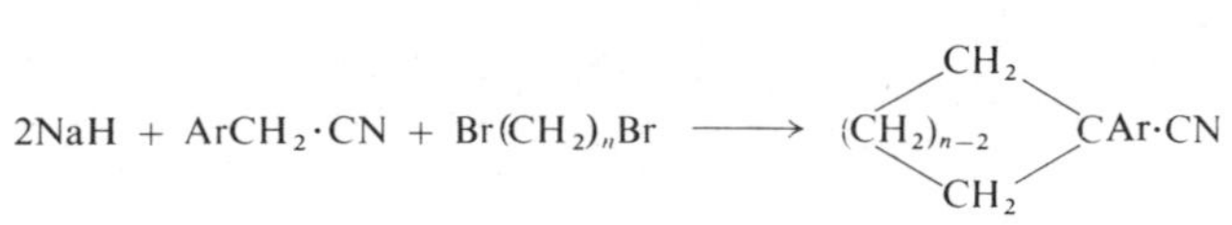

In another synthetic method good yields of 1-arylcycloalkanecarbonitriles are reported in the sodium hydride–dimethyl sulphoxide alkylation of arylacetonitriles with α,ω-dibromoalkanes, Br(CH$_2$)$_n$Br (n = 3–5)[20]. Halogen substitution in the aryl group does not affect the course of the reaction, since the reagent is unaffected by aryl halides (as would be the conventionally used sodium amide). Also the reaction is extremely rapid.

Organoboranes, readily available by hydroboration[21], undergo facile reactions with dichloroacetonitrile to yield the mono- or di-alkylated nitrile in high yields[22].

$$R_3B + Cl_2CH \cdot CN \underset{2:1}{\overset{1:1}{<}} \begin{array}{l} RCHCl \cdot CN \\ \\ R_2CH \cdot CN \end{array}$$

The products of successive alkylation of dichloroacetonitrile by two different alkyl groups were also obtained in good yield without isolation of the intermediate.

$$Cl_2CH \cdot CN \xrightarrow{R_3^1B} [R^1CHCl \cdot CN] \xrightarrow{R_3^2B} R^1R^2CH \cdot CN$$

Trichloroacetic acid ($R = CCl_3 \cdot CH_2 \cdot$) and nitroacetic acid ($R = NO_2CH_2 \cdot$) react with enamines, and the products on decarboxylation yield the corresponding α-amino compounds[23, 24]. However, treatment of 1-morpholino-cyclohex-1-ene with cyanoacetic acid 'gave an immediate exothermic reaction and a crystalline salt separated in almost quantitative yield'[25]. Conversion into the free acid and comparison with an authentic sample

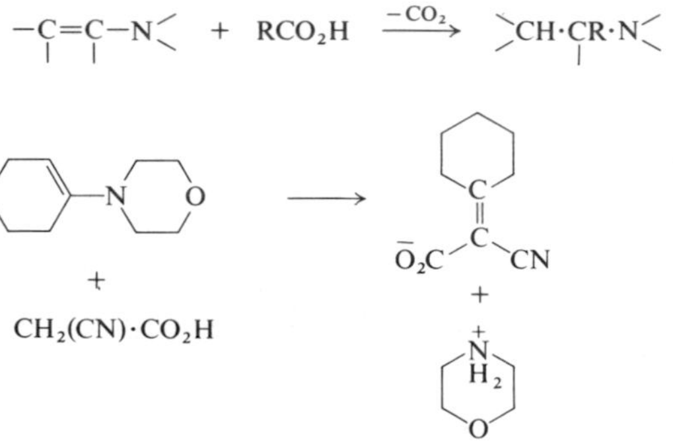

$$-\underset{|}{C}=\underset{|}{C}-N{<} \; + \; RCO_2H \; \xrightarrow{-CO_2} \; {>}CH \cdot \underset{|}{C}R \cdot N{<}$$

established that the product was the morpholine salt of α-cyanocyclo-hexylideneacetic acid. Clearly the cyanoacetic acid did not undergo decarboxylation but added to the enamine by way of the α-carbanion. Enamines from other aldehydes and ketones also yield the α-cyanoalkylideneacetic acid salts.

When diphenylcyclopropenone and cyanoacetic acid were allowed to react, the anticipated product, 4-cyano-1,2-diphenyltriafulvene-4-carboxylic

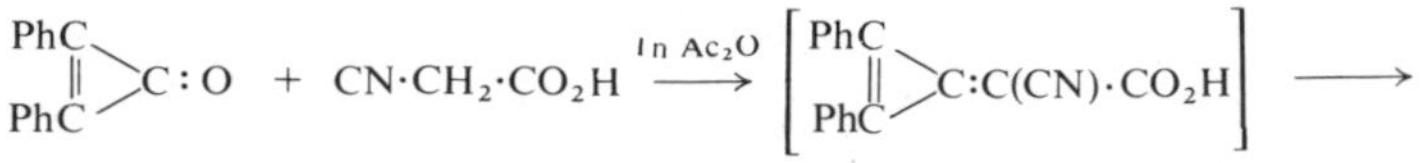

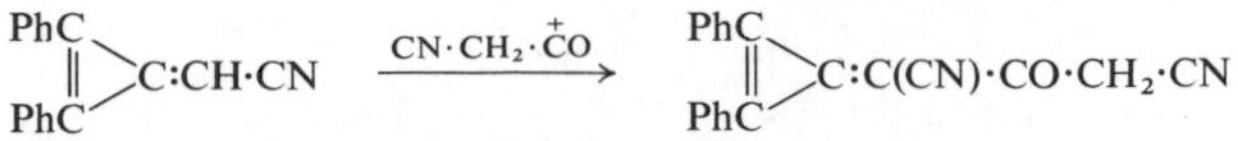

PhC
$\quad\Vert$ $\quad$C:CH·CN $\quad\xrightarrow{\text{CN·CH}_2\cdot\overset{+}{\text{C}}\text{O}}\quad$ PhC
PhC $\quad\qquad\qquad\qquad\qquad$ $\Vert$ $\quad$C:C(CN)·CO·CH$_2$·CN
$\qquad\qquad\qquad\qquad\qquad\qquad$ PhC

acid, was not isolated. The product obtained was 4-cyano-4-cyanoacetyl-1,2-diphenyltriafulvene[26]. It has been proposed that the initially formed acid undergoes decarboxylation to afford 4-cyano-1,2-diphenyltriafulvene, which then reacts with the cyanoacetyl cation generated by reaction with the solvent.

The Thorpe–Ziegler method of cyclisation has been reported as a feasible method for the preparation of substituted cyclopenta[c]thiophenes[27].

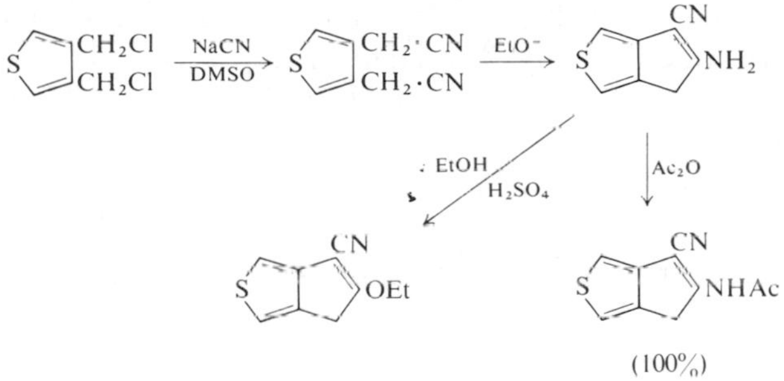

(100%)

4.2.2.2 Amidation reactions

Hydrogen fluoride has proved to be an excellent solvent for the preparation of N-alkylamides by the Ritter reaction[28]. Linear alkenes react smoothly with nitriles in hydrogen fluoride containing 0–10% of water to produce N-substituted amides. High yields (80–85%) of the three N-oct-2-ylacet-

$$R^1CH{:}CHR^2 + R^3CN \xrightarrow[\text{H}_2\text{O}]{\text{HF}} R^1CH(NH{\cdot}COR^3){\cdot}CH_2R^2$$

amides are obtained with oct-1-ene ($R^1 = C_6H_{13}$, $R^2 = H$) and acetonitrile ($R^3 = Me$). Branched alkenes also give high yields of the amides, but required 20–45% water in the reaction medium. Cycloalkenes behaved in a manner similar to linear systems with no evidence of ring contraction, even with cycloheptene. The amide isolated from norbornene (74% yield) appeared to be the exo-isomer.

Imidoyl fluorides ($R^1N{:}CFR^2$) analogous to the iminosulphonate intermediate, $R^1N{:}CR^2{\cdot}SO_3H$, proposed for the Ritter reaction in H_2SO_4[29], have been isolated in the pure state from reactions in anhydrous liquid hydrogen fluoride[30]. Addition of water to the reaction mixture instantly yields the amide.

The mechanism for the conversion of nitriles into amides by the action of H_2O_2 first proposed by Wiberg[31] has now been refined by using double-isotope labelling[32]. Wiberg's mechanism predicts that for doubly labelled hydrogen peroxide $H^{18}O—^{18}OH$, all the double label would be retained and

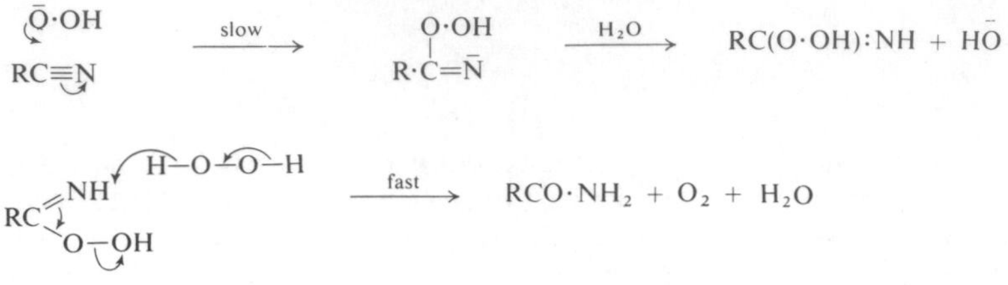

$$\underset{RC\equiv N}{\overset{\bar{O}\cdot OH}{\,}} \quad\xrightarrow{\text{slow}}\quad \underset{R\cdot C=\bar{N}}{\overset{O\cdot OH}{\,}} \quad\xrightarrow{H_2O}\quad RC(O\cdot OH)\!:\!NH + H\bar{O}$$

$$\underset{RC}{\overset{NH}{\underset{O-OH}{\,}}} \;\; H\!-\!O\!-\!O\!-\!H \quad\xrightarrow{\text{fast}}\quad RCO\cdot NH_2 + O_2 + H_2O$$

would appear as $^{36}O_2$. However, only 81 % of the double label did appear as $^{36}O_2$ in the experiment, an observation suggesting the presence of at least one additional mechanism. Analogous to the decomposition of peroxy-carboxylic acids, three pathways have been suggested for the fast step. The first predicts no scrambling. The labelling experiment did not allow for

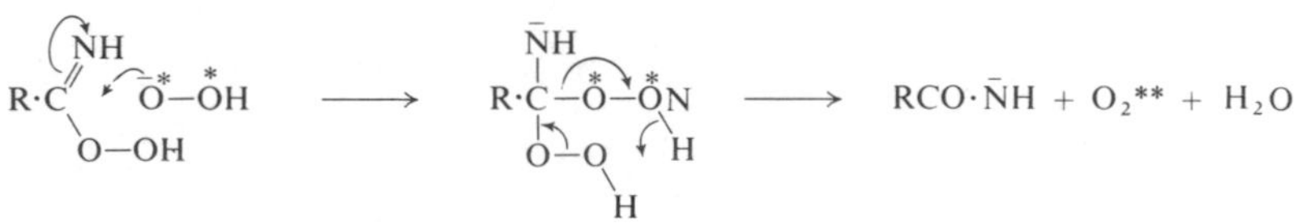

$$\underset{R\cdot C}{\overset{NH}{\underset{O-OH}{\,}}}\overset{*}{O}-\overset{*}{OH} \longrightarrow \underset{R\cdot C}{\overset{\bar{N}H}{\underset{\overset{|}{O}-O}{\,}}}\overset{*}{O}-\overset{*}{O}N \;\; \overset{|}{\underset{H}{O}}\;\; H \longrightarrow RCO\cdot \bar{N}H + O_2{}^{**} + H_2O$$

distinction between the two other possibilities, one of which predicts 50 % scrambling,

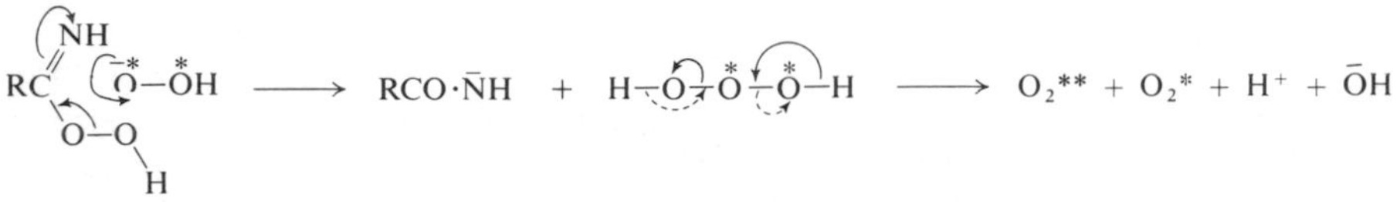

$$\underset{RC}{\overset{\bar{N}H}{\underset{\overset{|}{O}-O}{\,}}}\overset{*}{O}-\overset{*}{OH} \longrightarrow RCO\cdot \bar{N}H \;+\; H\!-\!\overset{*}{O}\!-\!O\!-\!\overset{*}{O}\!-\!H \longrightarrow O_2{}^{**} + O_2{}^{*} + H^+ + \bar{O}H$$

and the second, complete scrambling.

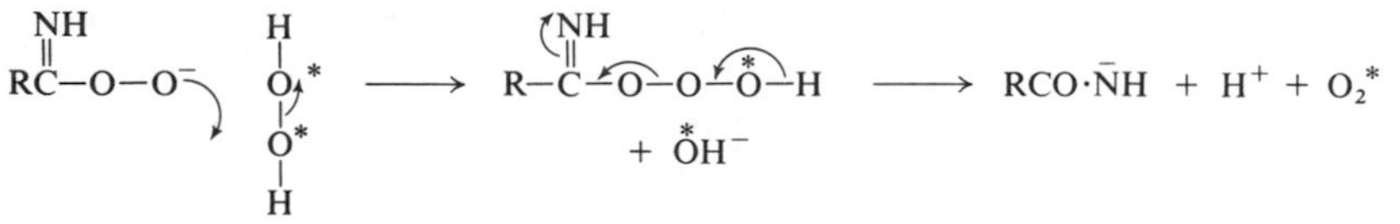

$$\underset{RC-O-O^-}{\overset{NH}{\,}}\;\; \overset{H}{\underset{\overset{|}{\underset{H}{O^*}}}{O_2{}^{*}}} \longrightarrow R-\overset{NH}{\underset{}{C}}\!-\!O\!-\!o\!-\!\overset{*}{O}\!-\!H \longrightarrow RCO\cdot \bar{N}H + H^+ + O_2{}^{*}$$
$$+ \,\overset{*}{O}H^-$$

4.2.2.3   *At a β-unsaturated carbon*

Diethylaminopropiolonitrile (R = NEt$_2$), when treated with allyl alcohol (X = O) in the presence of boron trifluoride etherate gave the Claisen rearrangement. The Claisen rearrangement, involving the conversion of allyl vinyl ethers into the homoallylic carbonyl compounds, is best known in the allyl aromatic and heterocyclic ether systems. To obtain information about the rearrangement in aliphatic compounds, the addition of several allyl alcohols and allylamines, and of prop-2-ene-thiol to a propiolonitrile derivative was studied[33]. No amino-Claisen rearrangement was observed

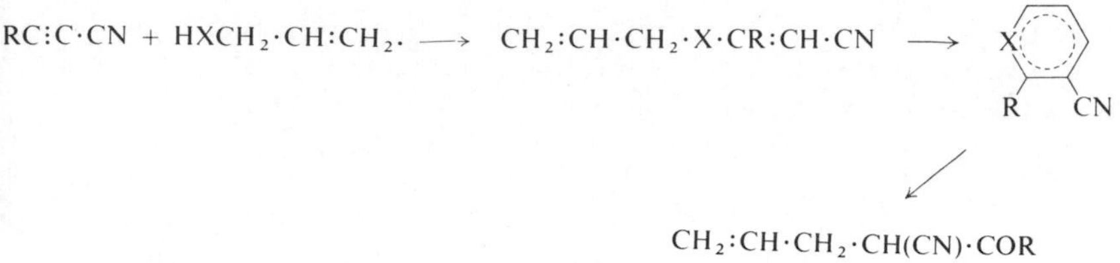

$$RC:C \cdot CN + HXCH_2 \cdot CH:CH_2. \longrightarrow CH_2:CH \cdot CH_2 \cdot X \cdot CR:CH \cdot CN \longrightarrow$$

$$CH_2:CH \cdot CH_2 \cdot CH(CN) \cdot COR$$

in any of the reactions (R = H, Cl, or NEt$_2$; X = NH or *N*-allyl). However, thio-Claisen rearrangements did occur in the reactions of chloropropiolo-nitrile (R = Cl) and diethylaminopropiolonitrile (R = NEt$_2$) with prop-2-ene-1-thiol (X = S).

Hydrogen chloride adds to cyclohexene-1-carbonitrile in an alcohol to give a 2-chlorocyclohexanecarboxylate and a small amount of the correspond-ing amide and unsaturated ester[34]. The orientation of the chlorine was found to be axial in both the chloro-amide and chloro-ester and the *cis*-configuration

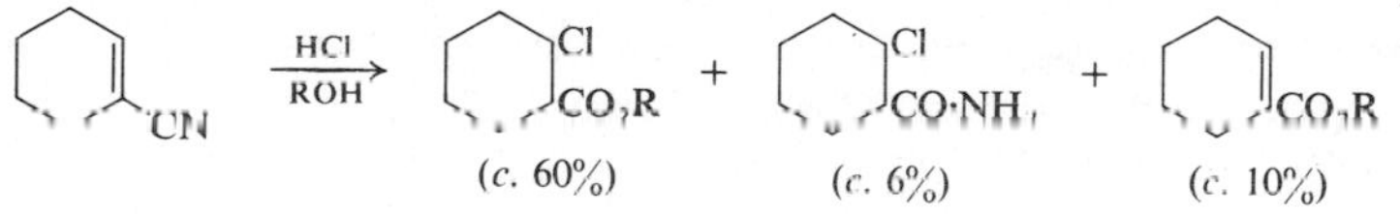

was assigned to the ester and the amide function. When the reaction was carried out in anhydrous ether, only one isomer of the β-chloronitrile (presumably the *cis*-isomer) was detected. This is evidence that the con-figuration of the products may not be equilibrium-controlled in alcoholic media.

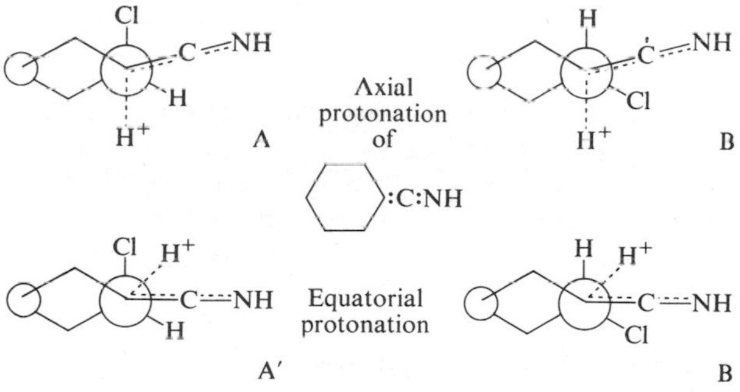

An examination of steric factors involved in protonation indicated that A was the most likely transition state since A' suffered from eclipsing of a solvated proton by axial chlorine, and B and B' exhibited interaction between the chlorine atom and the linear enamine group. Thus, transition state A is favoured for protonation leading to a *cis* product, if interaction between the —C:N— group and the chlorine atom is large. Such an interaction was designated as a pseudo-A$^{(1,3)}$ interaction[35].

A novel synthesis in good yield of cyanoketene acetals has been observed in the reaction of 2,2-dichloroacrylonitrile, prepared in 50–60% yield by the thermolysis of a mixture of acetonitrile and carbon tetrachloride under reduced pressure at 800–1000 °C, with aliphatic alcohols or phenol, in the presence of 2 mol. equiv. of base.

$$Cl_2C{:}CH{\cdot}CN \begin{cases} \xrightarrow{R^1OH \text{ (2 moles of base)}} & (R^1O)_2C{:}CH{\cdot}CN \\ \xrightarrow{R^1OH \text{ (3 moles of base)}} & (R^1O)_3C{\cdot}CH_2{\cdot}CN \\ \xrightarrow{R^2S^-} & (R^2S)_2C{:}CH{\cdot}CN \\ \xrightarrow{R^3R^4NH} & (R^3R^4N)_2C{:}CH{\cdot}CN \end{cases}$$

With 3 mol. equiv. of sodium ethoxide, triethyl orthocyano acetate was obtained in 72% yield. Reaction with thiols in the presence of bases gave the cyanoketene dithioacetals. With some amines 2,2-dichloroacrylonitrile gave the 2,2-diaminoacrylonitrile derivatives, although reaction with lower molecular weight primary aliphatic amines did not proceed smoothly, and gave in some instances unidentified products.

Treatment of chlorocyanoacetylene with secondary amines (Me$_2$NH, Et$_2$NH, piperidine, morpholine, or pyrrolidine) afforded cyano-ynamines in fair to good yield[37].

$$CN{\cdot}C{:}CCl \xrightarrow{HNR^1R^2} CN{\cdot}C{:}C{\cdot}NR^1R^2 \xrightarrow{H^+} R^1R^2N{\cdot}CO{\cdot}CH_2{\cdot}CN$$

$$\downarrow \begin{array}{l}(i)\ BF_3{-}Et_2O \\ (ii)\ HNR^3R^4\end{array}$$

$$CN{\cdot}CH{:}C(NR^1R^2){\cdot}NR^3R^4$$

These ynamines readily underwent hydration in almost quantitative yield to give substituted cyanoacetamides. Treatment of the cyanoynamines in an aprotic solvent with secondary amines (and benzylamine) in the presence of boron trifluoride etherate afforded the corresponding cyano-enamines.

4.2.2.4 *Reactions of* C:NX *compounds*

Nitrile imines (isoelectronic with nitrile oxides) are produced in the reaction of α-halogenophenylhydrazones with bases. By analogy with the nitrile

$$R^1CX{:}N{\cdot}NHR^2 \xrightarrow{NEt_3} R^1C{:}\overset{+}{N}{\cdot}\bar{N}R^2 \longleftrightarrow R^1\overset{+}{C}{:}N{\cdot}\bar{N}R^2$$

oxides, which upon reaction with acetylene afforded both acetylenic oximes and isoxazoles[38], nitrile imines have been found to yield pyrazoles and hydrazones when exposed to alkynes[39].

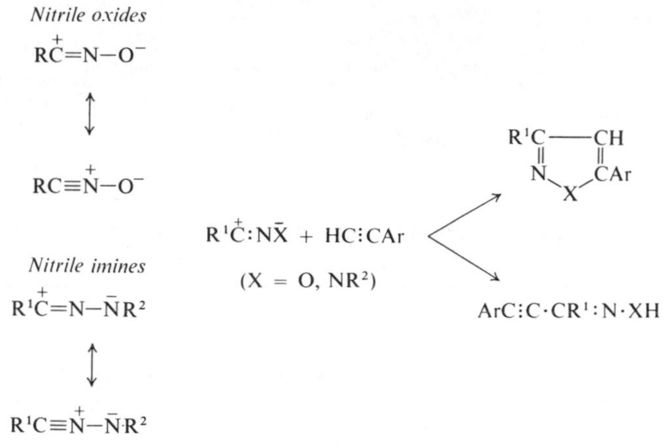

Nitrile oxides

$$\overset{+}{R\overset{.}{C}}=N-O^-$$

↕

$$RC\equiv\overset{+}{N}-O^-$$

$$R^1\overset{+}{\overset{..}{C}}:N\overline{\underset{..}{X}} + HC:CAr$$

$$(X = O, NR^2)$$

Nitrile imines

$$R^1\overset{+}{\overset{.}{C}}=N-\overline{\underset{..}{N}}R^2$$

↕

$$R^1C\equiv\overset{+}{N}-\overline{\underset{..}{N}}R^2$$

$$R^1C\text{——}CH$$
$$\underset{N}{\|}\qquad\underset{CAr}{\|}$$
$$N\diagdown\underset{X}{\diagup}CAr$$

ArC:C·CR¹:N·XH

4,5′-Di-isoxazoles have been isolated among the products of the cyclo-additions of benzonitrile oxide and 1-*NN*-dialkylaminobuta-1,3-dienes[40].

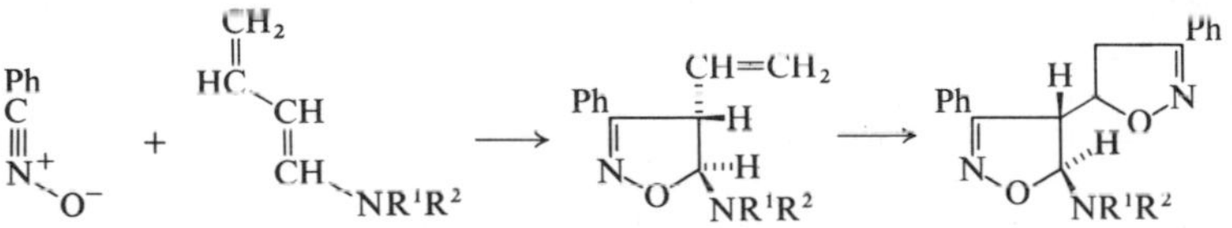

The mono-adducts can also be isolated.

Reaction of 3-methyl-2-phenyl-1-azirine with 2,4,6-trimethylbenzonitrile oxide furnishes carbodi-imides in > 80% yield early hydrolysed to the urea[41].

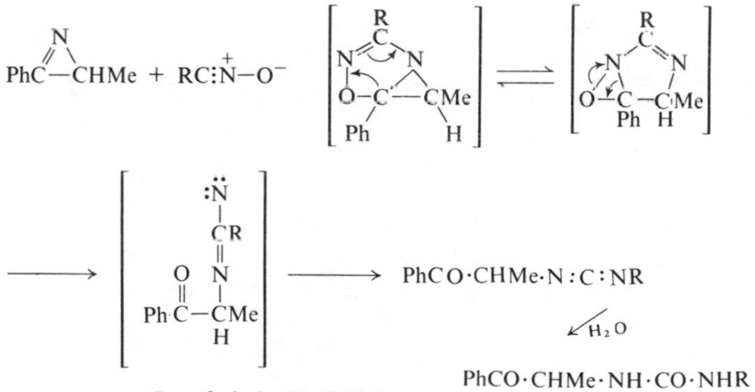

PhCO·CHMe·N:C:NR

↙ H₂O

PhCO·CHMe·NH·CO·NHR

(R = 2,4,6- Me₃C₆H₂)

The adduct initially formed undergoes ring cleavage followed by an intra-molecular migration of the aryl group. The corresponding carbodi-imides are also obtained with 2,3-diphenyl-1-azirine and 2-phenyl-1-azirine.

Extremely facile base-induced rearrangements have been observed for the bridgehead nitrile oxide shown below, the product being the oxime[42].

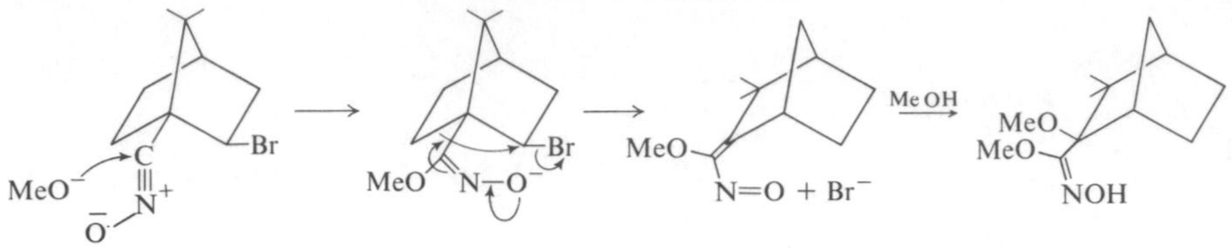

With ammonia the bridgehead nitrile oxide gave the amidoxime, which was converted into camphenilone cyanohydrin as shown below.

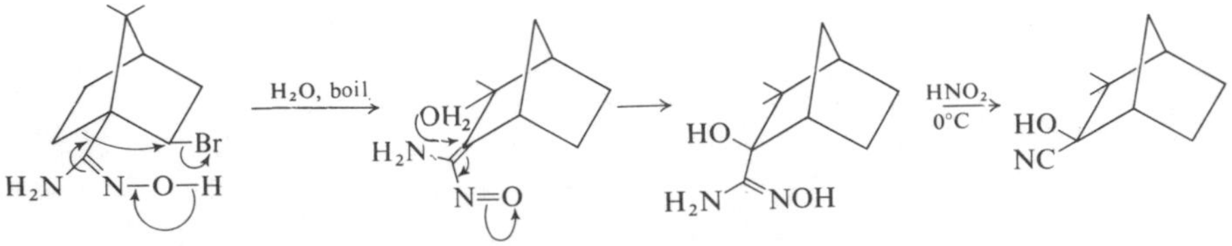

The rearranged hydroxyamidoxime expected on the basis of oxime participation could be isolated, and then gave the cyanohydrin on treatment with nitrous acid.

4.2.2.5 Sulphonyl cyanides

Reaction of sulphonyl cyanides with nucleophiles results in an efficient transfer of the cyano group from tosyl cyanide, for example, to the nucleophile[43].

$$p\text{-Tol SO}_2\text{·CN} \xrightarrow{\text{Nu}^-} p\text{-Tol SO}_2\text{H} + \text{NuCN}$$

Nucleophile	Product	Yield %
PhOLi	PhO·CN	40
p-MeC$_6$H$_4$SNa	p-MeC$_6$H$_4$S·CN	65
PhMgBr	PhCN	40
(Et)$_2$NH	Et$_2$N·CN	70

These 'activated' cyano groups of sulphonyl cyanides have also been found to act as dienophiles and dipolarophiles[44]. Tosyl cyanide was found to react with diazomethane and tetracyclone, as shown below.

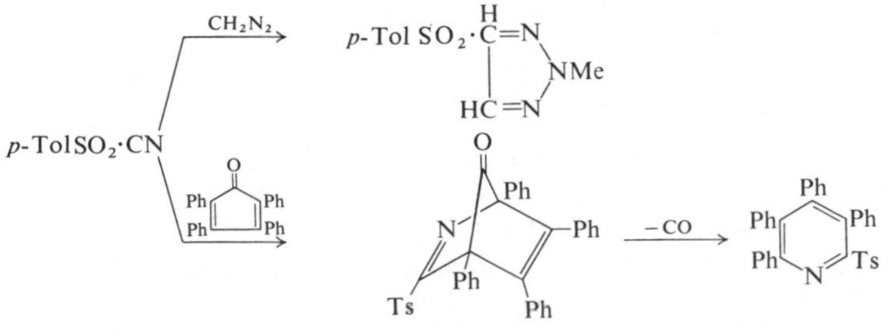

As for other activated cyano groups[45], alkyl sulphonyl cyanides also react with chlorine to provide substituted N-chloroalkylsulphonylformimidoyl chlorides[46].

$$RSO_2 \cdot CN + Cl_2 \longrightarrow RSO_2CCl:NCl$$

Thionyl chloride and salts of dicyanamide yield the structurally similar N-chlorothio(dichloroformimido) cyanide, a stable but extremely reactive compound[47].

$$NaN(CN)_2 + SOCl_2 \longrightarrow ClS \cdot N(CN) \cdot CCl:NCl \xrightarrow{H_2O-HCl} H_2N \cdot CCl:N \cdot CN$$

4.2.3 Difunctional compounds

Thermal cleavage of 2,5-diazido-3,6-di-t-butyl-1,4-benzoquinone (1 mol) in refluxing ethanol yields 2 moles of t-butylcyanoketene[48]. The ketene which is formed in nearly quantitative yield is stable in solution, but surprisingly reactive in 2 + 2 cyclo-additions. Decomposition of the diazide in cyclohexene gives the cyclobutanone in 63% yield. Reaction with dicyclohexyl carbodiimide yields the imino-β-lactam in 84% yield. The ketene also reacts instantaneously and virtually quantitatively with methanol to give methyl 2-cyano-3,3-dimethylbutanoate.

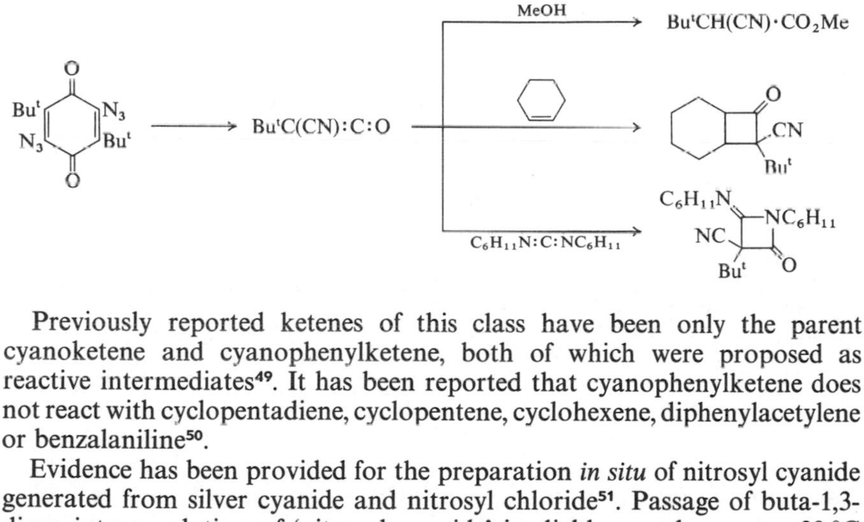

Previously reported ketenes of this class have been only the parent cyanoketene and cyanophenylketene, both of which were proposed as reactive intermediates[49]. It has been reported that cyanophenylketene does not react with cyclopentadiene, cyclopentene, cyclohexene, diphenylacetylene or benzalaniline[50].

Evidence has been provided for the preparation in situ of nitrosyl cyanide generated from silver cyanide and nitrosyl chloride[51]. Passage of buta-1,3-diene into a solution of 'nitrosyl cyanide' in dichloromethane at −20 °C gave the Diels–Alder 1,4-adduct.

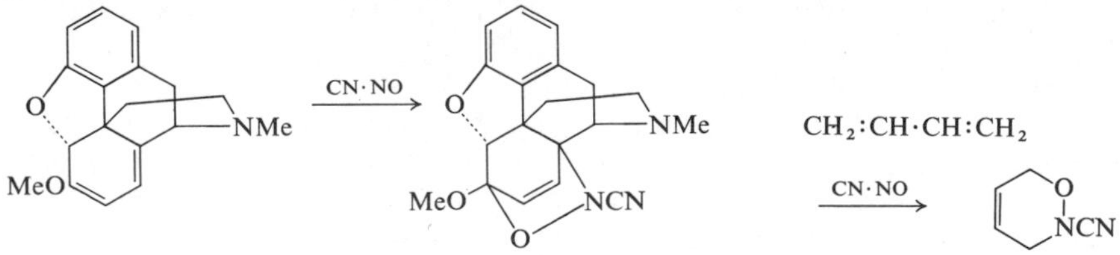

Vicinal cyanamidohalogenoalkenes, prepared by the addition of halogen (from NBS, dichlorourethan, t-butyl hypochlorite) and cyanamide to an alkene, are converted quantitatively in base into the N-cyanoaziridine[52]. N^1-Substituted-N^2-cyano-S-methylisothioureas, which are conveniently

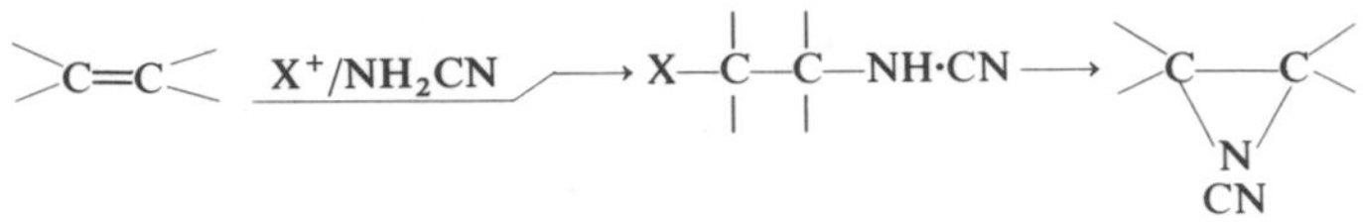

$$\text{>C=C<} \quad \underline{X^+/NH_2CN} \longrightarrow X\text{—}\overset{|}{C}\text{—}\overset{|}{C}\text{—NH·CN} \longrightarrow \text{>C——C<} \quad \underset{CN}{\overset{N}{\diagdown}}$$

prepared by the reaction of ammonia (or an appropriate amine) with dimethyl cyanoimidodithiocarbonate, readily lose methyl thiol at their melting points

$$(\text{MeS})_2\text{C:N·CN} + \text{RNH}_2 \rightarrow \text{RNH·C(SMe):N·CN} \rightarrow \text{RN:C:N·CN}$$

to yield the N-cyanocarbodi-imides[53], trapped as cyanoquanidines.

Di-iminosuccinonitrile prepared by base-catalysed addition of hydrogen cyanide to cyanogen has proved to be a valuable intermediate in the synthesis of diaza heterocyclic compounds[54].

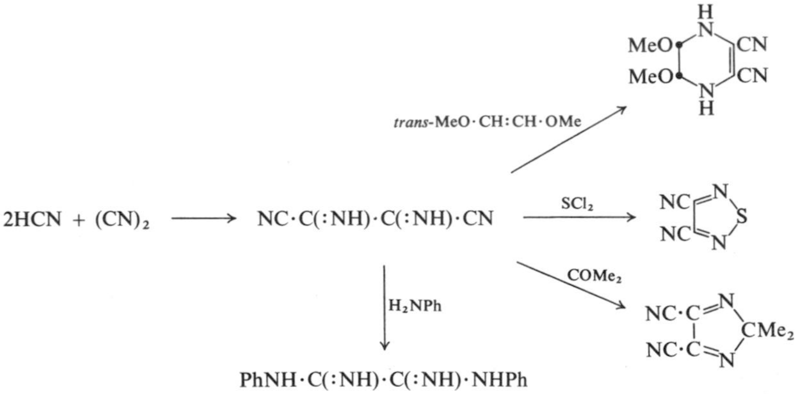

For example, reaction with electron-rich alkenes yielded the tetrahydro-pyrazine, product of a 4+2 cyclo-addition. Sulphur dichloride with di-iminosuccinonitrile gave dicyano-1,2,5-thiadiazole in 96% yield. Condensation with low molecular weight ketones yielded dicyanoisoimidazoles. Nucleophiles normally displace the nitrile groups as seen in the reaction with aniline to give N^2N^4-diphenyloxamidine.

A series of N-cyanotrialkylammonium salts which have previously been postulated only as reactive intermediates in the von Braun reaction have now been isolated in good yield as their tetrafluoroborates[55]. These stable salts gives the same cyanoamines as the postulated von Braun intermediates

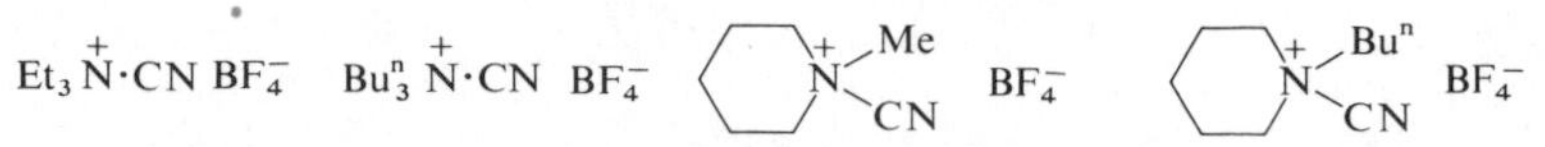

$$Et_3 \overset{+}{N}\cdot CN \; BF_4^- \quad Bu_3^n \overset{+}{N}\cdot CN \; BF_4^-$$

when subjected to the action of nucleophiles. The reactions were found to be first order both in nucleophile (NN-dimethylformamide) and in cyano-ammonium salt.

N.M.R. studies of cyanation of N-methyl-trans-decahydroquinoline with cyanogen bromide suggested first-order kinetics, consistent with rapid formation of the adduct followed by slower decomposition to the cyanide[56].

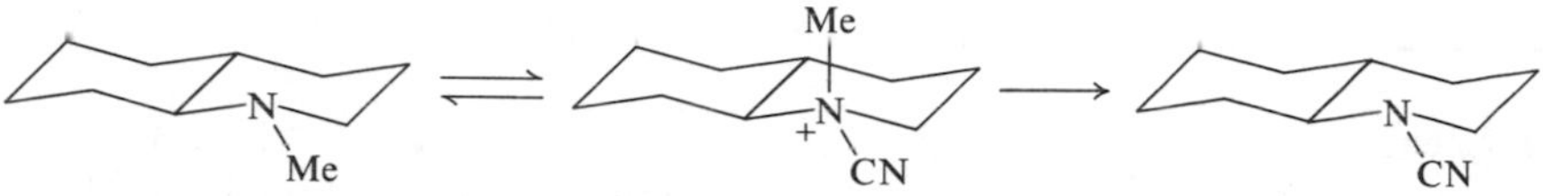

The n.m.r. trace of the reaction mixture immediately after initiation of the reaction at $-20\,^\circ C$ showed only $\overset{+}{N}Me$, $\overset{+}{N}CH_2$ and $\overset{+}{N}CH$ signals, confirming the extremely fast formation of the adduct. A sharp signal for $\overset{+}{N}Me$ also indicated steric selectivity in the reaction with cyanogen bromide. Using $CNCl\cdot SbCl_5$ as the cyanating agent, two $\overset{+}{N}Me$ resonances were observed.

4.2.4 Photochemically induced cyclo-additions

Simple unsaturated nitriles have been shown to undergo well-defined photochemical reactions. The isolation of 1-cyanobicyclobutane as the major

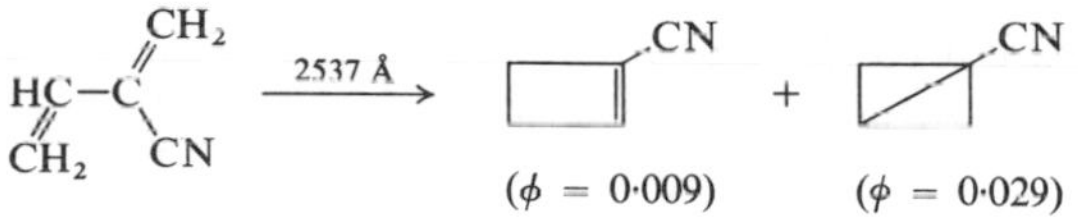

$$(\phi = 0\cdot009) \qquad (\phi = 0\cdot029)$$

product from photolysis of 2-cyanobuta-1,3-diene suggests a diradical intermediate[57].

$$S_0^T \xrightarrow{h\nu} S_1^T \longrightarrow$$

Acrylonitrile photodimerises to a mixture of cis- and trans-1,2-dicyanocyclobutane in the presence of triplet sensitisers.

The product ratio was found to vary with the medium but not with the sensitiser, suggesting the intermediacy of triplet acrylonitrile.

In contrast to the thermal 4+2 cyclo-addition of buta-1,3-diene to arylonitrile, the photosensitised addition of the diene to the nitrile, with acetophenone as the sensitiser, afforded considerable amounts of the 2+2

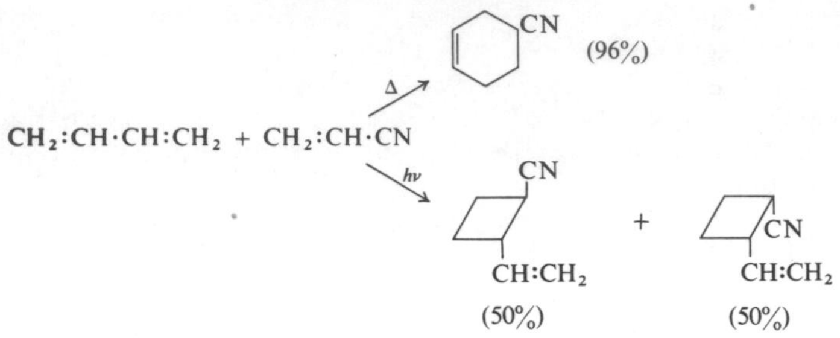

$$\text{CH}_2\text{:CH}\cdot\text{CH:CH}_2 + \text{CH}_2\text{:CH}\cdot\text{CN} \quad \overset{\Delta}{\underset{h\nu}{\rightleftarrows}}$$

(top, via Δ): cyclohexene-CN (96%)

(bottom, via $h\nu$): cyclobutane with CN and CH:CH$_2$ (50%) + cyclobutane with CN and CH:CH$_2$ (50%)

adducts, and virtually none (1.5%) of the cyanocyclohexene[58]. Irradiation of α-acetoxyacrylonitrile and buta-1,3-diene gave a mixture of three cross-adducts (plus three buta-1,3-diene dimers)[59]. The cyclobutane isomers were isolated in yields of up to 60%, while the cyclohexene derivative was a minor

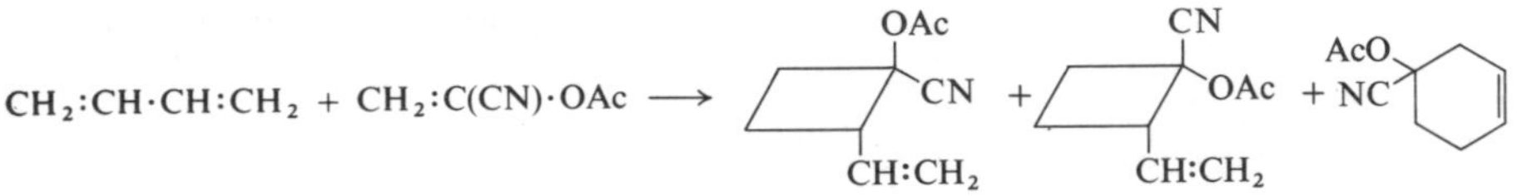

$$\text{CH}_2\text{:CH}\cdot\text{CH:CH}_2 + \text{CH}_2\text{:C(CN)}\cdot\text{OAc} \longrightarrow$$

(cyclobutane with OAc, CN, CH:CH$_2$) + (cyclobutane with CN, OAc, CH:CH$_2$) + (cyclohexene with AcO and NC substituents)

product (1–12%). The photosensitised cyclo-addition of cyclopentadiene to α-acetoxyacrylonitrile gave a mixture of cyanoacetoxybicyclo[2.2.1]- and cyanoacetoxybicyclo[3.2.0]-heptanes.

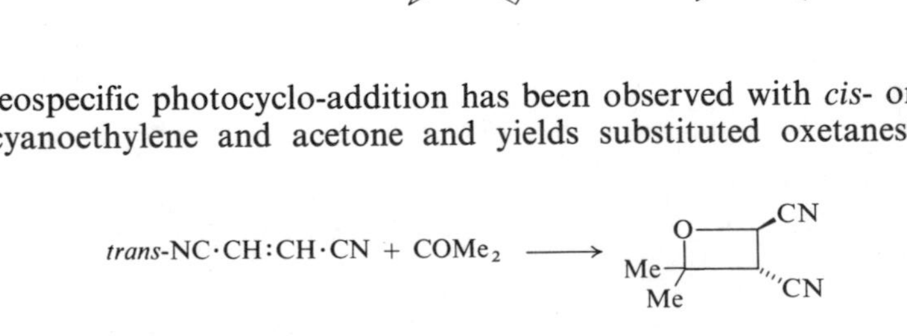

$$\text{(cyclopentadiene)} + \text{CH}_2\text{:C(CN)}\cdot\text{OAc} \longrightarrow$$

(bicyclo[3.2.0] with CN, OAc) + (bicyclo[3.2.0] with OAc, CN)

+

(bicyclo[2.2.1] with CN, OAc) + (bicyclo[2.2.1] with OAc, CN)

Stereospecific photocyclo-addition has been observed with cis- or trans-1,2-dicyanoethylene and acetone and yields substituted oxetanes[60]. The

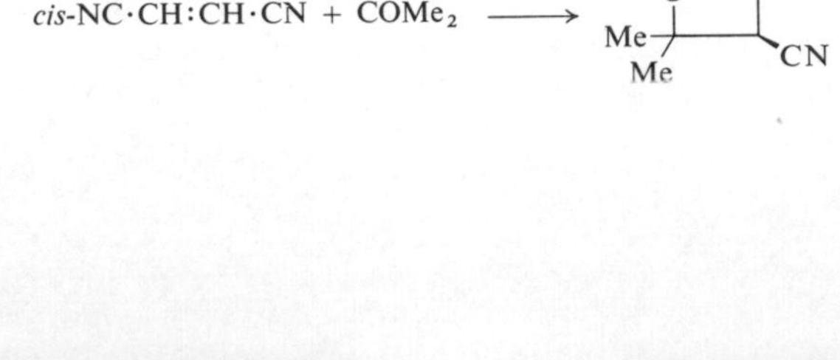

$$\textit{trans}\text{-NC}\cdot\text{CH:CH}\cdot\text{CN} + \text{COMe}_2 \longrightarrow$$

(oxetane: O, Me, Me, CN, ""CN)

$$\textit{cis}\text{-NC}\cdot\text{CH:CH}\cdot\text{CN} + \text{COMe}_2 \longrightarrow$$

(oxetane: O, Me, Me, CN, CN)

reaction proceeds by reaction of singlet acetone with the ground-state alkene.

4.2.5   Synthesis of heterocyclic compounds

Retrodiene reactions involving elimination of hydrogen cyanide or a nitrile have been observed in the Diels–Alder reactions of oxazoles and diphenyl-cyclopropenone[61]. In no case was the bicyclic adduct isolable, the major

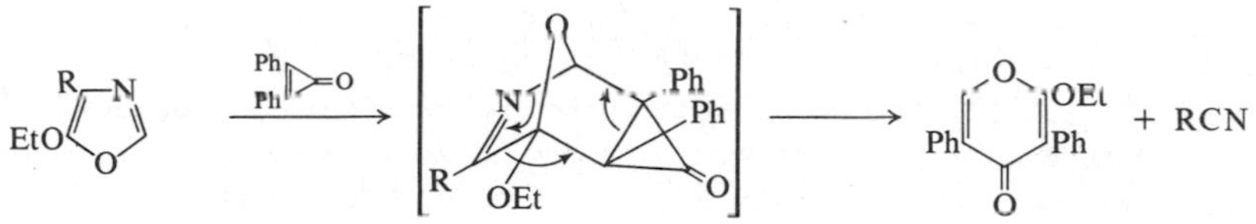

products being a γ-pyrone or a furan.

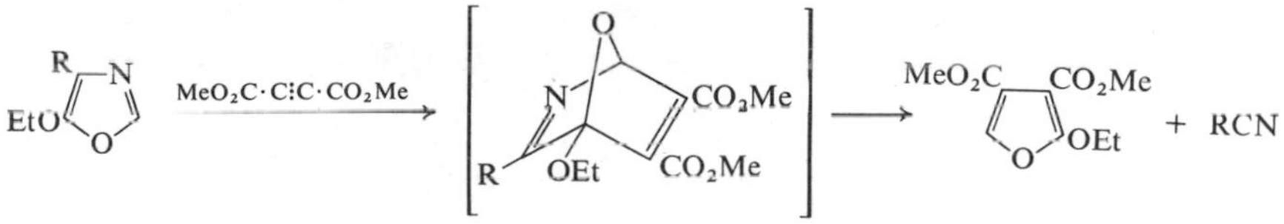

The reaction of nitriles in ethanol with hydrazine hydrate and catalytic amounts of Raney nickel yields primary amines, aldazines, or 3,6-disubstituted 1,2,4,5-tetrazines depending on the structure of the nitrile[62].

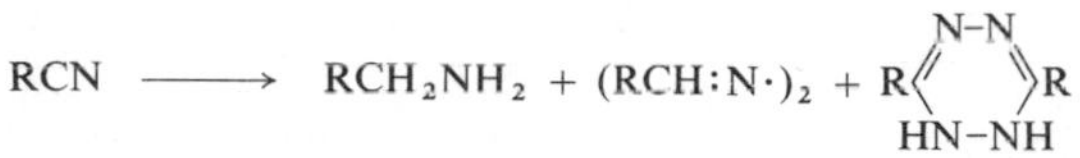

However, as a general synthesis for aldehydes (isolated as the aldazine), the reaction appears to be limited to benzonitrile and substituted benzonitriles.

The unusual ring opening of α-cyano-β-2-furylacrylic esters by morpholine has been shown to yield β-(4-alkoxycarbonyl-5-amino-2-furyl)-acrolein[63]

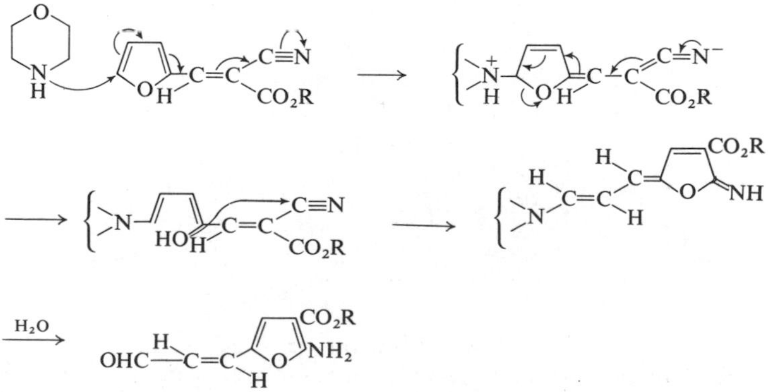

Opening of furan rings with ammonia and amines has been known for some time[64]. Specifically, α-cyano-β-2-furylacrylic esters with primary arylamines afford β-1-aryl-2-pyrrolyl-α-cyano-acrylic esters[65], probably through an intermediate similar to that proposed for the rearrangement above. In contrast to the reaction of epichlorohydrin in unbuffered

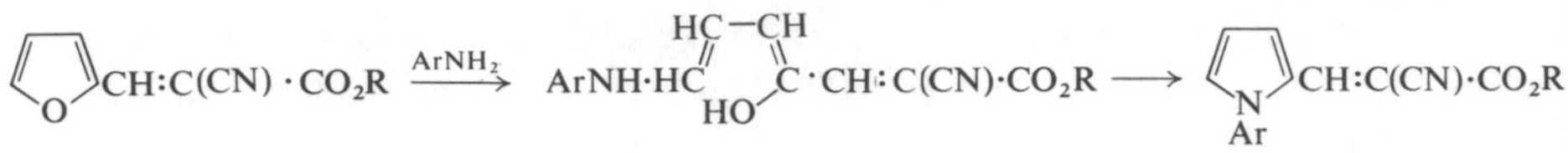

sodium cyanide solution, which yields the substituted cyclohexadiene, 2-methylepichlorohydrin under the same conditions leads to an entirely different product identified as 4,9-dimethyl-2,6-diazaspiro[4,5]dec-3-ene-1,7-dione[66]. The latter product appears to arise from two C-5 moieties. Either *trans*-1-cyano-2-methylpropen-3-ol or its self-condensation product (a

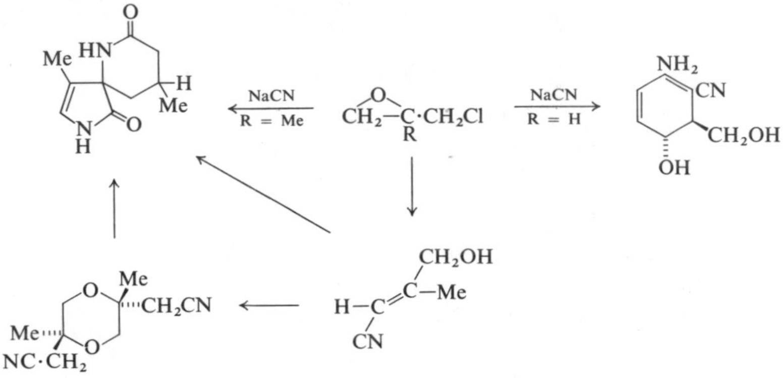

dioxan derivative) is converted into the azadione under the influence of base.

With sodium cyanide in a polar solvent, trifluoroacetonitrile reacts to form a 3:1 adduct which has been shown to be a salt of a dihydrotetra-azapentalene[67]. On acidification the water-soluble salt gives the acid. The reaction was presumed to proceed first through 1:1 and 2:1 adducts which

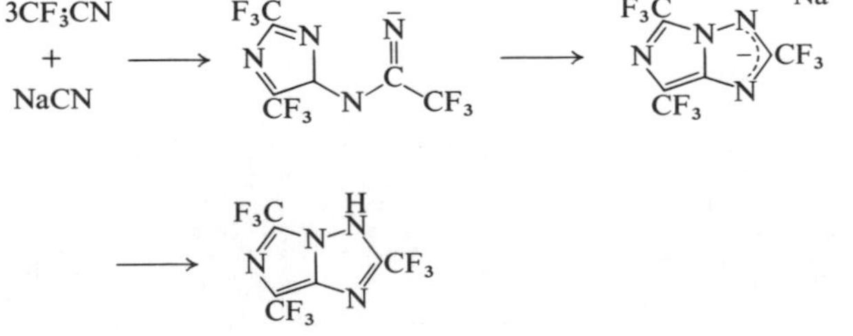

could not be isolated. The final cyclisation, in which a nitrogen anion adds to a carbon–nitrogen double bond to form a nitrogen–nitrogen bond, is unusual and occurs in a direction opposite to that normally observed. This addition of sodium cyanide closely parallels a similar addition to 3 mol. equiv. of cyanogen in which cyano groups replace the trifluoromethyl groups in the dihydrotetra-azapentalene product[68].

Alkenes react with nitrosyl tetra fluoroborate, $NOBF_4$, in nitrile media to yield heterocyclic compounds[69]. The salts first formed may be reduced by

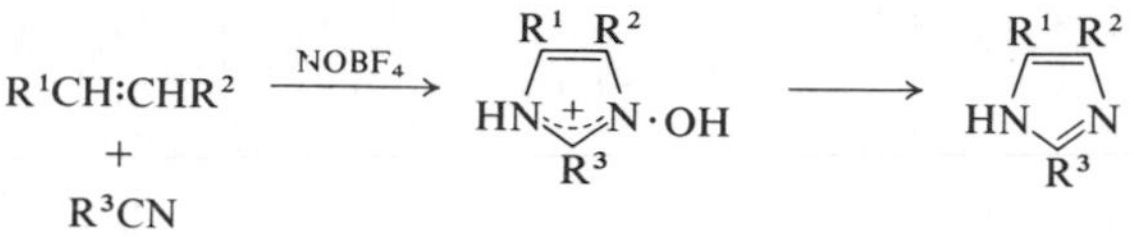

hydride reagents to imidazoles. For example, propene with acetonitrile gives 2,4(5)-dimethylimidazole ($R^2 = R^3 = Me$, $R^1 = H$) in 80% yield and both cis- and trans-but-2-ene give 2,4,5-trimethylimidazole ($R^1 = R^2 = R^3 = Me$) (80%). Styrene yields 2-methyl-4(5)-phenylimidazole ($R^1 = Ph$, $R^2 = H$, $R^3 = Me$) in 50% yield, and cyclohexene gives 2-methyltetrahydro-benzimidazole ($R^1, R^2 = -(CH_2)_4-$, $R^3 = Me$) also in 50% yield.

The cyanoimidodithiocarbonate anion undergoes chlorination to give 3-chloro-1,2,4-thiadiazol-5-ylsulphenyl chloride in yields of 85–100%, while bromination gives bis-(3-bromo-1,2,4-thiadiazol-5-yl) disulphide, also in

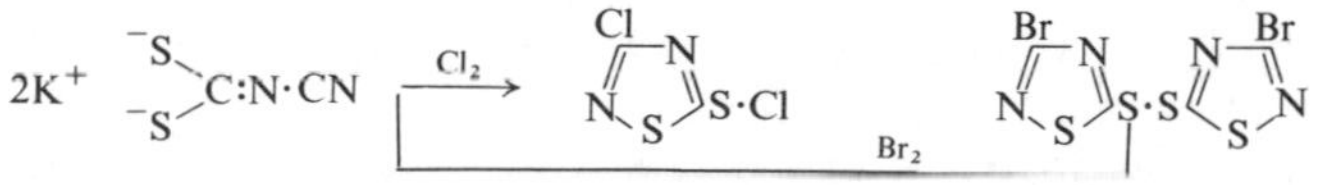

excellent yield[70]. The sulphenyl chlorides reacted rapidly with alkenes, the products being the corresponding chlorosulphides. Different products were obtained from such reactions involving cis- and trans-but-2-ene, presumably the threo- and erythro-chlorosulphides. Single products were also obtained in reactions of unsymmetrical alkenes.

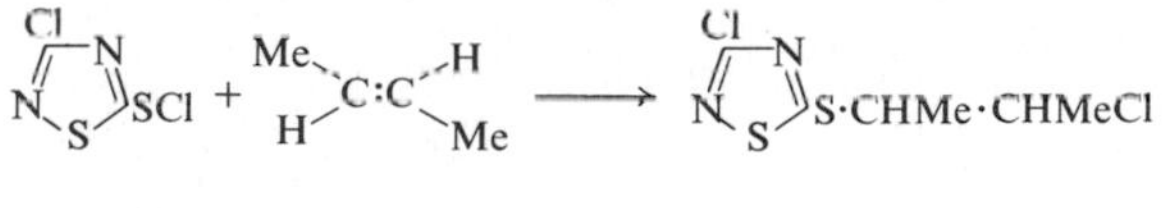

Transition-metal promoted reductive decyanation of nitriles has been achieved by using $Fe(acac)_3$ [71]. The best yields (58–100%) were obtained with saturated nitriles, while allyl and phenyl cyanide gave lower yields (40–46%).

$$RCN + Fe(C_5H_7O_2)_3 + 2Na \rightarrow RH + Fe(C_5H_7O_2)_2C_5H_6O_2^- + CN^- + 2Na^+$$

4.3 ISOCYANIDES

4.3.1 Synthesis

Recent reports by Meyers and Adickes[72] and Walborsky and Niznik[73] of base induced α-elimination of the elements of ROH from formimidate derivatives as a potentially general route to the isonitrile function, prompted

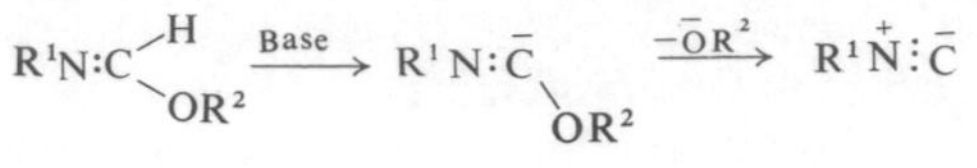

$$R^1N{:}C{\overset{H}{\underset{OR^2}{}}} \xrightarrow{\text{Base}} R^1N{:}\bar{C}{\underset{OR^2}{}} \xrightarrow{-\bar{O}R^2} R^1\overset{+}{N}{:}\bar{C}$$

Anselime and his co-workers to investigate the action of n-butyl-lithium on ethyl N-phenylformimidate[74]. With three equivalents of n-butyl-lithium, PhN:CHOEt gave phenyl isocyanide along with N-5-nonylaniline (PhNHCHBu$_2^n$) as the major product. With methyl-lithium, N-phenyl-formimidate gave phenyl cyanide (30%) and as second product (25%), N^1N^2-diphenylformimidate (PhNH·CH:NPh). Unfortunately this approach is not considered useful as a general method for the synthesis of isocyanides.

4.3.2 Reactions

4.3.2.1 Cyclisation to heterocyclic compounds

Several interesting cyclo-addition reactions of potential value in the synthesis of heterocyclic compounds have been reported. A novel route to Δ^1- (or Δ^2-)pyrroline derivatives in high yield has been found in the reactions of isonitriles with an acidic hydrogen at the α-carbon, such as benzyl isocyanide or carbethoxymethyl isocyanide, with the carbon–carbon double bond of α,β-unsaturated cyanides and carbonyl compounds in the presence of Cu$_2$O as catalyst[75]. Also in the presence of Cu$_2$O, these isocyanides react with the C=O bond of the carbonyl compound to provide Δ^2-oxazolines. An organocopper–isocyanide complex is formed initially by abstraction of the α-hydrogen and then the organocopper complex adds to the C=C or the C=O double bond respectively.

(X = CN or CO$_2$Me)

4.3.2.2 Insertion reactions

New reactions of isocyanides characterised by their nucleophilicity have been reported for α,β-unsaturated carbonyl and cyano compounds and with acetylenemonocarboxylates[76]. The reaction with an α,β-unsaturated compound gives two products, the first being produced exclusively in t-butyl

$$\overset{+}{R N} \overset{-}{\vdots} C + CH_2 : CHX \longrightarrow RN:CH \cdot CH:CHX + RN:C(OMe) \cdot CH_2 \cdot CH_2 X$$

alcohol. The second is found when the reaction is carried out in methanol. The reaction with an acetylenemonocarboxylate proceeds as follows.

$$R^1 \overset{+}{N} \overset{-}{\vdots} \overset{-}{C} + R^2 C \vdots C \cdot CO_2 Me \xrightarrow{MeOH} R^1 N:C(OMe) \cdot CR^2 :CH \cdot CO_2 Me$$

In another report[77] Saegusa describes the reaction of vinyl isocyanide in the presence of Cu_2O as catalyst with amines, alcohols, thiols, N-alkyl derivatives of carboxamides, carbonates, urea and thiourea to form the products shown below:

RN(CII:N·CII:CII$_2$)$_2$ RZ·CH:N·CH:CH$_2$ R^{1}CO·NR^2·CH:N·CH:CH$_2$
  (from the amine)   (RZ)$_2$CH·N:CHMe   (from the amide)
           (from the alcohol or thiol, Z = O or S)

4.3.2.3   With carbenes and nitrenes

Although carbenes and isocyanides might be expected to interact readily to give ketenimines, only the second such reaction[78] has recently been reported by Ciganek. Methoxycarbonylphenylketene N-t-butylimine was prepared in 51% yield by the thermolysis of methyl diazophenylacetate in t-butyl isocyanide.

$$PhC(:N_2) \cdot CO_2 Me + Bu^t NC \xrightarrow{N_2} PhC(CO_2 Me):C:NBu^t$$

The most likely route for the reaction involves loss of nitrogen to form the carbene followed by α-addition to t-butyl isocyanide. Another process similar to this, but involving an intermediate nitrene, has been reported by Saegusa in the reaction of an azide with an isocyanide[79]. The reaction, which is a highly acceptable synthetic method for unsymmetrical carbodi-imides,

$$R^1 N_3 + R^2 \overset{+}{N} \overset{-}{\vdots} \overset{-}{C} \xrightarrow{catalyst} R^1 N:C:NR^2 + N_2$$

proceeds most efficiently in the presence of pentacarbonyliron, rather than with Cu_2O, which has been found effective in similar reactions.

4.3.2.4   Synthetic reactions catalysed by metals

Two significant reactions of isonitrile with synthetic possibilities have been described by Walborsky[80] and by Saegusa[81]. Walborsky has found that aliphatic Grignard reagents, like lithium alkyls, add to 2,4,4-trimethylpent-2-yl isocyanide to yield the corresponding metallo-aldimine, which undergoes hydrolysis with H_2O or D_2O, or carbonation in good to excellent yields to

$$RMgX + Bu^t CH_2 \cdot CMe_2 \cdot \overset{+}{N} \overset{-}{\vdots} \overset{-}{C} \longrightarrow Bu^t CH_2 \cdot CMe_2 N:CR \cdot MgX$$

provide aldehydes and α-keto-acids, respectively. The lithium aldimine has been found to be a particularly versatile synthetic intermediate.

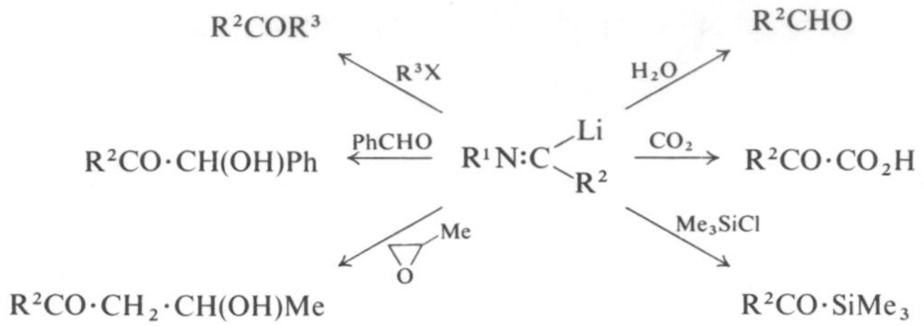

$$R^2COR^3 \qquad\qquad\qquad R^2CHO$$

$$R^2CO \cdot CH(OH)Ph \xleftarrow{PhCHO} R^1N{:}C{\overset{Li}{\underset{R^2}{\diagup}}} \xrightarrow{CO_2} R^2CO \cdot CO_2H$$

$$R^2CO \cdot CH_2 \cdot CH(OH)Me \qquad\qquad\qquad R^2CO \cdot SiMe_3$$

In another report of synthetic reactions catalysed by Cu^I, Saegusa has described two groups of reactions of allyl isocyanide, namely its isomerisation to propenyl isocyanide, and the formimidation of amides, amines, and alcohols[81].

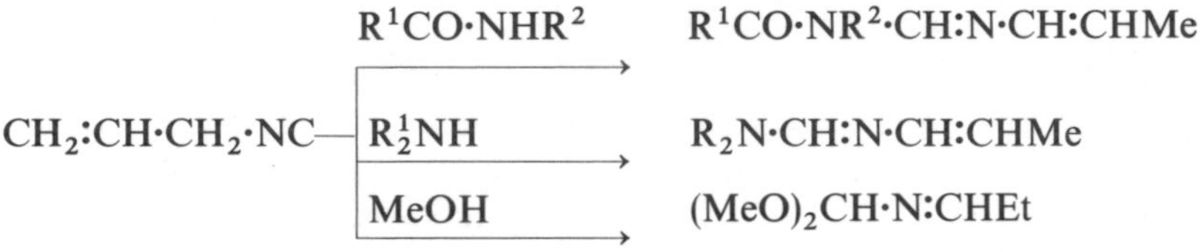

$$R^1CO \cdot NHR^2 \qquad\qquad R^1CO \cdot NR^2 \cdot CH{:}N \cdot CH{:}CHMe$$

$$CH_2{:}CH \cdot CH_2 \cdot NC\text{---}R^1_2NH \qquad\qquad R_2N \cdot CH{:}N \cdot CH{:}CHMe$$

$$MeOH \qquad\qquad (MeO)_2CH \cdot N{:}CHEt$$

4.3.2.5 Additional reactions

A few isolated reactions of isocyanides of general importance have been reported. Cyclohexyl and t-butyl isocyanide react with 1-halogenoalkynes to afford N-substituted β-halogenoacrylamides in good, though not excellent yields[82].

$$R^1NC + R^2C{:}CX \longrightarrow R^1N{:}\overset{+}{C} \cdot CR^2{:}\bar{C}X \longrightarrow R^1NH \cdot CO \cdot CR^2{:}CHX$$

Two examples of radical reactions have been reported[83, 84]. In the first the reaction of isocyanides with thiols is described. There are two courses observed, one of which gives the thioformimidate and the other the isothiocyanate from the isocyanide, and an alkane. With cyclohexyl isocyanide and 2-propylthiol, the reaction can be represented as follows.

$$C_6H_{11}NC + Pr^iSH \to Pr^iS \cdot CH{:}NC_6H_{11} + C_6H_{11}N{:}C{:}S + C_3H_8$$

Experimental results support a mechanism involving the thiyl radical. Radical reactions of primary and secondary thiols follow the course which gives thioformimidates, while those of benzylthiol and tertiary thiols afford isothiocyanates and the hydrocarbon derived from the thiol. Secondary thiols pursue both reaction routes, but the Cu^I catalysed reaction takes the first course almost exclusively.

The reaction of isocyanides with disubstituted phosphines in the presence of azobisisobutyronitrile (AIBN) produced disubstituted cyanophosphines and the hydrocarbon derived from the isocyanide.

$$RNC + Et_2PH \xrightarrow{AIBN} Et_2P \cdot CN + RH$$

$$(R = Bu^t \text{ or } PhCH_2)$$

When cyclohexyl or n-hexyl isocyanide was employed, the course of the reaction with diethylphosphine was different and the product was diethyl-formimidoylphosphine.

$$RNC + Et_2PH \xrightarrow{AIBN} Et_2PCH:NR$$

In the absence of AIBN, starting materials were recovered quantitatively. With the Cu^I catalyst in common usage in Saegusa's laboratory, only the corresponding formimidoylphosphine was obtained, regardless of the nature of the isocyanide.

4.4 NITRO COMPOUNDS

4.4.1 Synthetic methods

4.4.1.1 From acids and acid derivatives

The addition of n-propyl nitrate to the α-anion of straight-chain carboxylic acids offers a direct method for the preparation of nitroalkanes containing

$$RCH_2 \cdot CO_2H \xrightarrow[HMPA]{2LiNPr_2^i} RCHLi \cdot CO_2Li \xrightarrow{Pr^nO \cdot NO_2}$$

$$[RC(NO_2)Li \cdot CO_2Li \rightleftharpoons RC(CO_2Li):\overset{+}{N}(OLi) \cdot \bar{O}] \xrightarrow[-CO_2]{H^+} RCH_2NO_2$$

one less carbon atom[85]. The intermediate α-nitrocarboxylate salt is not isolated. Acidification of the reaction mixture causes decarboxylation to the nitroalkane in moderate to good yields (45–68 %).

Acyl nitrates, prepared by a number of different methods from the corresponding anhydrides, undergo thermolysis to give carbon dioxide and

$$(RCO)_2O \xrightarrow[\text{agent}]{\text{Nitrating}} RCO_2 \cdot NO_2 \xrightarrow{-CO_2} RNO_2$$

the nitroalkane[86]. The decarboxylation is accomplished satisfactorily with little or no danger of explosion in the presence of an excess of the nitrating agent, the carboxylic acid derivative, or an inert solvent. Yields are in the range 20–60%. A free radical process is thought to be involved.

4.4.1.2 From other nitro compounds

Terminal nitroalkanes have been prepared in good yields (58–94 %) by sodium borohydride reduction of β-nitroalkyl nitrates[87]. The β-nitroalkyl nitrates in turn can be obtained in high yield from alk-1-enes, nitrogen oxides, and oxygen. The reduction requires only mixing of the reactants at room tem-

$$O_2N \cdot OCR^1R^2 \cdot CH_2NO_2 \longrightarrow [R^1R^2C:CHNO_2] \longrightarrow R^1R^2CH \cdot CH_2NO_2$$

perature and the reaction probably proceeds through a nitroalkene intermediate which results from the elimination of nitric acid.

Potassium salts (di and mono) of α,α'-dinitrocyclanones have been found to undergo ring cleavage and to give the α,ω-dinitroalkanes in high yield[88].

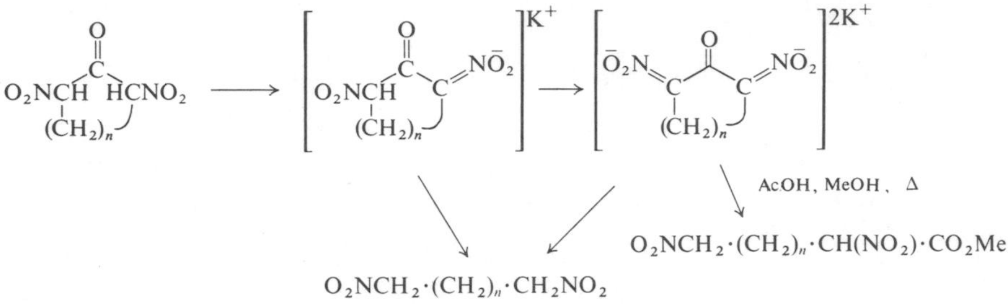

$$O_2NCH_2 \cdot (CH_2)_n \cdot CH_2NO_2$$

$$O_2NCH_2 \cdot (CH_2)_n \cdot CH(NO_2) \cdot CO_2Me$$

The reaction is pH dependent. For example, at pH 5, dipotassium 2-keto-1,3-cyclopentanedinitronate ($n = 2$) gave only the monopotassium salt (95.5% yield), while dipotassium-2-keto-1,3-cyclohexanedinitronate ($n = 3$) was cleaved to 1,5-dinitropentane in 88% yield. The di-anions yield the same products as the monopotassium salts, but two molar equivalents of acid are required. Two of the dipotassium salts ($n = 2$) and ($n = 3$), also 2,6-dinitrocyclohexanone, were cleaved in refluxing methanolic acetic acid without decarboxylation to give methyl α,ω-dinitro alkanoates in yields of c. 65%.

4.4.1.3 From alkenes

Synthesis of nitromethyl derivatives by cis-addition of the elements of nitromethane to an ethylenic linkage has been reported for 3,4,5,6,7,8-hexahydroquinazolin-2(1H)-one by reaction with nitroacetic acid at the

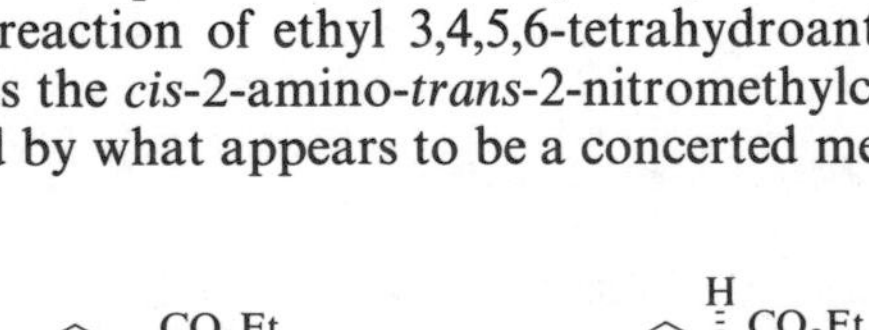

decarboxylation temperature (c. 60 °C)[89]. Quantitative conversion was obtained. Also reaction of ethyl 3,4,5,6-tetrahydroanthranilate with nitroacetic acid yields the cis-2-amino-trans-2-nitromethylcyclohexane carboxylate in 70% yield by what appears to be a concerted mechanism[90].

The addition of acetyl nitrate to 1,2,3,4,4a,10a-(*trans*-4a,10a)-hexahydro-phenanthrene gives, in low yield, the four isomeric 9-acetoxy-10-nitro-octahydrophenanthrene addition products, along with the 10-nitro derivative of the starting hexahydrophenanthrene[91]. The products were separated and

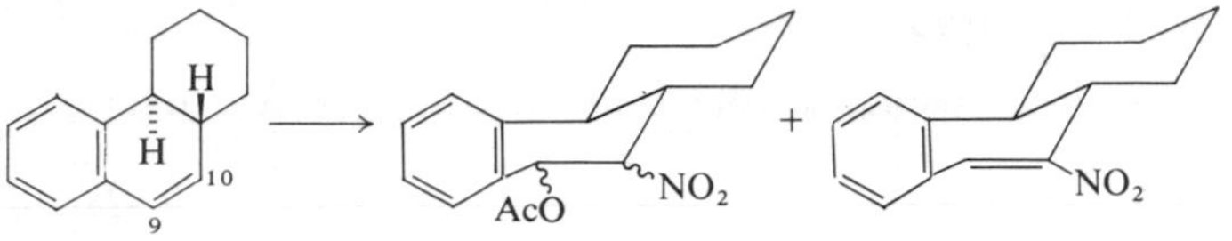

their characterisation was achieved by analysis of n.m.r. data, and x-ray diffraction studies. The product in the greatest yield had a C-10 equatorial nitro-group and an axial group at C-9.

4.4.1.4 Electrophilic nitration

Electrophilic nitration (substitution of hydrogen by the nitro group) and nitrolysis (nitrolytic cleavage of C—C (σ) bonds) of alkanes and cyclo-alkanes have been observed by using stable nitronium salts ($NO_2^+PF_6^-$, $NO_2^+SbF_6^-$, $NO_2^+BF_4^-$) in solution in methylene chloride–tetramethylene sulphone[92]. The solvents employed and the nitro products themselves are lone pair (n) donors. Since the σ basicity of alkanes is lower than the n basicity of these lone pair molecules, nitration is generally slow and only low yields of nitroalkanes are obtained. At 25 °C, 0.1 % of nitromethane was obtained in the nitration of methane. Higher yields have been obtained in HF and HSO_3F indicating that the protonitronium ion NO_2H^{2+} may be the nitrating species. Higher alkanes and isoalkanes gave yields of 2–5% and adamantane was nitrated in 10% yield. Tertiary C—H bonds were highest

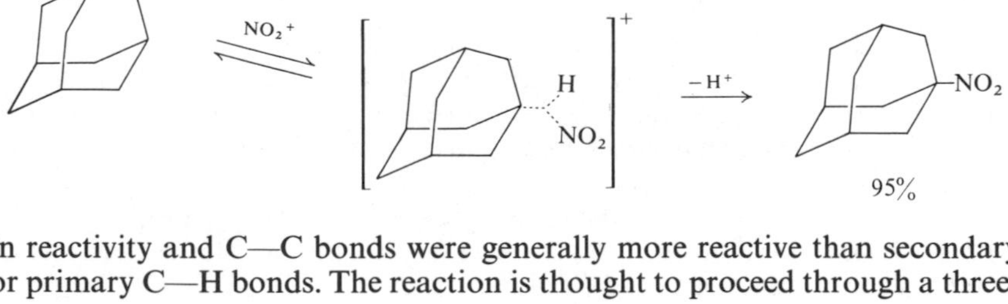

in reactivity and C—C bonds were generally more reactive than secondary or primary C—H bonds. The reaction is thought to proceed through a three-centred bond transition state formed by attack of the nitrating agent on the two-electron covalent (σ) bond.

4.4.2 Reactions

4.4.2.1 At an α-carbon atom

A free-radical chain mechanism has been proposed for the coupling reactions of substituted nitroalkanes with tertiary carbanions. Reaction of the lithium salt of 2-nitropropane with 2-bromo-2-nitro-propane in the dark, or with illumination, produces a nearly quantitative yield of 2,3-dimethyl-2,3-dinitrobutane[93]. Reaction of the lithium salt with 2,2-dinitropropane at 60 °C

$$Me_2\bar{C}\cdot NO_2 + Me_2CBr\cdot NO_2 \longrightarrow O_2N\cdot CMe_2\cdot CMe_2\cdot NO_2$$

in the dark resulted in the facile formation of the coupled product. The failure of 2-bromo-2-nitropropane to give the mixed dimer when treated with diethyl ethylmalonate anion is thought to be due to bromine transfer between the anion and the bromine compound. 2,2-Dinitropropane reacts with

$$Me_2C(NO_2)_2 + [EtC(CO_2Et)_2]^- \longrightarrow O_2NCMe_2\cdot CEt(CO_2Et)_2$$

the anion of diethyl ethylmalonate to yield only the mixed dimer in which no atom or group appears to be transferred. Various cyano-substituted anions also react with 2-bromo-2-nitropropane without extensive bromine transfer. The authors suggest that the 'key to successful utilisation of these reactions involves the formation of an intermediate radical anion by trapping of a radical by an ion'[93]. Catalysis of these reactions by light or by radical anions (sodium naphthenide or potassium nitrobenzenide) also lends support to a radical process.

The nitro-group of α-nitro-esters, -ketones, and -nitriles and α,α-dinitro compounds is readily replaced by a variety of anions, thus providing an easy method for the preparation of some highly branched compounds[94].

$$R^1CMe_2\cdot NO_2 + R_2^2\bar{C}\cdot NO_2 \quad \overset{+}{Li} \longrightarrow R^1CMe_2\cdot CR_2^2NO_2$$

Reaction of ethyl α-nitroisobutyrate ($R^1 = CO_2Et$) with the lithium salt of 2-nitropropane ($R^2 = Me$) gave a 95% yield of product. When the anion is that of nitrocyclohexane ($R^3-R^2 = -(CH_2)_5$), the corresponding alkylate is obtained in 82% yield. Also readily replaced are the nitro groups of α-nitroketones ($R^1 = PhCO$, $R^2 = Me$; 80–85% yield) and α-nitronitriles ($R^1 = CN$, $R^2 = Me$; 90% yield). The displacement of a nitro group from α,α-dinitro compounds was also found to take place readily. This displacement of a nitro group by nitroalkane salts is believed to have a mechanism involving a radical-anion and a free radical, rather than being an S_N2 displacement[95].

Direct fluorination of salts of nitro-compounds[96, 97] has been used to prepare 1-fluoro-1,1-dinitroalkanes. This method is also applicable to the salts of 2-nitro-alcohols and -carboxylic esters.

$$Me(CH_2)_nCX{:}NO_2^- \xrightarrow[H_2O]{F_2} Me(CH_2)_n{\cdot}CFX{\cdot}NO_2 \xrightarrow[H_2O]{NaOBr} Me(CH_2)_n{\cdot}CBrF{\cdot}NO_2$$

$(n = 1)$

$$(X = CH_2OH \text{ or } CO_2Et)$$

Direct fluorination of a series of nitroalcohols $(X = CH_2OH)$, 2-nitro-butan-1-ol $(n = 1)$, -pentan-1-ol $(n = 2)$, -hexan-1-ol $(n = 3)$, and -heptan-1-ol $(n = 4)$ afforded the corresponding 2-fluoro-2-nitroalcohols in yields of 21–42% [98]. The fluorination of the nitronate salt from ethyl 2-nitropenta-noate $(n = 2, X = CO_2Et)$ gave ethyl 2-fluoro-2-nitropentanoate in 85% yield. Reaction of 2-fluoro-2-nitrobutan-1-ol with sodium hypobromite solution gave 1-bromo-1-fluoro-1-nitropropane and oxidation of 2-fluoro-2,2-dinitroethanol gave fluorodinitromethane.

Reaction of t-butyl hypochlorite with nitromethane in the presence of styrene yielded chloronitromethane and t-butyl alcohol in nearly quantitative

$$Me_3C{\cdot}OCl + Me_3NO_2 \xrightarrow{PhCH{:}CH_2} ClCH_2{\cdot}NO_2 + Bu^tOH$$

yield [99]. The reaction is accomplished only in the presence of an alkene which evidently serves as a catalyst. Other alkenes which have been used are hex-1-ene, tetramethylethylene, and 1,2-dichloroethylene, and the indications are that the rate of reaction increases with increasing basicity of the alkene. Active aromatic compounds also catalyse the reaction. The reaction of nitroethane is similar to that of nitromethane. The mechanism proposed involves formation of a π-complex between the alkene and the hypochlorite

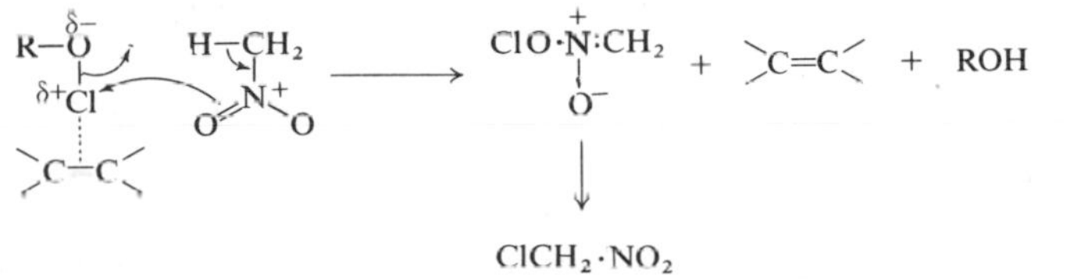

followed by reaction of the polarised bond with nitromethane to give an intermediate which easily rearranges to the chlorinated product.

Reaction of methyl $(R^1 = Me)$ and phenyl $(R^1 = Ph)$ 2-(2-nitroethyl)-benzoate $(R^2 = H)$ in base leads to a Dieckmann-type cyclisation which affords 2-nitroindanone [100]. However, when methyl 2-(1-methoxy-2-nitro-

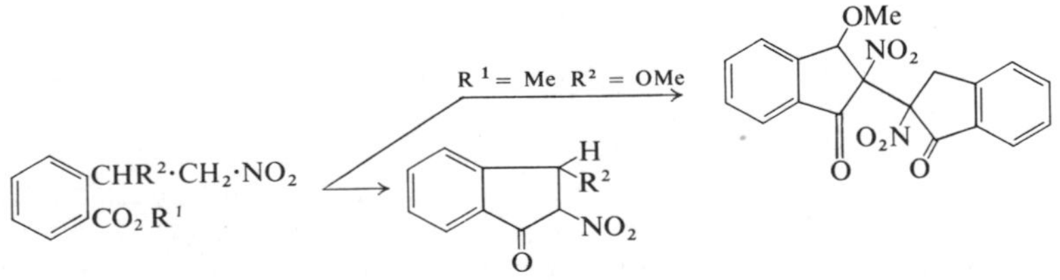

ethyl)benzoate $(R^1 = Me, R^2 = OMe)$ was treated with methanolic sodium methoxide, the product proved to be 3-methoxy-2-nitro(2-nitroindanone-2-yl)indanone rather than the expected cyclised product. This reaction was explained by postulating elimination of methanol from the normal cyclisation product to give the α,β-unsaturated ketone, which then underwent Michael addition with the surviving product of the normal cyclisation.

4.4.2.2　Reactions of α, β-unsaturated compounds

The reaction of nitro-alkenes with basic peroxides leads to a rapid transformation into α-nitroepoxides[101]. With β-methyl-β-nitrostyrene (R^1 = Ph, R^2 = Me), the α-nitroepoxide was isolated in 67% yield.

Higher yields were obtained with cis-α-nitrostilbene (R^1 = R^2 = Ph, 85% yield) and 1-cyclohexyl-2-nitroethylene (R^1 = cyclohexyl, R^2 = H, 91% yield). Reaction of the epoxide from β-methyl-β-nitrostyrene with nucleophiles (PhONa, PhSNa, Me_2NH, $PhCS \cdot NH_2$ and $LiAlH_4$) resulted in each case in attack at C-1. In dilute aqueous sulphuric acid, the nitroepoxide reacted to give 1-hydroxy-1-phenylacetone along with small amounts of the diketone. In catalytic reduction with platinum oxide in ethyl acetate, a mixture of 1-hydroxy-1-phenylacetone oxime and 1-hydroxy-2-nitro-1-phenylpropane was obtained, while reduction with zinc in aqueous acetic acid led only to the oxime.

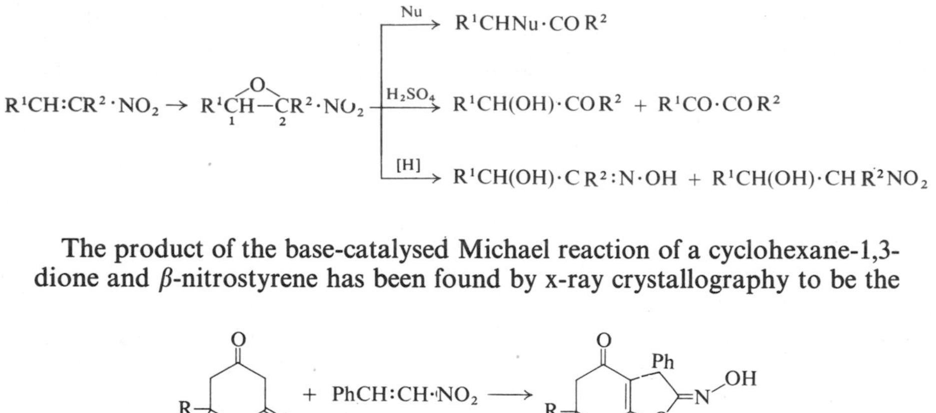

$$R^1CH:CR^2 \cdot NO_2 \rightarrow R^1\overset{O}{\overset{\frown}{CH}}-\underset{1}{\overset{}{C}}R^2 \cdot NO_2 \underset{2}{}$$

$$\xrightarrow{Nu} R^1CHNu \cdot COR^2$$

$$\xrightarrow{H_2SO_4} R^1CH(OH) \cdot COR^2 + R^1CO \cdot COR^2$$

$$\xrightarrow{[H]} R^1CH(OH) \cdot CR^2:N \cdot OH + R^1CH(OH) \cdot CHR^2NO_2$$

The product of the base-catalysed Michael reaction of a cyclohexane-1,3-dione and β-nitrostyrene has been found by x-ray crystallography to be the

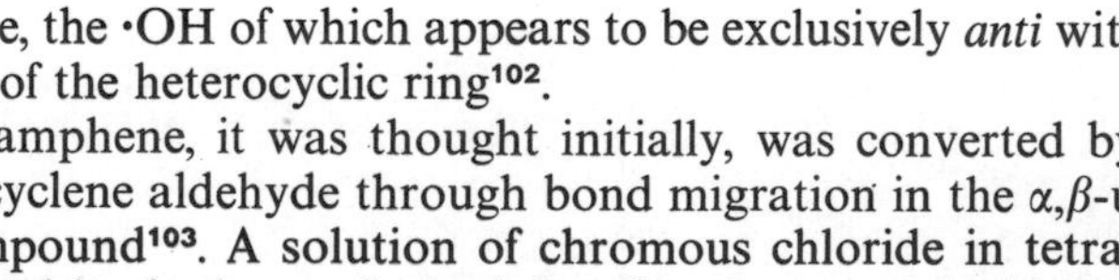

(R = H, Me)

above oxime, the ·OH of which appears to be exclusively *anti* with respect to the oxygen of the heterocyclic ring[102].

ω-Nitrocamphene, it was thought initially, was converted by reduction into the tricyclene aldehyde through bond migration in the α,β-unsaturated nitroso-compound[103]. A solution of chromous chloride in tetrahydrofuran was employed in the hope of obtaining the unsaturated nitroso-compound and, subsequently, the tricyclene aldehyde[104]. The product isolated, however, was the acyloin derivative in 78% yield. The nitroso-compound, rather than undergoing the expected bond migration, gave the ring expanded product by a reaction sequence including hydrolysis and ring expansion.

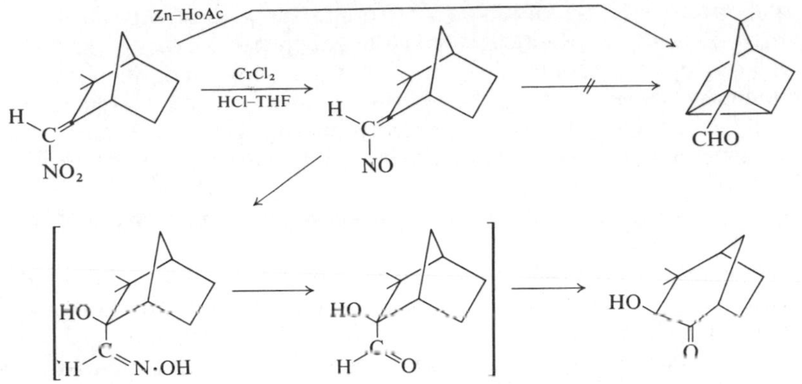

4.4.2.3 Oxidation-reduction

When heated in alkaline high boiling ($>200\,°C$) alcohols such as glycerol or diethylene glycol, the salts of 2-nitronorbornanes are largely reduced to

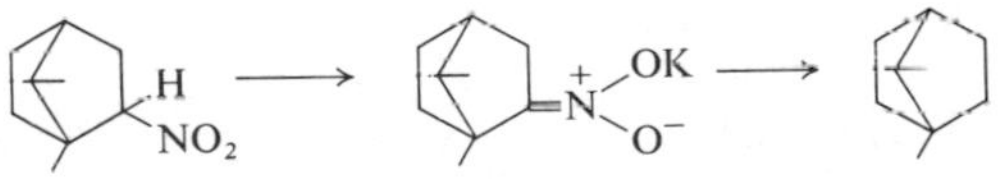

the parent hydrocarbon in good yield (60 %)[105]. Camphor (10–13 %) and camphor oxime (7–9 %) are formed as by-products in the reaction.

Persulphate oxidation of nitro-compounds has been found to yield secondary vicinal dinitro-compounds and carbonyl compounds[106]. Reaction of phenylnitromethane anion ($R^1 = Ph$, $R^2 = H$) with ammonium persul-

$$R^1R^2C{:}NO_2^- \xrightarrow{S_2O_8^{2-}} R^1R^2C(NO_2){\cdot}CR^1R^2NO_2 + R^1R^2CO$$

phate led to the meso- (15–20 %) and the (±)-1,2-dinitro-1,2-diphenyl ethane (40–65 %) and benzaldehyde (15 %). Rates of oxidative dimerisation of several alkane nitronates were greatly increased by small amounts of silver ion, although yields were essentially identical with those from the uncatalysed reactions. Chloroform used as a heterogeneous extractant was found to minimise transformation of the dimers into conjugated nitro-alkenes and their addition products.

The preparation of 1,4-diketones (valuable intermediates in furan and cyclopentanone synthesis) from γ-nitroketones has been accomplished by using titanium trichloride under relatively mild conditions[107]. γ-Nitroketones

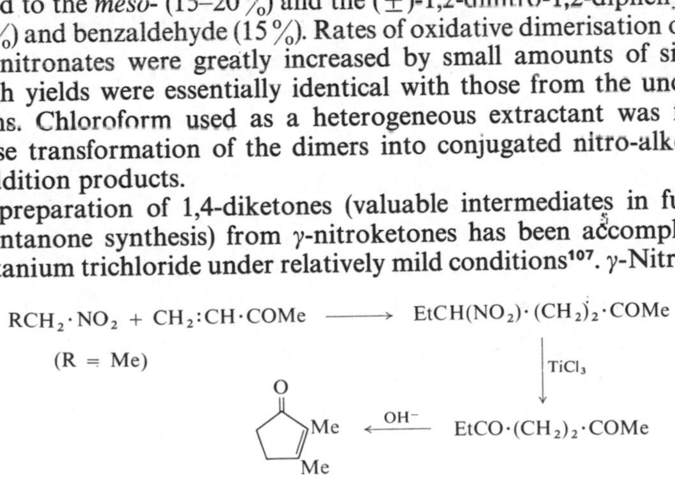

$$RCH_2{\cdot}NO_2 + CH_2{:}CH{\cdot}COMe \longrightarrow EtCH(NO_2){\cdot}(CH_2)_2{\cdot}COMe$$

$$(R = Me)$$

are easily prepared by addition of a nitronate anion to an α,β-unsaturated carbonyl compound. Reduction of 5-nitroheptane-2-one to give heptane-2,5-dione was accomplished in 85% yield with titanium trichloride. 'Although for simple cases it is convenient simply to use unbuffered aqueous $TiCl_3$ (pH < 1), it is significant synthetically that in delicate cases the reaction can be carried out at pH $\approx$ 4, making this a mild method indeed'[107]. Previous methods of converting nitro-groups directly into carbonyl groups (Nef reaction, permanganate oxidation) require harsh conditions and often result in poor yields.

Permanganate oxidation of alkane nitronate anions has been followed by spectrophotometric stopped-flow techniques: the reaction is first order with respect both to the permanganate ion and to the nitronate anion, and zeroth order with respect to the hydroxide ion[108]. The kinetic results are consistent with a rate determining attack of permanganate ion at the carbon of the C=N bond, with a synchronous movement of a pair of electrons to the

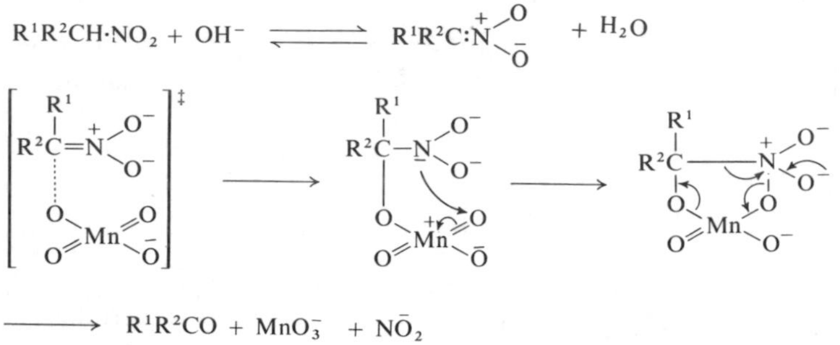

nitrogen atom, forming an activated complex. Carbon–oxygen bond formation and closure to the cyclic complex are followed by rearrangement to the ketone.

Kinetic studies on the inverse effect of water in the hydrolysis of nitro-aliphatic compounds reveal the existence of two reaction pathways for the

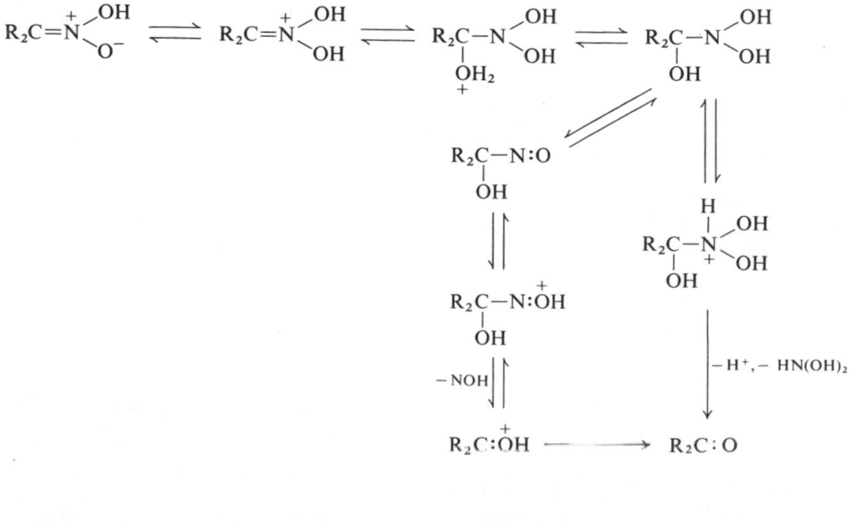

Nef reaction[109]. The relative amounts of reaction by these pathways depends upon the concentration of water.

The conversion of bromonitro-compounds in sulphuric acid into ketones has been found to proceed not through an initial loss of the nitro-group but via loss of Br^+ followed by a transformation closely resembling the Nef reaction[110].

$$Me_2CBr \cdot \overset{+}{N}(:O) \cdot OH \xrightarrow{-Br^+} Me_2C:\overset{+}{N}(OH) \cdot \bar{O} \longrightarrow Me_2CO$$

This mechanism was suggested by the isolation of p-dibromobenzene in 63% yield when a mixture of 2-bromo-2-nitropropane and bromobenzene was shaken with sulphuric acid. This confirmed the presence of a brominating species in the reaction mixture.

4.4.2.4   Additional reactions

The rate of proton abstraction for cis-trans isomers of certain nitrocyclo-alkanes has been shown to be strongly diminished by substitution of a trans-2-phenyl group in nitrocyclohexanes: c. 350-fold relative to that for the cis isomer[111]. Comparison of p-values for 1-aryl-2-nitropropanes (0.87), 2-aryl-1-nitrocyclopentanes (cis 0.89, trans 1.45), and 2-aryl-1-nitrocyclo-hexanes (cis 0.84, trans 1.23) reveals that the aryl groups are much closer to the acidic proton in the trans isomer than in the cis, thus indicating that the chair is deformed in trans-2-aryl-1-nitrocyclohexanes and accounting for the large differences in ground-state energies in 2-substituted nitrocyclo-hexanes.

If interference of the aryl group with the removal of the acidic hydrogen in 2-aryl-1-nitrocyclohexanes causes retardation of deprotonation in the trans-isomer[111], this same factor must then cause retardation of protonation to form the trans-isomer in the microscopic reverse reaction[112]. It was proposed originally that the preference for equatorial attack and thus formation of the less stable isomer in 2-substituted cyclohexane nitronate ions was due to greater hindrance to axial attack by the two axial hydrogens, compared to the extent to which the three axial hydrogens hinder equatorial

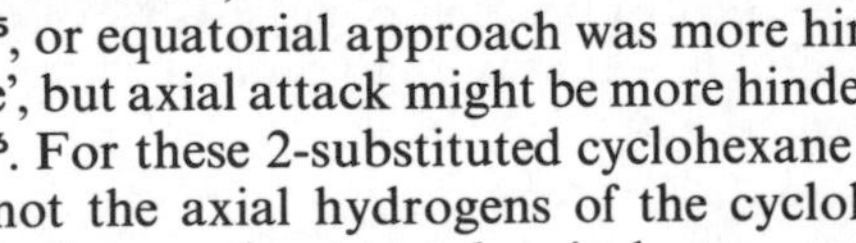

trans   cis

attack[113, 114]. In analogous situations, attack at carbonyl groups in cyclo-hexanones has been observed, for which it was concluded that axial approach was less hindered[115], or equatorial approach was more hindered if the transition state came 'late', but axial attack might be more hindered if the transition state came 'early'[116]. For these 2-substituted cyclohexane nitronate ions, it is the 2-substituent, not the axial hydrogens of the cyclohexane ring, which exerts the principal effect on the stereochemical outcome.

When nitroethane was photolysed in the presence of cyclohexane, the product isolated was acetamidocyclohexane in 10% yield[117]. The reaction can be rationalised in terms of initial hydrogen abstraction from cyclohexane by the excited nitro group, combination of the two radical species,

$$EtNo_2 \xrightarrow[RH]{hv} Et\overset{+}{N}R(OH)\bar{O} \xrightarrow{-H_2O} MeCH\!:\!\overset{+}{N}R\!\cdot\!\bar{O} \xrightarrow{hv}$$

$$MeCH\!\!-\!\!NR \longrightarrow AcNHR$$
$$\diagdown O \diagup$$

elimination of water, and photorearrangement of the resulting nitrone to the amide by way of the oxaziridine. The oxaziridine was synthesised independently from the corresponding amine, and on photolysis it yielded the amide. A similar reaction occurred with nitroethane in diethyl ether $(R = CH_2\!\cdot\!CH_2\!\cdot\!OEt)$.

Primary nitro compounds with acetic anhydride–sodium acetate at the temperature of the steam bath, afford triacylhydroxylamines in high yields[118].

$$RCH_2NO_2 \xrightarrow{Ac_2O,\ NaOAc} RCO\!\cdot\!NAc\!\cdot\!OAc$$

4.5 C- AND N-NITROSO COMPOUNDS

4.5.1 Synthesis

With an amine containing no α-hydrogen, hydrogen peroxide oxidation with a sodium tungstate catalyst yields the nitroso-compound rather than the oxime, the product isolated when α-hydrogens are present[119]. Thus 2-methyl-2-nitrosopropane has been prepared from t-butylamine in 24% yield, while 2,4,4-trimethylpent-2-ylamine gives the nitroso-compound in 36% yield[120]. The other product of the oxidation of Bu^tNH_2 is 2-methyl-2-

$$RNH_2 + H_2O_2 \xrightarrow{Na_2WO_4} RNO \longrightarrow RNO_2$$

nitropropane, the yield of which increases with increased concentration of peroxide. Unlike most aliphatic nitroso-compounds which exist as dimers, these highly hindered t-alkyl nitroso-compounds are monomeric in solution.

Studies of i.r. and n.m.r. spectra indicate strongly that nitrosation of aldehyde arylhydrazones occurs at nitrogen (X = H; Y = NO) and not at carbon (X = NO; Y = H)[121].

$$RCH\!:\!N\!\cdot\!NHAr \xrightarrow{HONO} RCX\!:\!N\!\cdot\!NYAr \longrightarrow RC(\!:\!N\!\cdot\!OH)\!\cdot\!N\!:\!NAr$$
$$\searrow$$
$$RC(NO_2)\!:\!N\!\cdot\!NHAr$$

This conclusion arises from the fact that the electronic absorption spectrum of the product from benzophenone phenylhydrazone, which could not contain a C-nitroso group, was amazingly similar to that obtained from the product of nitrosation of anisaldehyde.

$$Ph_2C\!:\!N\!\cdot\!NPh\!\cdot\!NO \qquad\qquad p\text{-MeO}\!\cdot\!C_6H_4CH\!:\!NPh\!\cdot\!NO$$

Thus, the formation of C-nitrohydrazones and the azo-oximes from Nnitrosohydrazones must involve a rearrangement, which, in the case of nitrohydrazones, is also accompanied by oxidation.

It is well known that the addition of nitrosyl chloride to an alkene yields the chloronitroso addition product, which may then react by one or both of two pathways: (a) dimerisation of the nitroso group, (b) oxidation of the nitroso group to a nitro group[122]. Recently a third pathway (c) has been observed, namely, isomerisation to an oxime followed by oxidation to a nitrimine; this reaction has been accomplished by using nitrosyl chloride[123].

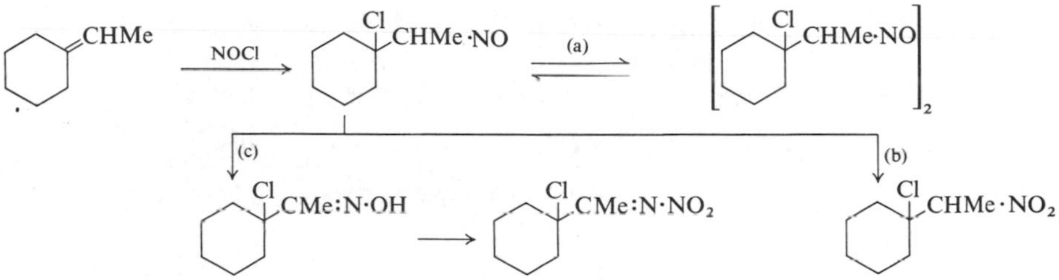

For example, addition of NOCl to ethylidenecyclohexane in ether gives 75% of the equilibrium mixture of monomer and dimer, 16% of the chloronitro-compound and 7% of the chloronitrimine[124].

4.5.2 Reactions

4.5.2.1 Photo-additions to alkenes

The photochemical reactions of N-nitrosopiperidine in mineral acid including (a) photo-elimination, (b) photo-addition, (c) fragmentation–recombination in the presence of methanol to give piperidine, formaldehyde, and N piperi dinoformamide, all appear to involve a common reactive transient species derived from a singlet excited state of the N-nitrosopiperidine–acid complex[125].

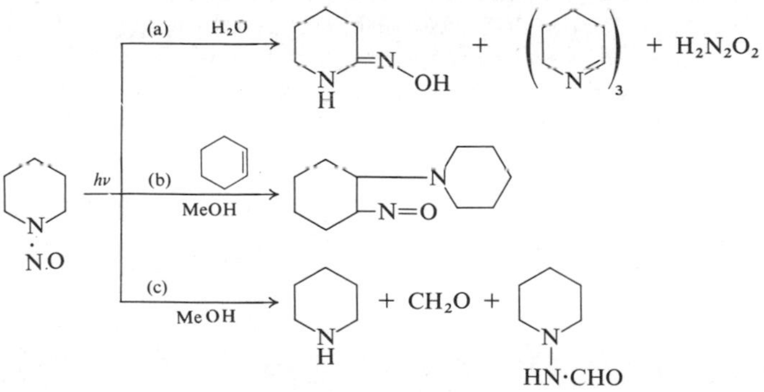

Photo-addition of N-nitrosopiperidine to symmetrical conjugated dienes, such as buta-1,3-diene, cyclopentadiene, cyclohexa-1,3-diene, and cis, cis-cyclo-octa-1,3-diene, leads to the 1,4-adduct as the major product, with smaller amounts of the 1,2-adduct also being produced[126].

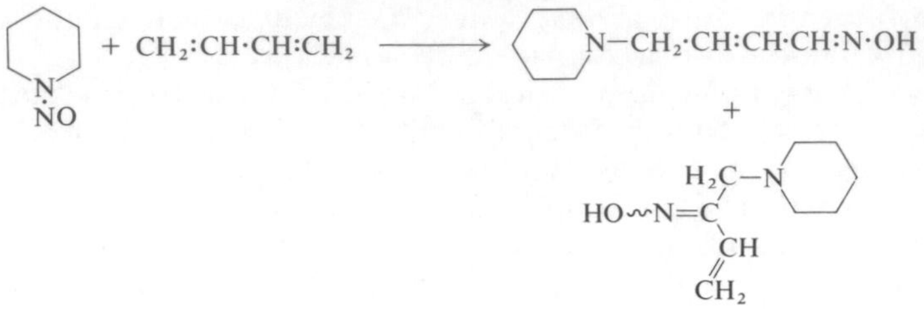

Mixtures of *syn*- and *anti*-isomers were also detected, except for the 1,4-adducts of cyclopentadiene and cyclohexa-1,3-diene which gave only the *anti*-isomer. The typical blue colour suggests that the *C*-nitroso compounds

$$R^1R^2\overset{+}{N}H\cdot CH_2\cdot CH{:}CH\cdot CH_2\cdot NO \qquad R^1R^2\overset{+}{N}H\cdot CH_2\cdot CH(NO)\cdot CH{:}CH_2$$

above are the primary photo-adducts during the photo-addition to cyclopentadiene.

Instead of the anticipated oxime, photoaddition of *N*-nitrosopiperidine to 3,3-dimethylbut-1-ene gave 3,3-dimethyl-1-piperidino-2-(*N*-nitrosohydroxylamino)butane[127].

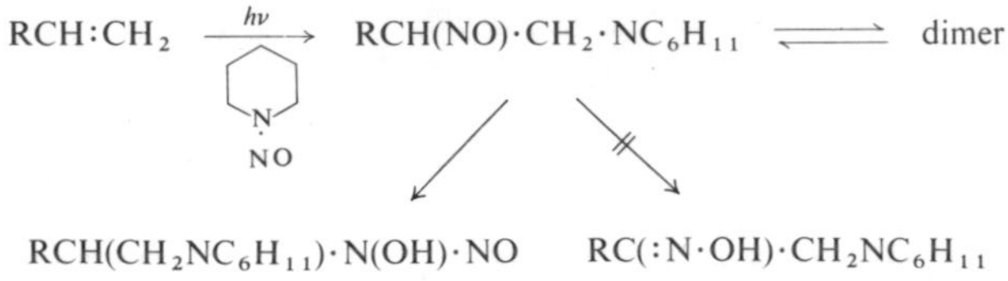

$$RCH{:}CH_2 \xrightarrow{h\nu} RCH(NO)\cdot CH_2\cdot NC_6H_{11} \rightleftharpoons dimer$$

$$RCH(CH_2NC_6H_{11})\cdot N(OH)\cdot NO \qquad RC({:}N\cdot OH)\cdot CH_2NC_6H_{11}$$

$$(R = Bu^t)$$

The formation of this product is credited to the high concentration (and therefore longer life time) of the monomer, which can then react with a continuous supply of HNO from the reaction. The t-butyl group is thought to have a steric effect which accelerates the dissociation of the dimer and also retards tautomerisation of the monomer to the oxime. Also, the enamine from the photoaddition of *N*-nitrosopiperidine to diphenylacetylene yielded on hydrolysis the expected benzil monoxime (R = Ph) in 61 % yield[128]. Also obtained in good yield was phenylgloxal ketoxime (R = H) which tautomerised reversibly to the *C*-nitroso compound, isolated as the dimer.

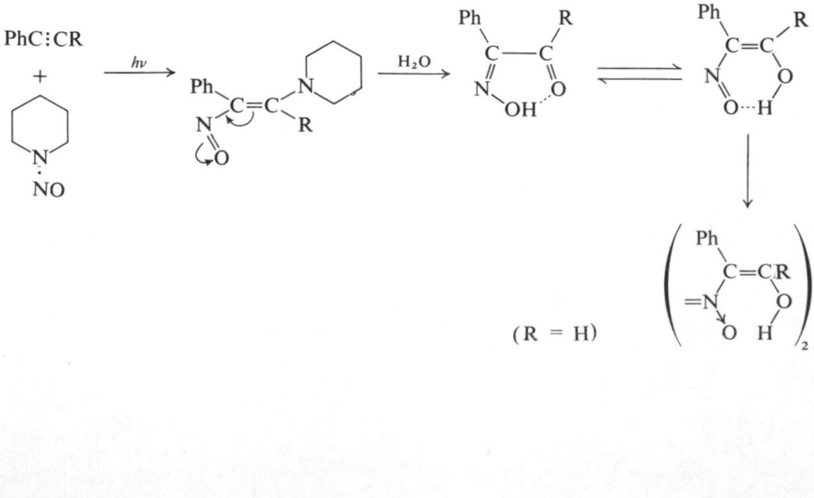

Radical-initiated intramolecular addition of nitrosamines appears to favour five-membered over six-membered rings[129, 130].

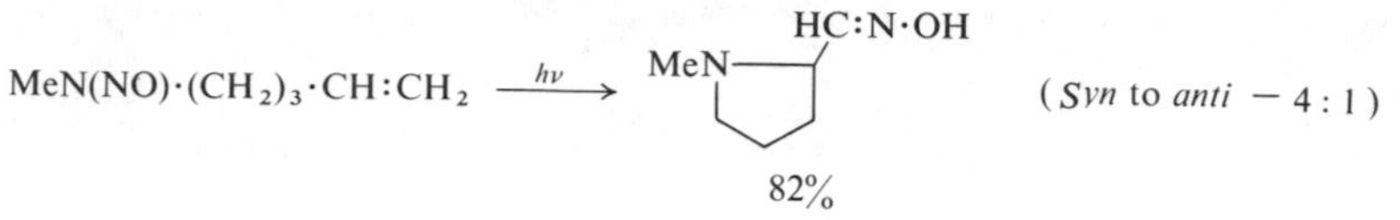
MeN(NO)·(CH2)3·CH:CH2
hν
MeN
HC:N·OH
82%
(Syn to anti – 4 : 1)

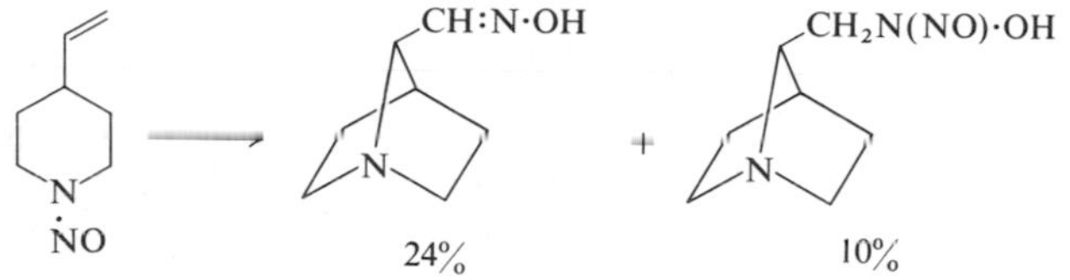
N
NO
CH:N·OH
N
24%
+
CH2N(NO)·OH
N
10%

This reaction provides a direct and efficient method for the synthesis of 5-membered azacyclic compounds from nitrosamines.

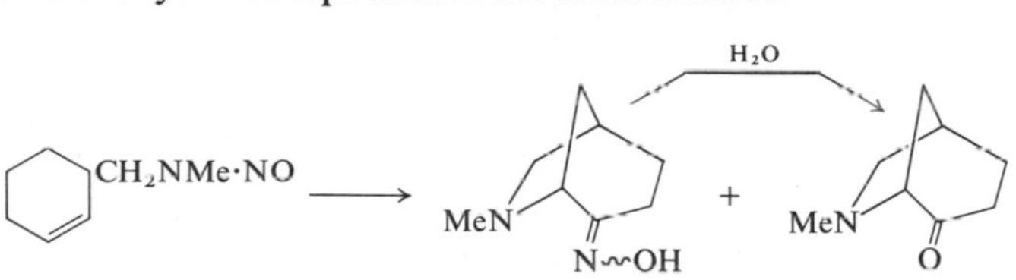
CH2NMe·NO
H2O
MeN
N~OH
+
MeN
O

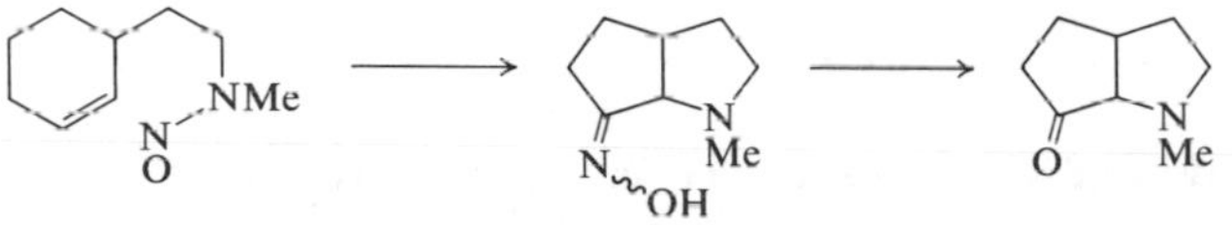
NMe
N
O
N
Me
N~OH
O
N
Me

4.5.2.2 *Azacyclic compound formation*

A new route to the synthesis of hexahydrotetrazines from aliphatic nitrosamines has been found[131]. The reaction of phenyl lithium with dimethyl- and diethyl-nitrosamine and *N*-nitrosopiperidine leads to the corresponding 'head-to-tail' dimers in fair yield (30–40 %). It is suggested that the reaction proceeds via an azomethine imine intermediate.

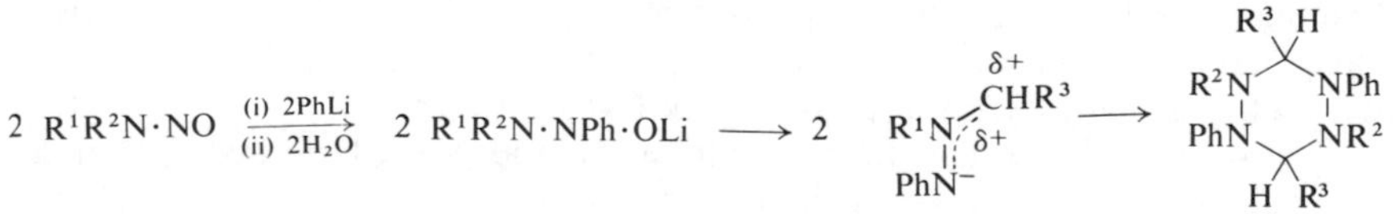
2 R1R2N·NO
(i) 2PhLi
(ii) 2H2O
2 R1R2N·NPh·OLi
2
δ+
CHR3
R1N
δ+
PhN–
R3 H
R2N
NPh
PhN
NR2
H R3

Reaction of the benzonorbornadiene nitrosochloride dimer with lithium aluminium hydride followed by acylation with toluene-*p*-sulphonyl chloride leads to the *exo*-aziridine as the major product with smaller amounts of the *exo*- and *endo*-tosylamides[132]. Treatment of the 7-oxa-analogue under similar conditions yielded two isomeric chlorine-containing tosylamides in approximately equal amounts, to which have been assigned the *exo-cis* and *endo-trans* configurations. This difference in behaviour can be rationalised by assuming that with the 7-oxa-compound the organometallic reagent complexes on the oxygen (and nitroso group) rather than displacing the chloride ion.

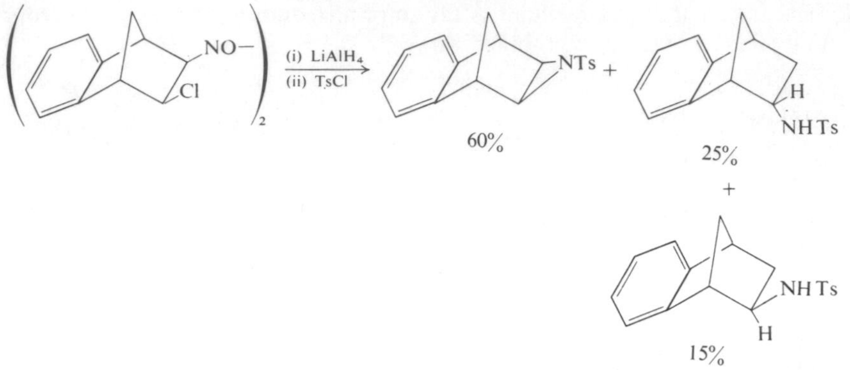

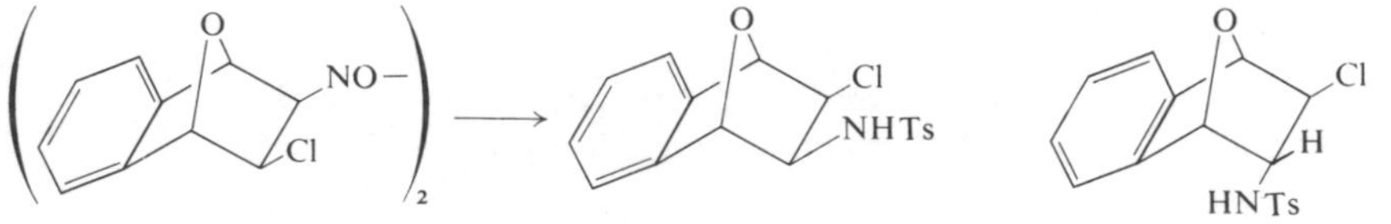

4.5.2.3 N-*nitroso compounds*

Acyloxy–t-butoxy radical pairs have been reported as generated in a deamination process involving decomposition of N-benzoyl-N-nitroso-O-t-butyl-hydroxylamine[133].

$$PhCO \cdot NH \cdot OBu^t \longrightarrow PhCO \cdot N(NO) \cdot OBu^t \longrightarrow PhCO \cdot O \cdot N:N \cdot OBu^t$$

$$PhCO_2Bu^t \longleftarrow PhC\overset{O}{\underset{O}{<}} \cdot \; + \; \dot{O}Bu^t$$

A similar reaction has been employed by White in the stereospecific intramolecular conversion of N-alkylamines into esters, followed by hydrolysis into alcohols, with partial retention of configuration[134-136].

N-Nitrosamides have also been employed in the photolytic generation of N-arylalkylacetamido radicals[137].

$$ArCH_2N(NO)Ac \xrightarrow{h\nu} ArCH_2\dot{N}Ac$$

Alkyl azides react rapidly and efficiently with stable nitrosonium salts to produce carbonium ions[138].

$$RN_3 + NO^+ \longrightarrow RN(NO) \cdot \overset{+}{N}:N \xrightarrow{-N_2}$$

$$RN\overset{+}{:}\overset{..}{N}\overset{..}{:}O \longleftrightarrow R\overset{+}{N}\cdot N\overset{..}{:}O \longrightarrow R^+ \text{ rearranged products}$$

For example, the organic products from the reaction of diphenylmethyl azide in acetonitrile included diphenylmethanol (26%), N-diphenylmethyl-acetamide (53%), benzophenone (13%), benzaldehyde (8%), and benzene diazonium tetrafluoroborate (8%).

$$Ph_2CHN_3 \xrightarrow[\text{MeCN}]{NO^+BF_4^-} \xrightarrow{H_2O} Ph_2CHOH + Ph_2CH\cdot NHAc + PhCHO +$$

$$\lceil Ph\overset{+}{N_2} \rceil BF_4^- + Ph_2CO$$

4.5.3 α-Nitrosoamino carbanions

Evidence for an α-nitrosoamino carbanion has been provided by studies of 4-methyl1-nitroso-piperazine in a base catalysed exchange with deuterium[139].

Also alkylation of NN-dimethylnitrosoamine in c. 15% yield has been effected.

$$Me_2N\cdot NO \xrightarrow[\text{NaH}]{MeI} MeN(NO)Et$$

$$\Big\downarrow \begin{smallmatrix}OD^-\\D_2O\end{smallmatrix}$$

$$(CD_3)_2N\cdot NO \xrightarrow[\text{NaH}]{MeI} CD_3\cdot N(NO)\cdot CD_2Me$$

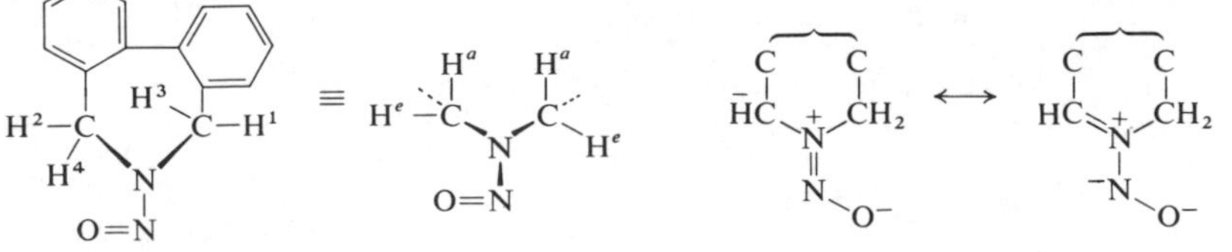

'It is noteworthy', the authors report, 'that canonical structures involving resonance delocalisation of the formal negative charge cannot be formulated, and the facility of the reaction must be attributed entirely to inductive effects'[139], because of the polarisation of the N—N—O group to effect ylid formation $R^1\overset{-}{C}H\cdot N\overset{+}{R^2}:NO$. On the other hand, an investigation of the relative rates of hydrogen–deuterium exchange of the four α-protons of N-nitroso-6,7-dihydro-1,11-dimethyl-5H dibenz[c,e]azepine has shown that the rate of exchange of H^4 by deuterium was 30 times that of H^2 [140]. Exchange of H^3 was also favoured over that of H^1 by c. 3.5 to 1. Thus, both quasi-equatorial protons exchanged faster than the quasi-axial protons. These rates are believed to reflect the stereochemical requirements for carbanion stabilisation by resonance interaction, and, in addition, the carbanions fomed by loss of protons H^3 and H^4 would have a more favourable geometry for resonance stabilisation. Further evidence against the ylid structure is provided by the fact that the quaternary salts of N-nitroso compounds exchange H for D more slowly than the compound itself, even though in this case the nitrogen would already be positive[141].

The tosylates and O-alkyl derivatives of some N-nitrosohydroxylamines have been prepared. Reaction of these tosylates with base produced the corresponding aldehyde[142].

$$RCH_2 \cdot N(NO) \cdot OH \longrightarrow RCH_2 \cdot N \overset{+}{:} \overset{-}{N}(OTs) \overset{-}{O} \xrightarrow{Bu^tO^-} RCHO + N_2 + TsO^-$$

O-Alkyl derivatives decomposed much more slowly in base and only when the α-methylene group was suitably activated (with Ph for example). Exchange studies in a suitable O-deuteriated alcohol indicated that O-alkyl derivatives undergo complete exchange of the benzylic protons (R = Ph), while the tosylates undergo decomposition since no deuteriated compound was recovered.

4.5.4   The C-nitroso chromophore

Some success has been reported in the application of the techniques of c.d. and o.r.d. to reactions in which steroidal and terpenoidal amines are converted into their C-nitroso derivatives and in which this chromophore is employed in studying the environment in the parent amine[143]. The long wavelength band (660–700 nm) of the nitroso function is readily distinguished and may be measured in the presence of any other organic chromophore.

Careful oxidation of the parent secondary amine with peroxyacetic acid yielded the nitroso monomer, with presumably the same configuration as the parent amine mixture. The spectra taken during reaction demonstrate that the nitroso chromophore is influenced by a disymmetric environment. In practice it proved difficult to obtain spectra with reproducible intensities. This problem can be overcome to some extent by employing the α-halogeno-C-nitroso-compounds derived from the oxime of the amine, since the bulky halogen suppresses dimerisation and there is no hydrogen available for tautomerisation.

$$R^1R^2C{:}N \cdot OH \xrightarrow[X = Br, Cl \text{ or } NO_2]{X_2} R^1R^2CX \cdot NO$$

Bromination of oximes of several steroids and terpenes gave blue nitroso compounds whose absorption band was readily discernible.

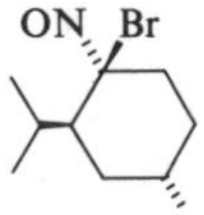

Bromonitrosomenthane

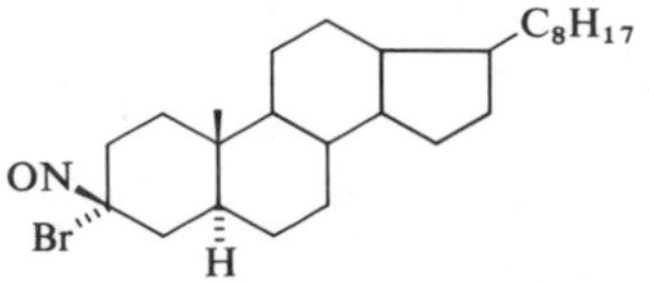

3α-Bromo-3β-nitroso-5α-cholestane

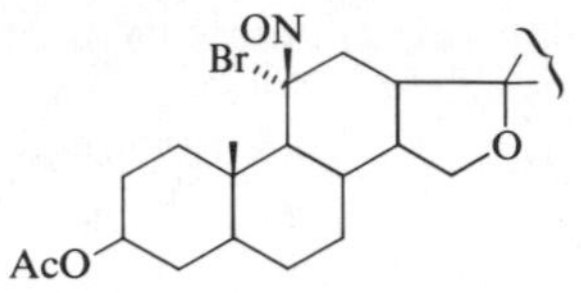

12α-Bromo-12β-nitrosotigogenin acetate 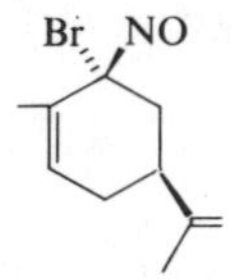   3-Bromo-2-nitroso-carvo-6,8-diene

In these compounds the halogen is assumed to be axial. Sign and frequency of absorbance are given for c. 20 steroids and terpenoids[143].

4.6 OXIMES

4.6.1 Synthesis

The addition of nitrosyl chloride to cyclopentadiene has been found to yield the nitroso dimer in 92% yield[144]. The transient existence of the monomeric

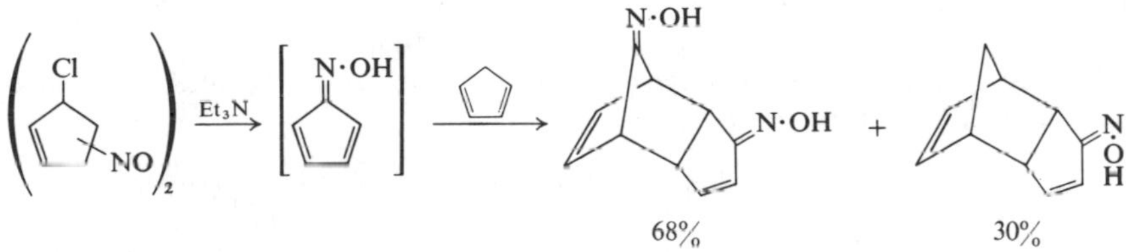

68%   30%

oxime was indicated by trapping experiments which afforded both the mono and the dioximino adduct, while a quantitative yield of dioxime was obtained in the absence of cyclopentadiene. Reaction of alkenes with dinitrogen trioxide affords 2-nitronitroso dimers which were thermally rearranged to nitroximes[145].

$$R^1CH{:}CHR^2 \xrightarrow{N_2O_3} (R^1CH(NO){\cdot}CHR^2NO_2)_2 \rightarrow R^1C({:}N{\cdot}OH){\cdot}CHR^2NO_2$$

The dinitrogen trioxide reaction product with *endo*-dicyclopentadiene, after isomerisation, affords the norbornyl addition product in high yield. Nitroximes have also been obtained from the reaction of dinitrogen trioxide with

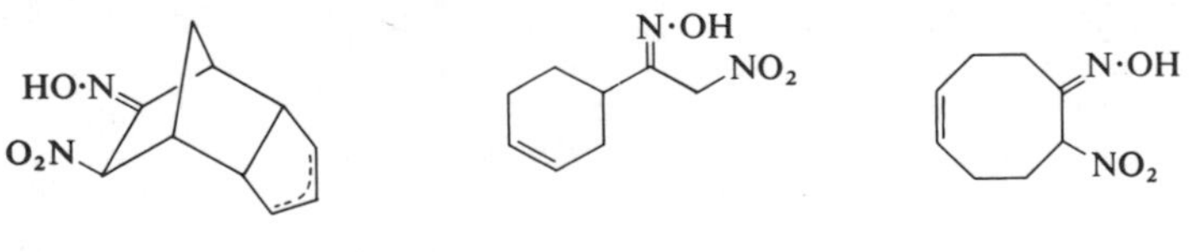

4-vinylcyclohexene and cyclo-octa-1,5-diene. These 1,2-nitroximes can be selectively reduced with palladium-on-carbon to give 1-hydroxylamino-2-oximes, which can react further with hydride to yield vicinal diamines, or

react with another molecule of the hydroxylamino-oxime to afford pyrazine NN-dioxides[146].

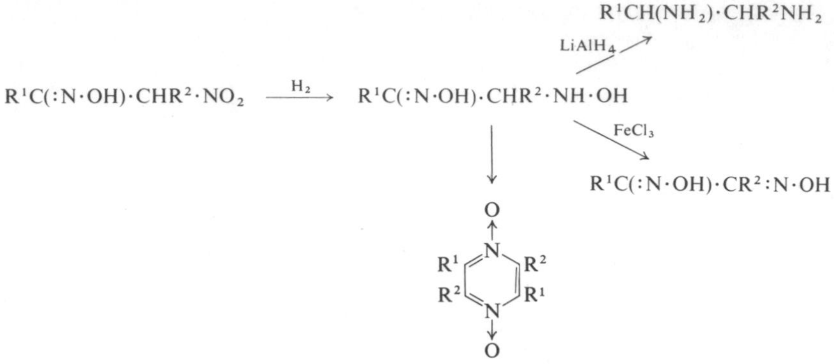

$$R^1C(:N \cdot OH) \cdot CHR^2 \cdot NO_2 \xrightarrow{H_2} R^1C(:N \cdot OH) \cdot CHR^2 \cdot NH \cdot OH$$

$$\xrightarrow{LiAlH_4} R^1CH(NH_2) \cdot CHR^2NH_2$$

$$\xrightarrow{FeCl_3} R^1C(:N \cdot OH) \cdot CR^2 : N \cdot OH$$

4.6.2   Reactions

4.6.2.1   With Grignard reagents

Investigations of the reaction of oximes with Grignard reagents indicate that the aziridines formed depend upon the configuration of the oxime[147, 148]. For example, the pure *anti*-benzyl oxime below gave the less substituted aziridine as the major product.

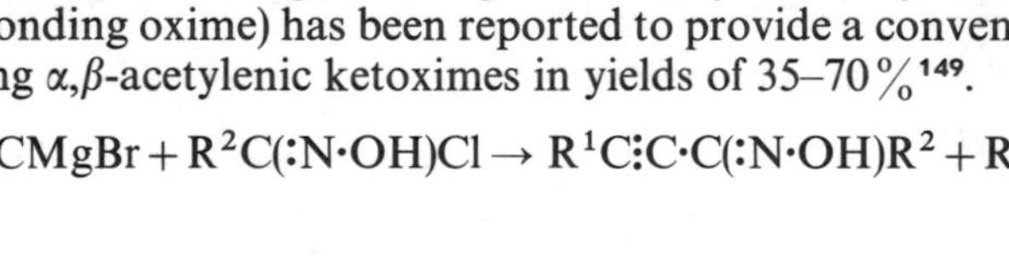

Reaction of acetylenic Grignard reagents with hydroximoyl chlorides (from the corresponding oxime) has been reported to provide a convenient method for preparing α,β-acetylenic ketoximes in yields of 35–70 %[149].

$$2R^1C \vdots CMgBr + R^2C(:N \cdot OH)Cl \rightarrow R^1C \vdots C \cdot C(:N \cdot OH)R^2 + R^1C \vdots CH$$

4.6.2.2 Cyclisations

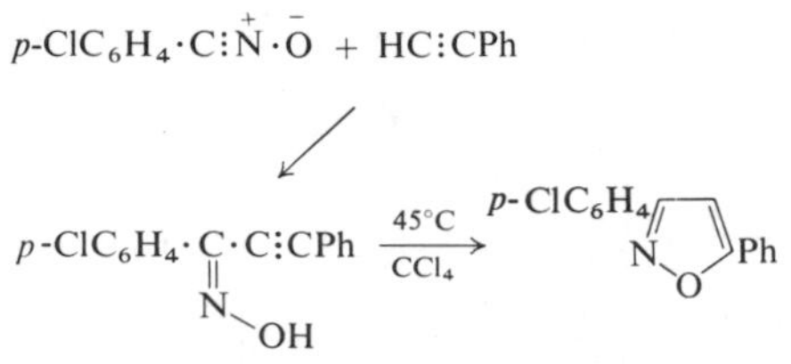

One such α,β-acetylenic ketoxime has been observed to undergo a slow, first-order isomerisation to an isoxazole[150]. This isoxazole was the major product (80%) from the reaction of *p*-chlorobenzonitrile *N*-oxide with a tenfold excess of phenylacetylene, while the oxime was obtained from the same reaction in only modest yield (16%).

5-Hydroxy-3,4-diphenyl-Δ^2-isoxazoline has been prepared (42%) by the reaction of dimethylsulphonium methylide with α-benzil monoxime, presumably through an epoxide, which, upon rearrangement to an aldehyde,

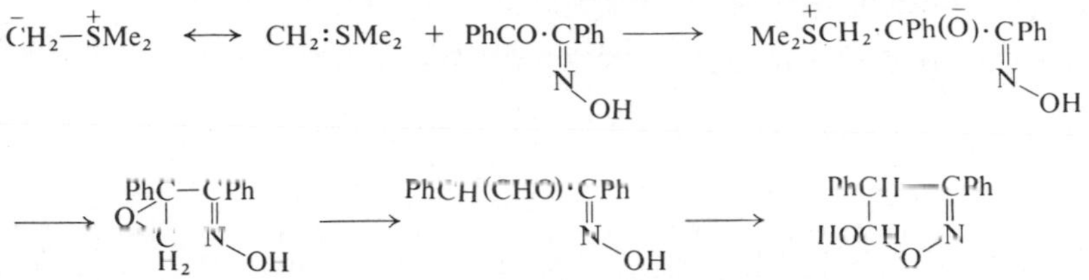

gave the product by nucleophilic attack of the oxime upon the carbonyl carbon[151]. Another example of ring closure by reaction of oximes and carbonyl compounds, this time to give 6-membered rings, is seen in the formation of 6-hydroxy-5,6-dihydro-1,2,4*H*-oxazines[152].

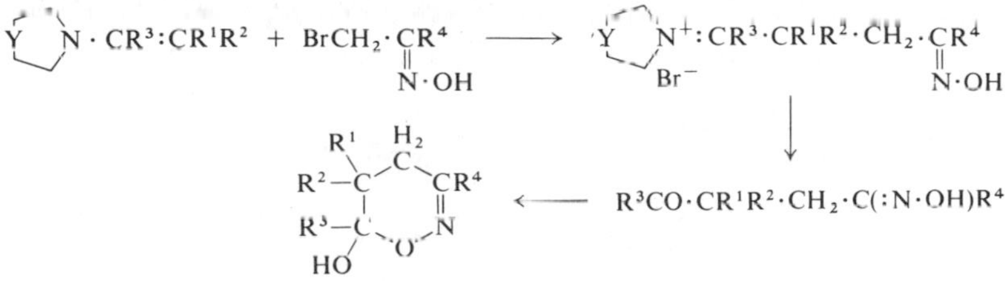

$$Y = -, CH_2 \text{ or } O$$

α-Halogenoketoximes were found to react easily with enamines to give salts which undergo hydrolysis to the oxazines, very probably through oximes. Attempted acylation of N^1N^2-trimethylene-N^3-hydroxyguanidine with ethyl chloroformate in the presence of potassium carbonate readily gave 5,6,7,8-tetrahydro-[1,2,4]oxadiazolo[3,4-*a*]pyrimidine-3-one[153]. The reac-

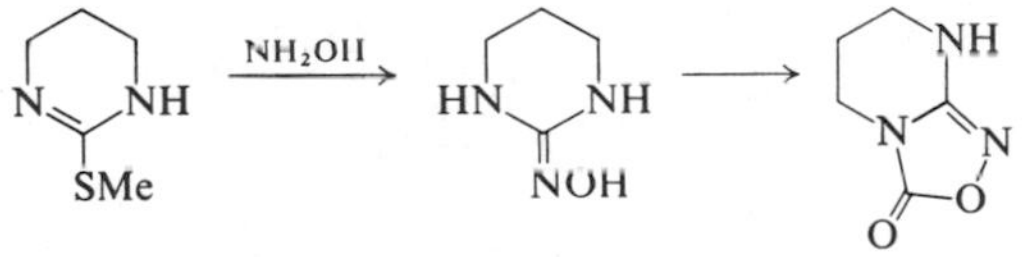

tion of 'activated' esters with amidoximes offers what appears to be a convenient and general route to 1,2,4-oxadiazoles and is particularly useful for the preparation of lower boiling dialkyl derivatives[154]. The reaction involves an initial *O*-acylation of the amidoxime followed by cyclisation to the oxadi-

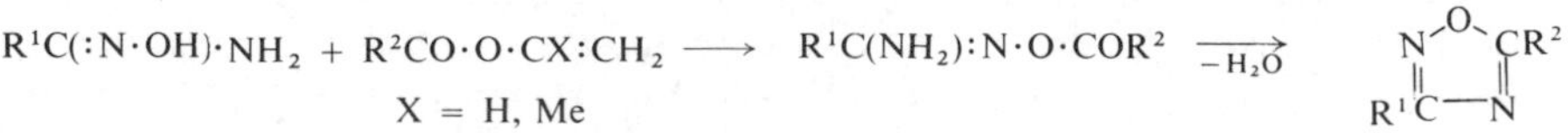

$$R^1C(:N\cdot OH)\cdot NH_2 + R^2CO\cdot O\cdot CX:CH_2 \longrightarrow R^1C(NH_2):N\cdot O\cdot COR^2 \xrightarrow{-H_2O}$$

$$X = H, Me$$

azole. Yields range from 21 to 69%.

Alkylation of the potassium and silver salts of alkyl benzohydroxamates with primary alkyl halides has been found to yield a mixture of N-alkyl-benzohydroxamates (major product) and *syn*- and *anti*-O-alkylbenzo-hydroximates, while the use of isopropyl halides led to mixtures in which the latter predominate[155]. When potassium benzohydroxamates were allowed to react with α,ω-dihalogenoalkanes cyclised products were obtained.

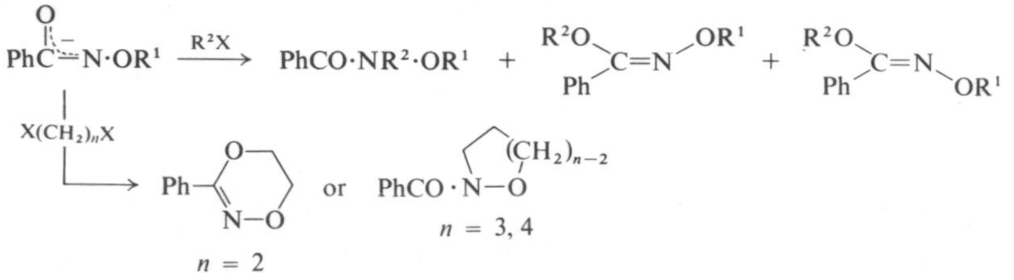

Addition of nitrosyl chloride to several oximes (X = OMe) results in oxidation to a nitrimine and in replacement of bromine by chlorine (Y = Cl)

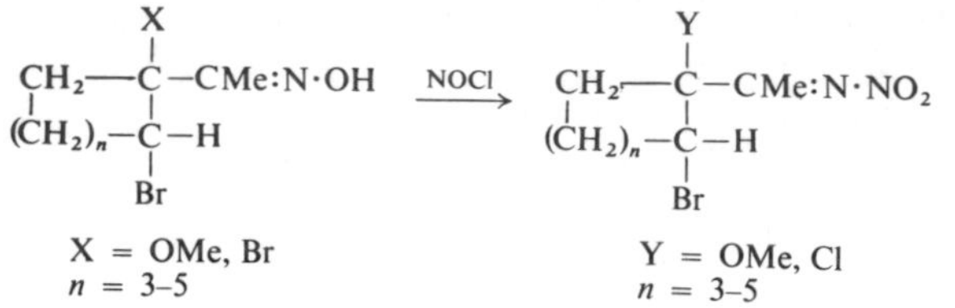

in the case of α-bromoketoximes (X = Br)[156].

4.6.3 Beckmann rearrangements

Reaction of *anti*-homoadamantan-4-one oxime with polyphosphate ester (PPE) occurred by a stereospecific Beckmann rearrangement to give 4-aza-tricyclo[5.3.1.1^{3,9}]dodecan-5-one in 58% yield with 32% recovery of the

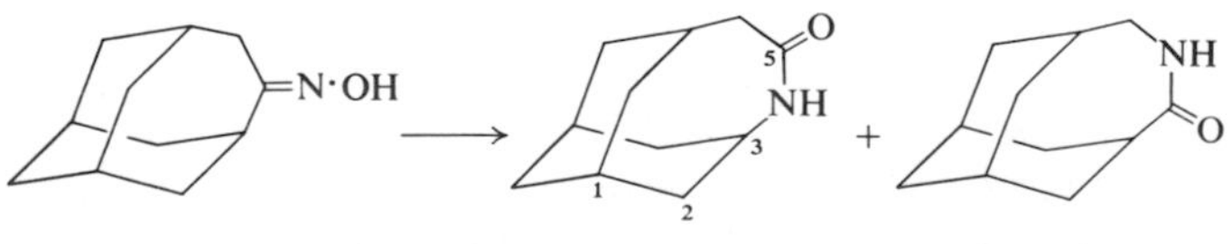

starting oxime[157]. In 85% H_2SO_4, however, a mixture of the 4-aza-product and the isomeric lactam, 5-azabis-homoadamantan-4-one was obtained (1:4 ratio). No abnormal Beckmann (fission) product was observed even in concentrated sulphuric acid. This had been observed for adamantanone oxime under similar conditions[158], and is attributed to the difference in the spatial arrange-ments of the participating bonds in the adamantane and homoadamantane derivative. Rearrangement of adamantanone oxime under photochemical conditions by irradiation in acetic acid failed to yield fission products, but 4-azatricyclo[4.3.1.1^{3,8}]undecan-5-one was isolated in 89% yield along with the starting ketone $(8\%)^{159}$.

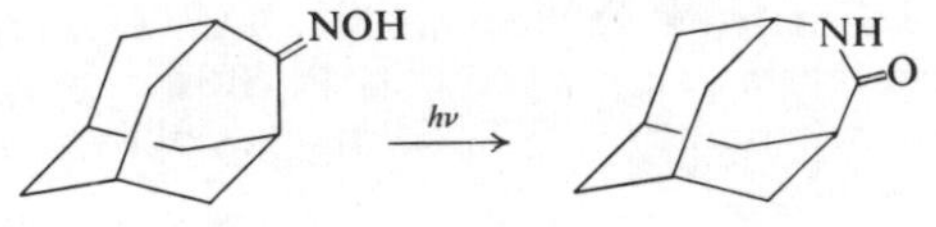

The photolysis of *anti-α-oximinocyclododecanone*, in addition to iso-merisation, has been found to yield abnormal Beckmann products[160]. The

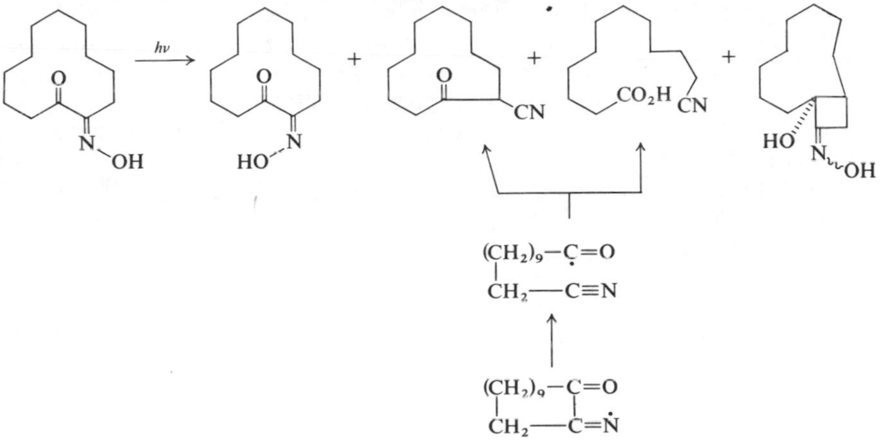

$$(CH_2)_9-\overset{\cdot}{C}=O$$
$$CH_2-C\equiv N$$

$$\uparrow$$

$$(CH_2)_9-C=O$$
$$CH_2-\overset{\cdot}{C}=\overset{\cdot}{N}$$

reaction is thought to proceed by an initial photochemical fission of the N—O bond, followed by fission of the bond adjacent to the carbonyl group to produce the cyano group and the carbonyl radical. Loss of hydrogen α to the cyano group would afford the ketonitrile, while recombination with ·OH radicals from N—O bond cleavage would yield the ω-cyanoundecanoic acid.

Generally in the Beckmann rearrangement, the group which migrates is the one *anti* to the hydroxy group. Several investigations of this rearrange-ment in α,β-unsaturated systems have suggested that oximes with the allyl group *syn* undergo facile rearrangement to the lactam, but that the *anti*-isomer resists rearrangement, suggesting that alkeneic are not so effective migrating groups as alkyl groups[161-163].

It has been recently observed, however, that when several cyclic α,β-unsaturated ketoximes (or their tosylates) were subjected to the conditions of the Beckmann rearrangement the group *anti* to the hydroxy group migra-ted efficiently, whether alkyl or alkenic[164]. The exception was the oxime of *anti*-3-methylcyclohex-2-enone (R = Me, n = 3); this was attributed to

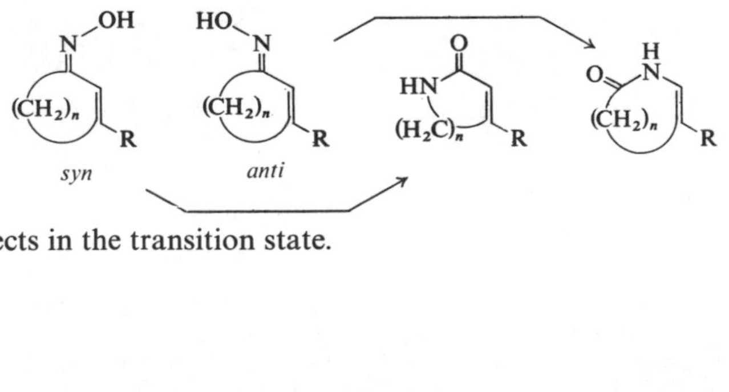

steric effects in the transition state.

Complexes of sulphur trioxide with various Lewis bases have been used as reagents for Beckmann rearrangements[165]. Studies with cyclohexanone oxime indicated that when amide–SO_3 complexes of five carbons or less are

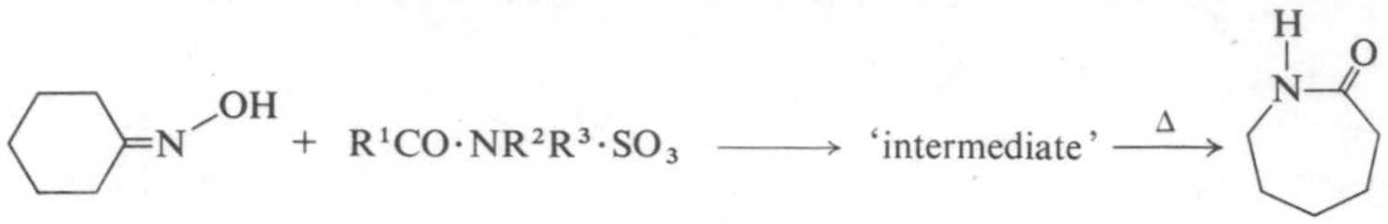

used to form the intermediate, the oxime is regenerated upon hydrolysis, while those intermediates formed from amide complexes containing six carbons or more will proceed to the lactam. All cyclohexanone oxime intermediates investigated afforded the lactam in good yield under thermolytic conditions, a process which is highly exothermic and once initiated is essentially spontaneous.

The oximes of the substituted cyclohexanones, 2-cyclohexylcyclohexanone and 2-(1-chlorocyclohexyl)cyclohexanone, with PCl_5 in diethyl ether give

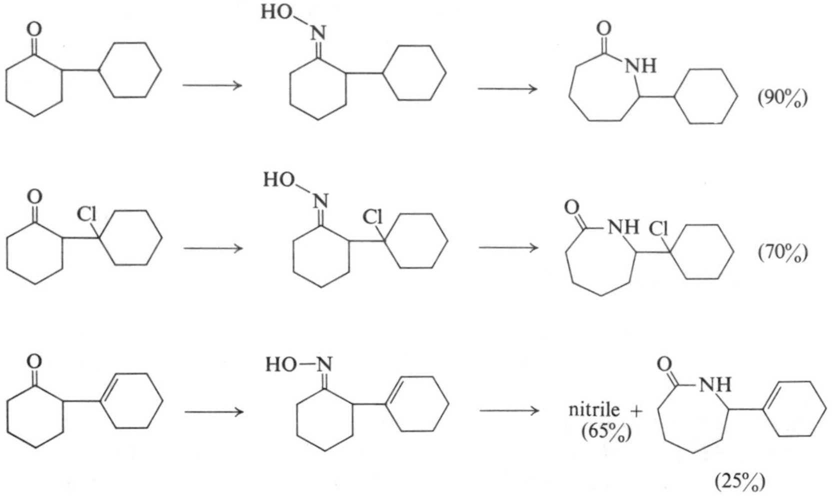

rise to Beckmann rearrangement products in high yields[166]. The rearrangement of 2-(cyclohex-1-enyl)cyclohexanone oxime when carried out under similar conditions gave in addition to the normal Beckmann product a large amount of nitrile or mixture of nitriles (fission products), possibly derived from the stable allylic carbonium ion, and formulated as either or both of the two conjugated double bond nitriles.

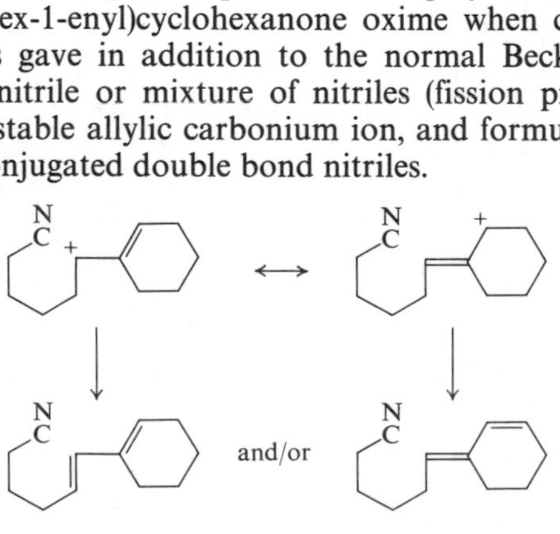

In the Beckmann rearrangement effected by photolysis of 5α-cholestan-6-one oxime, the migration was found to occur with retention of configuration

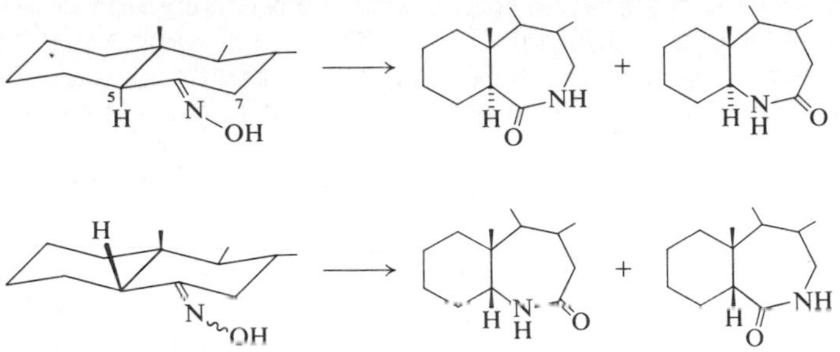

at C_5 suggesting that on cleavage, the C_5—C_6 bond of the oximes gives neither a radical pair nor an ion pair, and that the N—O bond is cleaved with concerted C_5 or C_7 migration[167]. Similar results were obtained with the 5β-cholestan-6-one oximes (*syn* and *anti*) in which the lactams retained the *cis* ring fusion.

The normal Beckmann products anticipated from the rearrangement of some NN-disubstituted oxime thiocarbamates were not obtained[168]. Instead, isomerisation in dry carbon tetrachloride under reflux yielded the thio-oxime carbonate.

$$R^1R^2C:N\cdot O\cdot CS\cdot NR_2^3 \longrightarrow R^1R^2C:N\cdot S\cdot CO\cdot NR_2^3$$
$$\updownarrow$$
$$R^2N:CR^1\cdot S\cdot CO\cdot NR_2^3$$

4.6.4 Determination of configuration

The classical, although at times unreliable, method for determining the stereochemistry of an oxime has been based upon the preferential migration of the group *anti* to the hydroxy group. Recent n.m.r. techniques using induced changes in shifts by solvent or contact shift reagents have provided more generally reliable methods for configurational assignment. By using benzene solutions of dialkyl, cycloalkyl, alkyl phenyl and α,β-unsaturated ketoximes of known configuration, it has been found that the addition of small amounts of concentrated hydrochloric acid vapour to the n.m.r. sample caused the α-protons *syn* to the hydroxy group to shift to higher field and the *anti* α-protons to lower field[169]. In all cases studied the induced chemical shift changes were large enough to be unambiguously recognised and measured (the normal range appears to be from c. 3 to 10 Hz). That the aromatic solvent was important was seen when chloroform solutions of the ketoximes showed virtually no induced shift upon addition of hydrogen chloride.

It has been pointed out that bonding in these complexes of benzene with oximes is very weak, probably not exceeding van der Waals interaction and that the aromatic-solvent-induced shifts observed for benzene solutions of the oxime are most likely to be only 'time-average values resulting from very rapid processes of formation and breakdown of favourably oriented groups of one solute and one or more solvent molecules drawn at random from the medium. Still the effect of these interactions may be considered as a statistical result of a single centre of anisotropy"[170]. The relative positions of benzene and oxime have been determined such that a nodal surface may be drawn through the oxime molecule separating upfield and downfield shift regions ($\Delta\delta = \nu(CCl_4) - \nu(C_6D_6)$). This benzene induced shift could be annihilated

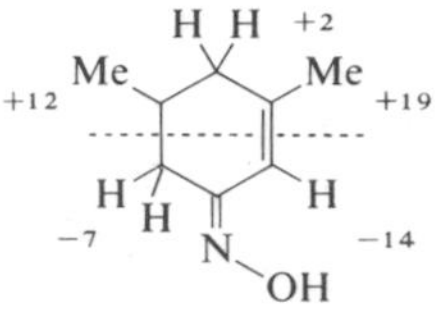

(a) by dilution with $CDCl_3$ or CCl_4, (b) by protonation of the oxime, in this case with $CF_3 \cdot CO_2H$, when oxime protons with large upfield shifts give downfield shifts and vice versa.

Another n.m.r. technique for investigation of isomeric oximes involves the use of tris-dipivalomethanato(europium) $[Eu(DPM)_3]$ [171]. Significant shift differences have been noted for protons in the *syn*- and the *anti*-isomer. In general the *syn*-protons are more strongly shifted than the *anti*. For example, the methyl resonances in acetoxime, which give a unique signal under standard conditions, have been separated by almost 6 p.p.m. by using this shift

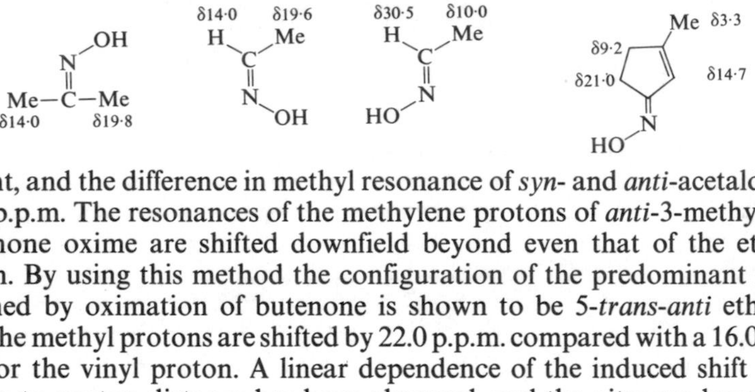

reagent, and the difference in methyl resonance of *syn*- and *anti*-acetaldoxime is 9.6 p.p.m. The resonances of the methylene protons of *anti*-3-methylcyclopentenone oxime are shifted downfield beyond even that of the ethylene proton. By using this method the configuration of the predominant isomer obtained by oximation of butenone is shown to be 5-*trans-anti* ethylenic, since the methyl protons are shifted by 22.0 p.p.m. compared with a 16.0 p.p.m. shift for the vinyl proton. A linear dependence of the induced shift on the oxygen-to-proton distance has been observed, and the nitrogen lone pair is

Table 4.2   Percentage of *syn* and *anti* forms of $R^1R^2C{:}NOH$

R^1	R^2	% *syn*	% *anti*
Me	Et	72	28
Me	Pr^n	73	27
Me	n-Pentyl	75	25
Me	n-Hexyl	74	26
Me	Pr^i	86	14
Me	Bu^i	71	29

thought to play an insignificant role in complexation. In a related investigation, rather different conclusions are drawn. N.M.R. chemical shifts induced by Eu(DPM)$_3$ have also been used to determine the ratios of *syn* to *anti* isomer formation in the oximation of a series of alkyl methyl ketones[172]. In this series the *syn* forms were found to be present in larger amounts. It was observed that all protons of the *anti* forms were more deshielded than the corresponding protons of the *syn* forms, and therefore co-ordination through the nitrogen lone pair is reasonable on the basis of better coordination between Eu(DPM)$_3$ and the nitrogen lone pair in the *anti* configuration, because of the smaller steric hindrance of the methyl group[172].

The configurations for a series of α-hydroxyiminoketones were assigned when it was found that upon reaction with toluene-*p*-sulphonyl chloride in pyridine, the *syn*-isomers formed stable toluene-*p*-sulphonates while the *anti*-isomers underwent a Beckmann fission to yield the oxime ester[173].

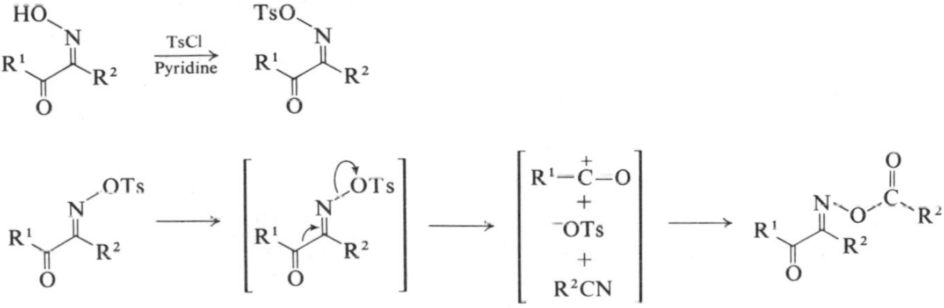

On the basis of acidity and spectral studies, the configuration of 1,3-dioximino-acetone (DIA) prepared by nitrosation of acetone dicarboxylic acid has been

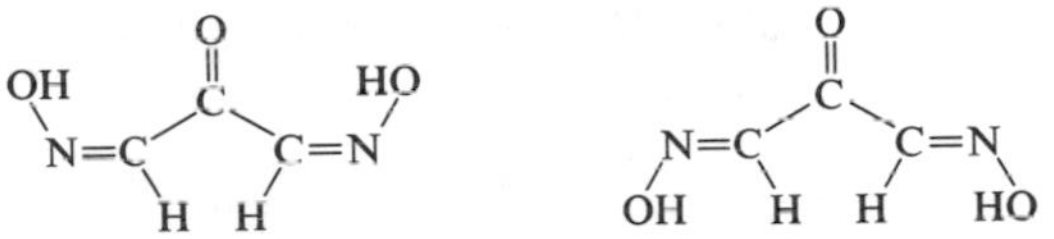

anti-anti   *syn-syn*

determined to be *syn-syn*[174]. The n.m.r. spectrum shows only one aldehydic proton resonance, which effectively excludes the *syn-anti* isomer. Comparison of acidity characteristics and spectral data of model compounds (1-oximino-acetone and ethyl oximinoacetoacetate) has led to the conclusion that DIA exists in the configuration in which intramolecular hydrogen bonding is not present.

In an investigation of arylazoketoximines, in which three configurations are possible, it has been found again that the hydrogen bonded structure is not

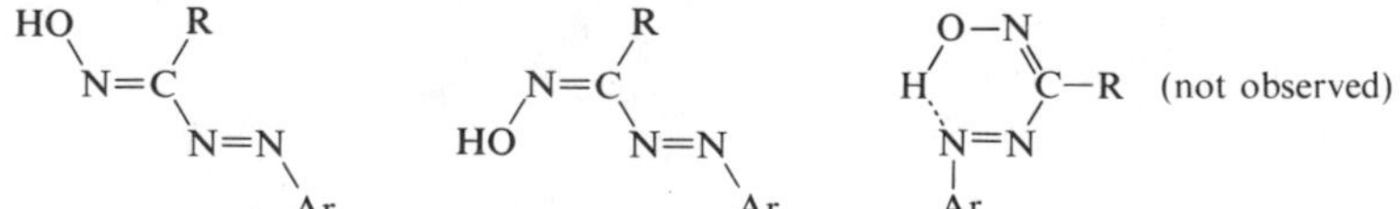

trans-anti   *trans-syn*   (not observed)

preferred[175]. Phenylazoformaldoxime (R = H) and phenylazoacetaldoxine (R = Me) were found to be pure *trans-anti*-isomers, while phenylazobenzaldoxime (R = Ph) is a mixture of the *trans-anti-* and the *trans-syn-* isomer in the ratio of 2:1.

Assignment of *syn* and *anti* configurations in substituted acylferrocene oximes is based upon the shift to lower field of the α-proton(s) of the ferrocene ring, very likely the result of steric compression of the oxime oxygen in the *syn* isomer[176]. Also, in the i.r. spectrum, the C=N stretching band of the *syn*-isomer appears at higher frequency than that for the *anti*-isomer.

Although the barrier to *syn-anti* isomerisation in oximes and oxime ethers has been found to be large[177], evidence has been reported for much lower barriers in the thio-oxime ethers, where the *syn-anti* isomerisation is estimated

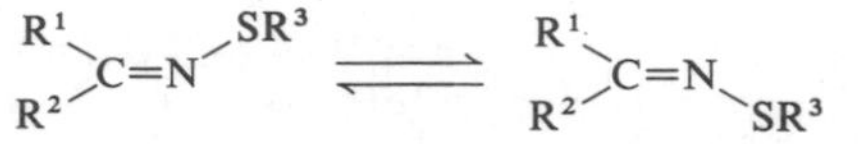

to be 10^{11} times faster than for the corresponding oxime ethers[178].

4.7 HYDRAZINE DERIVATIVES

4.7.1 Synthesis

The synthesis of NN-disubstituted hydrazines in good yields by reduction of N-nitrosamines with zinc–ammonia–ammonium–carbonate has been reported[179]. Cyclohexanone enamines have been found to react with

$$R^1R^2N{\cdot}NO \longrightarrow R^1R^2N{\cdot}NH_2$$

Table 4.3

	Reduction of Nitrosoamines	
R^1	R^2	Yield of Hydrazine, %
Me	Me	38
Ph	Me	70
Ph	Ph	82
Ph	CH_2Ph	72
CH_2Ph	CH_2Ph	67
$CH_2C_6H_4OMe$-p	$C_6H_4CO_2H$-p	63

dibenzoyldi-imide at room temperature to provide a quantitative yield of the diazetidine by a 1,2-cyclo-addition[180]. Upon mild hydrolysis the adduct yields 2-(N^1N^2-dibenzoyl)hydrazinocyclohexanone.

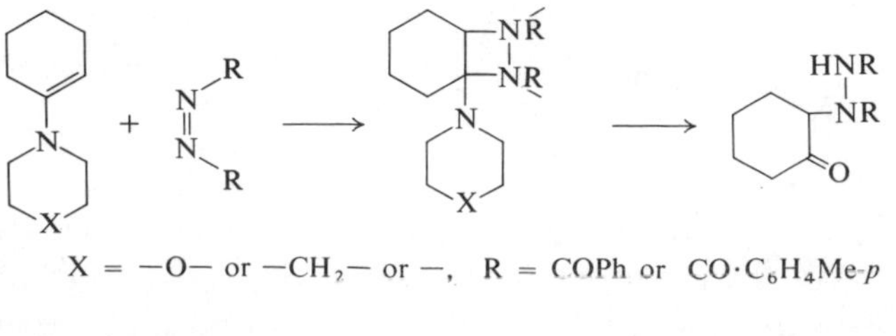

X = —O— or —CH$_2$— or —, R = COPh or CO·C$_6$H$_4$Me-p

The oxidation of NN-disubstituted hydrazones with lead tetra-acetate has been found to yield diacylhydrazines and an aldehyde[181]. This oxidation

$$p\text{-}XC_6H_4\cdot CH{:}N\cdot NR^2CH_2R^1 \xrightarrow{\text{Pb(OAc)}_4} p\text{-}XC_6H_4\cdot CH{:}N\cdot NHR^2 + R^1CHO$$

$$\downarrow \text{Pb(OAc)}_4$$

$$p\text{-}XC_6H_4\cdot CO\cdot NH\cdot NR^2Ac$$

requires two equivalents of lead tetra-acetate; the first equivalent yields the monosubstituted hydrazone and the aldehyde; the second oxidises the monosubstituted hydrazone to the N^2-acetyl-N^1-aroylhydrazine. Different alkylhydrazines are obtained depending upon the course of the de-alkylation step. The oxidation of benzaldehyde NN-dibenzylhydrazone ($X = H$, $R^1 = Ph$, $R^2 = CH_2Ph$) afforded the diacylhydrazine (75%) and benzaldehyde (92%). When benzaldehyde N-benzyl-N-p-methoxybenzylhydrazone ($X = H$, $R^1 = p\text{-}MeOC_6H_4$, $R^2 = CH_2Ph$) was oxidised it was found that the p-methoxybenzyl group was cleaved much more readily than the benzyl group, and benzaldehyde (30%) and p-methoxybenzaldehyde (62%) were isolated along with the corresponding hydrazines. A number of other benzaldehyde hydrazone NN-dialkyl compounds have been examined to determine their relative ease of cleavage. The order reported was p-methoxybenzyl > benzyl > p-chlorobenzyl ⩾ methyl. The reaction of lead tetra-acetate with the benzothiazolylhydrazone of some aromatic aldehydes again produced the acylated derivative of the hydrazone rather than the cyclised thiazolobenzothiazoles[182]. By heating diacylhydrazines under reflux in

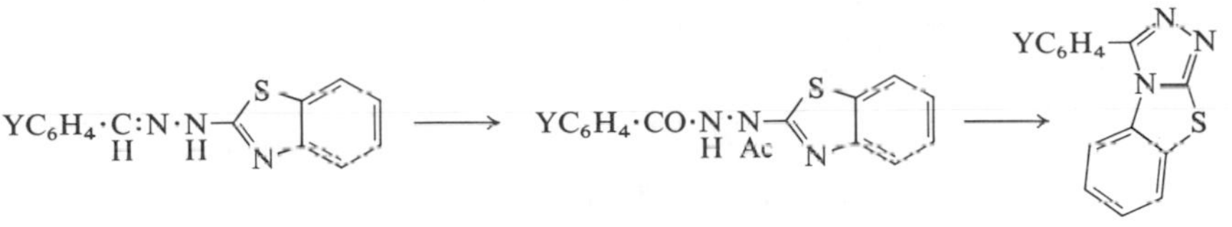

Y = H, p-Cl, p-Br, p-Me or p-Pr^i

phenol, the cyclic product was obtained in yields up to 60%.

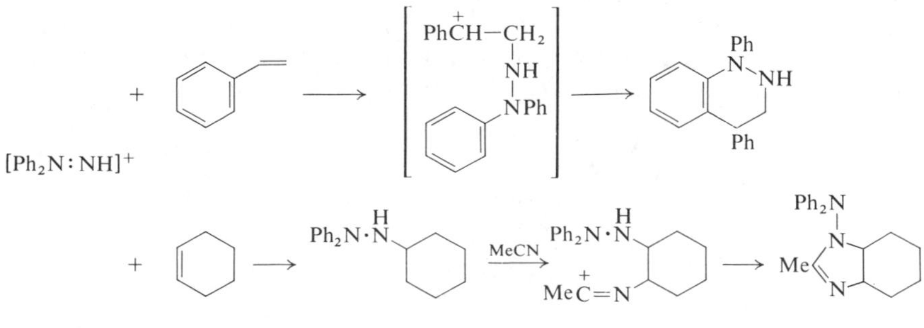

Reaction of the diphenyldiazonium cation (from electrochemical oxidation of acidic solutions of 1,1-diphenylhydrazine in acetonitrile which is 0.1 M in $LiClO_4$) with styrene gives the tetrahydrocinnoline in high yield[183]. With cyclohexene, 1-diphenylamino-2-methylhexahydrobenzimidazole is isolated in 80% yield. Upon reaction with the isomeric but-2-enes, viscous liquids were obtained corresponding to the *cis*- and the *trans*-1-diphenylamino-2,3-dimethylaziridines.

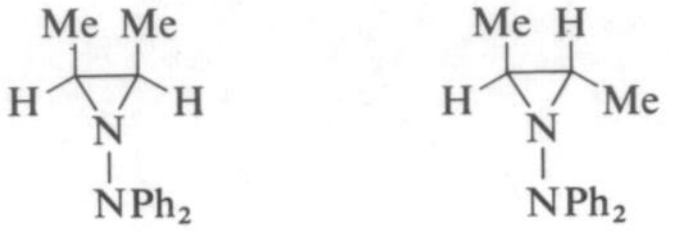

$$ \text{Me} \quad \text{Me} \qquad\qquad \text{Me} \quad \text{H} $$

$$ H \diagdown \diagup H \qquad\qquad H \diagdown \diagup Me $$

$$ \underset{\overset{|}{\text{NPh}_2}}{\text{N}} \qquad\qquad \underset{\overset{|}{\text{NPh}_2}}{\text{N}} $$

The *cis*-product is easily formed (95 % yield) and is stable. The *trans*-product, however, is difficult to obtain pure since it is unstable under acidic conditions and rearranges to the open-chain hydrazine.

4.7.2   Reactions

4.7.2.1   With α, β-unsaturated compounds

Reaction of hydrazine or arylhydrazines with dimethyl acetylenedicarboxylate affords 1:1 adducts which have been shown by n.m.r. studies to exist as previously unreported imine–enamine tautomers[184]. Reaction of an alcoholic

$$ \text{RNH·NH}_2 + \text{MeO}_2\text{C·C:C·CO}_2\text{Me} \longrightarrow \text{RNH·N:C(CO}_2\text{Me)·CH}_2\text{·CO}_2\text{Me} \quad \begin{array}{l}\text{imine}\\ \text{(hydrazone)}\end{array} $$

$$ + $$

$$ \text{RNH·NH·C(CO}_2\text{Me):CH·CO}_2\text{Me} \quad \begin{array}{l}\text{enamine}\\ \text{(enehydrazine)}\end{array} $$

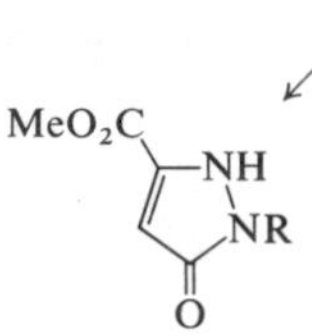

solution of hydrazine hydrate (R = H) with dimethyl acetylenedicarboxylate gave the hydrazone (40 %) and the pyrazolinone (60 %) and on being kept at room temperature the hydrazone was converted quantitatively into the pyrazolinone. The adducts resulting from the reaction of arylhydrazines (R = Ar) with the ester could be isolated as discrete tautomeric forms which were stable toward cyclisation. Isomerisation of an enehydrazine to a hydrazone (or an enamine to an imine) required 45 min in refluxing methanol. This isomerisation was not sufficiently rapid to explain the ratio of isomers observed in the direct reaction (4:1, imine:enamine) and so both isomers appear to arise directly from interaction of the reactants.

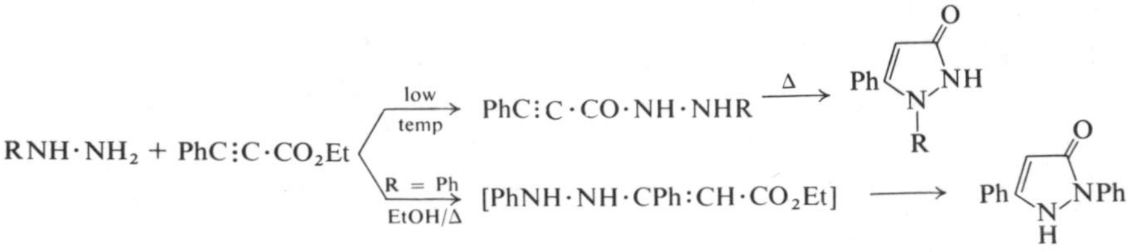

The reaction of hydrazine (R = H) and phenylhydrazine (R = Ph) with ethyl phenylpropiolate proceeded by a different route. When the reaction with hydrazine was carried out at low temperatures, the product was the phenyl propiolhydrazide which could be converted by heat into the 3-phenyl-pyrazol-5-one[185]. The reaction of the ester with phenylhydrazine (R = Ph) at room temperature yielded the intermediate phenylpropiolphenyl-hydrazide which on being heated gave 2,3-diphenylpyrazol-5-one. Reaction by Claisen addition followed by cyclisation at the triple bond is suggested. The isomeric 1,3-diphenylpyrazol-5-one was isolated in quantitative yield when the reaction between ethyl phenylpropiolate and phenylhydrazine was carried out under reflux in ethanol. The indication is that this reaction proceeds by an initial addition to the triple bond.

The Michael reaction of diethyl (substituted amino)methylenemalonate with hydrazine hydrate has been shown to be an effective method for the preparation of either 4-substituted-1,2,4-triazoles or 4-carbethoxy-5-hydroxy-4H-pyrazoles[186]. The hydrazine adduct initially formed may eliminate either diethyl malonate or the primary amine to form respectively the triazole or the pyrazole, the course of the reaction depending upon the nature of the leaving group.

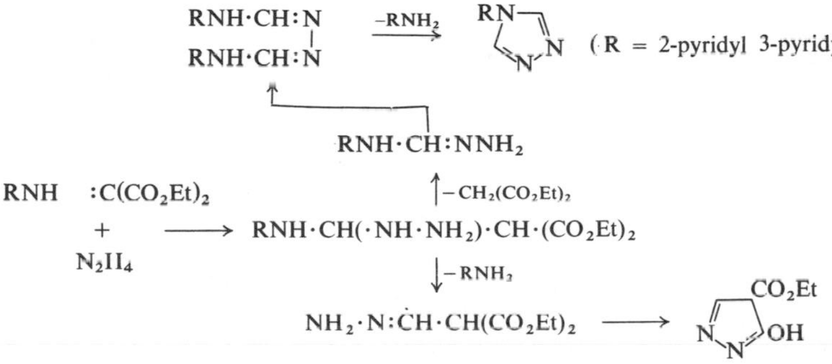

(R = 2-pyridyl 3-pyridyl , or 2-pyrazinyl)

(R = phenyl, p-tolyl, or N-phenylpiperazino)

Nucleophilic reaction of 2,3-diaryl-1-azirine-3-carboxamides with hydrazine leads to the formation of 1,2,4-triazin-6-one derivatives in moderate

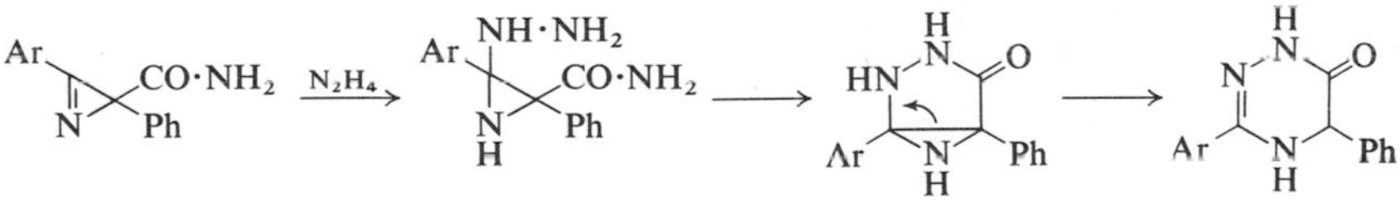

yield[187]. A bicyclic intermediate is postulated for this reaction.

4.7.2.2 Amidine and guanidine derivatives

Aminoguanidinium nitrate reacts with Mannich bases (as the free base) in 95% ethanolic solution at pH 10 to give good yields of the 1-formamidino-Δ^2-pyrazolines[188, 189]. At pH 2 the formamidinohydrazone dinitrates which were obtained could not be converted directly into the pyrazolines even under

$$NH_2 \cdot NH \cdot C \overset{\overset{+}{N}H_2}{\underset{NH_2}{\diagup}} \quad NO_3^- \; + \; ArCO \cdot (CH_2)_2 \cdot NMe_2 \xrightarrow{\text{pH 10}}$$

(structure: Ar-substituted pyrazoline)

$$\text{pH 2} \downarrow \qquad \qquad \nearrow\!\!\!/$$

$$[ArC(CH_2NHMe_2):N \cdot \overset{+}{N}H \cdot C(:\overset{+}{N}H_2)NH_2] \quad 2NO_3^-$$

$$\underset{NH_2}{\overset{C=\overset{+}{N}H_2}{|}} \quad NO_3^-$$

a variety of conditions, presumably because the hydrazine fragment was in the *anti*-configuration and hydrazine anchimerism could not occur. The mechanism proposed for production of a pyrazoline was (a) formation of the appropriate aryl vinyl ketone, (b) reaction of the ketone with hydrazine to form a *syn*-vinyl hydrazone, and (c) cyclisation of the hydrazone. It was observed, however, that *anti*-vinyl hydrazones did cyclise in good yield under acidic conditions because of acid-catalysed isomerisation to the *syn*-configuration. The reaction of triaminoguanidine with acetylacetone involves sequential attack of the diketone at the hydrazine groupings[190]. The reactants

$$\underset{\overset{|}{\overset{||}{N}}NH_2}{NH_2 \cdot NH \cdot \overset{|}{\underset{||}{C}} \cdot NH \cdot NH_2} \quad + \quad Ac_2CH_2 \longrightarrow (Pyr)_2C:N \cdot NH_2 \longrightarrow Pyr-\!\!\!\overset{N-N}{\underset{\underset{H\;\;H}{N-N}}{\diagdown\!\!\!/}}\!\!\!-Pyr$$

$$\downarrow$$

$$(Pyr)_2C:N \cdot N:CMe \cdot CH_2Ac$$

Pyr = 3,5-dimethylpyrazol-1-yl

may be varied at ambient temperatures to yield the dipyrazolyl ketone hydrazone (50%, with a 2:1 ratio of diketone to hydrazine), the dipyrazolyl-methylenehydrazono derivative (64–70%, with a 3:1 ratio of diketone to hydrazine), or the dehydro tetrazine (32.5%, with a 1:1.1 ratio). Amidinium chlorides are converted into mixtures of *meso*- and ($\pm$)-dihydrotetrazines by the reaction of hydrazine hydrate[191]. These dihydrotetrazines are readily oxidised by nitrous acid to the tetrazines.

$$R-C\overset{\overset{NH_2}{+}}{\underset{NH_2}{\diagup}} \quad Cl^- \xrightarrow[\text{EtOH}]{NH_2 \cdot NH_2} \quad R-\!\!\overset{N-N}{\underset{\underset{H\;\;H}{N-N}}{\diagdown\!\!\!/}}\!\!\!-R \longrightarrow R-\!\!\overset{N-N}{\underset{N=N}{\diagdown\!\!\!/}}\!\!\!-R$$

The R-($-$)-amidinium chlorides were found to yield the ($+$)-dihydrotetrazines and subsequently ($+$)-tetrazines of the (R,R) configuration.

4.7.2.3 Additional reactions

Thermal rearrangement of *N*-ammonio-amidates has been found to proceed by either of two pathways, a concerted suprafacial rearrangement (A) and a radical dissociation–recombination process (B)[192]. Hydrazinium salts (a)–(c) were found to rearrange thermally (3–5 °C) by path A, while acylated compounds (d)–(f) rearranged only by path B and only at higher temperature

(120 °C). It is thought that a fully developed negative charge at N′ is necessary for rearrangement by path A. In the reaction of compounds (d)–(f), extra stability is gained in the amidate ion from charge delocalisation, which contributes to the higher activation energy of path A and thus to competition from path B.

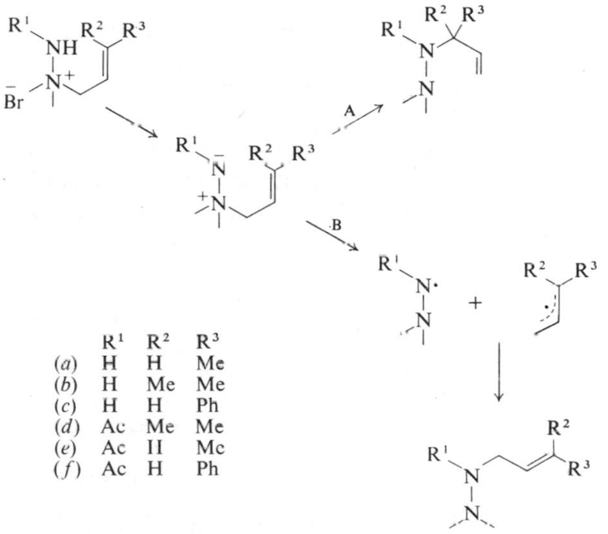

	R^1	R^2	R^3
(a)	H	H	Me
(b)	H	Me	Me
(c)	H	H	Ph
(d)	Ac	Me	Me
(e)	Ac	H	Me
(f)	Ac	H	Ph

Condensation of aldehydes, unsymmetrical hydrazines and cyanide leads to the formation of Mannich bases, which after acylation form 6-acylated 3-amino-5-imino-oxazolines upon treatment with an acid-acid anhydride mixture[193].

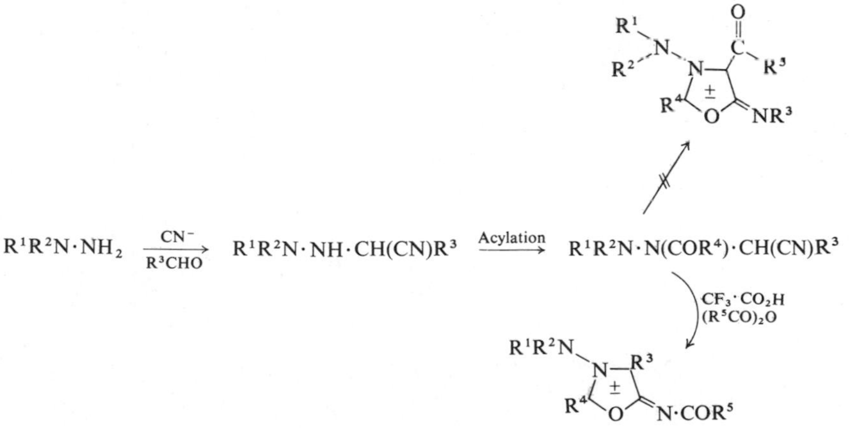

$$R^1R^2N\cdot NH_2 \xrightarrow[R^3CHO]{CN^-} R^1R^2N\cdot NH\cdot CH(CN)R^3 \xrightarrow{Acylation} R^1R^2N\cdot N(COR^4)\cdot CH(CN)R^3$$

These sydnone[194] and munchnone[195] relatives were assigned the N-acylated, as opposed to the C-acylated, structure primarily on the evidence that the proton in the 4-position (R^3 = H) was not exchangeable with water. Reexamination of the 3-aryl-5-imino-oxazolines indicated that these Dakin–West reaction products were C-acylated rather than the N-acylated products previously reported[196].

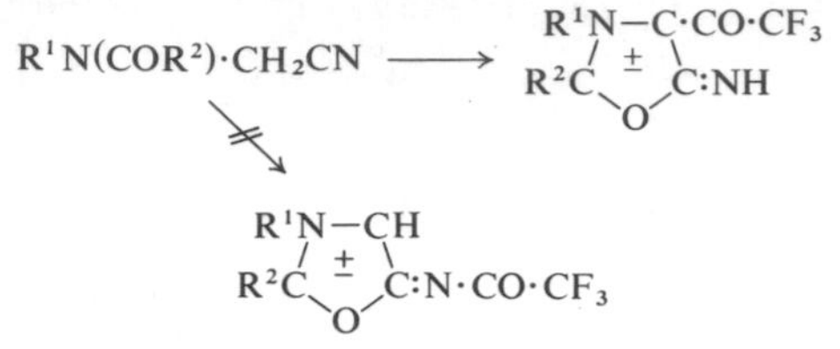

$$R^1N(COR^2) \cdot CH_2CN \longrightarrow \begin{matrix} R^1N - C \cdot CO \cdot CF_3 \\ \diagup \quad \pm \quad \diagdown \\ R^2C \quad \quad C{:}NH \\ \diagdown \quad \diagup \\ O \end{matrix}$$

$$\begin{matrix} R^1N - CH \\ \diagup \quad \pm \quad \diagdown \\ R^2C \quad \quad C{:}N \cdot CO \cdot CF_3 \\ \diagdown \quad \diagup \\ O \end{matrix}$$

The preparation of alkyldiazines by two routes has been reported[197]. Decarboxylation of diazinecarboxylate anions and elimination of toluene-p-sulphonic acid from 1-alkyl-1-tosyldiazanes. Alkenyldiazines can also be

$$RN{:}N \cdot CO_2^- \xrightarrow{BH^+} RN{:}\overset{+}{N}H \cdot CO_2^- \longrightarrow RN{:}NH$$

$$RNTs \cdot NH_2 \xrightarrow[\text{EtOH}]{\text{NaOEt}} RN{:}NH$$

synthesised by reaction of hydrazine with a chlorocarbonyl compound[198]. Vinyldiazine, prop-2-enyldiazine, and cyclohexenyldiazine have been pre-

$$ClCH_2 \cdot CHO + NH_2 \cdot NH_2 \longrightarrow ClCH_2 \cdot CH(OH) \cdot NH \cdot NH_2$$

$$\longrightarrow ClCH_2 \cdot CH{:}N \cdot NH_2 \longrightarrow CH_2{:}CH \cdot N{:}NH$$

pared in this manner. The reaction of hydrazones containing α-hydrogens with 3 mol. equiv. of n-butyl-lithium occurs through the $C(\alpha)NN$-trianion which with aromatic esters yields pyrazoles[199]. When steroidal tosylhydra-

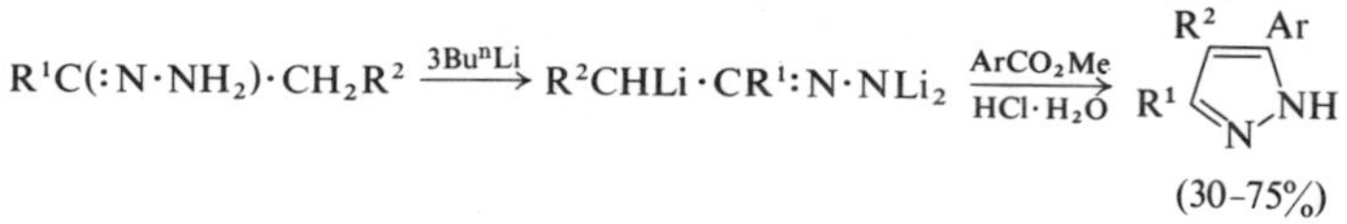

$$R^1C({:}N \cdot NH_2) \cdot CH_2R^2 \xrightarrow{3Bu^nLi} R^2CHLi \cdot CR^1{:}N \cdot NLi_2 \xrightarrow[\text{HCl} \cdot H_2O]{\text{ArCO}_2\text{Me}} \begin{matrix} R^2 \quad Ar \\ \diagdown \quad \diagup \\ \text{ring} \end{matrix}$$

(30–75%)

zones are treated with more than 2 mol equiv. of alkyl lithium, the less hindered (equatorial) alkyl-steroids are produced in good yield[200].

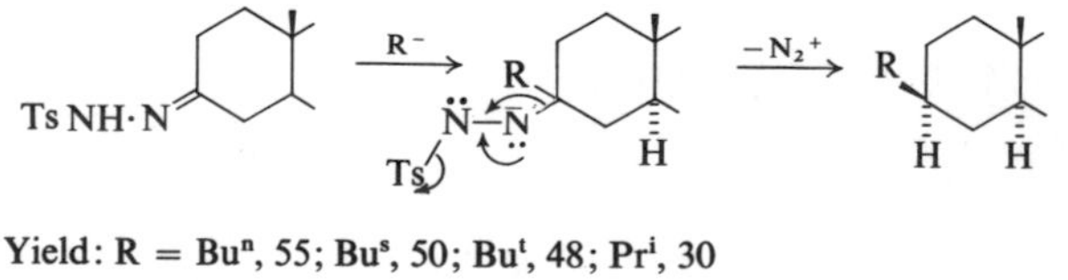

% Yield: R = Bu^n, 55; Bu^s, 50; Bu^t, 48; Pr^i, 30

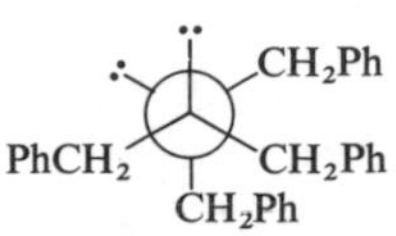

4.7.3 Other studies

In a study of the rotational and N-inversion barriers of tetrabenzylhydrazine, both rotation and inversion were found to be slow[201]. N.M.R. spectra at low temperatures (below 100 °C) indicated two pairs of non-equivalent benzyl groups. The methylene protons were also non-equivalent. It was concluded that rotation was slow relative to inversion.

Temperature dependence of the n.m.r. spectra of N^1N^2-dialkyl-N^1N^2-diarylhydrazines is reported to be consistent with non-planar conformations

$$\text{PhCH}_2\text{NR·NRCH}_2\text{Ph} \qquad \text{PhCH}_2\text{NR}^1\text{·OR}^2$$

for the hydrazines, having considerable barriers to rotation about the $N—N$ bond and with their rapid inversion at nitrogen[202]. However, similar temperature-dependent n.m.r. spectra obtained from NN-dialkylhydroxylamines may be the result of high barriers to N-inversion. The observed lack of a steric effect associated with replacement of an O-alkyl substituent by a much smaller hydrogen atom has led to this conclusion.

An infrared method for distinguishing between isomeric 1,1- and 1,2-disubstituted hydrazines has been found by examining the N–H stretching frequency of their hydrochlorides[203]. Monobasic salts of 1,1-disubstituted hydrazines show two or no v(NH) bands and 1,2-disubstituted hydrazine salts will show one v(NH) band above 3100 cm^{-1}.

$$\text{R}^1\text{R}^2\text{N·NH}_2 \xrightarrow{\text{HX}} \text{R}^1\text{R}^2\overset{+}{\text{N}}\text{H·NH}_2 \ \bar{\text{X}} \text{ or } \text{R}^1\text{R}^2\text{N·}\overset{+}{\text{N}}\text{H}_3 \ \bar{\text{X}}$$

Typical symmetric and antisymmetric v(NH); two bands

No bands

$$\text{R}^1\text{NH·NHR}^2 \xrightarrow{\text{HX}} \text{R}^1\overset{+}{\text{N}}\text{H}_2\text{·NHR}^2\bar{\text{X}} \text{ or } \text{R}^1\text{NH·}\overset{+}{\text{N}}\text{H}_2\text{R}^2\bar{\text{X}}$$

Both give one N–H band

4.8 AMINES

4.8.1 Preparation

4.8.1.1 Acyclic compounds

The condensation of lithium bis-benzenesulphenimide with alkyl bromides ($\text{R}^1 = \text{Bu}^n$, n-C_8H_{17}, PhCH_2, Bu^s, s-C_8H_{17}; X = Br) or alkyl toluene-p-sulphonates (X = OTs) yields the N-alkyl bis-benzenesulphenimide[204]. These imides are easily cleaved by hydrochloric acid or thiols to afford primary amines in good yields. Further, 2-aminopropionitrile ($\text{H}_2\text{N·CH}_2\text{·CH}_2\text{CN}$) was obtained in 91 % yield from acrylonitrile and the sulphenimide by using ethanethiol to obtain the amine. Likewise, ethyl 2-aminopropionate

$$(\text{PhS})_2\text{NH} \xrightarrow{\text{Bu}^n\text{Li}} (\text{PhS})_2\text{NLi} \xrightarrow{\text{R}^1\text{X}} (\text{PhS})_2\text{NR}^1 \xrightarrow[\text{or R}^2\text{SH}]{\text{HCl}} \text{R}^1\text{NH}_2$$

($\text{H}_2\text{N·CH}_2\text{·CH}_2\text{·CO}_2\text{Et}$) was obtained in 74% yield by using ethyl acrylate. These functional groups (nitrile, ester) would normally have been hydrolysed to the carboxy group under the conditions of the Gabriel synthesis[205].

The reduction of carboxylic acids with lithium in methylamine yields an imine which may be further reduced to give the secondary methylamine in good yield[206]. Amines are also produced by the reaction of enamines with mercuric acetate[207]. Following reduction of the organomercury salt (not

$$RCO_2H \xrightarrow[\text{MeNH}_2]{\text{Li}} RCH{:}NMe \xrightarrow[\text{or Li, MeNH}_2]{\text{H}_2\text{–catalyst}} RCH_2NHMe$$

isolated) with sodium borohydride, the tertiary amine is obtained in good yield. This reaction appears to be general, although sometimes N-mercuration

$$R_2^1NCH{:}CR_2^2 \xrightarrow{\text{Hg(OAc)}_2} R_2^1 \overset{+}{N}{:}CH{\cdot}CR_2^2{\cdot}HgOAc \xrightarrow{\text{NaBH}_4} R_2^1NCH_2{\cdot}CHR_2^2$$

(in which case the starting enamine is recovered) competes with C-mercuration.

Alkenes may be directly converted into amines without isolation of carbonyl intermediates by a reaction sequence which includes ozonolysis,

$$R^1CH{:}CHR^2 + \begin{array}{l}\text{(i) O}_3\text{, ROH} \\ \text{(ii) H}_2\text{, catalyst} \\ \text{(iii) H}_2\text{, R}^3\text{R}^4\text{NH, catalyst}\end{array} \longrightarrow \begin{array}{c} R^1CH_2NR^3R^4 \\ + \\ R^2CH_2NR^3R^4 \end{array}$$

partial reduction and reductive amination[208]. Overall yields ranged from 50 to 80%. Mixed amines can also be synthesised by the reaction of triphenylphosphinimines[209] with alkyl iodides[210]. The secondary amine is isolated

$$Ph_3\overset{+}{P}\overset{-}{N}R^1 \longleftrightarrow Ph_3P{:}NR^1 + R^2I \rightarrow [Ph_3\overset{+}{P}NR^1R^2]\ I^- \xrightarrow{\text{H}_2\text{O}} Ph_3P({:}O) + \\ HNR^1R^2$$

following hydrolysis of the intermediate dialkylaminophosphonium iodide. However, only methyl and ethyl iodide could be utilised in the reaction since higher alkyl iodides lose HI under the reaction conditions to give an alkene.

Direct amination of aliphatic compounds has been accomplished by the use of a trichloroamine–aluminium chloride reagent [$Cl_3N + AlCl_3 \rightarrow Cl^{\delta+}(AlCl_3NCl_2)^{\delta-}$][211]. The reaction has been described as involving chlorinium ion abstraction of hydride followed by nucleophilic attack by nitrogen. By using this reagent, 1-amino-1-methylcyclopentane was obtained

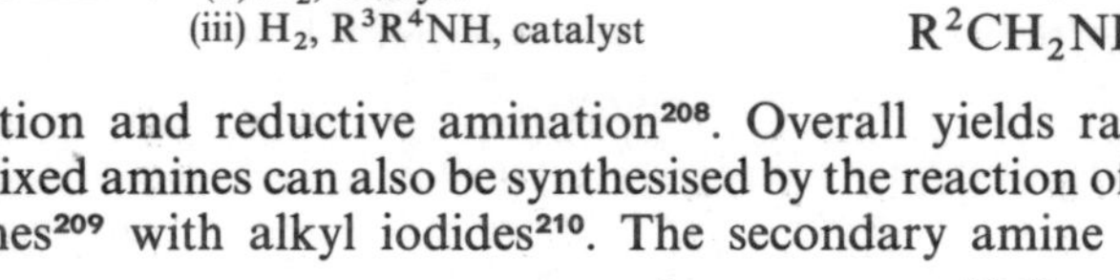

in 61% yield from methylcyclopentane. Amination of cis- or trans-decalin provided the same major product, cis-9-aminodecalin, in c. 50% yield. Reaction of cycloheptane with trichloroamine–aluminium chloride produced the ring-contracted product, 1-amino-1-methylcyclohexane. Cyclohexylamine was the major product in the low temperature (c. −10 °C) amination of cyclohexane, although at higher temperatures (10–15 °C) 1-amino-1-methylcyclopentane was formed preferentially. Amination of norbornane has also been carried out by using trichloroamine–aluminium chloride[212]: *exo*-2-aminobornane was isolated in 39% yield.

4.8.1.2 Cyclic and bicyclic compounds

Treatment of *exo*-2-chloronorbornane with trichloroamine–aluminium chloride resulted in ring expansion with nitrogen incorporation[212]. The major product after hydrogenation was 2-azabicyclo[3.2.1]octane (88%) with smaller amounts of the 3-isomer (6%) and *exo*-2-aminonorbornane (6%). The

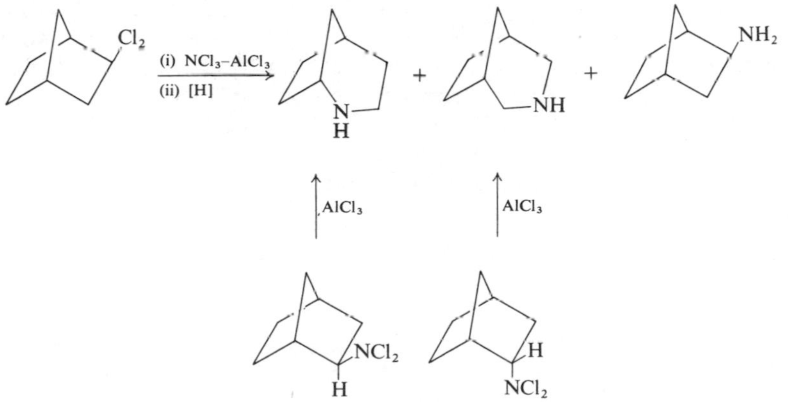

reaction is thought to proceed *via* NN-dichloroaminonorbornanes, since the *exo*-isomer rearranged in the presence of aluminium chloride and yielded only the 2-azabicyclo-octane (90%), while the *endo*-isomer gave predominantly the 3-aza-derivative.

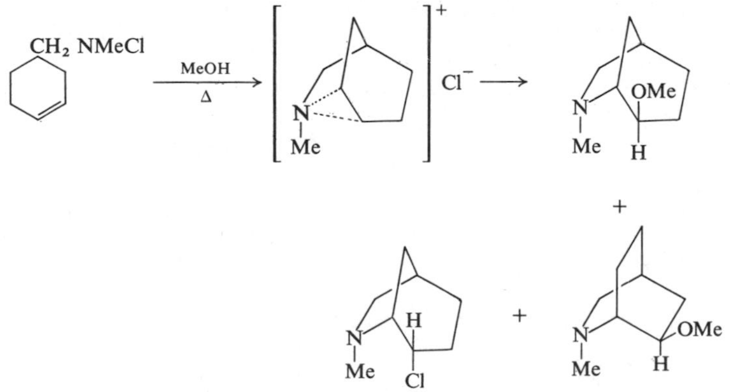

Routes to mono- and bi-cyclic nitrogen heterocyclic compounds may also involve reaction of nitrogen with a centre of unsaturation. Nitrenium ions generated through solvolysis of appropriate N-chloroamines have been employed in the synthesis of some aza-bicyclic compounds. Thus, treatment of 4-(N-chloro-N-methylaminomethyl)cyclohexene with methanol under reflux resulted in the formation of three cyclisation products, which may be rationalised as an intramolecular addition of a singlet nitrenium ion to the double bond[213]. Likewise, the 2-azabicyclo[3.3.1]nonanol could be prepared from the chloramine below[214].

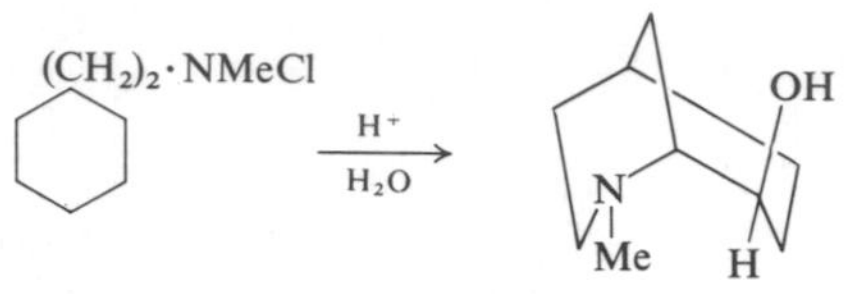

Another route to the nitrenium ion involves the deamination of dialkyl-hydrazines with nitrous acid. Reaction of nitrous acid with 2-amino-4,7,7-trimethyl-2-azabicyclo[2.2.1]heptane gave both the rearranged bridgehead nitrogen compound and the reduced compounds[215]. The large amount of this 'reductive deamination' compound (compared with a 3:1 ratio of rearranged to reduced products obtained from the N-chloro-analogue)[216] indicated decomposition via a diazoic acid.

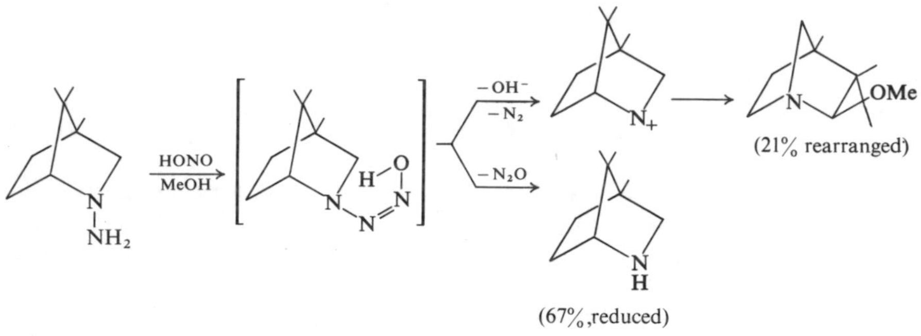

(21% rearranged)

(67%, reduced)

Similar results were obtained with the 2-amino-1,7,7-trimethyl-2-azabicyclo-[2.2.1]heptane; however, the rearranged products predominated.

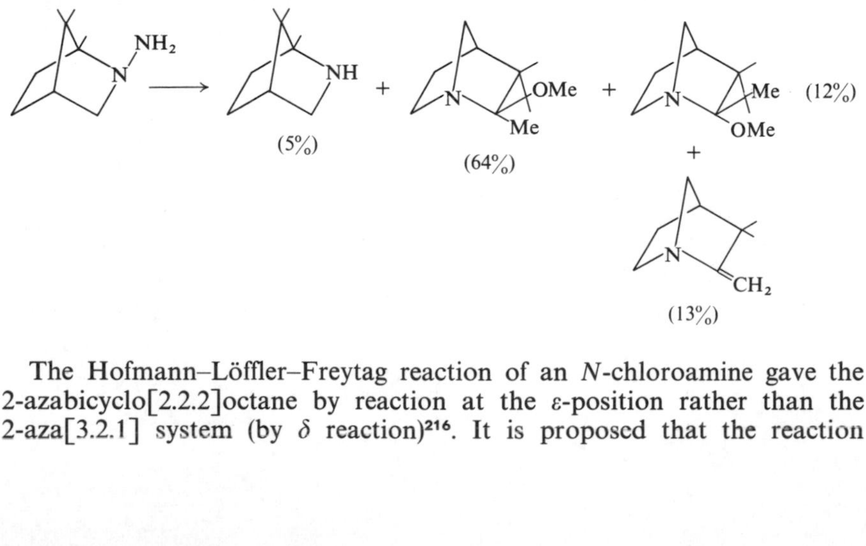

(5%)

(64%)

(12%)

(13%)

The Hofmann–Löffler–Freytag reaction of an N-chloroamine gave the 2-azabicyclo[2.2.2]octane by reaction at the ε-position rather than the 2-aza[3.2.1] system (by δ reaction)[216]. It is proposed that the reaction proceeds through a pseudo-boat intermediate. Lithium aluminium hydride reduction of the ethyl *cis*-4-aminocyclohexane carboxylate also leads to the 2-azabicyclo[2.2.2]octane system[217]. However, similar treatment of the

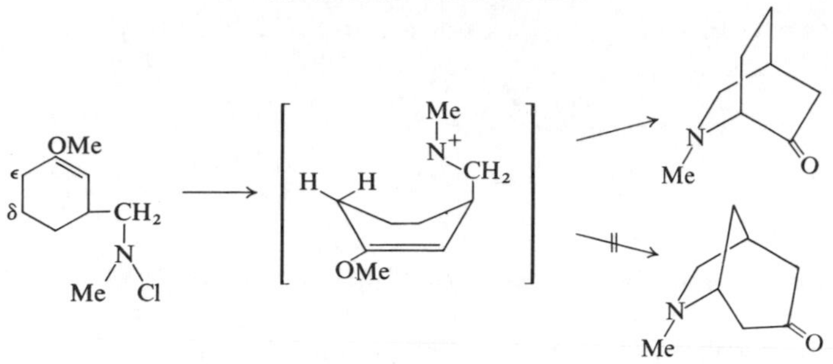

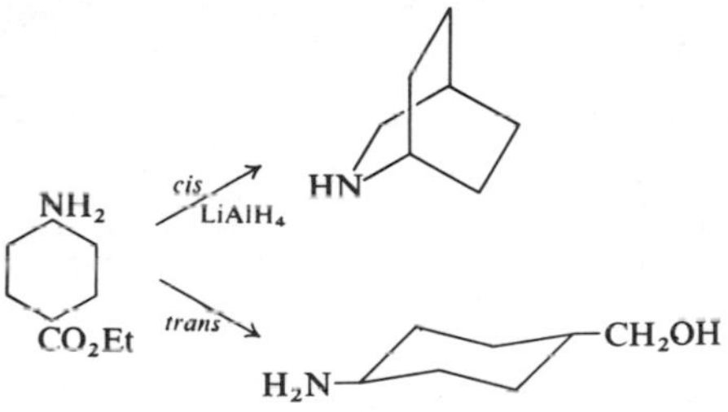

trans-isomer gave only the alcohol.

A photochemical cyclisation of N-propylpent-4-enylchloroamine has led to N-propylpyrrolidines[219]. The amine radical is thought to be an intermediate in this reaction. The chloromethylpyrrolidine was the only product isolated with acetic acid–water as solvent, while the methylpyrroli-

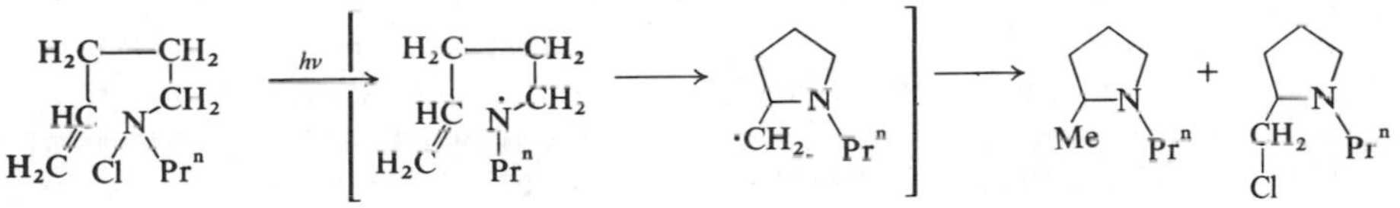

dine was the predominant product with methanol or propan-2-ol as solvent. The cyclisation of *trans*-2-(1-phenylcyclohexyl)cyclohexylamine upon treatment with NN-dichlorobenzenesulphonamide, it is proposed, proceeds through formation of the NN-dichloroamine, loss of HCl, then cyclisation

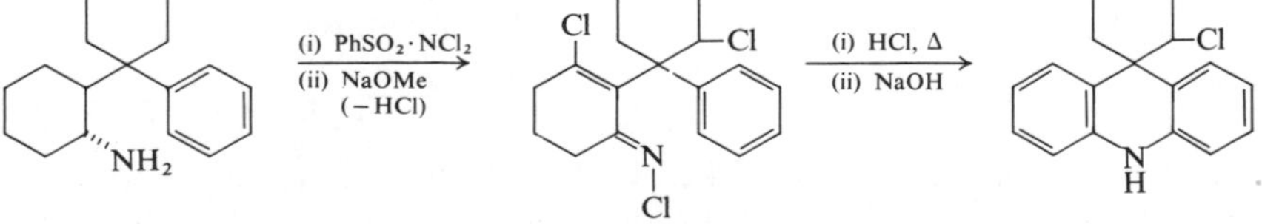

and aromatisation to the acridane derivative[220].

Aziridines with a formyl group at C_2 have been prepared by bromination of an α,β-unsaturated ester and then reaction with a primary amine[221].

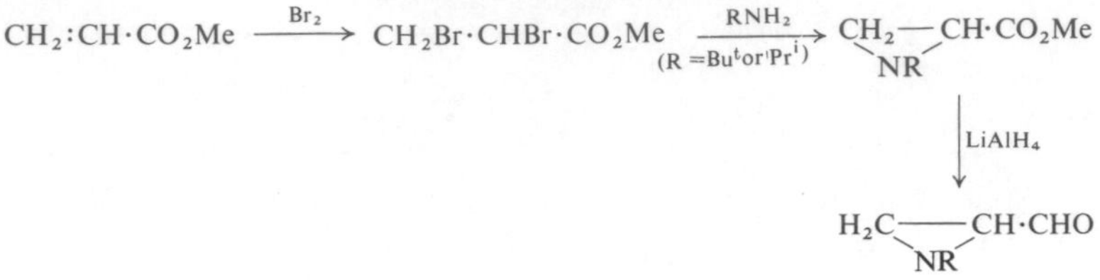

$$CH_2{:}CH{\cdot}CO_2Me \xrightarrow{Br_2} CH_2Br{\cdot}CHBr{\cdot}CO_2Me \xrightarrow[(R\,=Bu^t or^i Pr^i)]{RNH_2} \underset{NR}{CH_2\text{---}CH{\cdot}CO_2Me}$$

$$\downarrow LiAlH_4$$

$$\underset{NR}{H_2C\text{-----}CH{\cdot}CHO}$$

Hydride reduction of the ester at $-70\,^{\circ}C$ affords the aldehyde in 30% yield. Another method for the preparation of aziridines involves reaction of (dimethylamino)phenyl(2-phenylvinyl)oxosulphonium tetrafluoroborate with primary amines[222]. This Michael-type addition is unique in that the activator of the double bond is also the leaving group.

$$[PhCH{:}CH{\cdot}\overset{+}{S}({:}O)Ph{\cdot}NMe_2]BF_4{}^- \xrightarrow{Bu^tNH_2} \underset{CH\text{-----}NBu^t}{Ph\quad CH_2}$$

Addition of phenyl azide to norbornadienes followed by photolysis of the resulting $exo\text{-}\Delta^2$-triazole adducts gives the exo-3-aza-3-phenyltricyclo-[3.2.1.0]oct-6-enes[223]. The parent tricyclo-octene (R = H) was obtained in almost quantitative yield; the dicarboxylate (R = CO$_2$Me), however, was

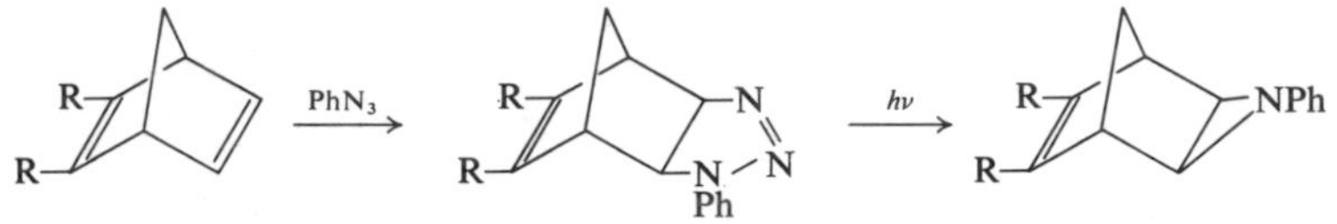

formed in considerably lower yield. A similar product was also isolated from the hydride reduction of benzonorbornadiene nitrosochloride dimer[132].

Alkyl-substituted α-chloroenamines (formed by the elimination of HCl from the product of the reaction of phosgene with tertiary amides) react with the azide ion to give 2-amino-1-azirines, the first representatives of three-membered ring amidines[224]. Hydride reduction of these azirines results in a highly exothermic reaction the product from which, after hydrolysis, furnishes 3,3,6,6-tetra-alkyldihydropyrazines (55%) probably via the α-amino-aldehyde. Reaction of 2-iodoalkyl azides[225] with several phosphorus-

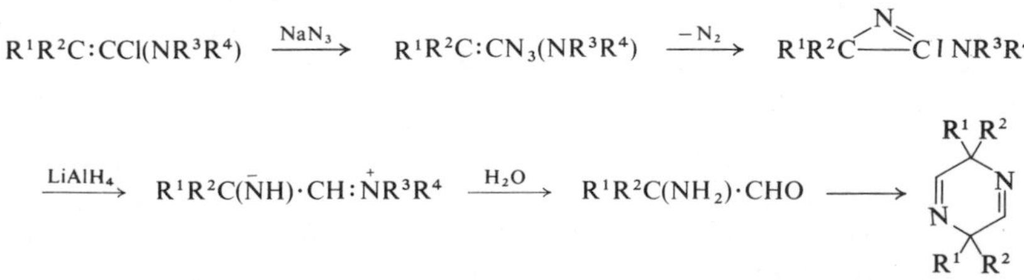

$$R^1R^2C{:}CCl(NR^3R^4) \xrightarrow{NaN_3} R^1R^2C{:}CN_3(NR^3R^4) \xrightarrow{-N_2} R^1R^2\underset{CINR^3R^4}{\overset{N}{C\text{----}C}}$$

$$\xrightarrow{LiAlH_4} R^1R^2C(\bar{N}H){\cdot}CH{:}\overset{+}{N}R^3R^4 \xrightarrow{H_2O} R^1R^2C(NH_2){\cdot}CHO \longrightarrow$$

containing nucleophiles (X$_3$ = Ph$_3$, (OMe)$_3$, (OEt)$_3$, Ph$_2$, Ome) produces aziridines in good yield[226]. Specifically, $threo$-iodo-azides derived from cis-butenes react with triphenylphosphine (X = Ph) to yield cis-aziridyl-phosphonium salts. $erythro$-Diastereoisomers produce the $trans$ salts. Lithium aluminium hydride reduction proceeded by cleavage of the P—N

threo   *cis*

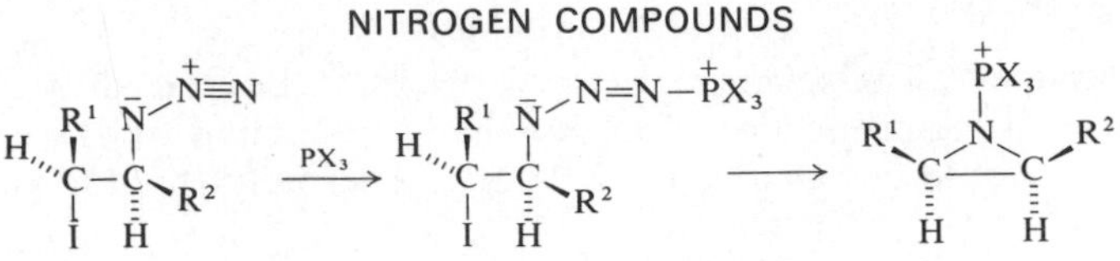

bond to give aziridines. A new class of organic compounds has been isolated
from the reaction of formaldehyde, a primary amine and hydroxylamine-*O*-
sulphonic acid[227]. The 2-alkyl-2,4-diazobicyclo[1.1.0]butanes have hereto-
fore been proposed only as reaction intermediates. l-Alkyldiazirines were
also produced in the reaction.

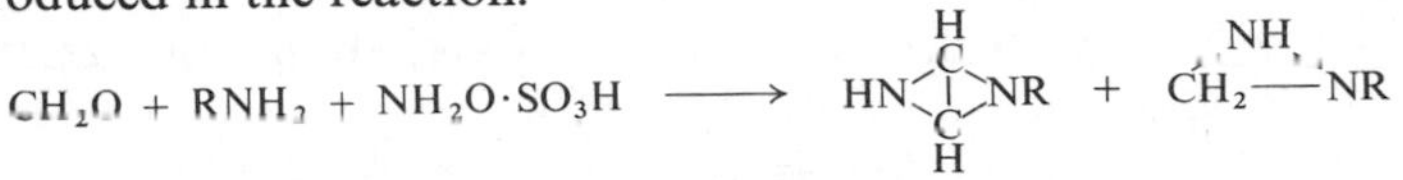

Chloroformates of 3- and 4-piperidinols, 3-pyrrolidinols, and 3-azetidinols
rearrange to oxazolidizones and oxazinones, conceivably via bicyclic
intermediates[228].

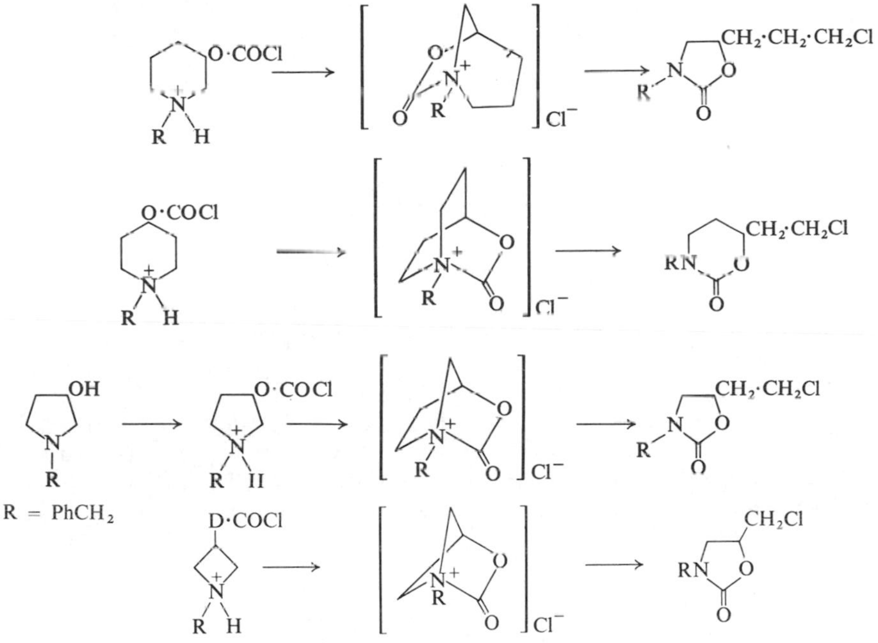

R = PhCH₂

Whenever it is possible, only the five-membered oxazolidin-2-ones are
isolated.

4.8.2   Reactions

4.8.2.1   Displacement by nitrogen

From the thermolysis of 1,1-bis-(2-dimethylaminoethyl)-3-phenylindene
monohydrochloride, the spiro-indenopiperidine was isolated along with

trimethylamine[229]. The spiro-compound appears to be formed by an initial intramolecular disproportionation involving *N*-methyl migration, then nucleophilic displacement of trimethylamine. Thermolysis of the analogous

1,3-compound however, gave not the bridged compound, but an indeno-pyrrolidine as the major product. The eight-membered ring system has

been postulated as an intermediate, which undergoes a transannular re-arrangement to give the observed product. A similar displacement is observed in the exchange reactions at nitrogen of piperid-4-one systems[230]. Reaction between a quaternary salt of 1,3-dimethylpiperid-4-one ($R^1 = H$) and Pr^iNH_2, Bu^sNH_2 or Bu^tNH_2 gave the corresponding exchanged *N*-substituted piperidones. The same acyclic ketone was isolated from each reaction indicating a ring-opening process, addition of the amine to this intermediate

and finally displacement of dimethylamine. The 1,2,5-trimethylpiperidone salts ($R^1 = Me$) also underwent the exchange reaction. The major products retained the *trans*-configuration of the starting amine salt. The 1-isopropyl and 1-s-butyl derivatives were in the preferred chair conformation while the *trans*-1-t-butyl analogue was shown to have a skew-boat conformation. Displacement by nitrogen of a halogen permits synthesis of endocyclic enamines by reaction of an α-chloro-ω-iodoalkane with the appropriate imine in lithium di-isopropylamide[231]. 2-Methyl-Δ^1-tetrahydropyridine and $\Delta^{1(9)}$-octahydroquinoline were both easily transformed into the corresponding cyclic compounds in excellent yields.

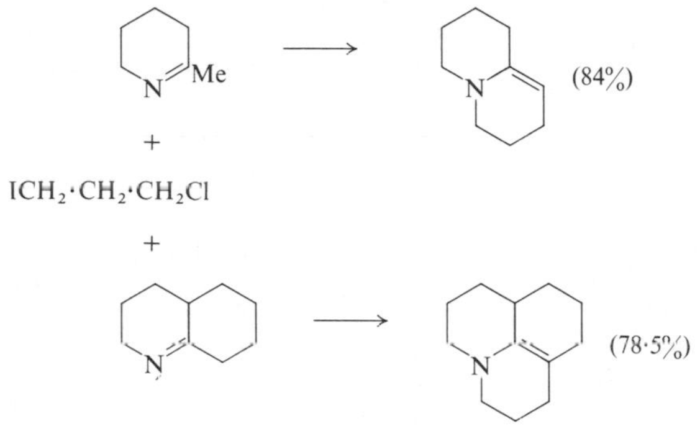

(84%)

+

$ICH_2 \cdot CH_2 \cdot CH_2Cl$

+

(78·5%)

4.8.2.2 Enamines and imines

Ketimines and enamines, products of the reaction of amines and ketones, can be conveniently prepared under mild conditions and in high yields by catalysis with molecular sieves[232]. This procedure appears to be completely general and has been extended to medium-size ring ketones and to camphor. The half-time for the condensation of aniline and acetophenone in the presence of molecular sieves is about 15 min; in the absence of the catalyst no ketimine was detectable even after 50 h. A novel synthesis of enamines in yields over 90% has been reported involving reaction of acetophenone anil (R = Ph) with one mol of $POCl_3$–DMF in DMF[233]. Treatment of the reaction mixture with perchloric acid allows isolation of the iminium salt, which gives the N-formylenamine upon alkaline hydrolysis.

$$PhN{:}CRMe \xrightarrow[DMF]{POCl_3} CH_2{:}CR{\cdot}(NPh{\cdot}CH{:}\overset{+}{N}Me_2)^- \xrightarrow{OH^-} PhN(CHO){\cdot}CR{:}CH_2$$
$$CH_2{:}CR{\cdot}(NPh{\cdot}CH{:}\overset{+}{N}Me_2)ClO_4$$

· Ethyl azodicarboxylate (EAD) has been used for chemical determination of the composition of mixtures of isomeric enamines from 2-substituted cyclohexanones[234].

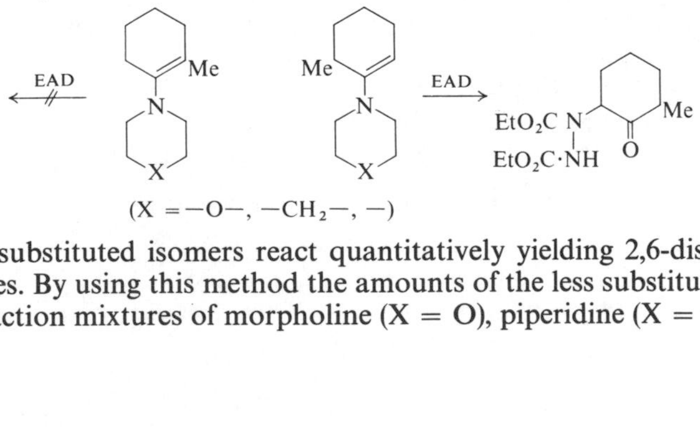

$(X = -O-, -CH_2-, -)$

The less substituted isomers react quantitatively yielding 2,6-disubstituted derivatives. By using this method the amounts of the less substituted isomer in the reaction mixtures of morpholine (X = O), piperidine (X = CH_2), and pyrrolidine (X = —) enamines were shown to be respectively 49%, 47%, and 81%, in agreement with results obtained by physical methods.

The product obtained from the reaction of cyclopropanone and piperidine undergoes reactions characteristic of a cyclopropyliminium salt[235]. Cyclic amines or enamines, acting like nucleophiles, substitute at carbon.

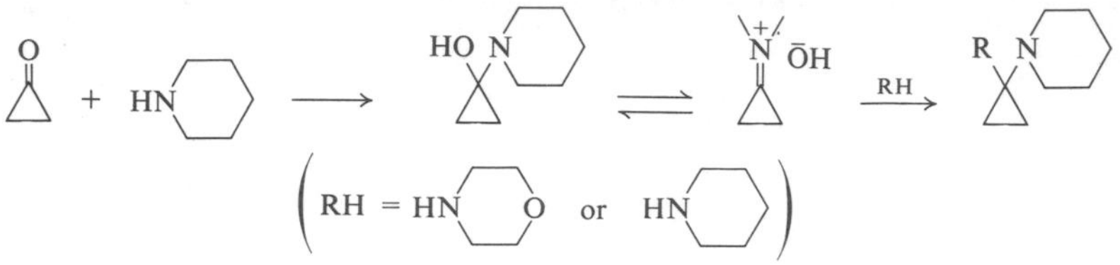

Stable methyleneiminium salts have been prepared in high yields from methylamines and di-p-anisylmethyl perchlorate. These salts reacted with

$$R^1R^2NMe \xrightarrow{(An)_2CH \ ClO_4^-} R^1R^2\overset{+}{N}{:}CH_2{\cdot}ClO_4^-$$

$$An = p\text{-}MeOC_6H_4$$

$$\xrightarrow{R^3MgBr} R^1R^2NCH_2CH_2R^3$$
$$\xrightarrow{PhCOMe} R^1R^2NCH_2{\cdot}CH_2{\cdot}COPh$$

nucleophiles at the methylene carbon to provide α-alkylated products[236]. Another method for α-alkylation of amines is the reaction of organo-lithium reagents with primary amines[237]. The process is thought to involve elimina-

$$PhCHMe{\cdot}NH_2 \xrightarrow{RLi} PhCHMe{\cdot}NLi_2^- \xrightarrow{LiH} PhCMe{:}NLi$$

$$\xrightarrow{RLi} PhCMeR{\cdot}NLi_2 \xrightarrow{H_2O} PhCMeR{\cdot}NH_2$$

tion of lithium hydride from the lithiated amine and then addition of the organo-lithium compound to the lithio-imine.

4.8.2.3 Mannich condensations

Primary and secondary amines have been found to yield Mannich condensation products upon reaction with 2-fluoro-2,2-dinitroethanol[238]. The 2-fluoro-2,2-dinitroethylamines are isolated in yields ranging from 64 to 97%. Allylamine ($R^1 = CH_2{\cdot}CH{:}CH_2$, $R^2 = H$) gave the 2:1 condensation

$$R^1R^2NH + CF(NO_2)_2{\cdot}CH_2{\cdot}OH \longrightarrow CF(NO_2)_2{\cdot}CH_2NR^1R^2$$

product $CH_2{:}CH{\cdot}CH_2{\cdot}N[CH_2{\cdot}CF(NO_2)_2]_2$ in 49% yield. Similar products are obtained when tris(propoxymethyl)amine reacts quantitatively with one equivalent of fluorodinitromethane, namely NN-bis(propoxymethyl)-2-fluoro-2,2-dinitroethylamine[239]. A second equivalent of fluorodinitromethane reacted more slowly to give the bis-derivative

$$(R^1OCH_2)_2NR^2 + FCH(NO_2)_2 \rightarrow [FC(NO_2)_2CH_2]NR^2{\cdot}CH_2OR^1$$

$$[R^2 = CH_2{\cdot}CF(NO_2)_2, \ R^1 = Pr^n]$$

(81% yield). The bis products were also obtained with NN-bis(ethoxymethyl)-t-butylamine ($R^1 = Et$, $R^2 = Bu^t$), NN-bis(propoxymethyl)benzylamine ($R^1 = Pr^n$, $R^2 = PhCH_2$) and NN-bis(ethoxymethyl)-2-amino-acetaldehyde diethyl ketal ($R^1 = Et$, $R^2 = (EtO)_2CH \cdot CH_2$). These products were dealkylated in strong acid to give bis(fluorodinitroethyl)amine in 85–88 % yield.

4.8.2.4 Transition metal catalysed reactions

gem-Diamines and enamines are the products when tris(dimethylamino)-stibine is treated with an aldehyde or ketone[240]. Thus benzaldehyde (R^1 = Ph, R^2 = H) gave benzylidene-bisdimethylamine quantitatively, and cyclohexanone and acetophenone could readily be converted into the correspond-

$$R^1R^2CO + Sb(NMe_2)_3 \begin{cases} \longrightarrow R^1R^2C(NMe_2)_2 \\ \xrightarrow[\text{and } R^2 = CHR^3R^4]{\text{If } \alpha\text{-H is present}} R^3R^4C{:}CR^1NMe_2 \end{cases}$$

ing enamines. Also reaction with acetic anhydride gave the NN-dimethyl-acetamide.

Several additions of amines to alkenes catalysed by transition metals have been reported. Rhodium or iridium compounds ($RhCl_3$, $3H_2O$, Rh_3, H_2O, $Rh(NO_3)_3$, and $IrCl_3 3H_2O$) effectively catalyse the addition of secondary

$$CH_2{:}CH_2 + R_2NH \longrightarrow R_2NEt$$

$$[R = Me, R_2 = -(CH_2)_5-]$$

aliphatic amines to ethylene[241]. Also addition of benzylamine to the dichloro-(endo-dicyclopentadiene)platinum(II) complex and hydrogenation of the

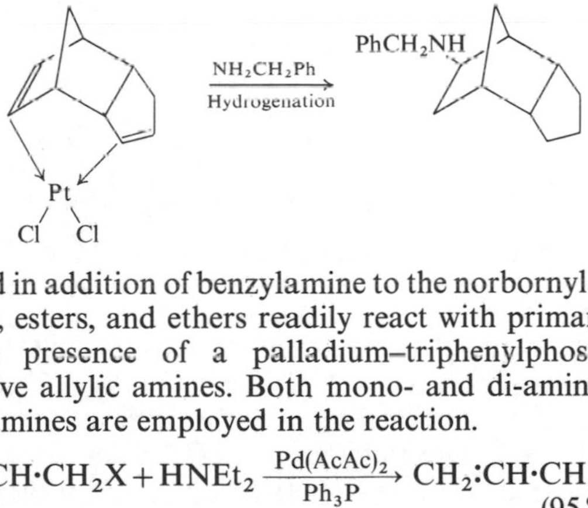

product resulted in addition of benzylamine to the norbornyl double bond[242]. Allylic alcohols, esters, and ethers readily react with primary or secondary amines in the presence of a palladium–triphenylphosphine complex catalyst[243] to give allylic amines. Both mono- and di-amines are obtained when primary amines are employed in the reaction.

$$CH_2{:}CH \cdot CH_2X + HNEt_2 \xrightarrow[Ph_3P]{Pd(AcAc)_2} CH_2{:}CH \cdot CH_2NEt_2$$
$$(95 \%)$$

An insertion reaction of chloramines with carbon monoxide is catalysed by palladium metal or platinum chlorides[244]. Thus dimethyl-, diethyl- and benzylmethyl-carbamoyl chloride were formed in good yields. The carbamoyl

$$R^1R^2NCl + CO \longrightarrow R^1R^2N \cdot COCl$$

chloride from N-chloropiperidine was also isolated. This carbonylation reaction can also be used for the preparation of monoalkyl compounds ($R^1 = H$).

In a similar reaction, rhodium complexes have been found to catalyse the carbonylation of some secondary amines[245]. By using this method, morpho-

$$[Ru(CO)_2(OAc)]_n + R_2NH \rightarrow [Ru(CO)_2(OAc)(HNR_2)]_2 \xrightarrow{CO} R_2N \cdot CHO$$

line, piperidine, and pyrrolidine were carbonylated in moderate yields (30–45%).

The first transition-metal complex, $[(C_5H_5)_2TiN_2]_2$, with a N_2 ligand capable of chemical modification has been employed in the synthesis of amines and nitriles[246]. When an excess of pentan-3-one ($R = Et$) is added to a tetrahydrofuran solution of the complex, a mixture of pent-3-ylamine and di(pent-3-yl)amine (2:1 ratio) is obtained in 25–50% yield. Benzoyl chloride

$$R_2CO \rightarrow R_2CHNH_2 + (R_2CH)_2NH$$

undergoes incorporation of nitrogen to yield benzonitrile.

4.8.2.5   Stereospecific reactions

Several stereospecific reactions of amines have been reported. The reaction of dimethylamino- and piperidino-alkyl-phenyl ketones ($R^2 = Ph$) with methylmagnesium iodide, and of the corresponding methyl ketones ($R^2 = Me$) with phenylmagnesium bromide have been found to be highly stereo-

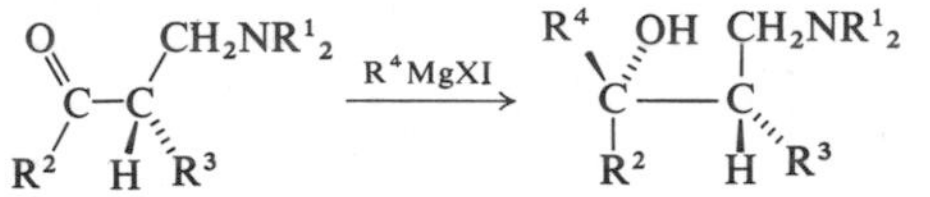

$R^3 = Me$, CH_2Ph or Ph; $NR^1_2 = NMe_2$ or $N(CH_2)_5$
$R^4 = Me$, $X = 1$; $R^4 = Ph$, $X = Cl$, Br, or I

yielded completely stereospecific reactions, while the methyl ketones gave high, though not complete stereospecificity. In both reactions, the organic group approaches from the same side of the C—CO—C plane. These facts are consistent with coordination of the Mg atom of the Grignard reagent with both the carbonyl oxygen and the amine nitrogen, and approach of the entering group to the carbonyl group from the side opposite to the R^3 group[248–250]. It is also interesting that in a preliminary study of the isomeric amino-ketones in which the asymmetric centre is β to the carbonyl group,

$$PhCO \cdot CH_2 \cdot CHR^1 \cdot NR^2_2$$

high stereospecificity is also observed[251]. Lithium aluminium hydride reduction of these same α-asymmetric-β-amino-ketones, though not so highly stereospecific as the Grignard reactions, still affords *erythro*-amino-alcohols as the predominant diastereoisomers[252]. Amine groups also promote Grignard reagent addition to double or triple bonds present in

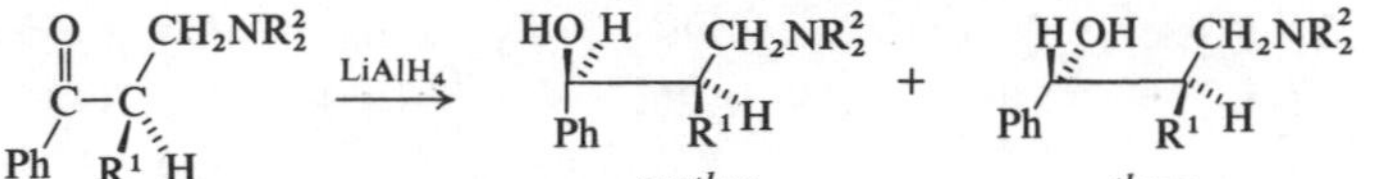

$$\text{erythro} \qquad \text{threo}$$

the same molecule[253]. Thus, reaction of an excess of allylmagnesium chloride

$$\text{PhCH:CH·CH}_2\text{NR}_2 \xrightarrow{\text{CH}_2\text{:CH·CH}_2\text{MgCl}} \text{CH}_2\text{:CH·CH}_2\text{·CH(CHPh)·CH}_2\text{NR}_2$$

$$\text{PhC:C·CH}_2\text{NMe}_2 \xrightarrow{\text{CH}_2\text{:CH·CH}_2\text{MgCl}} \text{CH}_2\text{:CH·CH}_2\text{·C(:CHPh)·CH}_2\text{NMe}_2$$

with some unsaturated amines led to significant amounts of the addition compounds (R = H or Me).

N-Bromobistrifluoromethylamine has been found to react with cis- and with trans-but-2-ene (under ionic conditions) to give the corresponding threo and erythro addition products respectively, indicating that a 'four-centred' mechanism is not operative in this case. Dehydrobromination

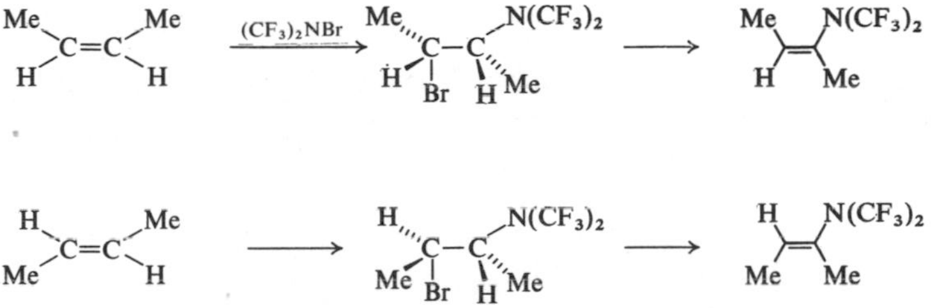

affords the trans- and the cis-but-2-ene[254]. Stereospecific addition of several secondary amines to buta-1,3-diene results in 1-NN-dialkylamino-cis-but-2-enes[255]. Reaction with diethylamine gives the cis-aminobutene in 98%

$$\text{R}_2\text{NLi} + \text{CH}_2\text{:CH·CH:CH}_2 \xrightarrow{\text{R}_2\text{NH}} \text{R}_2\text{NC}_4\text{H}_6\text{Li} \xrightarrow{\text{R}_2\text{NH}}$$
$$\text{R}_2\text{NCH}_2\text{·CH:CHMe}$$

yield. A complex formed from R_2NLi and R_2NH (1:2 mol ratio) appears to be the reacting species in the addition. Addition of an amine to a carbonyl group occurs when D($-$)-ephedrine and an aromatic aldehyde are refluxed in benzene and results in high yields of optically-pure oxazolidines[256]. The oxazolidines thus prepared have the R configuration at the induced asymmetric

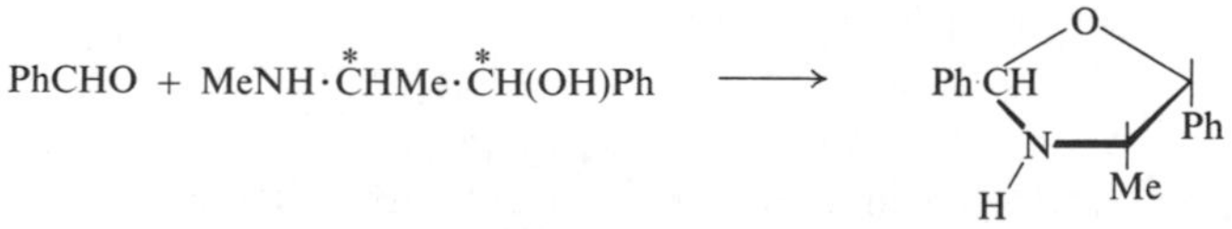

centre (the carbonyl carbon atom) and are designated 2R:4S:5R-oxazolidines. Although the stereochemical course of the asymmetric hydroboration of double bonds with ($-$)-di-3-pinanylborane can be generalised[21], these correlations have not been applicable to other types of alkenic systems, notably the carbocyclic analogues of cis-alkenes. The hydroboration of

1-methyl-1,2,3,6-tetrahydropyridine with $(-)$-di-3-pinanylborane would be expected to give the S-1-methylpiperidin-3-ol if the reaction course is the

same as that with acyclic *cis*-alkenes. The formation of R-1-methyl-piperidin-3-ol in the hydroboration is consistent with results obtained for other hindered cyclic alkenes and suggests that hydroboration occurs with the Lewis salt of the amine rather than the free base[257]. Asymmetric reduction of imines with lithium alkyl(hydro)dipinan-3-ylborates has been found to yield the corresponding amines of higher optical purity than an opposite

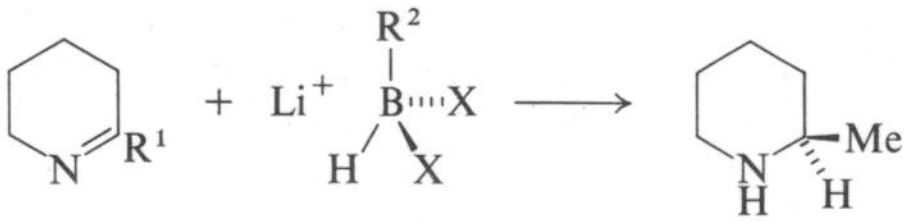

preferred chirality to those obtained from dipinan-3-ylborane[258]. Thus, 2-methylpiperidine $(R^2 = Me)$ was obtained with 25% optical purity from the corresponding imine when treated with lithium butyl(hydro)dipinan-3-ylborate $(R^2 = Bu^n)$[259].

4.8.2.6 Protecting groups

α-Methyl-α(4-methyl-2-phenylazophenoxy)propionic acid $(R^1 = H, R^2 = Me)$ and the dimethyl derivative $(R^1 = R^2 = Me)$ provide effective protection for amino groups, with the advantage of an internal nucleophile to assist in removal of the protecting moiety[260]. The group is introduced by acylation and removed by reduction of the azo portion with potassium borohydride

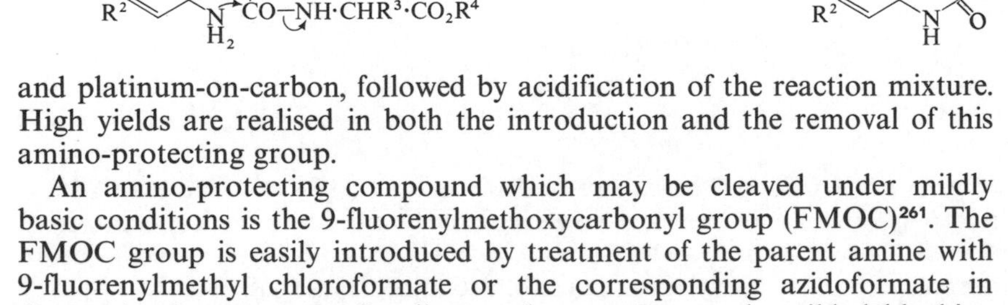

and platinum-on-carbon, followed by acidification of the reaction mixture. High yields are realised in both the introduction and the removal of this amino-protecting group.

An amino-protecting compound which may be cleaved under mildly basic conditions is the 9-fluorenylmethoxycarbonyl group (FMOC)[261]. The FMOC group is easily introduced by treatment of the parent amine with 9-fluorenylmethyl chloroformate or the corresponding azidoformate in dioxan in the presence of sodium carbonate. Extremely mild deblocking conditions include dissolution in liquid ammonia for several hours or in ethanolamine, morpholine or a similar amine. Yields are essentially quantitative.

Phenacylsulphonyl chloride provides a protecting group which is resistant

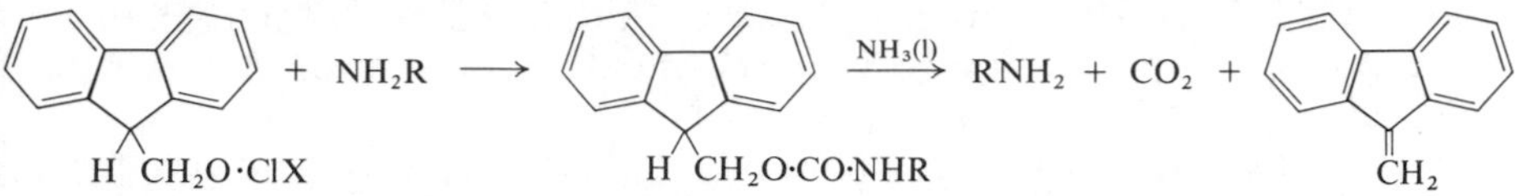

$$\text{H} \quad \text{CH}_2\text{O·ClX} \qquad\qquad \text{H} \quad \text{CH}_2\text{O·CO·NHR} \qquad\qquad \text{CH}_2$$

to both acidic or basic hydrolysis, but is easily removed by zinc in acetic acid[262]. In addition, sufficient acidity is conferred on the methylene-hydrogen to permit facile dialkylation of the sulphonamide. Phenacylsulphonyl chloride reacts readily with amines at ice temperature. The alkylation

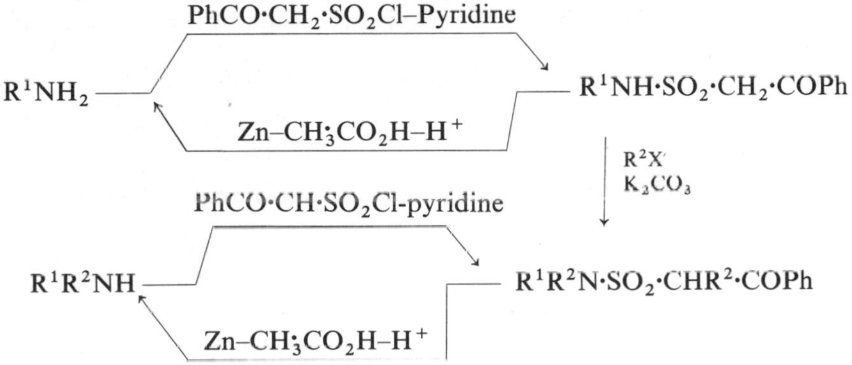

proceeds normally with dry potassium carbonate in acetone at room temperature, and all reactions give good yields.

The resolution of the enantiomers of *trans*- and *cis*-2-(*o*-bromophenyl)-cyclohexylamine was followed by n.m.r. through the diastereoisomeric (−)-menthoxyacetamide derivatives[263]. The methylene hydrogens of the

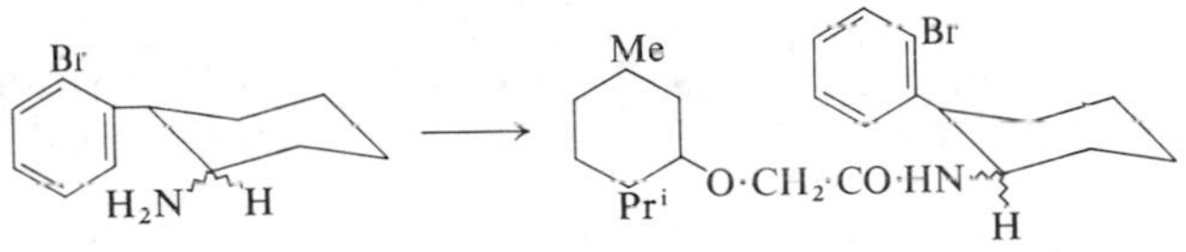

acetamide portion of the (−)-menthoxyacetamides were distinctly different for each of the four stereoisomers and these differences allowed each to be distinguished readily by its n.m.r. spectrum. Thus, by following the changing peak intensities a complete assessment of the progress of the resolution of the enantiomers is effected.

4.8.2.7   Other reactions

When *p*-nitrocumyl chloride is treated with quinuclidine in DMSO at room temperature, the product is the pure quaternary ammonium chloride in 90% yield[264].

$$p\text{-}NO_2 \cdot C_6H_4 \cdot CMe_2Cl \;+\; \overset{\displaystyle\bigcirc}{N} \longrightarrow \; p\text{-}NO_2 \cdot C_6H_4 \cdot CMe_2 \cdot \overset{+}{N} \quad \overset{-}{Cl}$$

Reactions with other aliphatic amines proceed smoothly, and generally in good yield to give the quaternary salts. The observation of a profound decrease of reaction rate in the presence of oxygen or absence of light has indicated a radical-anion chain mechanism.

The preparation of gem-bis(difluoroamino)-compounds has been accomplished by using difluoroamine with halogenoketones[265].

$$AcCH_2 \cdot CH_2Cl \xrightarrow[\text{H}_2\text{SO}_4]{\text{HNF}_2} MeC(NF_2)_2 \cdot CH_2 \cdot CH_2Cl \xrightarrow{\text{KOH}} MeC(NF_2)_2 \cdot CH{:}CH_2$$

Addition of amines to the carbon–carbon triple bond of dialkyl alk-1-ynylthiophosphonates[266] gives enamine thiophosphonates in essentially quantitative yields[267]. Reaction with sodium hydride followed by an aldehyde

$$(EtO)_2P({:}S) \cdot C{:}CR^1 \xrightarrow{\text{R}^2\text{R}^3\text{NH}} (EtO)_2P({:}S) \cdot CH{:}CR^1 \cdot NR^2R^3$$

$$\xrightarrow[\text{(ii) R}^4\text{CHO}]{\text{(i) NaH}} R^4CH{:}CH \cdot CR^1{:}NR^2 \xrightarrow{\text{H}^+} R^4CH{:}CH \cdot COR^1$$

produces the α,β-ethylenic ketimines which may be hydrolysed to the α,β-ethylenic ketones.

Triphenylphosphine has been used in the synthesis of amides. Treatment of an organic acid and amine with the adduct of triphenylphosphine and

$$Ph_3P + CCl_4 + R^1CO_2H + 2R^2R^3NH \longrightarrow$$
$$R^1CO \cdot NR^2R^3 + Ph_3P({:}O) + HCCl_3 + R^2R^3\overset{+}{N}H_2\overset{-}{Cl}$$

carbon tetrachloride (or bromotrichloromethane) affords good yields (80–95%) of the amide[268].

4.8.3   Rearrangements

The Wolff, Beckmann, Hofmann, Curtius and Schmidt rearrangements which involve cleavage of a C—C bond and migration of an electron-rich group to an electron deficient centre have now been found to occur with B—C bond cleavage in appropriate o-carborane derivatives[269]. For example, treatment of 3-o-carboranecarbonyl chloride with an excess of diazomethane affords a diazoketone, which undergoes a Wolff rearrangement to give

$$\underset{B_{10}H_9 \cdot COCl}{\overset{\text{CH}\!-\!\!-\!\text{CH}}{\bigtriangleup\!\!O}} \xrightarrow{\text{CH}_2\text{N}_2} {>}B \cdot CO \cdot CHN_2 \xrightarrow[\text{Ag}_2\text{O}]{-\text{N}_2} {>}B\!-\!CO\!-\!\ddot{C}H \longrightarrow$$

$${>}B \cdot CH{:}C{:}O \longrightarrow {>}B \cdot CH_2 \cdot CO_2H$$

o-carborane-3-ylacetic acid. In like manner, the amide and azide of 3-o-carboranecarboxylic acid gave the amines by the Hofmann–Curtius rearrangement. The parent acid also gave the amine by the Schmidt rearrangement when first treated with N_3H. Heating of 1-methyl-o-carborane-3-yl-phenyl ketoxime with PCl_5 in benzene followed by hydrolysis gave only the Beckmann rearrangement product, 1-methyl-3-N-benzoylamino-o-carborane.

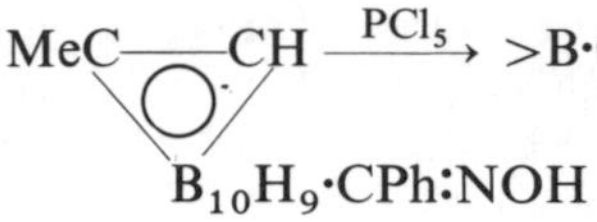

$$\text{MeC}\underset{\text{B}_{10}\text{H}_9\cdot\text{CPh:NOH}}{\overline{\quad\text{CH}\quad}} \xrightarrow{\text{PCl}_5} >\text{B·CPh:N}^+ \longrightarrow >\text{B·NH·COPh}$$

The Schmidt reaction has also been shown to occur with diethyl p-anisoyl-phosphonate to yield diethyl N-(p-anisoyl)phosphoramidate (30%), which represents the first example of migration of a dialkoxyphosphino-group to an electron deficient nitrogen[270]. A study of the Schmidt reaction of adamantan-2-one has indicated that the product distribution may be a function of

$$\text{ArCO·P(:O)(OEt)}_2 + \text{HN}_3 \xrightarrow{\text{H}^+} \text{ArCO·NHP(:O)(OEt)}_2$$

$$(\text{Ar} = p\text{-MeO·C}_6\text{H}_4)$$

the catalyst–solvent system employed, and may yield any one of several products as the major product[271]. Thus in methanesulphonic acid, the 4-substituted adamantan-2-one (R = SO$_2$Me) was the major product (88%).

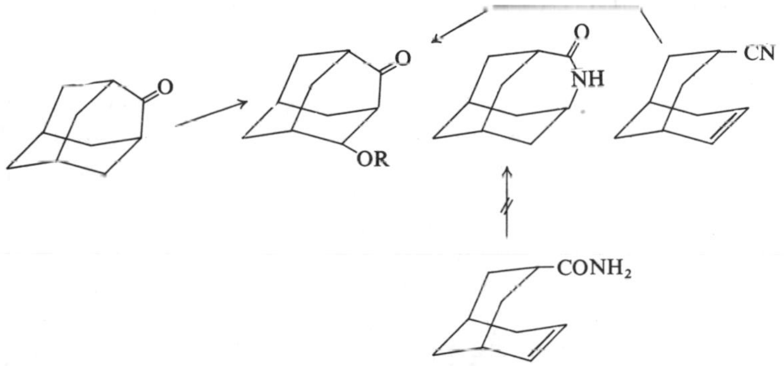

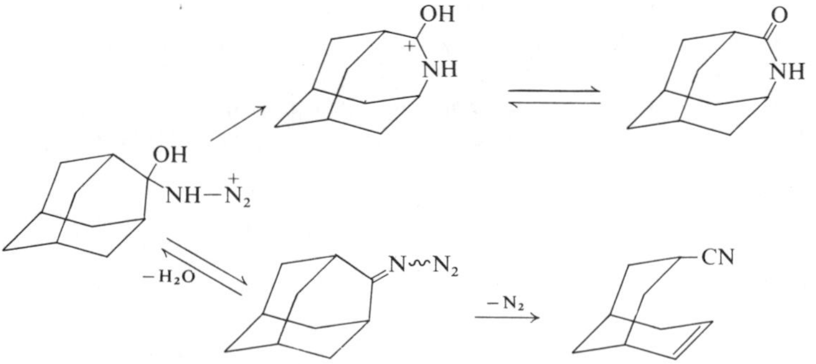

In methanesulphonic acid–acetic acid, in water, and in trichloro- or trifluoro-acetic acid, the nitrile was obtained (40–61%), along with the lactam (27–60%). Although the 4-substituted adamantanone was found to be derived exclusively from the nitrile, the lactam was not isolated when the amide from the nitrile reacted in appropriate solvents. Instead, the lactam is thought to be produced by a pathway involving loss of N_2 from the azidohydrin, with bond migration.

The reaction of adamantan-2-one with diazomethane generated *in situ* in methanol yielded pure homoadamantan-4-one (87–92 %)[272]. Hydride reduction afforded the ($\pm$)-homoadamantan-4-ol (92–98 %) which was dehydrated

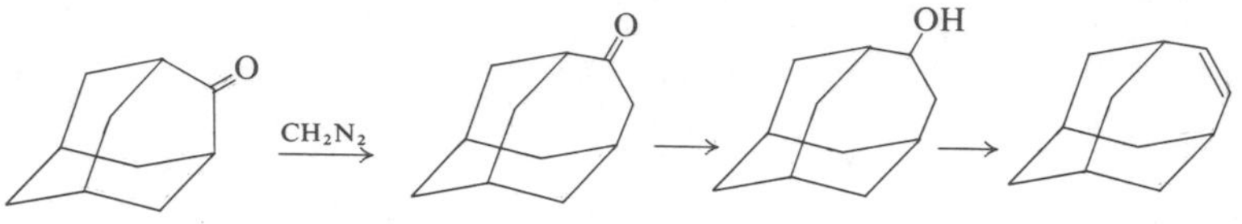

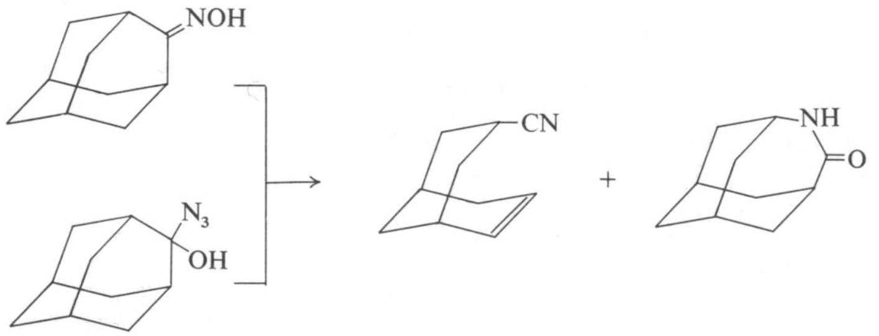

to the cycloalkene (68–85 %). Attempts to prepare the homolactam by a Beckmann (oxime) or a Schmidt (azide) rearrangement, gave large amounts of the anomalous product *endo*-7-cyanobicyclo[3.3.1]non-2-ene in addition to the expected lactam. Conversion of the adamantanone or its oxime into the lactam was achieved without contamination in polyphosphoric acid (yields 23 and 52 % respectively).

The 'Wawzonek rearrangement'[273] (a Stevens rearrangement involving an aminimide), was observed to give 1-carbethoxy-1,2,2-trimethylhydrazine (R = Me) in the thermolysis of trimethylamine carbethoxyimide[274]. The

$$\text{EtO·CO·}\bar{\text{N}}\cdot\overset{+}{\text{N}}\text{RMe}_2 \xrightarrow[175°C]{\Delta} \text{EtO·CO·NR·NMe}_2$$

more electrophilic benzyl group was found to migrate exclusively in the benzyl dimethyl derivative (R = CH_2Ph). This rearrangement is believed to proceed in part, if not entirely, by a diradical pathway. The diradical nature of the Stevens rearrangement of aminimides has been demonstrated by the use of chemically-induced dynamic nuclear polarisation (CIDNP)[275]. A further investigation of the mechanism using 1-benzyl-1,4,4-trimethyl-3-oxapyrazolidinium inner salt and 1,1-dimethyl-1-benzylamine-2-acetimide has been reported[276]. The heterocyclic aminimide rearranged 240 times faster than the acyclic compound. The rate difference is thought to reflect

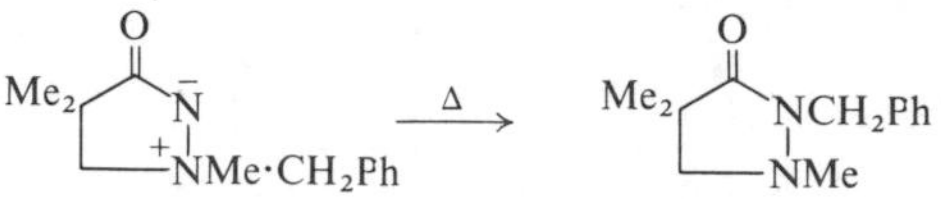

$$\text{PhCH}_2\overset{+}{\text{N}}\text{Me}_2\cdot\bar{\text{N}}\text{Ac} \xrightarrow{\Delta} \text{Me}_2\text{N}\cdot\text{NAc}\cdot\text{CH}_2\text{Ph}$$

relief of strain associated with the change in hybridisation of the α-nitrogen ($sp^2 \to sp^3$) and therefore the transition state must include a partially-bonded benzyl radical at the α-nitrogen atom (and consequently a partially rehybridised α-nitrogen atom). In addition to CIDNP studies, trapping experiments have verified the radical nature of this rearrangement.

Rearrangement of 3-dimethylamino-3-methylbut-1-yne methiodide with sodium amide has also been observed to give products indicative of a Stevens rearrangement, by a homolytic cleavage–recombination mechanism[277].

$$Me_3\overset{+}{N}\cdot CMe_2\cdot C\vdots CH \xrightarrow{2NaNH_2} Me_2N(CH_2)_2NMe_2 \text{ (dimer, 4\%)} +$$

$$[CH\vdots C\cdot CMe_2\cdot]_2 \text{ (dimer, 7.2\%)} + Me_2N\cdot CH_2\cdot CMe_2\cdot C\vdots CH$$

Expected Stevens product 24%

However, the rearrangement of NNN-trimethylneopentylammonium iodide to give the three rearranged products is considered to proceed by an ion-pair pathway[278]. This route was indicated by the isolation of minor by-products,

$$Bu^tCH_2\overset{+}{N}Me_3 \ \bar{I} \xrightarrow{RLi} \left[\overset{|}{-}\overset{}{\underset{+}{N}}\overset{/}{-}\overset{}{C} \right] \longrightarrow Bu^tCH_2\cdot CH_2NMe_2 +$$
$$(80\%)$$

$$Bu^tCH_2NMeEt + Bu^tCHMeNMe_2$$
$$(3\%) \qquad\qquad (1\%)$$

NN-dimethylbenzylamine (from phenyl lithium), NN-dimethylpent-1-yl-amine (from n-butyl lithium), and neopentane. Also consistent with this

$$Bu^tCH_2\cdot\overset{+}{N}Me_2\cdot CH_2 \longrightarrow Me_3C\cdot\bar{C}H_2 + CH_2\vdots\overset{+}{N}Me_2$$

solvent	PhLi	Bu^{n}Li
↓	↓	↓
Me$_4$C	PhCH$_2$NMe$_2$	Bu^{n}CH$_2$NMe$_2$

mechanism was (i) the absence of the dimer, $Bu^tCH_2\cdot CH_2Bu^t$, expected from a diradical-pair intermediate and (ii) unsuccessful attempts to observe CIDNP during product formation.

4.8.4 Deaminations

Sodium nitroprusside reacts with primary amines to give deaminated products, but with secondary amines to give N-nitrosamines[279]. The reagent is unique in that it is a deaminating agent which is so stable in aqueous

$$CH_2\vdots CH\cdot CH_2NH_2 \xrightarrow{NaFe(CN)_5NO_2} CH_2\vdots CH\cdot CH_2OH$$

$$Et_2NH \longrightarrow Et_2N\cdot NO$$

alkaline solution that deaminations may be carried out at initial pH values as high as 12.7. Intermediates in deamination reactions have been observed to be quite generally unstable, but the reaction of salts of nitrosohydroxyl-amines with toluene-p-suphonyl chloride gave a stable crystalline solid

$$R\overset{+}{N}(:N\cdot\bar{O})\cdot\bar{O} \xrightarrow{\text{TsCl}} R\overset{+}{N}(\cdot\bar{O}){:}N\cdot O\cdot SO_2\cdot C_6H_4Me\text{-}p$$

(R = Ph, Bu$^\text{i}$, or PhCH$_2$)[280]. The structures were determined by x-ray diffraction techniques. The stability is attributed to (i) the lack of resonance contributors with a full octet on each atom following loss of the tosyl group and/or (ii) the absence of an available electron pair to assist the departing

$$RN^+(O^-){:}N\cdot O\cdot SO_2\cdot C_6H_4Me\text{-}p \xrightarrow[-OSO_2Ar]{} R\overset{+}{N}(\bar{O}){:}\overset{+}{\ddot{N}} \longleftrightarrow R\overset{+}{N}(:O)\cdot\ddot{N}{:}$$

tosyl group. Studies of the stereochemistry of the elimination reactions of quaternary ammonium salts have shown a wide variation in the *trans* to *cis* ratio among the alkenic products obtained[281, 282]. It has been postulated that in acyclic systems, 'the formation of *trans* olefin by *anti* elimination is sterically hindered, and that formation of *trans* olefin by *syn* elimination becomes more important as the hindrance increases'[283]. Thus, the bulky trimethylammonio-group forces alkyl groups attached to the β and the γ carbon into positions where they hinder access to the *anti*-β-hydrogen. It has

been found that, with very few exceptions, the percentage of *cis*-alkene increases with the decreasing steric congestion about the β-carbon. However, as steric hindrance to the *anti*-β-hydrogen increases, the importance of *syn*-elimination increases, and *trans*-alkenes are preferred.

Yields of alkenes obtained by Hofmann elimination have been improved by carrying out the reaction under reduced pressure and at lower temperatures than normally required[284]. Thus, the yield of elimination product on thermolysis of NN-dimethyl-1,2,3,4-tetrahydroquinolinium hydroxide ranged from 5% (160°C at 760 mmHg) to 75% (60°C at 0.005 mmHg). Improved yields of alkenes were invariably obtained by using these experimental conditions and in no case was it necessary to raise the temperature above 100 °C during the thermolysis. The first example of a 1,3 bromine migration in a saturated system has been reported in the deamination of 3-bromo-2,2-bis-(bromo-methyl)-1-d$_2$-propylamine perchlorate[285]. The 1,3 bromine shift was demon-

$$(BrCH_2)_3C\cdot CD_2\overset{+}{N}H_3ClO_4^- \xrightarrow[CH_3\cdot CO_2H]{NaNO_2} (BrCH_2)_3C\cdot CD_2OAc +$$

$$(BrCH_2)_2C(CH_2OAc)\cdot CD_2Br$$

strated by the appearance of the CH$_2$OAc signal at δ 4.21 in the n.m.r. spectrum (1:6.5 ratio compared to the CH$_2$Br signal).

4.8.5 Configurational stability at nitrogen

Compounds possessing configurational stability at nitrogen have been the subject of much recent discussion[286-290]. Indeed, several pairs of diastereoismers, isolated as a result of a high barrier to nitrogen inversion, have been characterised[291, 292]. Recently reported has been the isolation of two N-chloro-aziridine invertomers[293]. Reaction with $AgNO_3$ in an attempt to form the

PhCO·CH—CHPh   PhCO·CH—CHPh
     N   N
     Cl   Cl

nitrenium ion resulted in recovery of the N-chloroaziridines unchanged from the reaction mixture. At ambient temperatures N-methoxytetramethyl-aziridine was found[294] to have a barrier to inversion $(\Delta G^{\ddagger}) > 22 \text{ kcal mol}^{-1}$.

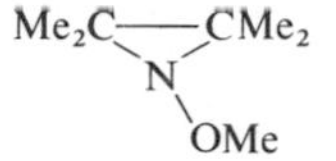

Me$_2$C——CMe$_2$
      N
      OMe

This value was suggested by the observation in the n.m.r. spectrum of the non-equivalent geminal methyl proton resonances, which did not coalesce even at temperatures up to 130 °C.

A study of the electronic effects on the barrier to pyramidal inversion in the series of substituted N-aryl-aziridines has shown a decrease in the energy of activation $(\Delta G^{\ddagger})$ from 12.5 to 8.2 kcal mol^{-1} (R = OMe > H > 4-Cl > 3-CF$_3$ > 4-CF$_3$ > 4-NO$_2$) paralleling the ability of the substituent to withdraw electrons from the aromatic ring[295]. Substituents which withdraw electrons from the aromatic ring could permit direct conjugation between

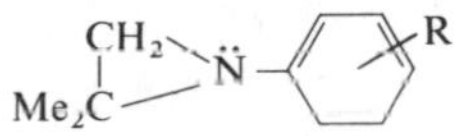

CH$_2$
    N——⟨ ⟩—R
Me$_2$C

the inversion centre (nitrogen) and the electron withdrawing groups. A satisfactory correlation of the magnitude of the barrier to inversion with σ constants was obtained. Among aziridine derivatives possessing configurational stability at nitrogen, a series of aziridine esters bearing 2,3-dihydro-2-oxobenzoxazol-3-yl, phthalimido, 3,4-dihydro-2-methyl-4-oxo-quinazolin-3-yl and 1,2-dihydro-2-oxoquinolin-1-yl groups as N-substituents has been investigated[296]. It was found that these aziridines showed slow inversion on the n.m.r. time scale at room temperature and that as the size of the ester alkyl group was increased, the proportion of invertomer with the N-substituent cis to the ester group increased. For example, the ratio of

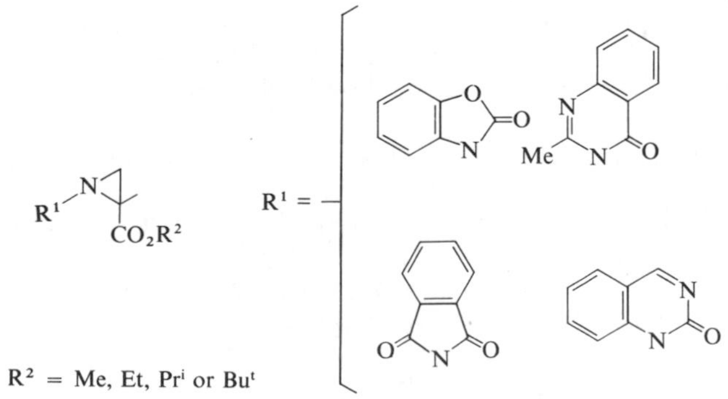

R^1—N⟨ ⟩CO$_2$R^2 R^1 =

R^2 = Me, Et, Pr^i or Bu^t

cis- to trans-isomer is approximately doubled as the ester alkyl group is changed from methyl to t-butyl. On steric grounds the reverse trend would be anticipated. The cause is believed to be a favourable interaction between the carbonyl oxygen atom and the heterocyclic lactam carbonyl carbon atom[296]. An interesting case of non-equivalence of the protons of a methylene group is found in the spectrum of 3,5-diethyltetrahydro-1,3,5-thiadiazine at −70 °C, in which the methylene protons of the ethyl group appear as two sextuplets[297]. This chirality results almost certainly from the slow chair–chair inversion of the six-membered ring in which the nitrogen atoms are centres of chirality.

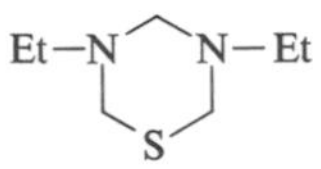

$$Et-N\underset{S}{\overset{\frown}{}}N-Et$$

An investigation of the barriers to N-inversion in several mono- and bi-cyclic compounds has indicated that the barrier in cyclic tertiary amines depends on ring size[298]. Another study of nitrogen inversion v. ring reversal

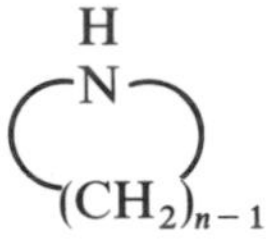

Table 4.4

n	Barrier, $\Delta G^{\ddagger}$, kcal mol^{-1}
3	20
6	8.5
4	8.5
5	8
7	6.5

in a series of cyclic amines has indicated that the former is the rate determining process when $n = 3, 4$ or $6.$ $(\Delta G^{\ddagger} = 13.4, 10.3,$ and 9.0 kcal mol^{-1} respectively,

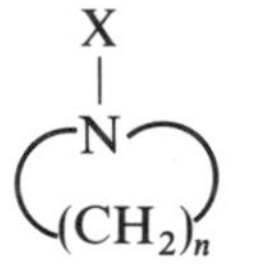

$$X = Cl \text{ or } Me$$
$$n = 3, 4, 5 \text{ or } 6$$

for the N-chloro series and 10.0, 8.3 and 6.8 kcal mol^{-1}, respectively, for the N-methyl series). However, ring reversal was found to be the rate determining process in the piperidine derivative $(n = 5)$ $(\Delta G^{\ddagger} = 13.5$ kcal mol^{-1} when $X = Cl)$[299]. N-Methylhomopiperidine $(X = Me, n = 6)$ was believed to furnish the simplest unambiguous example of nitrogen inversion in trialkyl-amines yet observed.

The barrier to N-inversion $(\Delta G^{\ddagger})$ is reported to be a multiplicative function of the N-substituent (as opposed to additive as in the Hammett and the Taft equation)[300]. Calculations of barriers to nitrogen inversion have been attempted by using two substituent parameters, X which represents the free energy of activation necessary for inversion of a reference N-substituent (R), and Z, which represents the variable N-substituent (Y), defined as $Z_y = \Delta G^{\ddagger}_y/\Delta G^{\ddagger}_R$. These parameters are related to the barrier to pyramidal inversion by the following equation:

$$\Delta G_y^{\ddagger} = XZy$$

From a series of N-methyl- and the corresponding N-chloro-compounds, a value of $Z_{Cl} = 1.28$ is obtained, which permits calculation of the barrier to nitrogen inversion of the N-chloro-compounds ($\Delta G_{Cl}^{\ddagger}$) when $\Delta G_{Me}^{\ddagger}$ (X) is known. Also, the barrier for pyramidal inversion is increased (compared to

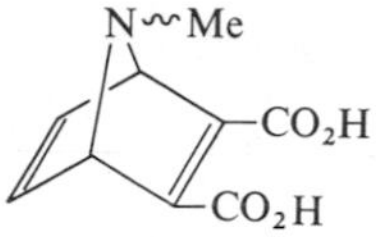

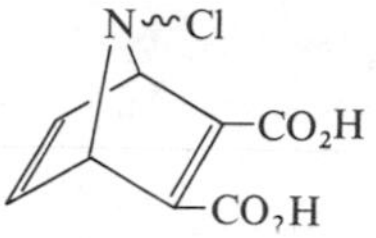

Predicted	
$\Delta G^+ \approx 24\ \text{kcal mol}^{-1}$	$\Delta G^+ = 18\cdot4\ \text{kcal mol}^{-1}$	$\Delta G^+ = 23\cdot5\ \text{kcal mol}^{-1}$

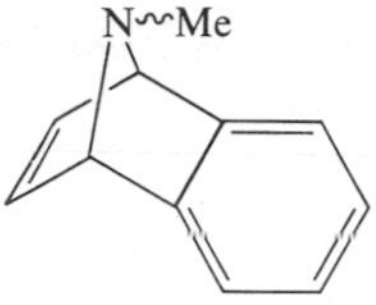

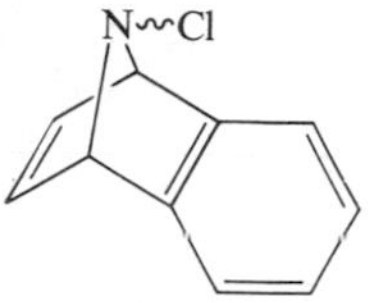

N-methyl) by N-oxygen substitution ($Z_{OMe} = 1.75$)[301], or by replacing a carbon adjacent to nitrogen by a second nitrogen atom ($Z = 1.45$)[301], but decreased by replacing a N-methyl by aN-tosyl substituent ($Z = 0.57$)[302]. Observation of slow nitrogen inversion on the n.m.r. time scale in the 7-methyl-7-azatetrahalogenobenzonorbornadienes has permitted a reasonable approximation of $\Delta G^+ = 14\ \text{kcal mol}^{-1}$ to the free energy of activation[303].

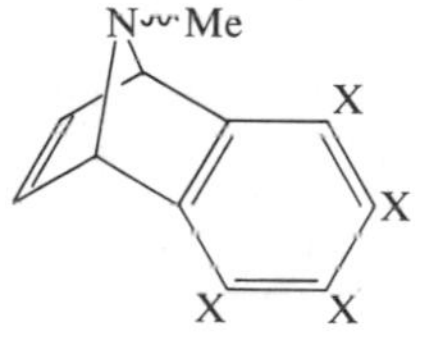

$(X = F, Cl)$

References

1. Blum, J. and Fisher, A. (1970). *Tetrahedron Lett.*, 1963
2. Dennis, W. E. (1970). *J. Org. Chem.*, **35**, 3253
3. Binkley, R. W. (1970). *Tetrahedron Lett.*, 2085
4. Rhee, I., Ryang, M. and Tsutsumi, S. (1970). *Tetrahedron Lett.*, 3419
5. Hanson, R. B., Foley, P., Jr., Anderson, E. L. and Aldridge, M. H. (1970). *J. Org. Chem.*, **35**, 1753
6. Klein, D. A. (1971). *J. Org. Chem.*, **36**, 3050
7. Wilt, J. W. and Ho, A. J. (1971). *J. Org. Chem.*, **36**, 2026
8. Ingold, C. K. (1953). *Structure and Mechanism in Organic Chemistry*, 562 (New York: Cornell University Press)
9. Powell, R. L. and Hall, C. D. (1971). *J. Chem. Soc. C*, 2336
10. Moriconi, E. J. and Jalandoni, C. C. (1970). *J. Org. Chem.*, **35**, 3796
11. Pande, K. C. and Trampe, G. (1970). *J. Org. Chem.*, **35**, 1169
12. Hashimoto, N., Matsumura, K., Saraie, T., Kawano, Y. and Morita, K. (1970). *J. Org. Chem.*, **35**, 675
13. Nagata, W., Yoshioka, M., Okumura, T. and Murakami, M. (1970). *J. Chem. Soc. C*, 2355
14. Kuehne, M. E. and Nelson, J. A. (1970). *J. Org. Chem.*, **35**, 161
15. Wenkert, E., Afonso, A., Bredenberg, J. B., Kaneko, C. and Tahara, A. (1964). *J. Amer. Chem. Soc.*, **86**, 2038; Spencer, T. A., Weaver, T. D., Villarica, R. M., Friary, R. J., Posler, J. and Schwartz, M. A. (1968). *J. Org. Chem.*, **33**, 712; Spencer, T. A., Friary, R. J., Schmiegel, W. W., Simeone, J. F. and Watt, D. S. (1968). *J. Org. Chem.*, **33**, 719

16. Kuehne, M. E. (1970). *J. Org. Chem.*, **35**, 171
17. Nagata, W., Yoshioka, M. and Okumura, T. (1970). *J. Chem. Soc. C*, 2365
18. Hallsworth, A. S. and Henbest, H. B. (1957). *J. Chem. Soc.*, 4604
19. Kofron, W. O. and Hauser, C. R. (1970). *J. Org. Chem.*, **35**, 2085
20. Butler, D. E. and Pollatz, J. C. (1971). *J. Org. Chem.*, **36**, 1308
21. Brown, H. C. (1962). *Hydroboration.* (New York: W. A. Benjamin Inc.)
22. Nambu, H. and Brown, H. C. (1970). *J. Amer. Chem. Soc.*, **92**, 5790
23. Alt, G. H. (1968). *J. Org. Chem.*, **33**, 2858; Alt, G. H. and Spezae, A. J. (1966). *J. Org. Chem.*, **31**, 1340
24. Armarego, W. L. F. (1969). *J. Chem. Soc. C*, 986
25. Alt, G. H. and Gallegos, G. A. (1971). *J. Org. Chem.*, **36**, 1000
26. Bergman, E. D. and Agranat, I. (1971). *J. Chem. Soc. C*, 1541
27. Seitz, G. and Mönnighoff, H. (1971). *Tetrahedron Lett.*, 4889
28. Norell, J. R. (1970). *J. Org. Chem.*, **35**, 1611
29. Glikmans, G., Torck, B., Hellin, M. and Coussemant, F. (1966). *Bull. Soc. Chim. Fr.*, 1376
30. Norell, J. R. (1970). *J. Org. Chem.*, **35**, 1619
31. Wiberg, K. B. (1953). *J. Amer. Chem. Soc.*, **75**, 3961; Wiberg, K. B. (1955). *J. Amer. Chem. Soc.*, **77**, 2519
32. McIsaac, J. E., Jr., Ball, R. E. and Behrman, E. J. (1971). *J. Org. Chem.*, **36**, 3048
33. Sasaki, T., Kojima, A. and Ohta, M. (1971). *J. Chem. Soc. C*, 196
34. Balasubrahmanyam, S. N. and Balasubramanian, M. (1971). *J. Chem. Soc. C*, 827
35. Johnson, F. and Malhotra, S. K. (1965). *J. Amer. Chem. Soc.*, **87**, 5492
36. Hashimoto, N., Kawano, Y. and Morita, K. (1970). *J. Org. Chem.*, **35**, 828
37. Sasaki, T. and Kojima, A. (1970). *J. Chem. Soc. C*, 476
38. Morrocchi, S., Ricca, A., Zanarotti, A., Bianchi, G., Gandolfi, R. and Grünanger, P. (1969). *Tetrahedron Lett.*, 3329
39. Morrocchi, S., Ricca, A. and Zanarotti, A. (1970). *Tetrahedron Lett.*, 3215
40. Carmella, P. and Bianchessi, P. (1970). *Tetrahedron, 26*, 5773
41. Nair, V. (1971). *Tetrahedron Lett.*, 4831
42. Ranganathan, S., Singh, B. B. and Panda, C. S. (1970). *Tetrahedron Lett.*, 1225
43. van Leusen, A. M. and Jagt, J. C. (1970). *Tetrahedron Lett.*, 967
44. van Leusen, A. M. and Jagt, J. C. (1970). *Tetrahedron Lett.*, 971
45. Carpenter, W. R. and Armstrong, P. (1964). *J. Org. Chem.*, **29**, 2772; Hagemann, N., Arlt, D. and Ugi, I. (1969). *Angew. Chem.*, **81**, 572
46. Vrijland, M. S. A. and Hackmann, J. Th. (1970). *Tetrahedron Lett.*, 3763
47. Geevers, J., Hackmann, J. Th. and Trompen, W. P. (1970). *J. Chem. Soc. C*, 875
48. Moore, H. W. and Weyler, W., Jr. (1970). *J. Amer. Chem. Soc.*, **92**, 4132
49. Bose, A. K., Anjaneyulu, B., Bhattacharya, S. K. and Manhas, M. S. (1967). *Tetrahedron, 23*, 4769; Böhme, H., Ebel, S. and Hartke, K. (1965). *Chem. Ber.*, **98**, 1463; Moore, H. W., Weyler, W., Jr. and Shelden, H. R. (1969). *Tetrahedron Lett.*, 3947
50. de Selms, R. C. (1969). *Tetrahedron Lett.*, 1179
51. Horsewood, P. and Kirby, G. W. (1971). *Chem. Commun.*, 1139
52. Ponsold, K. and Ihn, W. (1970). *Tetrahedron Lett.*, 1125
53. McCarty, C. G., Parkinson, J. E. and Wieland, D. M. (1970). *J. Org. Chem.*, **35**, 2067
54. Begland, R. W., Cairncross, A., Donald, D. S., Hartter, D. R., Sheppard, W. A. and Webster, O. W. (1971). *J. Amer. Chem. Soc.*, **93**, 4953
55. Paukstelis, J. V. and Kim, M. (1970). *Tetrahedron Lett.*, 4731
56. Fodor, G. and Abidi, S. (1971). *Tetrahedron Lett.*, 1369
57. Gale, D. M. (1970). *J. Org. Chem.*, **35**, 970
58. Dilling, W. L. and Kroening, R. D. (1970). *Tetrahedron Lett.*, 695
59. Dilling, W. L., Kroening, R. D. and Little, J. C. (1970). *J. Amer. Chem. Soc.*, **92**, 928
60. Dalton, J. C., Wriede, P. A. and Turro, N. J. (1970). *J. Amer. Chem. Soc.*, **92**, 1318
61. Grigg, R. and Jackson, J. L. (1970). *J. Chem. Soc. C*, 552
62. Zajac, W. W., Jr., Siuda, J. F., Nolan, M. J. and Santosusso, T. M. (1971). *J. Org. Chem.*, **36**, 3539
63. Yasuda, H., Hayashi, T. and Midorikawa, H. (1970). *J. Org. Chem.*, **35**, 1234
64. Bosshard, P. and Eugster, C. H. (1966). *Advan. Heterocycl. Chem.*, **7**, 377
65. Leditschke, H. (1952). *Chem. Ber.*, **85**, 483
66. Moppett, C. E., Johnson, F. and Dix, D. T. (1971). *Chem. Commun.*, 1560
67. Middleton, W. J. and Metzger, D. (1970). *J. Org. Chem.*, **35**, 3985

68. Webster, O. W. (1963). *U.S. Patent* 3 093 653
69. Scheinbaum, M. L. and Dines, M. B. (1971). *Tetrahedron Lett.,* 2205
70. Thaler, W. A. and McDivitt, J. R. (1971). *J. Org. Chem.,* **36,** 14
71. van Tamelen, E. E., Rudler, H. and Bjorklund, C. (1971). *J. Amer. Chem. Soc.,* **93,** 7113
72. Meyers, A. I. and Adickes, H. W. (1969). *Tetrahedron Lett.,* 5151
73. Walborsky, H. M. and Niznik, G. E. (1969). *J. Amer. Chem. Soc.,* **91,** 7778
74. Koga, N., Koga, G. and Anselme, J. -P. (1970). *Tetrahedron Lett.,* 3309
75. Saegusa, T., Ito, Y., Kinoshita, H. and Tomita, S. (1971). *J. Org. Chem.,* **36,** 3316
76. Saegusa, T., Ito, Y., Tomita, S., Kinoshita, H. and Taka-ishi, N. (1971). *Tetrahedron,* **27,** 27
77. Saegusa, T., Murase, I. and Ito, Y. (1971). *J. Org. Chem.,* **36,** 2876
78. Ciganek, E. (1970). *J. Org. Chem.,* **35,** 862
79. Saegusa, T., Ito, Y. and Shimizu, T. (1970). *J. Org. Chem.,* **35,** 3995
80. Walborsky, H. M., Morrison, W. H., III and Niznik, G. E. (1970). *J. Amer. Chem. Soc.,* **92,** 6675
81. Saegusa, T., Murase, I. and Ito, Y. (1971). *Tetrahedron,* **27,** 3795
82. Johnson, F., Gulbenkian, A. H. and Nasutavicus, W. A. (1970). *Chem. Commun.,* 608
83. Saegusa, T., Kobayshi, S. and Ito, Y. (1970). *J. Org. Chem.,* **35,** 2118
84. Saegusa, T., Ito, Y., Yasuda, N. and Hotaka, T. (1970). *J. Org. Chem.,* **35,** 4238
85. Pfeffer, P. E. and Silbert, L. S. (1970). *Tetrahedron Lett.,* 699
86. Bachman, G. B. and Biermann, T. F. (1970). *J. Org. Chem.,* **35,** 4229
87. Larkin, J. M. and Kreuz, K. L. (1971). *J. Org. Chem,* **36,** 2574
88. Feuer, H., Hall, A. M. and Anderson, R. S. (1971). *J. Org. Chem.,* **36,** 140
89. Armarego, W. L. F. (1971). *J. Chem. Soc. C,* 1812
90. Armarego, W. L. F. and Kobayashi, T. (1971) *J. Chem. Soc. C,* 3222
91. Nelson, W. L., Miller, D. D. and Shefter, E. (1970). *J. Org. Chem.,* **35,** 3433
92. Olah, G. A. and Lin, H. C. (1971). *J. Amer. Chem. Soc.,* **93,** 1259
93. Russell, G. A., Norris, R. K. and Panek, E. J. (1971). *J. Amer. Chem. Soc.,* **93,** 5839
94. Kornblum, N. and Boyd, S. D. and Stuchal, F. W. (1970). *J. Amer. Chem. Soc.,* **92,** 5783
95. Kornblum, N. and Boyd, S. D. (1970). *J. Amer. Chem. Soc.,* **92,** 5784
96. Grakauskas, V. and Baum, K. (1968). *J. Org. Chem.,* **33,** 3080; Kamlet, M. J. and Adolph, H. G. (1968). *J. Org. Chem.,* **33,** 3073
97. Eremenko, L. T. and Natsibullin, F. Ya. (1968). *Izv. Akad. Nauk. S.S.S.R.,* 912
98. Baum, K. (1970). *J. Org. Chem.,* **35,** 846
99. Heasley, V. L., Hensley, G. E., McConnell, M. R., Martin, K. A., Ingle, D. M. and Davis, P. D. (1971). *Tetrahedron Lett.,* 4819
100. Baer, H. H. and Naik, S. R. (1970). *J. Org. Chem.,* **35,** 2927
101. Newman, H. and Angier, R. B. (1970). *Tetrahedron,* **26,** 825
102. Dominani, S. J., Chaney, M. O. and Jones, N. D. (1970). *Tetrahedron Lett.,* 4735
103. Lipp, P. and Braucker, H. (1939). *Chem. Ber.,* **72,** 2079
104. Ranganathan, S. and Singh, B. B. (1970). *Chem. Commun.,* 218
105. Toivonen, H., Laurema, S. A. and Ilvonen, P. J. (1971). *Tetrahedron Lett.,* 3203
106. Pagano, A. H. and Shechter, H. (1970). *J. Org. Chem.,* **35,** 295
107. McMurry, J. E. and Melton, J. (1971). *J. Amer. Chem. Soc.,* **93,** 5309
108. Freeman, F. and Lin, D. K. (1971). *J. Org. Chem.,* **36,** 1335
109. Sun, S. F. and Folliard, J. T. (1971). *Tetrahedron,* **27,** 323
110. Ranganathan, S. and Raman, H. (1970). *Tetrahedron Lett.,* 3331
111. Bordwell, F. G. and Yee, K. C. (1970). *J. Amer. Chem. Soc.,* **92,** 5933
112. Bordwell, F. G. and Yee, K. C. (1970). *J. Amer. Chem. Soc.,* **92,** 5939
113. Zimmermann, H. E. (1955). *J. Org. Chem.,* **20,** 549
114. Zimmerman, H. E. and Nevins, T. E. (1963). *Molecular Rearrangements,* **1,** (New York: Interscience)
115. Richer, J. C. (1965). *J. Org. Chem.,* **30,** 324
116. Marshall, J. A. and Carroll, R. D. (1965). *J. Org. Chem.,* **30,** 2748
117. Reid, S. T. and Tucker, J. N. (1970). *Chem. Commun.,* 1286
118. Stermitz, F. R. and Norris, F. A. (1970). *J. Org. Chem.,* **35,** 527
119. Kahr, K. and Berther, C. (1960). *Chem. Ber.,* **93,** 132
120. Stowell, J. C. (1971). *J. Org. Chem.,* **36,** 3055
121. Buckingham, J. (1970). *Tetrahedron Lett.,* 2341
122. Hassner, A. and Heathcock, C. (1964). *J. Org. Chem.,* **29,** 1305

123. Shiue, C. Y., Park, K. P. and Clapp, L. B. (1970). *J. Org. Chem.*, **35**, 2063
124. Shiue, C. and Clapp, L. B. (1971). *J. Org. Chem.*, **36**, 1169
125. Lau, M. P., Cessna, A. J., Chow, Y. L. and Yip, R. W. (1971). *J. Amer. Chem. Soc.*, **93**, 3808
126. Chow, Y. L., Colón, C. J. and Chang, D. W. L. (1970). *Can. J. Chem.*, **48**, 1664
127. Chow, Y. L., Chen, S. C. and Chang, D. W. L. (1970). *Can. J. Chem.*, **48**, 157
128. Chow, Y. L., and Chang, D. W. L. (1971). *Chem. Commun.*, 64
129. Chow, Y. L., Perry, R. A., Menon, B. C. and Chen, S. C. (1971). *Tetrahedron Lett.*, 1545
130. Chow, Y. L., Perry, R. A., and Menon, B. C. (1971). *Tetrahedron Lett.*, 1549
131. Farina, P. R. (1970). *Tetrahedron Lett.*, 4917
132. Dominianni, S. J. and Demarco, P. V. (1971). *J. Org. Chem.*, **36**, 2534
133. Koenig, T., Deinzer, M. and Hoobler, J. A. (1971). *J. Amer. Chem. Soc.*, **93**, 938
134. White, E. H. (1955). *J. Amer. Chem. Soc.*, **77**, 6011, 6014
135. White, E. H. and Aufdermarsh, C. A., Jr. (1958). *J. Amer. Chem. Soc.*, **80**, 2597
136. White, E. H. and Aufdermarsh, C. A., Jr. (1961). *J. Amer. Chem. Soc.*, **83**, 1174, 1179
137. Chow, Y. L. and Tam, J. N. S. (1970). *J. Chem. Soc. C*, 1138
138. Doyle, M. P. and Wierenga, W. (1970). *J. Amer. Chem. Soc.*, **92**, 4999
139. Keefer, L. K. and Fodor, C. H. (1970). *J. Amer. Chem. Soc.*, **92**, 5747
140. Fraser, R. R. and Wigfield, Y. Y. (1971). *Tetrahedron Lett.*, 2515
141. Doering, W. von E. and Hoffmann, A. K. (1955). *J. Amer. Chem. Soc.*, **77**, 521; Saunders, M. and Gold, E. H. (1966). *J. Amer. Chem. Soc.*, **88**, 3376
142. Freeman, J. P. and Lillwitz, L. D. (1970). *J. Org. Chem.*, **35**, 3107
143. Vietmeyer, N. D. and Djerassi, C. (1970). *J. Org. Chem.*, **35**, 3591
144. Ponder, B. W. and Walker, P. L. (1970). *J. Org. Chem.*, **35**, 4283
145. Scheinbaum, M. L. (1970). *J. Org. Chem.*, **35**, 2785
146. Scheinbaum, M. L. (1970). *J. Org. Chem.*, **35**, 2790
147. Alvernhe, G. and Laurent, A. (1970). *Bull. Soc. Chim. Fr.*, 3003
148. Alvernhe, G. and Laurent, A. (1971). *Tetrahedron Lett.*, 1913
149. Hamlet, Z., Pampersad, M. and Shearing, D. J. (1970). *Tetrahedron Lett.*, 2101
150. Battaglia, A. and Dondoni, A. (1970). *Tetrahedron Lett.*, 1221
151. Bravo, P., Gaudiano, G. and Ticozzi, C. (1970). *Tetrahedron Lett.*, 3223
152. Bravo, P., Gaudiano, G., Ponti, P. P. and Umani-Ronchi, A. (1970). *Tetrahedron*, **26**, 1315
153. Befzecki, C. and Trojnar, J. (1970). *Tetrahedron Lett.*, 1879
154. Durden, J. A., Jr. and Heywood, D. L. (1971). *J. Org. Chem.*, **36**, 1306
155. Johnson, J. E., Springfield, J. R., Hwang, J. S., Hayes, L. J., Cunningham, W. C. and McClaugherty, D. L. (1971). *J. Org. Chem.*, **36**, 284
156. Shiue, C., Park, K. P. and Clapp, L. B. (1970). *J. Org. Chem.*, **35**, 2063
157. Sasaki, T. Eguchi, S. and Toru, T. (1971). *J. Org. Chem.*, **36**, 2454
158. Korsloot, J. G. and Keizer, V. G. (1969). *Tetrahedron Lett.*, 3517
159. Sasaki, T. Eguchi, S. and Toru, T. (1970). *Chem. Commun.*, 1239
160. Stojiljković, A. and Tasovac, R. (1970). *Tetrahedron Lett.*, 1405
161. Horning, E. C., Stromberg, V. L. and Lloyd, H. A. (1952). *J. Amer. Chem. Soc.*, **74**, 5153
162. Donat, F. J. and Nelson, A. L. (1957). *J. Org. Chem.*, **22**, 1107
163. Shoppee, C. W., Krüger, G. and Mirrington, R. N. (1962). *J. Chem. Soc.*, 1050; Shoppee, C. W., Lack, R. E., Mirrington, R. N. and Smith, L. R. (1965). *J. Chem. Soc.*, 5868
164. Sato, T. Wakatsuka, H. and Amano, K. (1971). *Tetrahedron*, **27**, 5381
165. Kelly, K. K. and Matthews, J. S. (1971). *J. Org. Chem.*, **36**, 2159
166. Kelly, K. K. and Matthews, J. S. (1970). *Tetrahedron*, **26**, 1555
167. Suginome, H. and Takahashi, H. (1970). *Tetrahedron Lett.*, 5119
168. Cross, B., Searle, R. J. G. and Woodall, R. E. (1971). *J. Chem. Soc. C*, 1833
169. Fox, B. L. and Reboulet, J. E. (1970). *J. Org. Chem.*, **35**, 4234
170. Wolkowski, Z. W., Thoai, N. and Wiemann, J. (1970). *Tetrahedron Lett.*, 93
171. Wolkowski, Z. W. (1971). *Tetrahedron Lett.*, 825
172. Berlin, K. D. and Rengaraju, S. (1971). *J. Org. Chem.*, **36**, 2912
173. Danilewicz, J. C. (1970). *J. Chem. Soc. C*, 1049
174. Mosher, W. A., Hively, R. N. and Dean, F. H. (1970). *J. Org. Chem.*, **35**, 3689
175. Neugeybauer, F. A. (1970). *Tetrahedron Lett.*, 2345
176. Yamakawa, K. and Hisatome, M. (1970). *Tetrahedron*, **26**, 4483
177. Curtin, D. Y., Grubbs, E. J. and McCarty, C. (1966). *J. Amer. Chem. Soc.*, **88**, 2775
178. Brown, C., Grayson, B. T. and Hudson, R. F. (1970). *Tetrahedron Lett.*, **56**, 4925

179. Hayes, B. T. and Stevens, T. S. (1970). *J. Chem. Soc. C*, 1088
180. Marchetti, L. and Tosi, A. (1971). *Tetrahedron Lett.*, 3071
181. Aylward, J. B. (1970). *J. Chem. Soc. C*, 1494
182. Butler, R. N., O'Sullivan, P. and Scott, F. L. (1971). *J. Chem. Soc. C*, 2265
183. Cauquis, G. and Genies, M. (1971). *Tetrahedron Lett.*, 3959
184. Heindel, N. D., Kennewell, P. D. and Pfau, M. (1970). *J. Org. Chem.*, **35**, 80
185. Al-Jallo, H. N. (1970). *Tetrahedron Lett.*, 875
186. Gupta, C. M., Bhaduri, A. P. and Khanna, N. M. (1970). *Tetrahedron*, **26**, 3069
187. Nishiwaki, T. and Saito, T. (1970). *Chem. Commun.*, 1479
188. Scott, F. L., Houlihan, S. A. and Fenton, D. F. (1971). *J. Chem. Soc. C*, 80
189. Scott, F. L., Houlihan, S. A. and Fenton, D. F. (1970). *Tetrahedron Lett.*, 1991
190. Butler, R. N., Scott, F. L. and Scott, R. D. (1970). *J. Chem. Soc. C*, 2510
191. Fahey, J. L., Foster, P. A., Neilson, D. G., Watson, K. M. and (in part) Brokenshire, J. L. and Peters, D. A. V. (1970). *J. Chem. Soc. C*, 719
192. Baldwin, J. E., Brown, J. E. and Cordell, R. D. (1970) *Chem Commun.*, 31
193. Götz, M. and Zeile, K. (1970). *Tetrahedron*, **26**, 3185
194. Daeniker, H. U. and Druey, J. (1962). *Helv. Chim. Acta*, **45**, 2426, 2441
195. Huisgen, R. (1965). *Bull. Chim. Soc. Fr.*, 3431
196. Roesler, P. and Fleury, J. -P. (1968). *Bull. Chim. Soc. Fr.*, 631
197. Tsuji, T. and Kosower, E. M. (1971). *J. Amer. Chem. Soc.*, **93**, 1992
198. Tsuji, T. and Kosower, E. M. (1971). *J. Amer. Chem. Soc.*, **93**, 1999
199. Beam, C. F., Foote, R. S. and Hauser, C. R. (1971). *J. Chem. Soc. C*, 1658
200. Herz, J. E. and Ortiz, C. V. (1971). *J. Chem. Soc. C*, 2294
201. Dewar, M. J. S. and Jennings, W. B. (1970). *Tetrahedron Lett.*, 339
202. Fletcher, J. R. and Sutherland, I. O. (1970). *Chem. Commun.*, 687
203. Blair, J. A. and Gardner, R. J. (1970). *J. Chem. Soc. C*, 2707
204. Mukaiyama, T. and Taguchi, T. A. (1970). *Tetrahedron Lett.*, 3411
205. Gabriel, S. (1887). *Chem. Ber.*, **20**, 2224
206. Bedenbaugh, A. O., Bedenbaugh, J. H., Bergin, W. A. and Adkins, J. D. (1970). *J. Amer. Chem. Soc.*, **92**, 5774
207. Bach, R. D. and Mitra, D. K. (1971). *Chem. Commun.*, 1433
208. White, R. W., King, S. W. and O'Brien, J. L. (1971). *Tetrahedron Lett.*, 3591
209. Zimmer, H. and Singh, G. (1963). *J. Org. Chem.*, **28**, 483
210. Zimmer, H., Jayawant, M. and Gutsch, P. (1970). *J. Org. Chem.*, **35**, 2826
211. Field, K. W., Kovacic, P. and Herskovitz, T. (1970). *J. Org. Chem.*, **35**, 2146
212. Kovacic, P., Lowery, M. K. and Roskos, P. D. (1970). *Tetrahedron*, **26**, 529
213. Gassman, P. G. and Dygos, J. H. (1970). *Tetrahedron Lett.*, 4745
214. Gassman, P. G. and Dygos, J. H. (1970). *Tetrahedron Lett.*, 4749
215. Gassman, P. G. and Shudo, K. (1971). *J. Amer. Chem. Soc.*, **93**, 5899
216. Gassman, P. G. and Ryberg, R. L. (1969). *J. Amer. Chem. Soc.*, **91**, 2047, 5176
217. Furtoss, R., Teissier, P. and Waegell, B. (1970). *Tetrahedron Lett.*, 1263
218. Schneider, W. and Lehmann, K. (1970). *Tetrahedron Lett.*, 4285
219. Surzur, J. -M., Stella, L. and Tordo, P. (1970). *Tetrahedron Lett.*, 3107
220. Taguchi, T., Shimizu, Y. and Kawazoe, Y. (1970). *Tetrahedron Lett.*, 2853
221. Wartski, L., Wakselman, C. and Sierra Escudero, A. (1970). *Tetrahedron Lett.*, 4193
222. Johnson, C. R. and Lockard, J. P. (1971). *Tetrahedron Lett.*, 4589
223. Halton, B. and Woolhouse, A. D. (1971). *Tetrahedron Lett.*, 4877
224. Rens, M. and Ghosez, L. (1970). *Tetrahedron Lett.*, 3765
225. Fowler, F. W., Hassner, A. and Levy, L. A. (1967). *J. Amer. Chem. Soc.*, **89**, 2077
226. Hassner, A. and Galle, J. E. (1970). *J. Amer. Chem. Soc.*, **92**, 3733
227. Dudinskaya, A. A., Khmelnitski, L. I., Petrova, I. D., Baryshinikova, E. B. and Novikov, S. S. (1971). *Tetrahedron*, **27**, 4053
228. Li, J. P. and Biel, J. H. (1970). *J. Org. Chem.*, **35**, 4100
229. Matier, W. L. and Dykstra, S. J. (1971). *J. Org. Chem.*, **36**, 650
230. Hassan, M. M. A. and Casy, A. F. (1970). *Tetrahedron*, **26**, 4517
231. Evans, D. A. (1970). *J. Amer. Chem. Soc.*, **92**, 7593
232. Taguchi, K. and Westheimer, F. H. (1971). *J. Org. Chem.*, **36**, 1570
233. Kira, M. A., Nofal, Z. M. and Gadalla, K. Z. (1970). *Tetrahedron Lett.*, 4215
234. Colonna, F. P., Forchiassin, M., Risaliti, A. and Valentin, E. (1970). *Tetrahedron Lett.*, 571

235. Wasserman, H. H. and Baird, M. S. (1970). *Tetrahedron Lett.*, 1729
236. Volz, H. and Kiltz, H. H. (1970). *Tetrahedron Lett.*, 1917
237. Richey, H. G., Jr., Erickson, W. F. and Heyn, A. S. (1971). *Tetrahedron Lett.*, 2187
238. Grakauskas, V. and Baum, K. (1971). *J. Org. Chem.*, **36**, 2599
239. Gilligan, W. H. (1971). *J. Org. Chem.*, **36**, 2138
240. Koketsu, J. and Ishii, Y. (1971). *J. Chem. Soc. C*, 511
241. Coulson, D. R. (1971). *Tetrahedron Lett.*, 429
242. Stille, J. K. and Fox, D. B. (1970). *J. Amer. Chem. Soc.*, **92**, 1274
243. Atkins, K. E., Walker, W. E. and Manyik, R. M. (1970). *Tetrahedron Lett.*, 3821
244. Saegusa, T., Tsuda, T. and Isegawa, Y. (1971). *J. Org. Chem.*, **36**, 858
245. Byerley, J. J., Rempel, G. L. and Takebe, N. (1971). *Chem. Commun.*, 1482
246. van Tamelen, E. E. and Rudler, H. (1970). *J. Amer. Chem. Soc.*, **92**, 5253
247. Andrisano, R., Bizzari, P. C. and Tramontini, M. (1970). *Tetrahedron*, **26**, 3959
248. Cram, D. J. and Abd Elhafez, F. A. (1952). *J. Amer. Chem. Soc.*, **74**, 5828
249. Cram, D. J. and Kopecky, K. R. (1959). *J. Amer. Chem. Soc.*, **81**, 2748
250. Cram, D. J. and Wilson, D. R. (1963). *J. Amer. Chem. Soc.*, **85**, 1245
251. Cannata, V., Samori, B. and Tramontini, M. (1971). *Tetrahedron*, **27**, 5247
252. Andrisano, R. and Angiolini, L. (1970). *Tetrahedron*, **26**, 5247
253. Richey, R. G., Jr., Erickson, W. F. and Heyn, A. S. (1971). *Tetrahedron Lett.*, 2183
254. Barlow, M. G., Fleming, G. L., Haszeldine, R. N. and Tipping, A. E. (1971). *J. Chem. Soc. C*, 2744
255. Imai, N., Narita, T. and Tsuruta, T. (1971). *Tetrahedron Lett.*, 3517
256. Neelakantan, L. (1971). *J. Org. Chem.*, **36**, 2256
257. Lyle, R. E. and Spicer, C. K. (1970). *Tetrahedron Lett.*, 1133
258. Brown, H. C., Ayyangar, N. R. and Zweifel, G. (1964). *J. Amer. Chem. Soc.*, **86**, 397
259. Archer, J. F., Boyd, D. R., Jackson, W. R., Grundon, M. F. and Khan, W. A. (1971). *J. Chem. Soc. C*, 2560
260. Panetta, C. A. and Rahman, A. (1971). *J. Org. Chem.*, **36**, 2250
261. Carpino, L. A. and Han, G. Y. (1970). *J. Amer. Chem. Soc.*, **92**, 5748
262. Hendrickson, J. B. and Bergeron, R. (1970). *Tetrahedron Lett.*, 345
263. Cochran, T. G. and Huitric, A. C. (1971). *J. Org. Chem.*, **36**, 3046
264. Kornblum, N. and Stuchal, F. W. (1970). *J. Amer. Chem. Soc.*, **92**, 1804
265. Orlando, C. M., Jr., Engel, L. J., Catanes, F. C. and Gianni, M. H. (1971). *J. Org. Chem.*, **36**, 1148
266. Chattha, M. S. and Aguiar, A. M. (1971). *J. Org. Chem.*, **36**, 2720
267. Chattha, M. S. and Aguiar, A. M. (1971). *J. Org. Chem.*, **36**, 2892
268. Barstow, L. E. and Hruby, V. J. (1971). *J. Org. Chem.*, **36**, 1305
269. Zakharkin, L. I., Kalinin, V. N. and Gedymin, V. V. (1971). *Tetrahedron*, **27**, 1317
270. Kost, D. and Sprecher, M. (1970). *Tetrahedron Lett.*, 2535
271. Sasaki, T., Eguchi, S. and Toru, T. (1970). *J. Org. Chem.*, **35**, 4109
272. Black, R. M. and Gill, G. B. (1970). *J. Chem. Soc. C*, 671
273. Wawzonek, S. and Yeakey, E. (1960). *J. Amer. Chem. Soc.*, **82**, 5718
274. Sedor, E. A. (1971). *Tetrahedron Lett.*, 323
275. Jemison, R. W. and Morris, D. G. (1969). *Chem. Commun.*, 1226
276. Benecke, H. P. and Wikel, J. H. (1971). *Tetrahedron Lett.*, 3479
277. Hennion, G. F. and Shoemaker, M. J. (1970). *J. Amer. Chem. Soc.*, **92**, 1769
278. Pine, S. H., Catto, B. A. and Yamagishi, F. G. (1970). *J. Org. Chem.*, **35**, 3663
279. Maltz, H., Grant, M. A. and Navaroli, M. C. (1971). *J. Org. Chem.*, **36**, 363
280. White, E. H., Todd, M. J., Ribi, M., Ryan, T. J. and Sieber, A. A. F. (1970). *Tetrahedron Lett.*, 4467
281. Bailey, D. S. and Saunders, W. H., Jr. (1970). *J. Amer. Chem. Soc.*, **92**, 6904
282. Feit, I. N. and Saunders, W. H., Jr. (1970). *J. Amer. Chem. Soc.*, **92**, 5615
283. Bailey, D. S., Montgomery, F. C., Chodak, G. W. and Saunders, W. H., Jr. (1970). *J. Amer. Chem. Soc.*, **92**, 6911
284. Archer, D. A. (1971). *J. Chem. Soc. C*, 1327
285. Reineke, C. E. and McCarty, J. R., Jr. (1970). *J. Amer. Chem. Soc.*, **92**, 6376
286. Lehn, J. M. and Wagner, J. (1968). *Chem. Commun.*, 148
287. Dewar, M. J. S. and Jennings, B. (1969). *J. Amer. Chem. Soc.*, **91**, 3655
288. Brois, S. J. (1968). *J. Amer. Chem. Soc.*, **90**, 506
289. Kessler, H. (1970). *Angew. Chem. Int. Edn.*, **9**, 219

290. Rauk, A., Allen, L. C. and Mislow, K. (1970). *Angew. Chem. Int. Edn.*, **9**, 400
291. Brois, S. J. (1968). *J. Amer. Chem. Soc.*, **90**, 508
292. Felix, D. and Eschenmoser, A. (1968). *Angew. Chem. Int. Edn.*, **7**, 224
293. Padwa, A. and Battisti, A. (1971). *J. Org. Chem.*, **36**, 230
294. Brois, S. J. (1970). *J. Amer. Chem. Soc.*, **92**, 1079
295. Andose, J. D., Lehn, J. -M., Mislow, K. and Wagner, J. (1970). *J. Amer. Chem. Soc.*, **92**, 4050
296. Anderson, D. J., Horwell, D. C. and Atkinson, R. S. (1971). *J. Chem. Soc. C*, 624
297. Angiolini, L., Jones, R. A. Y. and Katritzky, A. R. (1971). *Tetrahedron Lett.*, 2209
298. Lehn, J. M. and Wagner, J. (1970). *Chem. Commun.*, 414
299. Lambert, J. B., Oliver, W. L., Jr. and Packard, B. S. (1971). *J. Amer. Chem. Soc.*, **93**, 933
300. Kessler, H. and Leibfritz, D. (1970). *Tetrahedron Lett.*, 4289
301. Kessler, H. and Leibfritz, D. (1970). *Tetrahedron Lett.*, 4293
302. Kessler, H. and Leibfritz, D. (1970). *Tetrahedron Lett.*, 4297
303. Gribble, G. W., Easton, N. R., Jr. and Eaton, J. T. (1970). *Tetrahedron Lett.*, 1075

5
Phosphorus Compounds

LOUIS D. QUIN
Duke University, Durham, N. Carolina

5.1 INTRODUCTION

Great interest is being shown in organophosphorus chemistry at present, and the last two years have seen the inauguration of an annual Specialist Periodical Report of The Chemical Society in this area, as well as a new journal, *Phosphorus*. The research literature is voluminous; even with a limitation to aliphatic derivatives, and a further limitation by the author that only compounds with a C—P bond would be included, there remain several hundred papers to be treated. In this two-year review, an attempt has been made to organise this literature into topics in which the greatest interest at present lies. This approach leaves unmentioned some valuable work in other areas, and regrettably, because of limitations of space, the review cannot be considered complete even for the topics treated.

5.2 REACTIONS FORMING THE C—P BOND

Methods for forming the C—P bond are legion; some of general use developed during the period under review are summarised here, while others are described in Section 5.3 where specific types of C-functional phosphorus compounds are treated.

Phosphorus pentachloride has found some use in constructing C—P compounds. Trialkylethylenes and olefinic compounds bearing electron-releasing groups, such as OR, SPh, form solid complexes which are readily hydrolysed and react with SO_2 to give phosphonic dichlorides[1, 2].

$$Me_2C{:}CHMe \xrightarrow{PCl_4^+ PCl_6^-} [Me_2C^+{\cdot}CHMe{\cdot}PCl_4] \xrightarrow[-HCl]{} Me_2C{:}CMe{\cdot}\overset{+}{P}Cl_3 \xrightarrow{SO_2}$$

$$Me_2C{:}CMe{\cdot}P({:}O)Cl_2$$

Methallyl chloride can lead to $ClCH_2{\cdot}CMe{:}CH{\cdot}P({:}O)Cl_2$ (40%) by a similar route[3]. With acetals and ketals, derivatives of vinylphosphonic acids may be obtained[4, 5], and α-halogenoethers give similar products[6, 7].

α,β-Unsaturated carbonyl compounds react with trivalent phosphorus species, and this reaction has been extensively employed to form the C—P bond in some novel structures. The reaction[8] of a phosphonous dichloride with an acrylamide is illustrative. Nucleophilic attack by phosphorus is followed by proton addition to the negative site.

$$MePCl_2 + CH_2{:}CH{\cdot}CO{\cdot}NEt_2 \xrightarrow{H^+} Cl_2\overset{+}{P}Me{\cdot}(CH_2)_2{\cdot}CO{\cdot}NEt_2$$

Acetic acid is present and reacts with the chlorophosphonium centre to form $MePCl({:}O){\cdot}(CH_2)_2{\cdot}CO{\cdot}NEt_2$. Acrylic esters form similar products[9]. The readily available trialkyl phosphites were shown several years ago to be useful in these reactions; more recent work has employed phosphites in reactions with unsaturated amides (but with poor yields[10]) and with unsaturated acids[11]. In both cases alkyl transfer from the phosphite to the carboxy group occurs; the product from an acid (e.g. acrylic) therefore has the structure $(RO)_2P(:O)\cdot(CH_2)_2\cdot CO_2R$. Attack at the carboxy group also occurs when a phosphinous amide reacts with an unsaturated acid[12].

$$Ph_2PNEt_2 + CH_2:CH\cdot CO_2H \rightarrow Ph_2P(:O)\cdot(CH_2)_2\cdot CO\cdot NEt_2 \ (73\%)$$

Spectral evidence suggests that a key step in this process involves formation of Ph_2PHO and an unsaturated amide, followed by their interaction.

Compounds with P—H bonds can be alkylated with alkenes. Secondary phosphines may be employed photochemically in such processes[13].

$$Me_2PH + MeCH:CH_2 \xrightarrow{h\nu} Me_2P\cdot CH_2\cdot CH_2\cdot Me \ (95\%)$$

Ethylene has also been used with nearly equal effectiveness[14], as have a number of fluorinated alkenes[13, 14]. Secondary phosphine oxides also react with alkenes; radical generating conditions are effective, although not necessary[15].

$$Me_2PHO + CH_2:CH\cdot(CH_2)_5Me \xrightarrow[195\,^\circ C]{Bu'O)_2} Me_2P(:O)\cdot(CH_2)_7Me \ (94\%)$$

Similarly, sodium hypophosphite adds to double bonds.

$$Na^+\bar{O}_2PH_2 + CH_2:CH\cdot CH_2\cdot O\cdot COMe \xrightarrow[140\,^\circ C]{(Bu'O)_2} Na^+\bar{O}_2PH\cdot(CH_2)_3\cdot O\cdot COMe$$

The actual product isolated was the phosphonic dichloride, $RPOCl_2$, formed by reaction with PCl_5–$POCl_3$ [16].

Phosphites $[(RO)_3P]$ and phosphinites (R_2POR) are ready participants in the classical Arbuzov rearrangement, and react with alkylating agents (halides, usually) to form phosphoryl derivatives. A new process which can be treated formally as of this type consists of the reaction of diethyl hydroxymethylmalonate with a phosphite[17].

$$(EtO)_3P + HOCH_2\cdot CH(CO_2Et)_2 \xrightarrow{150-210\,^\circ C} (EtO)_2P(:O)\cdot CH_2\cdot CH(CO_2Et)_2 \ (39.4\%)$$

A similar process has also been described wherein the hydroxymethyl compound is formed *in situ* from formaldehyde and diethyl malonate[18].

Alkylation of phosphorus by the electrophilic centre of a 1,3-dipole has been achieved[19]. The general process appears to have broad utility, yet to be explored.

$$(C_5H_{10}N)_3P + EtO_2C\cdot\overset{+}{C}:N\cdot\bar{N}Ph \rightarrow [(C_5H_{10}N)_3\overset{+}{P}\cdot C(CO_2Et):N\cdot\bar{N}Ph] \rightarrow$$

$$(C_5H_{10}N)_3P:C(CO_2Et)\cdot N:NPh$$

Another new alkylation process is that achieved when trialkylboranes react with trivalent phosphorus halides[20]. Only one example has been reported so far, but the process may prove to be valuable in attaching complex alkyl groups to phosphorus, for the trialkylboranes are readily prepared in diversity.

$$\text{Ph}_2\text{PCl} + (\text{C}_6\text{H}_{11})_3\text{B} \xrightarrow[\Delta]{\text{Bu}^n_2\text{O}} \text{Ph}_2\text{PC}_6\text{H}_{11} \ (57\%)$$

5.3 C-FUNCTIONAL ORGANOPHOSPHORUS COMPOUNDS

5.3.1 Carbonyl compounds

Phosphorus-containing aldehydes remain rather rare species, although dialkyl α-formylalkylphosphonates have been shown to be obtained by mild hydrolysis of appropriate enol ethers, $(\text{RO})_2\text{P}(:\text{O})\cdot\text{CR}:\text{CHOR}$ [21, 22]. These ethers, first prepared several years ago by the action of PCl_5 on dialkyl ethers[23], can be prepared by the reaction of triethyl phosphite with acetals of α-bromoaldehydes[22]. The lower formylalkylphosphonates, such as diethyl formylmethylphosphonate, are stable, distillable liquids, giving typical aldehyde reactions with, for example, Schiff's reagent and hydrazines.

   Ketonic phosphorus compounds are better known. Dialkyl α-ketoalkyl-phosphonates have been available for some time from the Arbuzov reaction of phosphites with acyl chlorides, as in the recently reported example[24] below.

$$\text{EtOCH}_2\cdot\text{CO}\cdot\text{Cl} + (\text{EtO})_3\text{P} \rightarrow \text{EtOCH}_2\cdot\text{CO}\cdot\text{P}(:\text{O})(\text{OEt})_2$$

α-Ketophosphine oxides may be obtained by the analogous reaction of a phosphinite with an acyl chloride[25, 26]. In the phosphonates, the C—P bond is readily cleaved. The acyl group is transferred to nucleophiles such as alcohols or amines, releasing a dialkyl phosphonate[27]; with NaBH_4, reduction of the carbonyl group occurs and the alcohol produced is cleaved into the aldehyde and a dialkyl phosphonate by base. The reaction represents a useful device for the conversion of an acid into an aldehyde[28].

$$\text{R}^1\text{CO}\cdot\text{Cl} \xrightarrow{\text{P(OR}^2)_3} \text{R}^1\text{CO}\cdot\text{PO(OR}^2)_2 \xrightarrow{\text{NaBH}_4} \text{R}^1\text{CH(OH)}\cdot\text{PO(OR}^2)_2 \xrightarrow{\bar{\text{O}}\text{H}}$$

$$\text{R}^1\text{CHO} + \text{HPO(OR}^2)_2$$

Photochemical rearrangement of α-ketophosphonates to β-ketophosphonates takes place, presumably via initial γ-hydrogen abstraction by carbonyl[25].

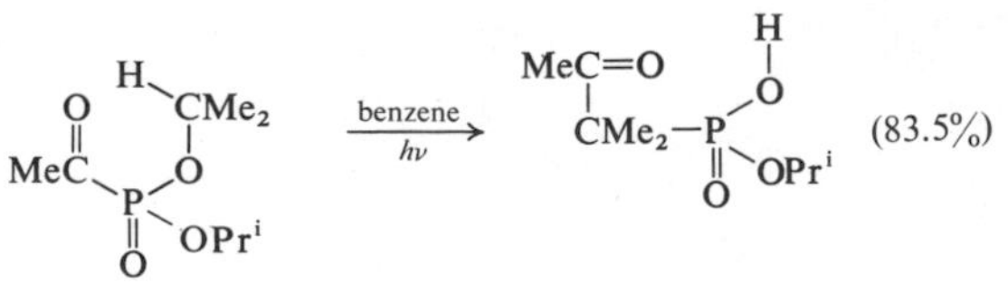

Yields are lower if the ester-alkyl possesses secondary or primary hydrogen at the γ-position.

   A promising new method for synthesising β-ketophosphine oxides consists of the formation of an organophosphorus enamine (see also Section 5.5.3) and its hydrolysis[29].

$$Ph_2P(:O){\cdot}C{:}CR^1 + R^2NH_2 \longrightarrow Ph_2P(:O){\cdot}CH{:}CR^1{\cdot}NHR^2 \xrightarrow[H^+]{H_2O} Ph_2P(:O){\cdot}CH_2{\cdot}COR^1$$

Yields from the acetylene derivatives are 83–94%; bis(alky-1-nyl)phosphines also form enamines, which are hydrolysed to diketones[30]. Direct acid-catalysed hydration of an acetylenic phosphine oxide, $Ph_2P(:O){\cdot}C{:}CPh$, also gives a β-keto-derivative[31], as does hydrolysis of alkoxyvinyl phosphonates[32]. γ-Ketophosphines have recently been prepared by Michael-like addition of metallic derivatives of primary or secondary phosphines to α,β-unsaturated ketones[250].

$$PhCH{:}CH{\cdot}COMe + C_6H_{11}\bar{P}H\ M^+ \rightarrow PhCH(PHC_6H_{11}){\cdot}CH_2{\cdot}COMe$$

Dialkyl γ-ketoalkylphosphonates have been obtained by alkylating phosphites with β-hydroxyalkyl ketones[33].

$$(EtO)_3P + HOCH_2{\cdot}CH_2{\cdot}COMe \xrightarrow{150\,^{\circ}C} (EtO)_2P(:O){\cdot}(CH_2)_2{\cdot}COMe\ (71\%)$$

Phosphonites $RP(OR)_2$ also undergo the reaction, providing γ-ketophosphinates, and phosphinites give γ-ketophosphine oxides.

Tri(ethoxycarbonyl)phosphine has been prepared[34] for the first time, by the reaction of Na_3P with ethyl chlorocarbonate (29% yield). The ester is a stable liquid, and fails to undergo various reactions characteristic of phosphines, such as oxidation with air or sulphur, quaternisation or complexation with CS_2. It has a very low basicity ($pK_a - 10.9$), and in every respect but for a surprisingly fast reaction with bromine, it seems to be characterised by very low electron density on phosphorus. It is, however, oxidised with C—P bond cleavage by stronger oxidising agents, and the ethoxycarbonyl groups exhibit normal behaviour. Hydrolysis results in complete disruption of C—P bonds, releasing PH_3.

It has been shown that carboxy derivatives of phosphine oxides can be obtained by ozonolysis of allyl substituents[35].

$$PhP(:O)(CH_2{\cdot}CH{:}CH_2)_2 \xrightarrow[\text{(ii) }H_2O]{\text{(i) }O_3} PhP(:O)(CH_2{\cdot}CO_2H)_2\ (83\%)$$

Carboxymethyl derivatives can also be obtained by lithiation followed by carbonation[36].

$$Ph_2P{\cdot}CH_2{\cdot}PPh_2 \xrightarrow[\text{(ii) }CO_2]{\text{(i) Li}} Ph_2P{\cdot}CH(CO_2Li){\cdot}PPh_2$$

Methylphosphonates may similarly be converted into carboxymethyl-phosphonates[37]. Unsaturated carboxylic acids result from reaction between phosphonous dichlorides and propiolic acid[38]. The reaction may proceed through nucleophilic attack of phosphorus on the β-carbon of propiolic acid.

$$HC{:}C{\cdot}CO_2H + PhPCl_2 \longrightarrow Cl_2\overset{+}{P}Ph{\cdot}CH{:}\bar{C}{\cdot}CO_2H \longrightarrow PhPCl(:O){\cdot}CH{:}CH{\cdot}CO{\cdot}Cl\ (50\%)$$

$$\xrightarrow{H_2O} HOPPh(:O){\cdot}CH{:}CH{\cdot}CO_2H\ (94\%)$$

Phosphoryl-substituted ketenes may be obtained in the usual way by the action of tertiary amines on the phosphoryl carbonyl chlorides, such as $(EtO)_2P(:O)\cdot CH_2\cdot CO\cdot Cl$ [39].

5.3.2 Hydroxy and alkoxy derivatives

α-Hydroxyalkyl derivatives of phosphorus compounds are easily obtained by reaction of carbonyl compounds with P—H containing structures. A new example of this reaction is that between phenylphosphinic (phenylphos-phonous) acid and aldehydes or ketones[40]. With α,β-unsaturated ketones, this reaction was followed by cyclisation to the oxaphospholene system.

$$PhCH:CH\cdot COMe + PhPH(:O)(OH) \longrightarrow PhCH:CH\cdot CMe(OH)\cdot PPh(:O)OH$$

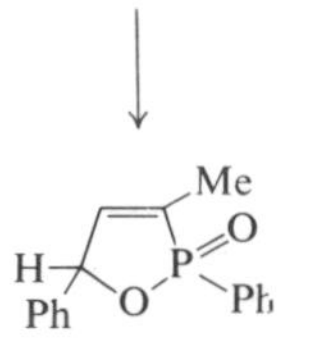

Another example of this type of alcohol synthesis is seen in the reaction of ethyl acetoacetate with diethyl phosphonate[41].

$$MeCO\cdot CH_2\cdot CO_2Et + (EtO)_2PH(:O) \xrightarrow{Et_2NH} (EtO)_2P(:O)\cdot CMe(OH)\cdot CH_2\cdot CO_2Et \ (95\%)$$

The above type of reaction is reversible, especially by base, and α-hydroxyalkyl derivatives can act in solution as donors of the P—H structure. This is a useful property; tris(hydroxymethyl)phosphine is a source of PH_3, obviating the direct use of this dangerous substance. A new example of this process is that in the reaction of bis(hydroxymethyl)methylphosphine with acrylonitrile; the product is the same as that from methylphosphine[42].

$$MeP(CH_2OH)_2 + 2\ CH_2:CH\cdot CN \rightarrow MeP(CH_2\cdot CH_2\cdot CN)_2$$

In another example, tetrakis (hydroxymethyl)phosphonium chloride reacted with ethylene oxide, with replacement of two hydroxymethyl groups[43].

$$(HOCH_2)_4\overset{+}{P}Cl^- + CH_2\overset{O}{-\!\!\!-\!\!\!-}CH_2 \xrightarrow{\bar{O}H} (HOCH_2)_2\overset{+}{P}(CH_2\cdot CH_2OH)_2\ Cl^-$$

This reaction probably proceeds through initial formation of tris(hydroxymethyl)phosphine.

Acetals containing phosphorus functions result from reactions of orthoformates with trivalent phosphorus halides. Presumably halogen–alkoxy interchange occurs initially and is followed by an Arbuzov reaction[44].

These processes are illustrated below, although the final product is the *P*-ethoxy derivative formed by additional interchange.

$$EtPCl_2 + CH(OEt)_3 \rightarrow [EtPClOEt + ClCH(OEt)_2]$$

$$\rightarrow Et\overset{+}{P}Cl(OEt){\cdot}CH(OEt)_2 \rightarrow EtPCl(:O){\cdot}CH(OEt)_2$$

Replacing one Cl by Et_2N results in a significantly slower reaction, and it was suggested that diminished electrophilicity at phosphorus (needed for the first step) through $p_\pi - d_\pi$ bonding of N with P was the cause. When an acetal is used with a phosphonous dihalide, an alkoxyalkylphosphinic acid derivative results.

$$EtPCl_2 + Me{\cdot}CH(OEt)_2 \rightarrow EtPCl(:O){\cdot}CHMe{\cdot}OEt \ (62\%)$$

An example has been reported of an epoxyalkyl halide participating in the Arbuzov reaction[45].

$$PhCO{\cdot}CH\!-\!\!CPh{\cdot}CH_2Br + (EtO)_3P \xrightarrow{120\,°C} PhCO{\cdot}CH\!-\!\!CPh{\cdot}CH_2PO(OEt)_2$$
$$\quad\quad\quad \overset{\backslash}{O}\!\!/ \quad\quad\quad\quad\quad\quad\quad\quad\quad\quad\quad\quad \overset{\backslash}{O}\!\!/$$

An epoxy group is also installed by treating an α-ketophosphonate with a diazoalkane[46] and double-bond epoxidation is effective for vinylphosphonates (Section 5.11).

5.3.3 Amino and other nitrogen functions

A new versatile synthesis of α-aminophosphonic acids, which proceeds in high yields, employs the following sequence[47].

$$RCH{:}NCH_2Ph + (EtO)_2PH(:O) \longrightarrow RCH(NHCH_2Ph){\cdot}P(:O)(OEt_2)_2 \xrightarrow[HCl]{H_2O}$$

$$RCH(NHCH_2Ph){\cdot}P(:O)(OH)_2 \xrightarrow{H_2} RCH(NH_2){\cdot}P(:O)(OH)_2$$

The phosphite–Schiff's base interaction has been known for many years[48]; the novel feature here is the use of the benzyl group on nitrogen, which is easily removed by hydrogenolysis. The Arbuzov reaction has also been used to prepare the *N*-acetyl derivatives of α-aminophosphonic acids, as follows[49].

$$MeCO{\cdot}NH{\cdot}CH_2OEt + (EtO)_3P \rightarrow [MeCO{\cdot}NH{\cdot}CH_2\overset{+}{P}(OEt)_3] \rightarrow$$

$$MeCO{\cdot}NH{\cdot}CH_2PO(OEt)_2$$

A process dubbed 'phosphonoaminomethylation' provides derivatives of

α-aminophosphonic acids. This process utilises the aminal of a formylphosphonic ester as a precursor of a reactive species, an iminium chloride[50].

$$(EtO)_2P(:O)\cdot CH(OMe)NMe_2 \xrightarrow{SOCl_2} (EtO)_2P(:O)\cdot CH:\overset{+}{N}Me_2\bar{C}l$$

$$\xrightarrow{PhCO\cdot NH_2} (EtO)_2P(:O)\cdot CH(NMe_2)\cdot NH\cdot COPh \ (95\%)$$

Several species with active hydrogen were added successfully to the quaternary salt; included were aryl methyl ketones, which gave moderate yields of C-phenacyl derivatives of the amino acid.

β-Aminophosphonic acids occur naturally and are treated in Section 5.11. A Michael addition of nitromethane to a vinylphosphonate has provided a 3-nitropropylphosphonate, reducible to the amino derivative[51].

$$MeNO_2 + CH_2:CHP(:O)(OEt)_2 \xrightarrow[\Delta]{EtONa} NO_2(CH_2)_3P(:O)(OEt)_2 \xrightarrow[\text{(ii) } H_2O, HCl]{\text{(i) } H_2, \text{ Raney Ni}}$$

$$NH_2(CH_2)_3P(:O)(OH)_2 \ (62\%)$$

A new class of nitrogen-containing phosphorus compounds consists of the (methyleneamino)phosphines, $R_2C:N\cdot PR_2$ [52].

$$Me_2PCl + HN{=}CPh_2 \xrightarrow{Et_3N} Me_2P\cdot N:CPh_2 \ (>80\%)$$

With olefinic compounds[53], these compounds form cyclo-adducts readily.

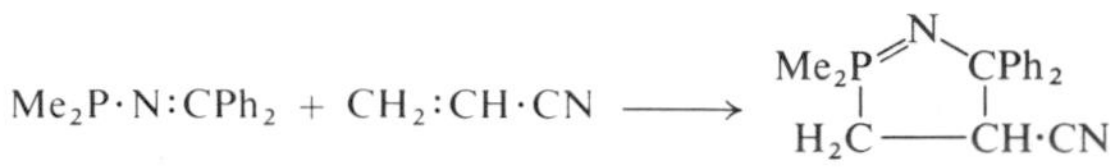

Their reactivity depends on nucleophilicity at P (seen also in ready quaternisation with halides) coupled with electrophilicity at C; both characteristics reside in the uncharged canonical form, and thus a formal resemblance to a 1,3-dipole is lacking. The formation of a p_π–d_π double bond between N and P assists in stabilisation of the cyclo-adduct.

A phosphine with an isocyanide function on an alkyl chain has been formed by the reaction of diphenylphosphine with vinyl isocyanide[54].

$$Ph_2PH + CH_2:CH\cdot NC \xrightarrow[\Delta]{KOBu^t} Ph_2P(CH_2\cdot)_2NC \ (53\%)$$

With phenylphosphine, the only product isolated (90%) had a novel cyclic structure, presumably formed from the initial linear adduct.

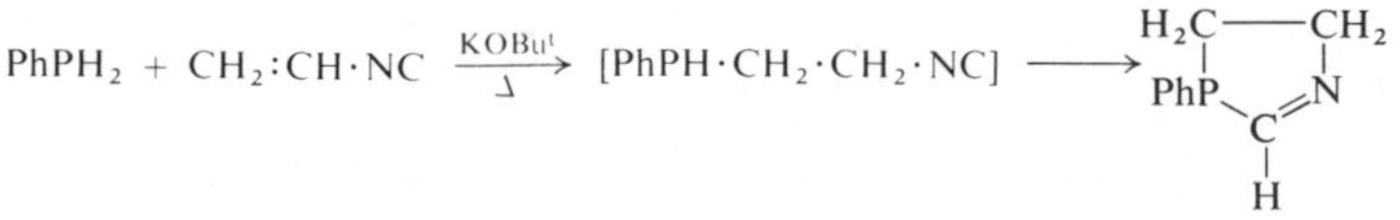

Phosphorus compounds containing diazo-groups have been under active study for the last few years, for considerable stability is imparted to the diazo-group by an α-phosphoryl function. Direct diazotisation of an amino group has been employed recently in the formation of diazophosphonates, $(RO)_2P(:O)\cdot CH:N_2$ [55-57], and a diazophosphine oxide, $Ph_2P(:O)\cdot CH:N_2$ [56]. Another approach involves formation of an appropriate hydrazone from an α-keto-phosphorus compound, followed by Bamford–Stevens elimination with a base[57, 58].

$$RCOP(:O)(OMe)_2 + H_2N\cdot NHTosyl \longrightarrow RC(:N\cdot NHTosyl)\cdot P(:O)(OMe)_2 \xrightarrow{Na_2CO_3}$$

$$RC(:N_2)\cdot P(:O)(OMe)_2$$

Yields in these processes can be very high and the diazo-compounds are obtained frequently as distillable liquids [e.g., $(MeO)_2P(:O)\cdot C(:N_2)\cdot Me$ has b.p. 50–52 °C at 0.20 mmHg [58]] or crystalline solids [$Ph_2P(:O)\cdot CH:N_2$ melts at 61–62 °C]. Some, however, undergo spontaneous decomposition. Typical diazo character resides in these compounds, and they have been used extensively for formation of phosphoryl-containing cyclopropanes (from alkenes, via the carbene formed with copper powder[55, 57, 58]) and pyrazolines (via 1,3-dipolar cyclo-addition to alkenes[57]). Carbenes are also generated by photolysis[59].

5.3.4 Halogenated compounds

Considerable interest has been expressed in C-halogeno-derivatives, mainly of phosphines and their oxides. Several useful reactions are available for the synthesis of such compounds, which are themselves quite valuable as precursors of other derivatives. Some of the more recent work includes development of the following reactions.

(a) (See Ref. 60) $\quad Cl(CH_2)_2PCl_2 + CH_2{=}CH_2 \xrightarrow{\;\;O\;\;} Cl(CH_2)_2P(OCH_2CH_2Cl)_2 \xrightarrow[Arbuzov]{\Delta}$

$$(ClCH_2\cdot CH_2)_2P(:O)\cdot OCH_2\cdot CH_2Cl \xrightarrow{PCl_3} (ClCH_2\cdot CH_2)_2POCl$$

The phosphinic chloride, as well as the thio-derivative made with P_4S_{10}, forms the usual phosphinate derivatives, readily dehydrohalogenated to form divinylphosphinic acid derivatives. The halogenoalkyl derivatives are also useful in the Arbuzov reaction.

$$(ClCH_2\cdot CH_2)_2P(:O)OEt + (EtO)_3P \rightarrow [(EtO)_2P(:O)\cdot CH_2\cdot CH_2]_2P(:O)OEt$$

(b) (See Ref. 61) $\quad (HOCH_2\cdot CH_2)_3\overset{+}{P}CH_2OH\ Cl^- \xrightarrow{Cl_2}$

$$(HOCH_2\cdot CH_2)_3P{:}O + (HOCH_2\cdot CH_2)_2P(:O)\cdot CH_2OH$$

$\qquad\qquad\qquad\quad \Big\downarrow PCl_5 \qquad\qquad\qquad\qquad\quad \Big\downarrow PCl_5$

$$(ClCH_2\cdot CH_2)_3P{:}O \qquad (ClCH_2\cdot CH_2)_2P(:O)\cdot CH_2Cl$$

As in the preceding study, the halogenoalkyl substituents of these phosphine oxides serve to introduce other groups via displacement of halide; they can also be converted into vinylphosphine oxides with tertiary amines, and undergo the Arbuzov reaction.

(c) (See Ref. 62) $(HOCH_2)_2P(:O)\cdot(OH) \xrightarrow[80\,°C]{\text{excess of } SOCl_2} (ClCH_2)_2POCl \xrightarrow{H_2S} (ClCH_2)_2PSCl$

$\Big\downarrow$ $(PhO)_3P, 170\,°C$

$(ClCH_2)_2PCl$

Of note here is the formation of a halogenoalkylphosphinous halide, which with Grignard reagents gives tertiary phosphines, $(ClCH_2)_2PR$. These interesting substances are unstable, possibly undergoing intermolecular quaternisation as well as more complicated reactions.

(d) (See Ref. 63) $(HOCH_2)_4P^+Cl^- \xrightarrow{PCl_5} (ClCH_2)_4\overset{+}{P}Cl^- \xrightarrow{NaOH} (ClCH_2)_3P$

With aqueous NaOH, or merely on being heated in water, tris-(chloromethyl)-phosphine underwent a useful transformation to a tertiary phosphine oxide, with loss of one chlorine atom.

$$(ClCH_2)_2PCH_2Cl \xrightarrow[\Delta]{24\,\% NaOH} (ClCH_2)_2PMe(:O)$$

Chloride displacements by various nucleophiles were possible with both the phosphine and the oxide.

(e) (See Ref. 64) $MeOCHCl_2 + (RO)_3P \xrightarrow{80\,°C} MeOCHCl\cdot P(:O)(OR)_2 \xleftarrow[\text{or } SO_2Cl_2]{Cl_2}$

$MeOCH_2\cdot P(:O)(OR)_2$

The adaptation of the Arbuzov reaction in the first approach to the chloro-methoxy-derivative has also been found applicable to the methylthio-derivative. The thio products reacted further in an Arbuzov process, yielding bisphosphonates, $[(RO)_2P(:O)]_2CH\cdot SMe$.

The characteristics of phosphorus functions can be greatly modified by the potent electron-attracting effect of a trifluoromethyl group, as exemplified in a study of the properties of some alkyl methyltrifluoromethylphosphin-ates[65]. The methyl ester undergoes normal Arbuzov reactions, but with HCl it is cleaved by a process unknown with a normal phosphinite.

$$MeOPMe\cdot CF_3 \xrightarrow{HCl} MeOH + MePCl\cdot CF_3$$

This is accompanied by some of the expected C—O cleavage to MeCl and the secondary phosphine oxide. The t-butyl ester gave predominantly C—O cleavage. The resulting oxide, $MePH(:O)CF_3$, was of some interest also; it was quite stable and underwent no disproportionation, and it did not rearrange to the isomeric phosphinous acid $(R_2\overset{..}{P}OH)$ as had been found for the bis(trifluoromethyl)-derivative. Thiophosphinites containing one CF_3 group were rather resistant to cleavage by HCl, and in this reaction only the P—S bond was broken.

Fluorinated alkylphosphines have been formed in good yield by a photo-chemical method[66].

$$(CF_3)_2PH + CHF:CF_2 \xrightarrow{h\nu} (CF_3)_2P\cdot CHF\cdot CHF_2 \ (91\,\%)$$

No reaction occurred in the dark even at 150 °C, and isomer formation was negligible. Dimethylphosphine in the same process gave a $1:1$ mixture of addition products.

$$Me_2PH + CF_2 \cdot CFH \xrightarrow{h\nu} Me_2P \cdot CF_2 \cdot CH_2F + Me_2P \cdot CHF \cdot CHF_2$$

$$(52\%) \qquad\qquad (48\%)$$

The $(CF_3)_2P \cdot$ radical is therefore seen to attack the CHF group more rapidly than the CF_2 group, while the $Me_2P \cdot$ radical reacts at nearly the same rate with both groups. The implication is that fluoromethyl greatly increases the electrophilicity of the phosphino radical, causing it to attack the more electron rich CHF centre.

Fluoroalkyl groups have also been attached to phosphorus in a reaction of hexafluoroacetone with dialkyl phosphonates[67].

$$(Bu^nO)_2PH(:O) + (CF_3)_2CO \xrightarrow{25\,°C} (Bu^nO)_2P(:O) \cdot C(OH)(CF_3)_2 \ (95\%)$$

However, an isomeric phosphate may also be formed in the process, sometimes as the predominant product.

$$(MeO)_2PH(:O) + O:C(CF_3)_2 \rightarrow (MeO)_2P(:O) \cdot OCH(CF_3)_2 \ (94\%)$$

A later publication[67a] revealed that the phosphate resulted from rearrangement of an initially formed hydroxyphosphonate, a reaction well known in phosphorus chemistry.

5.3.5 Compounds with tertiary alkyl substituents

Phosphorus compounds with bulky substituents are of special significance in revealing the possible operation of steric effects about the phosphorus atom. The greater size of this atom relative to those of the first-row elements tends to reduce the magnitude of some of the effects commonly encountered. For example, the basicity of phosphines is in the order expected from accumulation of inductive effects of the substituents, so that tertiary phosphines are considerably stronger bases than are secondary, themselves stronger than primary[68]. This, of course, is not the case with amines. Nevertheless, profound effects on reactivity at phosphorus may occur if bulky groups are attached thereto. These effects are the most dramatic when two t-butyl groups are attached to phosphorus. Recent work in this area includes that by Trippett[69], who has studied the properties of Bu_2^tPCl and its derivatives. While hydrolysis did proceed readily to give the expected secondary phosphine oxide, and ammonolysis has been reported elsewhere[70], no reaction occurred with diethylamine, alcohols, or alcoholic sodium alkoxides, indicating greatly reduced sensitivity to nucleophilic attack by these larger reagents. (Sodium phenoxide in dimethyl formamide did, however, give a 50% yield of the expected ester). Phosphorus in a phosphinous chloride or a phenylphosphinite was able to act as a nucleophilic centre, e.g. towards alkyl halides, chlorine, and sulphur. Di-t-butylphosphine oxide also showed some unusual properties attributable to steric effects. The oxide failed to exhibit the characteristic instability (disproportionation) towards refluxing base, and did not give adducts $[Bu_2^tP(:O)\cdot C(OH)R_2]$ with carbonyl compounds. This lack of reactivity was also noted by Crofts[71] who reported stability of the oxide towards bromine and $BrCl_3C$–aqueous alkali as well. Furthermore, a pronounced lack of reactivity of relevant phosphonium compounds towards nucleophiles was noted[69]. Thus, $PhCH_2P^+Bu_2^tCl$ Cl^-, while being readily hydrolysed to $PhCH_2P(:O)Bu_2^t$, failed to react with methanol or methanolic sodium methoxide. Also, the methiodide of phenyl di-t-butylphosphinite was surprisingly stable to water or methanol, and quaternary salts, such as $PhCH_2P^+PhBu_2^t$ Br^-, resisted alkaline hydrolysis[72]. The reluctance of the phosphonium species to undergo nucleophilic attack was taken to mean that phosphorus is reluctant to have two bulky t-butyl groups in the trigonal bipyramidal intermediate believed to be involved in this type of reaction. Similar reduced reactivity to nucleophiles has been encountered among di-t-butylthiophosphinic halides[73]; methylmagnesium iodide, for example, failed to displace the halide to form the tertiary phosphine sulphide.

An effect of a single large group on reactivity at phosphorus can also be realised. Recently reported[71] examples include (i) the unusually slow hydrolysis and alcoholysis of Bu^tPCl_2, (ii) the unsatisfactory nature of the Michaelis reaction between sodium ethyl t-butylphosphonite and ethyl iodide. A report that a conventional Arbuzov reaction failed with $Bu^tP(OEt)_2$ was later retracted[74] and this reaction remains untested. It was noted, however, that t-butylphosphonous dichloride participated readily in the alkyl-halide-aluminium chloride alkylation process, even leading to the di-t-butyl system in good yield.

$$Bu^tPCl_2 + Bu^tCl + AlCl_3 \longrightarrow [Bu_2^t\overset{+}{P}Cl_2\ AlCl_4^-] \xrightarrow{H_2O} Bu_2^tPOCl$$

Di-t-butylphosphinic chloride proved to be a useful precursor of the phosphinous chloride.

$$Bu_2^tPOCl \xrightarrow{P_4S_{10}} Bu_2^tPSCl \xrightarrow{Ph_3P} Bu_2^tPCl$$

A considerable amount of work has been done with derivatives of the t-butylthiophosphonic system[75, 76]. Pertinent to the present discussion is the observation of an order of reactivity for the methyl-t-butylthiophosphinic halides intermediate between that for the corresponding dimethyl- or di-t-butyl derivative[75].

Methyl-t-butylphosphinous chloride has been prepared by the action of a t-butylmagnesium halide on $MePCl_2$ [75, 77], as well as by the following sequence[78].

$$Bu^tPCl_2 + Et_2NH \longrightarrow Bu^tPClNEt_2 \xrightarrow{MeLi} Bu^tMePNEt_2 \xrightarrow{HCl} Bu^tMePCl\ (50-60\%)$$

The chloride readily formed a biphosphine $[(Bu^tMeP)_2]$ with sodium, and reacted with ammonia to give a mixture of Bu^tMePNH_2 and $(Bu^tMeP)_2NH$.

Dimethyl t-butylphosphonite has been subjected to a cyclo-addition reaction with 1-benzoyl-1-phenylethylene (methylenedeoxybenzoin)[79]. The 1:1 adduct has pentacovalent phosphorus, with a trigonal bipyramidal structure. The product is of interest in that n.m.r. spectral studies showed the t-butyl group to occupy the equatorial position.

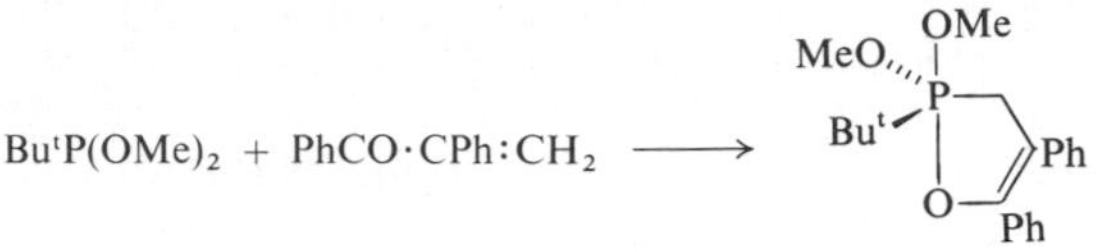

$$Bu^tP(OMe)_2 + PhCO \cdot CPh{:}CH_2 \longrightarrow$$

Mono- and di-t-pentyl derivatives have been prepared by the action of the appropriate Grignard reagent on PCl_3 [80]. Hydrolysis of the second derivative gave the secondary phosphine oxide, subsequently oxidised to the phosphinic acid. The former chloride gave a low yield of t-butyl-t-pentylphosphinous chloride with Bu^tMgCl.

Various phosphorus–fluorine compounds with one or two t-butyl groups present have been prepared[81], and evidence obtained for modified behaviour due to the presence of this large group.

5.4 YLIDS AND CARBANIONS IN PHOSPHORUS CHEMISTRY

Ylids prepared by abstraction of a proton alpha to a phosphonium group by bases are, of course, well-known for their alkene-forming reaction with carbonyl compounds. Emphasis in this review is, however, on the ylids as phosphorus compounds (phosphoranes) and not as alkene-forming reagents.

The general conditions of a Wittig reaction were found applicable to the synthesis of allylidenetrimethylphosphorane[82].

$$Me_3P + ClCH_2 \cdot CH{:}CH_2 \longrightarrow Me_3 \overset{+}{P}CH_2 \cdot CH{:}CH_2 \xrightarrow{BuLi} Me_3 \overset{+}{P} \bar{C}H \cdot CH{:}CH_2$$

As is characteristic of an alkylphosphorane, the product was very unstable, but was of interest because its n.m.r. spectrum indicated delocalisation of the negative charge into the π-system. In trimethylmethylenephosphorane, the CH_2 signals correspond to remarkable shielding (97 Hz *upfield* from $SiMe_4$), consistent with high electron density. This is dramatically changed by introduction of the vinyl group, the signal then appearing at 104 Hz *downfield* from $SiMe_4$. This is a strong indication of delocalisation of the charge from the α-carbon. A similar effect was seen for benzylidenetrimethylphosphorane.

Delocalisation also occurs into a carbonyl group, involving the enolate.

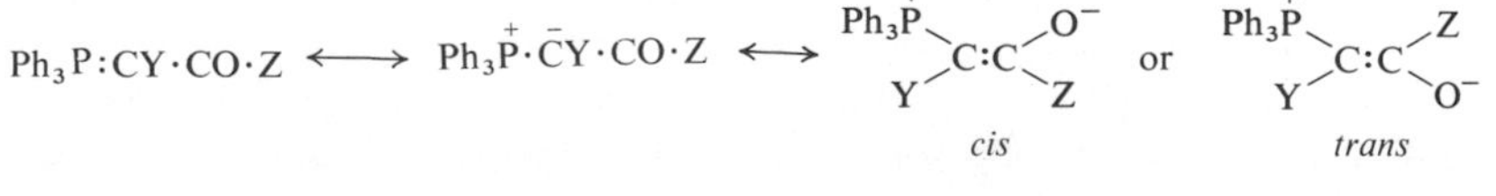

$$Ph_3P{:}CY \cdot CO \cdot Z \longleftrightarrow Ph_3 \overset{+}{P} \cdot \bar{C}Y \cdot CO \cdot Z \longleftrightarrow$$

For β-ketoalkylidenephosphoranes, n.m.r. evidence is strong that the *cis*-enolate[83, 84] is the exclusive structure. This preference may be attributed to steric effects, rather than electrostatic attraction between P^+ and O^-. Thus formylmethylenetriphenylphosphorane (Y = Z = H), lacking a steric effect, exists as a mixture of the *cis* and the *trans*-form; the *cis*-form pre-dominates[84, 85], although a substantial barrier hinders rotation, as is seen also for alkoxycarbonylphosphoranes[86].

In the Wittig reaction, betaines are formed as intermediates, and these decompose to phosphine oxides and alkenes. In protic solvents, the betaine would be converted into a hydroxyalkylphosphonium salt, which could participate in other reactions.

$$Ph_3\overset{+}{P}\overset{-}{C}HR^1 + O{:}CHR^2 \rightarrow Ph_3\overset{+}{P}CHR^1{\cdot}CHR^2O \rightarrow Ph_3PO + R^1CH{:}CHR^2$$
$$\downarrow H^+ \qquad\qquad\qquad \uparrow$$
$$Ph_3\overset{+}{P}CHR^1{\cdot}CH(OH)R^2 \underset{\overline{B}{:}}{\rightarrow} Ph_3\overset{+}{P}CR^1{:}CHR^2$$

One possibility is that water would be lost, forming the vinylphosphonium salt[87, 88]. If R^1 of the ylid is a group capable of conjugating with the double bond, the salt may undergo the elimination by base of triphenylphosphine oxide to give an alkene[87], a mechanism quite unlike that in aprotic solvents. If R^1 is alkyl, other processes occur, some involving the vinylphosphonium salt, to give complex mixtures of products[88]. In another study with protic media[89], the betaine, $Ph_3\overset{+}{P}CH_2{\cdot}CHPh\overset{-}{O}$, and related species were found to be in equilibrium with the hydroxyalkylphosphonium salt, the vinyl-phosphonium hydroxide, and the pentacovalent vinyl oxyphosphorane $[Ph_3P(O^-)CH{:}CHPh]$. Phenyl migration to give $Ph_2P({:}O)CHPh{\cdot}CH_2Ph$ is known to occur in protic media, and this study revealed the immediate precursor of this species to be the vinyl oxyphosphorane.

Carbanionic sites formed at positions alpha to phosphoryl functions receive some stabilisation by charge dispersal into this group. With additional stabilisation, as with CO_2R groups, the anions are quite stable and condense with carbonyl compounds in a useful variant of the Wittig olefin synthesis.

$$(R^1O)_2P({:}O){\cdot}\overset{-}{C}H{\cdot}CO_2R^2 + R^3_2CO \longrightarrow (R^1O)_2\overset{O}{\overset{\|}{P}}{-}CH{\cdot}CO_2R^2$$
$$\overset{}{\underset{O^-{-}CR^3_2}{\big|}}$$

$$\downarrow$$

$$(R^1O)_2PO^-_2 + R^3_2C{:}CH{\cdot}CO_2R^2$$

Recent work[90] has shown that the first step is reversible; addition of NaH to *erythro*- and to *threo*-$(EtO)_2P({:}O){\cdot}CH(CN){\cdot}CH(OH)Ph$ to form the anion, in the presence of *p*-chlorobenzaldehyde, gave *p*-chlorocinnamonitrile as well as cinnamonitrile. This result requires the oxyanion to regenerate the phosphonate carbanion by retroaldolisation. Other evidence suggested that direct interconversion of the diastereoisomeric oxyanions could also occur, but a mechanism for this could not be ascertained.

Carbanions can be made from α-thio-substituted phosphonates[91]; here sulphur assists in stabilising the adjacent negative charge. The anions are useful in synthesis, forming alkyl vinyl sulphides with carbonyl compounds.

$$MeS{\cdot}CHR^1{\cdot}P({:}O)(OEt)_2 \xrightarrow[\text{(ii) } R^2_2CO]{\text{(i) Bu}^n\text{Li}} R^2_2CO{\cdot}CR^1(SMe){\cdot}P({:}O)(OEt)_2 \rightarrow R^2_2C{:}CR^1{\cdot}SMe + \overset{-}{O}_2P(OEt)_2$$

With α,β-unsaturated ketones, the carbanions may add by the Michael reaction, or undergo the alkene-forming reaction. It has been shown[92], by using chalcone as the reactant, that either process can occur, depending on the conditions. With NaH as base and diglyme as solvent, triethyl phosphonacetate reacted to form the alkenic product (ethyl 2,4-diphenyl-butadiene-1-carboxylate). With NaOEt in benzene, the phosphonate reacted with two moles of chalcone in the Michael process. The 1:1 Michael product was formed with $NaNH_2$ in ether.

An example of alkene formation by elimination of a phosphinate from the oxyanion of a β-hydroxyphosphine oxide appears in a report by Lednicer[93].

$$Ar^1(CH\!:\!CH)_2\!\cdot\!CH_2P(\!:\!O)Me_2 \xrightarrow{Bu^tOK} Ar^1(CH\!:\!CH)_2\!\cdot\!\bar{C}HP(\!:\!O)Me_2 \xrightarrow{Ar^2CHO}$$

$$\underset{|}{\overset{Ar^2—C—O}{Ar^1(CH\!:\!CH)_2\!CHP(\!:\!O)Me_2}} \longrightarrow Ar^1(CH\!:\!CH)_3Ar^2 + \bar{O}_2/PMe_2$$

Here the carbanion is stabilised by the adjacent π-system.

Phosphonate carbanions also have utility in the synthesis of other organophosphorus compounds, and as seen in the following examples, participate in some familiar reactions.

$$(EtO)_2P(\!:\!O)\!\cdot\!\bar{C}H\!\cdot\!CO_2MeK^+ \xrightarrow{MeI} (EtO)_2P(\!:\!O)\!\cdot\!CHMe\!\cdot\!CO_2Me\ (85\%^{94})$$

$$(MeO)_2P(\!:\!O)\bar{C}Et\!\cdot\!COMe\ M^+ + ClCH\!:\!CH\!\cdot\!COMe \longrightarrow (MeO)_2P(\!:\!O)\!\cdot\!CEtAc\!\cdot\!CH\!:\!CH\!\cdot\!60\%^{95})$$

$$[(EtO)_2P(\!:\!O)\!\cdot\!\bar{C}H\!\cdot\!CN]_2\ Mg^{2+} + PhCHO \longrightarrow (EtO)_2P(\!:\!O)\!\cdot\!CH(CN)\!\cdot\!CH(OH)Ph$$
$$\text{(Ref. 90)}$$

$$[(EtO)_2P(\!:\!O)\!\cdot\!\bar{C}H\!\cdot\!CN]_2\ Mg^{2+} + Pr^nBr \longrightarrow (EtO)_2P(\!:\!O)\!\cdot\!CH(CN)Pr^n\ \text{(Ref. 96)}$$

The carbanions may be formed by direct metallation[94-96] or with Grignard reagents[90]. Butyl-lithium also formed carbanions from phosphonamides of structure $(Me_2N)_2P(\!:\!O)\!\cdot\!CH_2OEt$, which could then be alkylated[97]. Esters were not suitable for use in this process, since butylation on phosphorus occurred. Carbanions may be formed from certain benzylic phosphonates by the action of sodium amide[98, 99]; when generated in the presence of Schiff's bases, the carbanion reacts readily.

$$p\text{-}MeC_6H_4\ CH_2\!\cdot\!P(\!:\!O)(OEt)_2 + PhCH\!:\!NPh \xrightarrow[\text{ether, }-33°C]{NaNH_2} p\text{-}MeC_6H_4\!\cdot\!CHY\!\cdot\!CHPh\!\cdot\!NHPh\ (70\%)$$
$$[Y = P(\!:\!O)(OEt)]$$

A mixture of the *threo*- and the *erythro*-isomer is formed in a reversible process; the more stable *threo*-isomer is favoured under thermodynamic control, while at lower temperatures the *erythro*-isomer predominates under kinetic control[100]. In some processes of this type, however, alkene-formation was found to compete with the addition[98, 99].

In addition to the groupings displayed above as stabilising phosphonate carbanions, a second phosphoryl group is known to possess this property.

$$CH_2[P(\!:\!O)(OEt)_2]_2 + CH_2\!:\!CH\!\cdot\!CN \xrightarrow{NaOEt} (NC\!\cdot\!CH_2\!\cdot\!CH_2)_2C[P(\!:\!O)(OEt)_2]_2\ (54\%)^{101}$$

$$(RO)_2P(\!:\!O)\!\cdot\!CH_2\!\cdot\!P(\!:\!O)(OR)\!\cdot\!CH_2P(\!:\!O)(OR)_2 \xrightarrow[MeI]{K, \Delta}$$

$$\underset{(RO)_2P(\!:\!O)\!\cdot\!CHMe\!\cdot\!P(\!:\!O)(OR)\!\cdot\!CHMePO(OR)_2\ (58\%)^{102}}{KA\downarrow MeI}$$

Methylene flanked by two phosphine oxide groups is also activated.

$$Ph_2P(:O) \cdot CH_2 \cdot PPh_2(:O) \xrightarrow[RX]{Na, \Delta} Ph_2P(:O) \cdot CHR \cdot PPh_2(:O) \quad (46-70\%^{103})$$

5.5 UNSATURATED ORGANOPHOSPHORUS COMPOUNDS

5.5.1 Double-bonded structures

–A valuable new approach to vinylphosphonates has been devised[104]; vinyl halides can be caused to undergo the Arbuzov reaction at high temperatures in the presence of $NiCl_2$. From vinyl chloride and triethyl phosphite at 180–190 °C, a 70% yield of the vinylphosphonate is obtained. In another example, trans-dichloroethylene with an excess of phosphite gave a 99% yield of the diphosphonate.

$$trans\text{-}CHCl\text{:}CHCl + (EtO)_3P \xrightarrow{180°C} trans\text{-}CHX\text{:}CHX \quad [X = P(:O)(OEt)_2]$$

N.M.R. studies on these phosphonates reveal some valuable P–H coupling constant generalities (in Hz) as shown below.

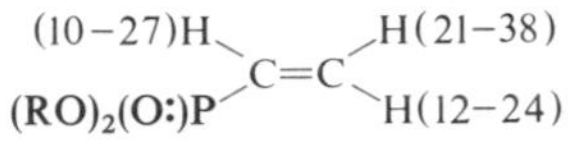

$$\begin{array}{c} (10-27)H \diagdown \qquad \diagup H(21-38) \\ \qquad\qquad C{=}C \\ (RO)_2(O\text{:})P \diagup \qquad \diagdown H(12-24) \end{array}$$

A much smaller value (7 Hz) for cis-P–H coupling is observed[105] when α-halogen is present, as in such structures as $Cl_2(O\text{:})P \cdot CX\text{:}CHOR$.

Some new examples of double-bond rearrangement from the β,γ- to the α,β-position have been reported[106]. Simply on passage through a basic alumina column, allyldiphenylphosphine oxide rearranged to the prop-1-enyl compound in 93% yield. Similar facile rearrangements in the phosphonate and in the phosphonium salt series were encountered. Phosphines were isomerised in boiling ethanolic sodium ethoxide (e.g. allyldiphenylphosphine gave the prop-1-enyl isomer in 87% yield). These rearrangements depend upon abstraction by base of a proton α to the phosphorus function; the high yields reveal the greater stability of the conjugated systems.

Another consequence of the tendency to rearrangement among allylphosphonium salts is their ready cleavage by aqueous KCN; methacrylonitrile is eliminated, and phosphines are formed[107]. Presumably the prop-1-enyl structure is first formed by action of the basic cyanide; Michael addition of cyanide then occurs and β-elimination follows,

$$P_3\overset{+}{P}CH_2 \cdot CH\text{:}CH_2 \rightarrow R_3\overset{+}{P}CH\text{:}CH \cdot Me \xrightarrow{HCN} R_3\overset{+}{P}CH_2 \cdot CHMeCN \rightarrow R_3P + CH_2\text{:}CMeCN$$

The propenyl group may also be cleaved in high yield by electrolysis[106].

Alkadienylphosphonic derivatives have been formed in a widely applicable process[108]. A phosphonodithionic anhydride is heated at 100 °C (autoclave) for several hours with an excess of a diene; the initial cyclic product is subjected to cleavage with base to give the dienyl derivative.

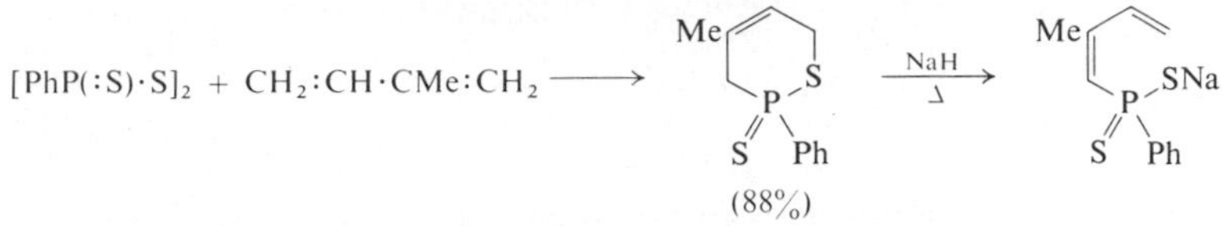

$$[PhP(:S)\cdot S]_2 + CH_2:CH\cdot CMe:CH_2 \longrightarrow \underset{(88\%)}{\text{(cyclic product)}} \xrightarrow[\Delta]{NaH} \text{(dienyl derivative)}$$

In recent years vinylphosphonium salts have been shown very effectively by Schweizer to be of considerable value in the synthesis of a variety of compounds. The process depends upon the susceptibility of the vinyl salts to nucleophiles in a Michael-type addition ('phosphonioethylation'). The process may be illustrated by the following new example[109], where the phosphonioethylation product is converted into an ylid, as usual, for further reaction.

$$Ph_3\overset{+}{P}CH:CH_2 + CH_2(COR)_2 \rightarrow Ph_3\overset{+}{P}CH_2\cdot CH_2\cdot CH(COR)_2 \xrightarrow{Base}$$

$$Ph_3\overset{+}{P}\overset{-}{C}H\cdot CH_2\cdot CH(COR)_2 \xrightarrow{\Delta} Ph_3P + CH_2\overset{CH_2}{=}C(COR)_2$$

The salts show high reactivity towards thiols and amines, and this reaction was used to label proteins. In this study[110], the vinyl grouping was formed in the salt by elimination reactions of 2-hydroxyethyl- and 2-chloroethyl- salts.

5.5.2 Acetylenic derivatives

A versatile synthesis of alk-1-ynylphosphonates, lacking to the present, has been developed[111].

$$MeC:CMgBr + ClP(:O)(OEt)_2 \xrightarrow{0^\circ C} MeC:CP(:O)(OEt)_2 \ (76\%)$$

The previously unknown thio counterparts may be prepared in a similar reaction with lithium alkynylides[112]. An older preparation of alkynylphosphonates, the reaction of phosphites with acetylenic halides, has been studied mechanistically[113] (see also Section 5.10) and found to involve the following steps as the most important processes.

$$(R^1O)_3P: \ X\!-\!C:C\!-\!R^2 \rightarrow (R^1O)_3\overset{+}{P}X + {}^-C:CR^2$$

$$(R^1O)_3\overset{+}{P}X + {}^-C:C\!-\!R^2 \rightarrow (R^1O)_3\overset{+}{P}C:CR^2 + X^-$$

$$X: \ R^1\!-\!O\!-\!\overset{+}{P}(OR^1)_2C:CR^2 \rightarrow X\!-\!R^1 + O:P(OR^1)_2\cdot C:CR^2$$

Attack of phosphorus on halogen to form a halogenophosphonium ion is a well-known process and was demonstrated here by trapping substantial amounts of this ion as a phosphate with an alcohol. Diacetylenic phosphonates are formed on oxidation of certain butyne phosphonates[113].

$$(EtO)_2(:O)PCHAc \cdot CH_2 \cdot C:CH \xrightarrow[\text{pyridine}]{O_2, \text{ CuCl}} [(EtO)_2(:O)PCHAc \cdot CH_2 \cdot C:C]_2 \ (66\%)$$

$$[Ac = Me \cdot CO]$$

Acetylenic phosphines are readily prepared by displacement of halogen from a phosphinous halide with an organometallic compound. Among some new examples is the synthesis[114] from Ph_2PCl and $LiC:CBu^t$ of $Ph_2PC:CBu^t$ in 68% yield. $Ph_2PC:CH$ also participated in this process; with $(Et_2N)_2PCl$ its anion gave the expected phosphine. The acetylenic group tends to stabilise the phosphine group to aerial oxidation.

Another common approach to acetylenic phosphines involves the action of a metallic derivative of a phosphine with a halogenoacetylene. 1,4-Dichlorobutyne gave $Ph_2PCH_2 \cdot C:C \cdot CH_2PPh_2$ on reaction with $Ph_2\bar{P}$ [115]. However, base-promoted isomerisation can occur in such processes where the halide is of general structure $XCHR^1 \cdot C:CR^2$, and the product with Ph_2PNa could be $Ph_2PCHR^1 \cdot C:CR^2$ or could contain $Ph_2PC:C \cdot C\equiv$, or $Ph_2PC:C:C$ [116]. If an excess of phosphide is avoided, isomerisation does not occur and the alk-2-ynylphosphine results. With some bromo-derivatives, however, the phosphide may attack the halogen, as in the following example.

$$Ph_2\bar{P}: \ \ \ \ ^{\frown}Br-CH_2 \cdot C:CH \xrightarrow[\text{NH}_3]{\text{liq.}} [Ph_2PBr + {}^-CH_2 \cdot C:CH] \xrightarrow{\text{NH}_3} Ph_2PNH_2 + MeC:CH$$

Acetylene bis-(phosphonous diamides) may be formed in the reaction of $BrMgC:CMgBr$ with amino-substituted phosphorus halides. Characteristically, the products undergo alkylation at phosphorus and are cleaved by dry HCl [117].

$$(Et_2N)_2PCl + BrMgC:CMgBr \rightarrow (Et_2N)_2PC:CP(NEt_2)_2 \ (71\%)$$

$$\overset{\text{MeI}}{\swarrow} \qquad\qquad \overset{\text{HCl}}{\searrow}$$

$$(Et_2N)_2\overset{+}{P}MeC:C\overset{+}{P}Me(NEt_2)_2 \qquad Cl_2PC:CPCl_2 \ (48\%)$$

The bis-(phosphonous dichloride) would appear to be a valuable material for synthesis of other acetylenic phosphorus compounds.

Alk-1-ynylphosphines with concentrated aqueous solutions of hydrogen halides add HX[118]. No hydration of the triple bond seems to occur. The reaction is probably preceded by protonation of phosphorus, and this product is then attacked by halide ion at the β-position. *Trans*-addition is suggested by the coupling constant (7.1 Hz) of the alkenic protons.

Acetylenic phosphinous esters can rearrange to allenes via an S_Ni process, as in the example below[119].

$$Ph_2PCl + HOCR^1R^2C:CBr \longrightarrow BrC\equiv C-CR^1R^2$$

$$\underset{Ph}{\overset{\displaystyle :P}{\diagdown}}\overset{\displaystyle O}{\diagup}_{Ph}$$

$$\downarrow$$

$$Ph_2P(:O)CBr:C:CR^1R^2 \ (60\%)$$

Allenic derivatives add methyl from Me_2CuLi at $-10\,°C$ on the β-carbon[120] and form pyrazole derivatives by 1,3-dipolar addition of diazomethane to the

α,β-carbons[121]. Among other results of a comprehensive study of allenic phosphine oxides[122] were the observations that various nucleophiles add to the α,β-double bond, and that lithiation takes place at the α-position, forming an anion which reacts with carbonyl compounds to give 1-hydroxy-alkylallenes.

5.5.3   Participation in cyclo-addition reactions

Vinylic and acetylenic phosphonates participate in the Diels–Alder reaction as moderately reactive dienophiles. Cyclic dienes have now been employed in this reaction and through n.m.r. spectral studies the stereochemistry of the products has been elucidated[124].

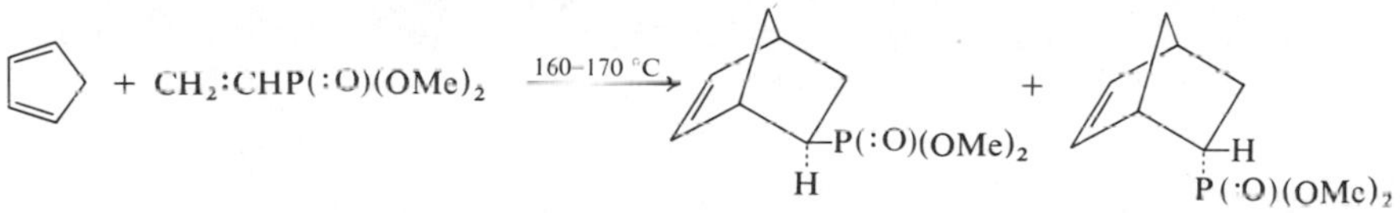

The isomers were separated by chromatography on silica gel; the *exo* isomer slightly predominated (1.2:1). With hexachlorocyclopentadiene or 5,5-dimethoxytetrachlorocyclopentadiene, however, only the *endo* adduct was obtained, presumably because of diminished steric interaction in the transition state relative to that forming the *exo*-isomer.

Diethyl vinylphosphonate has been used as dienophile with diethyl 1-(buta-1,3-dienyl)phosphonate, a compound shown in the same study to be an effective diene in other Diels–Alder reactions[125].

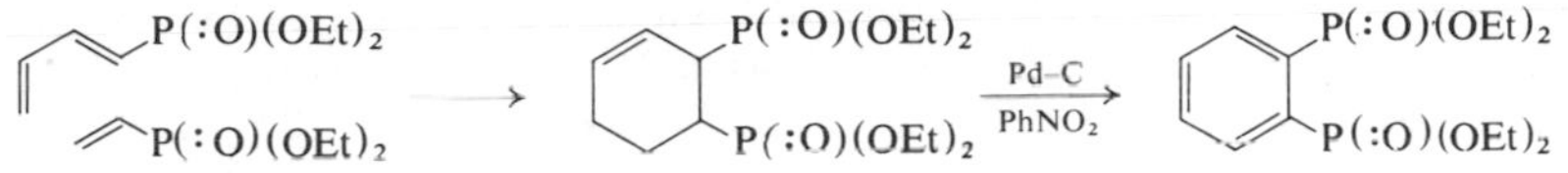

The dienylphosphonate also was shown to dimerise by the Diels–Alder route.

Much current interest is associated with the addition of 1,3-dipoles to unsaturated phosphorus compounds, as useful syntheses of heterocyclic compounds result. Diazoalkanes have been the most popular dipoles so far, as seen in the following examples.

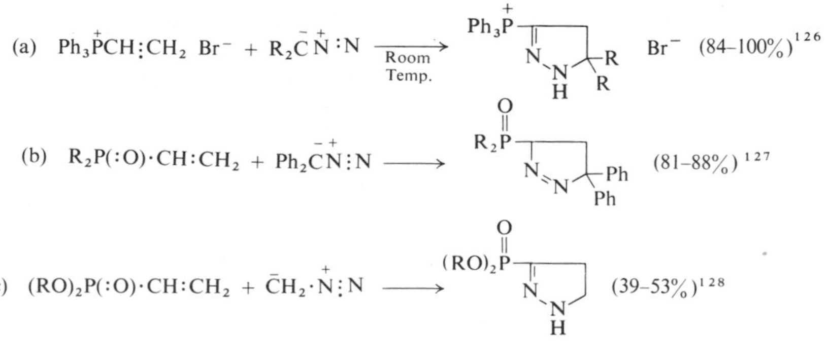

Nitrilimines condense readily with phosphoryl compounds to give the expected pyrazolines. A study[129] of structural influences on the reactivity

$$(R^1O)_2P(:O){\cdot}CH{:}CH_2 + R^2C{:}\overset{+}{N}{\cdot}\overset{-}{N}R^3 \longrightarrow$$

of diphenylnitrilimine towards α,β-unsaturated phosphoryl compounds showed that the rate of cyclo-addition increased with increasing electrophilic character of the unsaturated group. Thus $(MeO)_2P(:O){\cdot}CH{:}CH_2$ was much more reactive than $(Me_2N)_2P(:O){\cdot}CH{:}CH_2$, in which the electron-withdrawing effect of phosphoryl is moderated by electron release from nitrogen. However, the phosphonate group is not as activating as carbethoxy, as is usually the case when one compares electron withdrawing effects of groups having d_π–p_π conjugating ability rather than p_π–p_π. With vinylic phosphines, the cycloaddition with nitrilimines is of another type, and cyclic ylids result.

(made in situ)

With $EtO_2C{\cdot}C{:}\overset{+}{N}{\cdot}\overset{-}{N}Ar$, similar adducts were obtained[131]. An acetylenic phosphine $(Ph_2PC{:}CH)$ reacted with diphenylnitrilimine[130] to form the cyclic ylid in 98% yield. The departure from $3+2$ cyclo-addition when trivalent phosphorus is present is due to the high nucleophilicity of phosphorus in this function, and to the well-known tendency for positive phosphorus to stabilise an adjacent carbanionic centre in the ylid structure.

In other dipolar additions of the $3+2$ type, nitrile oxides react readily with vinylphosphonates[129] to form the expected isoxazoline ring and even a sydnone has reacted in the expected manner with an acetylenic phosphonate[132].

5.5.4   Enamines containing phosphorus functions

Enamines bearing a phosphorus-containing substituent have very recently become available and are proving to have some valuable properties. They are readily prepared by nucleophilic addition (trans) of amines to acetylenic compounds, and in Section 5.3.1 it was pointed out that hydrolysis of enamines having phosphine oxide groups gave β-keto-phosphine oxides in good yield[29]. Another synthetic application is their conversion into anions with strong bases, and subsequent reaction with an aldehyde or ketone[133]. The anions on reaction with carbonyl compounds behave as do phosphonate carbanions and give rise to α,β-unsaturated ketimines (and thence to corresponding ketones).

$$Ph_2P(:O)CH{:}CR^2{\cdot}NHR^1 \xrightarrow{\text{Na or BuLi}}_{\text{THF}} [Ph_2P(:O)CH{:}CR^2{\cdot}\overset{-}{N}R \longleftrightarrow Ph_2P(:O)\overset{-}{C}H{\cdot}CR^2({:}NR^1)]$$

$$Ph_2P(:O){\cdot}\overset{-}{C}H{\cdot}CR^2({:}NR^1) + R_2^3CO \longrightarrow Ph_2PO_2^- + R_2^3C{:}CH{\cdot}CR^2{:}NR^1 \xrightarrow{H_2O} R_2^3C{:}CH{\cdot}COR^2$$

$$(53\text{--}70\%)$$

Epoxides react with the carbanions to give good yields of cyclopropyl ketimines[134]. Enamine phosphonates, prepared similarly by adding amines to acetylenic phosphonates, were also useful in the ketimine synthesis[135], as were related thiophosphonates[136]. A new synthesis of enamine phosphonates has been reported from another laboratory[137], and involves reaction of ynamines with dialkyl phosphonates. *Cis* and *trans* isomers were formed; the predominating isomer was that revealed by n.m.r. studies to have H

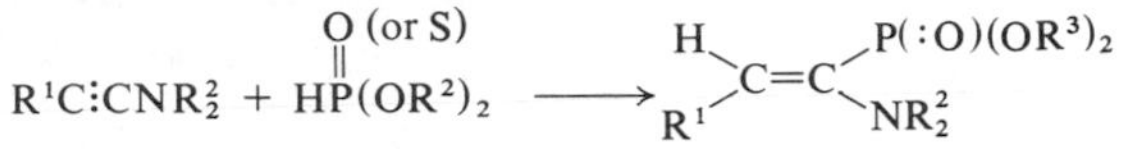

$$R^1C\!:\!CNR_2^2 + H\overset{\displaystyle O\ (\text{or } S)}{\overset{\|}{P}}(OR^2)_2 \longrightarrow \underset{R^1}{\overset{H}{\diagdown}}C\!=\!C\underset{NR_2^2}{\overset{P(:O)(OR^3)_2}{\diagup}}$$

and the phosphorus function *cis*.

5.6 COMPOUNDS WITH TWO OR MORE C—P BONDS

5.6.1 Phosphines

Great interest is presently associated with bisphosphines because of their value as multidentate ligands for metal ions. No attempt will be made to discuss here the chemistry of these complexes; the synthesis and properties of the phosphines only will be included.

A useful new method for synthesising polyphosphines is the Michael-like addition of phosphide ions to the double bond of vinylphosphines[138].

$$Ph_2PH + CH_2\!:\!CHPPh_2 \xrightarrow{PhLi} Ph_2PCH_2\!\cdot\!CH_2PPh_2 \ (80\%)$$

Selected examples of more complex phosphines prepared include those shown below.

$$PhPH_2 + 2CH_2\!:\!CHPPh_2 \xrightarrow{KOBu^t} (Ph_2PCH_2\!\cdot\!CH_2)_2PPh \ (87\%)$$

$$3Ph_2PH + (CH_2\!:\!CH)_3P \xrightarrow{PhLi} (Ph_2PCH_2\!\cdot\!CH_2)_3P \ (59\%)$$

$$PhPHCH_2PHPh + 2Ph_2PCH\!:\!CH_2$$
$$\downarrow KOBu^t$$
$$Ph_2P(CH_2)_2\!\cdot\!PPh(CH_2)_2PPh\!\cdot\!(CH_2)_2PPh_2 \ (51\%)$$

$$Ph_2PC\!:\!CH + Ph_2PH \longrightarrow \underset{Ph_2P}{\overset{H}{\diagdown}}C\!=\!C\underset{H}{\overset{PPh_2}{\diagup}} \ (72\%)$$

Arsines were also added to multiple bonds of phosphines. The orientation in these additions is of interest; it appears that the phosphino group stabilises an adjacent negative charge, as does the phosphoryl group[139].

$$(EtO)_2P(:O)CH\!:\!CH_2 + Ph_2PH \rightarrow (EtO)_2P(:O)CH_2\!\cdot\!CH_2PPh_2$$

This effect has been noted before in additions of amines to vinylphosphines, and has been taken to indicate that conjugation involving trivalent phosphorus depends on d-orbital acceptance of electrons[140]. While this is the generally accepted view for acyclic vinylphosphines, and recent n.m.r. evidence supports the concept for acetylenic phosphines[141], some spectral properties of *cyclic* vinylphosphines suggest that phosphorus can also act as an electron donor[142].

Alkylphenylphosphines, as their lithio derivatives, have been added to vinylphosphines[143], and free-radical additions have also been promoted by azo-bis-isobutyronitrile (AIBN)[144] of phosphines to simple alkenes[145].

$$H_2P(CH_2)_3PH_2 + 2Me\cdot(CH_2)_5\cdot CH\!:\!CH_2 \xrightarrow{\text{AIBN}} Me(CH_2)_7\cdot PH(CH_2)_3\cdot PH(CH_2)_7\cdot Me$$
$$(87\%)$$

A synthesis of a phosphino-phosphonite is as follows[146].

$$Ph_2PLi + \overset{O}{\overset{\frown}{CH_2CH_2}} \longrightarrow Ph_2PCH_2\cdot CH_2OLi \xrightarrow{Ph_2PCl} Ph_2PCH_2\cdot CH_2\,OPPh_2$$

Cleavage of phenyl groups from phosphorus with alkali metals occurs readily, and this reaction has been applied to bis-(diphenylphosphino)-alkanes to form bis-phosphides of synthetic value[147].

$$Ph_2PNa + Cl(CH_2)_nCl \rightarrow Ph_2P(CH_2)_nPPh_2 \xrightarrow{NA} Ph\bar{P}(CH_2)_n\bar{P}Ph \xrightarrow[\text{(ii) }H_2O_2]{\text{(i) }H_2O}$$

$$HOPPh(\!:\!O)\cdot(CH_2)_n\cdot PPh(\!:\!O)OH$$

Both phenyl groups could also be cleaved from diphenylphosphino-groups to form dichlorophosphine functions as shown below.

$$Ph_2P(CH_2)_nPPh_2 \xrightarrow[280\,^\circ C]{PCl_3,\ AlCl_3} Cl_2P(CH_2)_nPCl_2\ (55\text{--}65\%)$$

These products react normally as phosphonous dichlorides, and should prove of value as intermediates in various syntheses.

Tetramethylbiphosphine can serve as a source of the dimethylphosphino-radical, and gives 1,4 adducts with buta-1,3-diene[148].

$$Me_2P\cdot PMe_2 + CH_2\!:\!CH\cdot CH\!:\!CH_2 \xrightarrow{100\,^\circ C} Me_2PCH_2\cdot CH\!:\!CH\cdot CH_2PMe_2\ \ (cis\!:\!trans = 1\!:\!2.3)$$

The process may also be promoted by AIBN. Bisphosphines add to alkenes, and it has now been shown[149] that ultraviolet radiation promotes the process.

Ylids of structure $(Me_2N)_3\overset{+}{P}{}^-H_2$ may be obtained, as well as those in which one or two Me_2N groups are replaced by methyl; these ylids displace chlorine from trivalent phosphorus halides, forming new multi-phosphorus species[150].

$$(Me_2N)_3\overset{+}{P}\overset{-}{C}H_2 + MePCl_2 \rightarrow [(Me_2N)_3\overset{+}{P}\overset{-}{C}H]_2PMe$$

5.6.2 Phosphonates

Polyphosphonic acids are also receiving considerable attention at present, for they have marked chelating ability for polyvalent metal ions. A particularly simple new process[151] for preparing propane-1,2,3- and butane-1,2,3,4-polyphosphonic acid consists of the reaction of acetylenic alcohols with diethyl phosphonate in the presence of sodium. Presumably twofold addition of the phosphonate to the triple bond occurs, followed by dehydration to give a terminal alkene function and yet another addition of phosphonate.

$$HC\vdots C\cdot CH_2OH + HP(\vdots O)(OEt)_2 + NaP(\vdots O)(OEt)_2 \rightarrow (EtO)_2P(\vdots O)\cdot CH_2\cdot CH(CH_2OH)\cdot P(\vdots O)(OEt)_2$$
$$P(\vdots O)(OEt)_2$$

$$(EtO)_2P(\vdots O)\cdot CH_2\cdot C(\vdots CH_2)\cdot P(\vdots O)(OEt)_2 \xrightarrow[\text{(Michael reaction)}]{\text{HPO(OEt)}_2} (EtO)_2P(\vdots O)CH_2\cdot CH\cdot CH_2P(\vdots O)(OEt)_2$$
$$(72\%)$$

Hydrolysis with concentrated hydrochloric acid gave the triphosphonic acid, which may be conveniently re-esterified with trialkyl orthoformates[152].

Another type of polyphosphorus acid of current interest is that in which phosphorus and carbon are structured in chains, in various sequences, with phosphorus in the fully oxidised form, as exemplified below.

$$(HO)_2P(\vdots O)CH_2\cdot P(\vdots O)\cdot (CH_2)_5\cdot P(\vdots O)\cdot CH_2\cdot P(\vdots O)(OH)_2$$
$$\underset{OH}{|} \qquad \underset{OH}{|}$$

In one approach to such acids[153], pentamethylene-1,5-bis-phosphonite is prepared from $(EtO)_2PCl$ and the necessary di-Grignard reagent, and subjected to a double Arbuzov reaction (170°) with $(RO)_2P(\vdots O)CH_2Cl$ to give the ester in 87% yield. Another approach employed the following steps.

(a) $(EtO)_2P(\vdots O)(CH_2)_5P(\vdots O)(OEt)_2 \xrightarrow{\text{LiAlH}_4} H_2P(CH_2)_5PH_2 \xrightarrow[\text{HCl}]{\text{COCl}_2} Cl_2P(CH_2)_5PCl_2$

$(Et_2N)_2PCl + BrMg(CH_2)_5MgBr \rightarrow (Et_2N)_2P(CH_2)_5P(NEt_2)_2$

(b) $Cl_2P(CH_2)_5PCl_2 + CH_2O \rightarrow ClCH_2P(\vdots O)(CH_2)_5P(\vdots O)CH_2Cl \xrightarrow[\text{(ii) (EtO)}_3P]{\text{(i) EtOH}}$
$$\underset{Cl}{|} \qquad \underset{Cl}{|}$$

$$(EtO)_2P(\vdots O)CH_2P(\vdots O)(CH_2)_5P(\vdots O)CH_2P(\vdots O)(OEt)_2$$
$$\underset{OEt}{|} \qquad \underset{OEt}{|}$$

Esters with alternating C—P bonds have been formed by polymerisation of a chloromethylphosphinite via an Arbuzov mechanism[154].

$$ClCH_2P(OEt)_2 \xrightarrow{170°C} ClCH_2P(\vdots O)-[CH_2P(\vdots O)]_9-CH_2P(OEt)_2$$
$$\underset{OEt}{|} \qquad \underset{OEt}{|}$$

The terminal trivalent phosphorus function was then oxidised to the phosphonate. That the polymer had the length shown was revealed by ^{31}P n.m.r. studies; the monoester groupings had signals at lower field (-37.9 p.p.m.) than the di-ester (-17.0 p.p.m.), in the ratio 10:1. Other polyphosphinates were similarly prepared; the related polyphosphinic acids obtained by hydrolysis were excellent Ca^{2+} chelating agents.

The Arbuzov reaction can also be used to prepare other polyphosphorus esters.

$$(EtO)_3P + (ClCH_2)_2P(\vdots O)OEt \rightarrow [(EtO)_2P(\vdots O)CH_2]_2P(\vdots O)OEt \ (87\% \ [155])$$

$$(EtO)_3P + (ClCH_2\cdot CH_2)_2P(\vdots O)OEt \xrightarrow{160-180°C} (EtO)_2P(\vdots O)CH_2\cdot CH_2P(\vdots O)CH_2\cdot CH_2P(\vdots O)(OEt)_2$$
$$\underset{OEt}{|}$$

$$(\text{Ref. 156})$$

Nitrogen-containing polyphosphonic acids are likewise of current interest as chelating agents. Nitrilo-tri(methylenephosphonic acid), prepared several years ago[157], is an excellent chelating agent and very recently nitrilo-tri(ethylenephosphonic acid) has been synthesised[158]. While the Arbuzov reaction [between tri-(2-chloroethyl)amine and a phosphite] again proved of value for this purpose, the yield was low (23%). It was greatly improved by the Michaelis procedure.

$$N(CH_2 \cdot CH_2Cl)_3 + KP(:O)(OEt)_2 \xrightarrow[\text{toluene}]{\Delta} N[CH_2 \cdot CH_2P(:O)(OEt)_2]_3 \ (67\%)$$

Hydrolysis gave the acid (pentabasic on titration) as a high-melting solid. It proved to be a poor sequestering agent for calcium relative to the methylene derivative, an effect attributed to the increased distances between the phosphoryl groups. Phosphorus analogues of another familiar sequestering agent, EDTA, are also known and ethylenediamine tetra(methylenephosphonic) acid, recently obtained in especially pure form as a high-melting crystalline solid, has been found to have excellent chelating ability[159].

The base-catalysed addition of dialkyl phosphonates to the carbonyl group of α-ketophosphonates gives hydroxydiphosphonates[160].

$$PhCO \cdot P(:O)(OMe)_2 + HP(:O)(OMe)_2 \xrightarrow[0°C]{R_2NH} PhC(OH)[P(:O)(OMe)_2]_2 \ (96\%)$$

1-Hydroxyphosphonates are stable to secondary amines, but undergo rearrangement to phosphates with strong bases. 1-Hydroxyethane-1,1-diphosphonic acid, a strong chelating agent known for some years, can be obtained in excellent yield (94–97%) by acetylation of phosphorous acid in various ways[161]. One procedure calls for boiling a mixture of phosphorous acid with a 50% excess of acetyl chloride, and then passing steam through the mixture in a purification step. Presumably the α-ketophosphonic acid is first formed, and then reacts with phosphorous acid.

$$MeCOCl + HP(:O)(OH)_2 \rightarrow Me \cdot CO \cdot P(:O)(OH)_2 \xrightarrow{HP(:O)(OH)_2} Me \cdot C(OH)[P(:O)(OH)_2]_2$$

The process also provided other alkane derivatives, and in practice was conducted on a mixture of the carboxylic acid, water and PCl₃.

Other diphosphonic acid derivatives recently prepared include a series of halogenomethylene diphosphonates[162], formed by NaSH-reduction of the dihalogenomethylene compounds (from the action of hypohalite on the methylenediphosphonic acid[163]), and 1,2-dihydroxy-1,2-diphosphonosuccinic acid, obtained as follows[164].

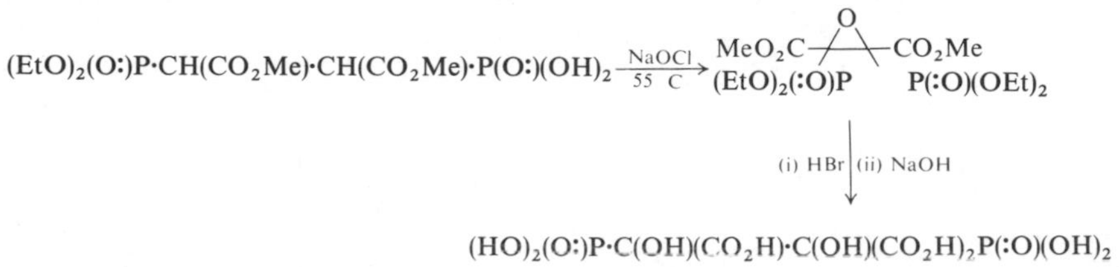

$$(HO)_2(O:)P \cdot C(OH)(CO_2H) \cdot C(OH)(CO_2H)_2 P(:O)(OH)_2$$

5.7 SPECIAL PROPERTIES OF PHOSPHINES

The remarkable discovery has been made that phosphines soluble in aqueous alkali are oxidised by that medium, with accompanying evolution of hydrogen[165]. Such phosphines as those with carboxyalkyl, hydroxyalkyl, aminoalkyl, and phenolic groups give this reaction; simple phosphines (e.g. tri-n-butyl phosphine) lack solubility in the medium and are unaffected by it.

Tertiary phosphines form salts ($R_3\overset{+}{P}NR_2$ Cl^-) on reaction with chloramines. Presumably p_π–d_π bonding between P and N assists in stabilising the S_N2-like transition state for displacement of chloride from N. Bulky groups on P tend to diminish the reaction rate[166]. Tertiary phosphines also react with halogenophosphines[167]; the addition products, not yet fully characterised, are unstable, decomposing by oxidation–reduction to elemental phosphorus and R_3PX_2. With phosphinous halides[168] the products from tertiary phosphines are crystalline solids, stable at 20 °C, but reverting to the reactants at 100 °C *in vacuo* and possibly in solution as well. The adduct is believed to have a P→P bond ($R_3\overset{+}{P}$—PPh_2 Cl^-); its ^{31}P n.m.r. spectrum supports the presence of the two types of phosphorus. A somewhat similar adduct was postulated several years ago to account for the formation of a diphosphine monoxide from a diarylphosphinous chloride on exposure to moisture and oxygen[169]. Apparently the higher nucleophilicity of a trialkylphosphine is needed for effective displacement of chlorine from phosphorus, for methyldiphenylphosphine fails to react with halogenophosphines[168]. An addition product is also formed from Me_2PH and Me_2PCl and appears to be the hydrochloride of $Me_2P\cdot PMe_2$ [170].

Tertiary phosphines are known to react with CCl_4 to form the species $R_3\overset{+}{P}CCl_3$ Cl^-; the adduct from triphenylphosphine is useful for converting alcohols into chlorides. It has now been found that Bu_3^nP and either CCl_4 or CBr_4 give an adduct useful for establishing the peptide bond, with high retention of optical purity[171]. The reaction may involve the following steps.

$$R_3^1\overset{+}{P}CX_3 \xrightarrow{R^2CO_2H} R_3^1\overset{+}{P}\cdot O\cdot COR^2 \xrightarrow{R^3NH_2} R^2CO\cdot NHR^3 + R_3^1PO$$

5.8 STEREOCHEMICAL CONSIDERATIONS

5.8.1 Pyramidal characteristics of phosphines

Followed the now classic work of Horner in 1961[172], which resulted in the first demonstration of the existence of stable, optically active forms among tertiary phosphines, there has been increasing interest in the pyramidal characteristics of trivalent phosphorus. Horner showed that an optically active acyclic phosphine had an activation energy of 28–30 kcal mol^{-1} for racemisation, a process which may be visualised as requiring passage through a transition state of planar configuration. The process is referred to as pyramidal inversion.

In more recent work[173], Horner has found that the solvent plays no role in this process, the energy of activation falling in the range 30 ± 3 kcal mol^{-1} for the racemisation (first-order kinetics) of (+)-methylphenyl-n-propyl-phosphine in a very wide variety of polar and non-polar solvents. The rate of pyramidal inversion is remarkably insensitive also to the steric charac-teristics of the groups about phosphorus[174]; for example, t-butylmethyl-phenylphosphine had an activation free energy ($\Delta G^{\ddagger}$) for racemisation of 32.7 kcal mol^{-1}, while that for methylphenyl-n-propylphosphine was 32.1 kcal mol^{-1}. Even the effects of a marked change of substituent on $\Delta G^{\ddagger}$ seem to be small (cf. cyclohexylmethyl-n-propylphosphine 35.6, with methyl-phenyl-n-propylphosphine, 32.1 kcal mol^{-1}). However, a definite, though slight, trend to lower activation free energies was detected as the electron-withdrawing character of substituents on phenyl increased. The direction is that expected for $p_{\pi}-p_{\pi}$ conjugation, which would stabilise the planar transition state. (This conjugative effect becomes very important for the phospholes, for which the planar transition state is extensively stabilised by cyclic conjugation in this heteroaromatic system. The inversion barrier in such compounds is only 15–16 kcal mol^{-1} [175].) This type of conjugation also appears to account for the low barrier to inversion detected in an acylphosphine[176] (e.g. for acetylisopropylphenylphosphine $\Delta G^{\ddagger}$ is c. 19.4 kcal mol^{-1}). In this case, dynamic n.m.r. studies (on the racemic form) were used to determine the barrier; near room temperature, the methyl signals of the isopropyl group are those of a conformationally restricted system, and appear as two well-separated four-line patterns of equal intensity. The signals broaden with increased temperature and coalesce at about 110 °C, thus allowing the determination of $\Delta G^{\ddagger}$. This observation is of general significance for α-ketophosphine chemistry. These substances are analogous to amides; as such $p_{\pi}-p_{\pi}$ conjugation is hardly unexpected, yet this is the first clear-cut demonstration of the phenomenon. The importance of $p_{\pi}-p_{\pi}$ conjugation is also revealed, as is true for other parameters, in semi-empirical calculations of the barrier by the Pople CNDO/2 method[177]. Mislow applied these calculations to a large number of compounds and obtained good agreement with experimental values only when proper consideration was given to the conjugative effect. As an example, acetyldimethylphosphine is calculated on the basis of no conjugation to have a barrier of 35.6 kcal mol^{-1}, while with conjugation the value (22.6 kcal mol^{-1}) is quite close to the one expected on the basis of the n.m.r. experiment described above.

Because of configurational stability at phosphorus, diphosphines exist in two diastereoisomeric forms, interconvertible through phosphorus inversion at high temperatures. The two forms, *meso* and *racemic*, appear to prefer a *transoid* conformation with respect to the lone pair on each phosphorus; their interconversion would require an inversion on one phosphorus atom, and a rotation about the P—P bond.

meso   *racemic*

These forms may be detected through differences in their ^{31}P n.m.r. spectra as well as in their ^1H spectra and these differences have been employed in a recent study on the effect of substituents on the inversion–rotation barrier[178]. From coalescence temperature measurements, the barrier was established to be in the range $22–24$ kcal mol^{-1} for several diphosphines of structure ArPMe·PMeAr. While $p_\pi–p_\pi$ bonding between phosphorus and the aromatic substituent may provide some stabilisation of the transition state for inversion, a more important effect appears to be stabilisation by $p_\pi–d_\pi$ bonding between the phosphorus atoms, and the net effect results from a balance between these factors. Inversion was tentatively presented as dominant over rotation for the rate-determining factor in the interconversion process. Values of $^1J_{pp}$ for the diastereoisomers of $(MePhP)_2$ differ by 19 Hz; assignments to structures could not be made, but this novel result accords with conformational differences in the isomers[179].

In diphosphine monosulphides, n.m.r. spectra are invariant up to 200 °C, implying a substantially higher barrier at the trivalent phosphorus, possibly of the level of that of a tertiary phosphine[178]. The thiophosphoryl group is itself involved in d-orbital bonding, and the contribution to stabilisation of the transition state for inversion through $p_\pi–d_\pi$ bonding of P atoms would be quite small.

Tertiary phosphorus with three different substituents acts as an intrinsic centre of asymmetry and as such brings about magnetic non-equivalence in an adjacent group. This is seen in $Pr^iPClNMe_2$ with non-equivalent C-methyls, and in $ClCH_2·PXNR_2$, with non-equivalent protons[180]. In racemic $(Bu^sO)_2PR$, the phosphorus atom is a centre of pseudo-asymmetry[181]. Another example is provided by the compound (2-cyanoethyl)(phenyl)-1-phenylethylphosphine, in which there are two asymmetric atoms present[182]. The methyl signal appears at a distinctly different position for the diastereoisomers, its environment being different regardless of conformational equilibration. In the formation of this substance, one of two cyanoethyl groups about phosphorus in the racemic phosphonium salt is eliminated; the fact that the

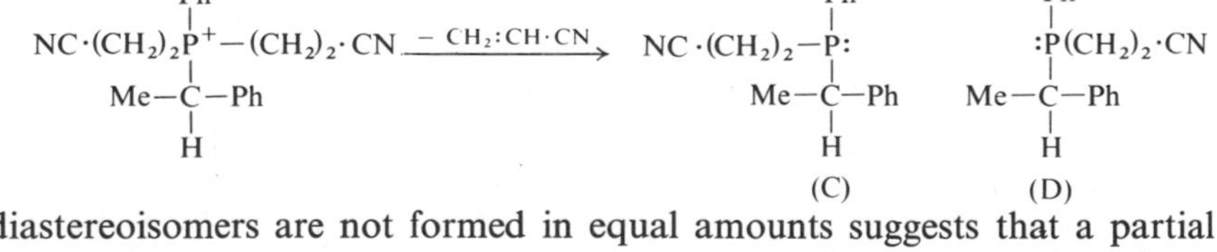

diastereoisomers are not formed in equal amounts suggests that a partial asymmetric synthesis is at hand, thought to be the first recorded* for a phosphine.

*Another example may be found in the report that racemic tertiary phosphines, on partial oxidation with an optically active peroxyacid, can lead to mixtures of phosphine oxides having optical activity[183].

The forms (A) and (D) constitute one ($\pm$) pair, (B) and (C) the other. The two pairs are formed in unequal amounts, which means that (B) is formed in preference to (A) (or vice versa), and (D) in preference to (C) (or vice versa), and a steric control by the asymmetric carbon atom of elimination of one of the cyanoethyl groups from the phosphonium salt is involved.

The stereochemistry of reactions occurring at trivalent phosphorus, or creating trivalent phosphorus, is of considerable importance to the development of mechanistic understanding of these processes. Some recent results may be summarised as follows. (i) It has been known for some years that racemisation occurs on reduction of phosphine oxides with LiAlH$_4$; this reagent has now been found to reduce phosphine sulphides with complete retention of configuration[184], possibly through hydride attack at S rather than at P, as in oxides. (ii) Selenium adds to phosphorus with retention, and as for the sulphides, may be removed with retention by LiAlH$_4$ reduction[185]. (iii) Oxidation of phosphines by H$_2$O$_2$ is known to occur with retention, and this is true also for KMnO$_4$ oxidation, but aqueous HNO$_3$ gives inversion of configuration[186] and N$_2$O$_4$ gives retention with much racemisation. (iv) Allyl groups of a phosphonium ion are displaced by cyanide with retention, while displacement with alkali proceeds with nearly complete inversion[187].

Optically active phosphonites have been shown to undergo the Arbuzov reaction with retention of configuration[188].

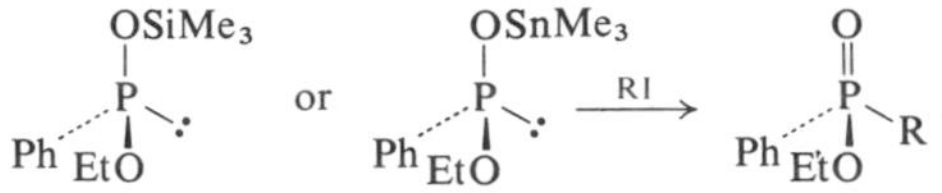

This result is in accord with the generally accepted two-step mechanism of the Arbuzov reaction, alkylation of phosphorus in an S$_N$2 process followed by dealkylation of an OR group by halide attack on R. It also agrees with an observation of retention at phosphorus in an Arbuzov reaction with a cyclic phosphite[189]. The anion derived from ethyl phenylphosphinate can also be treated in the present discussion as a trivalent derivative; it participates in the Michaelis–Becker reaction with retention.

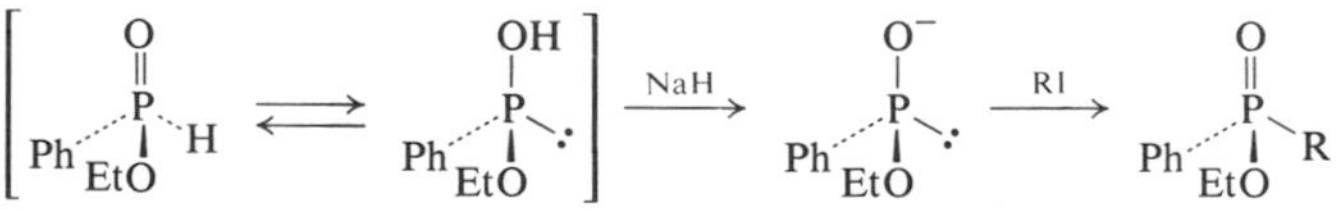

These observations are consistent with those of others[190], when retention was observed in the alkylation of the anion from active menthyl phenylphosphinate. The anion is, of course, an ambident species; attack could as well occur on oxygen, and the phosphonite so formed then rearranges to the phosphinate, but evidence supports direct P-alkylation. The silyl derivative[188] is obtained by the reaction of the phosphinate with triethylamine and trimethylchlorosilane, conditions which might also be visualised as providing the ambident anion as the reactive species.

5.8.2 Stereochemistry of reactions at phosphoryl and thiophosphoryl groups

Several years ago, Korpiun and Mislow[191] developed a highly useful device for obtaining optically active tertiary phosphine oxides. Menthyl esters of alkylarylphosphinic acids are prepared; the resulting mixture of diastereoisomeric esters is separated, and on reaction with a Grignard reagent each gives the phosphine oxide with high stereospecificity and inversion at phosphorus.

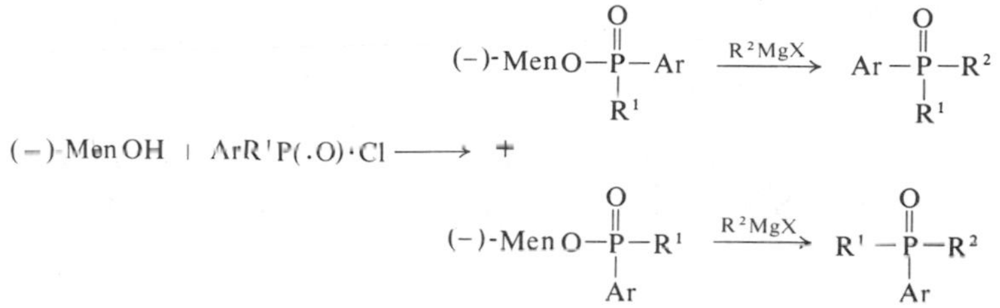

$$
(-)\text{-MenO}-\underset{\underset{R^1}{|}}{\overset{\overset{O}{\|}}{P}}-Ar \quad \xrightarrow{R^2MgX} \quad Ar-\underset{\underset{R^1}{|}}{\overset{\overset{O}{\|}}{P}}-R^2
$$

$(-)\text{-MenOH} + ArR^1P(.O)\cdot Cl \longrightarrow +$

$$
(-)\text{-MenO}-\underset{\underset{Ar}{|}}{\overset{\overset{O}{\|}}{P}}-R^1 \quad \xrightarrow{R^2MgX} \quad R^1-\underset{\underset{Ar}{|}}{\overset{\overset{O}{\|}}{P}}-R^2
$$

In more recent studies, the menthyl ester of phenylphosphinic acid has proved of value in similar reactions. The diastereoisomers obtained by fractional crystallisation were assigned[192] the R or S configuration at phosphorus by n.m.r. differences found also in the menthyl alkylarylphosphinate system (the doublets corresponding to the isopropyl group of the menthyl ring in the S-epimer are shifted up-field relative to those for the R-epimer); and were confirmed by comparisons of the Circular Dichroism (C.D.) curves of the two systems. As mentioned in Section 5.8.1, the anion of the phosphinate is alkylated with retention[190]. Another important discovery was that free-radical addition of the phosphinate to alkenes occurs with retention[192], suggesting that the phosphinyl radical is configurationally stable.

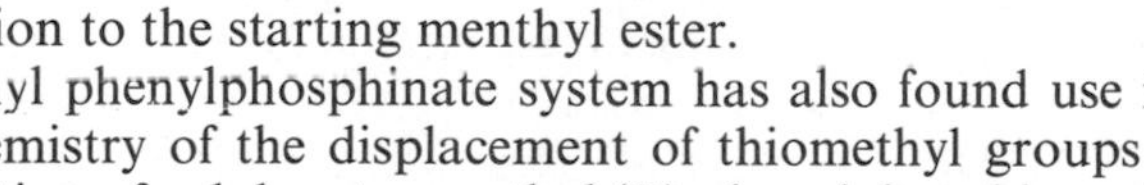

$$
O{=}P\overset{Ph}{\underset{OMen}{\cdots H}} + \bigcirc \quad \xrightarrow[\text{(PhCO·O)}_2]{10 \text{ mol \% of}} \quad O{=}P\overset{Ph}{\underset{OMen}{\cdots C_6H_{11}}} \quad (80\%)
$$

An earlier report[193] from another group had suggested that inversion occurred in the analogous reaction of menthyl methylphosphinate with alkenes, as well as with dimethyl disulphide; it was found[188, 194], however, that this conclusion was reached on the basis of an incorrect assignment[195] of configuration to the starting menthyl ester.

The menthyl phenylphosphinate system has also found use in exploring the stereochemistry of the displacement of thiomethyl groups from phosphorus. Addition of sulphur to menthyl (R)-phenylphosphinate, followed by methylation, gave the phosphonothioate, which reacted with Grignard reagents to form the phosphinate of overall retained configuration[194].

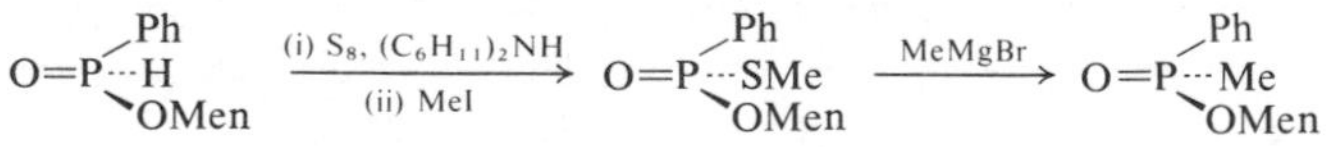

$$
O{=}P\overset{Ph}{\underset{OMen}{\cdots H}} \quad \xrightarrow[\text{(ii) MeI}]{\text{(i) S}_8, \text{ (C}_6\text{H}_{11}\text{)}_2\text{NH}} \quad O{=}P\overset{Ph}{\underset{OMen}{\cdots SMe}} \quad \xrightarrow{MeMgBr} \quad O{=}P\overset{Ph}{\underset{OMen}{\cdots Me}}
$$

An x-ray analysis of the phosphonothioate later showed[196] its absolute configuration to be S, as depicted above, hence showing that *all* steps in the process had gone with *retention*. This result is of profound importance to stereochemical studies in other phosphorus systems, for it has been assumed for years that displacement of RS-groups from phosphorus, as for RO-groups, proceeded with *inversion*. This assumption was employed in inferring the absolute configuration of isopropyl *S*-methyl methylphosphonothioate, the basis for numerous other steric correlations.

Menthyl methylphenylphosphinate has been used to show that displacement of RO from phosphorus by an amide (e.g. PhNHLi) occurs with inversion of configuration[197]. Cholesteryl esters were also used in this study; one of the diastereoisomeric forms could be obtained highly pure, and it was employed successfully in the displacements with metallic amides as well as with Grignard reagents.

Optically active methylphosphinates have also been obtained by a quite different approach[198]. (+)-Isopropyl methylphosphonothioate may be obtained in 100% optical purity by resolution; desulphurisation then gives the (−)-methylphosphinate.

$$\underset{\underset{\text{Me}}{|}}{\overset{\overset{\text{OH}}{|}}{\underset{Pr^iO}{\diagup}}\!\!P\!\!\diagdown S} \quad \xrightarrow{\text{Raney Ni}} \quad \underset{\underset{\text{Me}}{|}}{\overset{\overset{\text{OH}}{|}}{\underset{Pr^iO}{\diagup}}\!\!P\!\!\diagdown \!:} \quad \longrightarrow \quad \underset{\underset{\text{Me}}{|}}{\overset{\overset{\text{O}}{\|}}{\underset{Pr^iO}{\diagup}}\!\!P\!\!\diagdown H}$$

Here we are dealing with the system whose absolute configuration has been brought into question by the work of Mislow[196] and the assignment of the *R*-configuration to the (−)-isomer produced by desulphurisation of the (+)-phosphonothionate must be considered tentative for the present. The methylphosphinate added to alkenes under radical-forming conditions, giving with high stereoselectivity, as did the phenylphosphinate[192], a product presumably of retained configuration. Unlike the phenylphosphinate, however, the methyl compound on conversion into its anion racemised very rapidly, so that products of the Michaelis–Becker reaction were optically inactive. It is not immediately obvious why the two systems should differ drastically in this respect. Chlorination with *N*-chlorosuccinimide or with CCl_4–tri-n-butylamine both occurred with high stereoselectivity. The optically active methylphosphinate also served as a precursor of the active methylphosphinothionate[199]; reaction with P_4S_{10} gave the novel thionate $Pr^iOPHMe(:S)$ in 68% optical purity, with retention predominating. While it exhibited stereoselectivity in several reactions studied, it did not in general fare as well in this regard as did its oxygen counterpart. Application of the P_4S_{10} sulphurisation reaction to optically active phosphonothionates $[R^1OPR^2(:O)SR^3]$ also was found to be highly stereoselective, forming phosphonothionates $[R^1OPR^2(:S)SR^3]$ with presumed retention of configuration[200, 201]. Sulphuration of phosphine oxides, however, gave a racemic product[201]. Nitric acid has been reported to oxidise phosphine sulphides as well as phosphonothionates with inversion, rather unexpectedly, but N_2O_4 oxidation occurred with retention and considerable racemisation. In the presence of an acid, however, N_2O_4 oxidation also proceeded with net partial inversion[186]. An effect of acidity on peroxyacid oxidation of thiophosphoryl groups has also been observed[202]. In this study, the diastereoisomeric menthyl methylphenylphosphinothionates $[MePhP(:S)OMen]$ were oxidised with *m*-chloroperoxybenzoic acid in CH_2Cl_2 at 0 °C to give the phosphonate with a high degree of retention, but with trifluoroperoxyacetic acid inversion predominated. When a strong acid (e.g. $CF_3 \cdot CO_2H$) was present in the m-chloroperoxybenzoic acid oxidation, inversion predominated. Again protonation of the sulphur atom appears to be implicated in the inversion mechanism, but details remain obscure. Selenium may also be removed from phosphorus by oxidising agents; with $KMnO_4$, retention occurs, while inversion occurs with HNO_3 or N_2O_4 [185].

Optically active phosphine oxides are racemised by $LiAlH_4$, and it has now been established that this is true also for secondary phosphine oxides[203]. Diastereoisomers of phenyl-α-phenylethylphosphine oxide were obtained separately and found to be rapidly interconverted at room temperature by $LiAlH_4$. Attempts to repeat an earlier synthesis[204] of an optically active secondary phosphine oxide by $LiAlH_4$ reduction of an active phosphinate were unsuccessful, presumably because of stereomutation following the formation of the oxide.

In the attack of nucleophiles at phosphorus, trigonal bipyramidal intermediates are believed to be formed. Incoming nucleophiles generally take up an apical position, and departure of the displaced group also occurs at an apical position. The net result is inversion. An unusual example in which retention occurs has been reported[205]; a phosphonium salt was prepared by methylation of menthyl (S)-methylphenylphosphinothionate at sulphur with $Me_3O^+SbCl_6^-$, and on alkaline hydrolysis it gave menthyl (R)-methylphenylphosphinate, MePhP(:O)OMen.

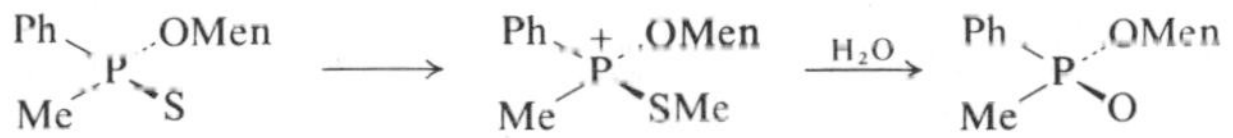

The unusual feature here is that in the initially formed trigonal bipyramid electronegativity considerations place the OH and OMen groups in apical positions, yet it is SMe which must be displaced. This may occur either from an equatorial position, or else from an apical position following pseudo-rotation.

A reaction resulting in the formation of the first stable penta-alkylphosphorane has been achieved[206]. It consists of reaction of a carbanion with a phosphonium ion of such strained character that relief of strain on conversion into the pentacovalent state directs the reaction away from the usual proton abstraction process. The phosphorane was distillable at room temperature and 10^{-6} mmHg; it had the high ^{31}P n.m.r. chemical shift (+90 p.p.m.) expected for this coordination state.

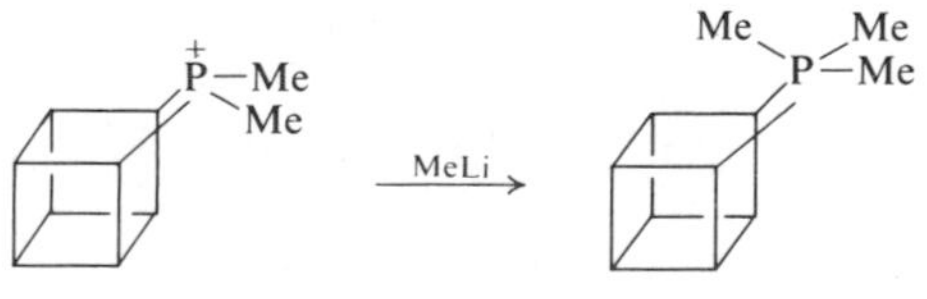

A new technique for the resolution of racemic isopropyl methylphosphinate, possibly applicable to other compounds as well, consists of the formation of inclusion complexes with cyclodextrins. The (−)-epimer was included preferentially, and was recovered in 66.5% optical purity. The non-included (+)–form was obtained 17% optically pure[207].

5.9   MECHANISTIC ASPECTS OF PHOSPHORUS CHEMISTRY

Displacement reactions at phosphoryl centres have been studied for some years, and some new stereochemical results were discussed in Section 5.8.2. Significant kinetic and catalytic studies have also been reported, and among the more important substrates used is isopropyl methylphosphonofluoridate, a highly toxic cholinesterase inhibitor known as Sarin or GB. Displacement of F by water[208] or by alcohols[209] is catalysed by amines. These processes are first-order in the amine, which appears to act as a general base catalyst. Calcium and magnesium ions also catalyse this hydrolysis, which accounts for the greater rate of destruction of the substance in sea water than in water[210]. Yttrium(III), as a polymeric hydrate in solution, is also a potent catalyst for the hydrolysis of p-nitrophenyl methylphosphonate[211]. In another study[212] on this ester, as well as on the phenyl ester, second-order kinetics were observed for hydrolysis in basic media, with attack of OH^- on the phosphorus of the anion. Neutral hydrolysis involves nucleophilic attack by H_2O; hydrolysis in strongly acidic media has a rate maximum, possibly to be attributed to diminished water activity in such media. Displacement reactions of the anion of the nitrophenyl ester with a variety of nucleophiles[213], including a collection of amines[214], were later studied, and found to proceed generally by attack at phosphorus, and with second-order kinetics. Steric bulk near the nucleophilic centre retards the reaction; the α-effect is operative, making nucleophiles bearing an electron-pair on an atom adjacent to the nucleophilic centre more reactive. Only with piperidine was a significant amount (35%) of cleavage of the aromatic carbon–oxygen bond detected.

Neighbouring group participation can occur in phosphonate hydrolysis; acidic hydrolysis of alkyl phosphonates normally occurs with C—O cleavage, but anchimeric assistance, as with the oximino-group, can bring about P—O cleavage instead[215].

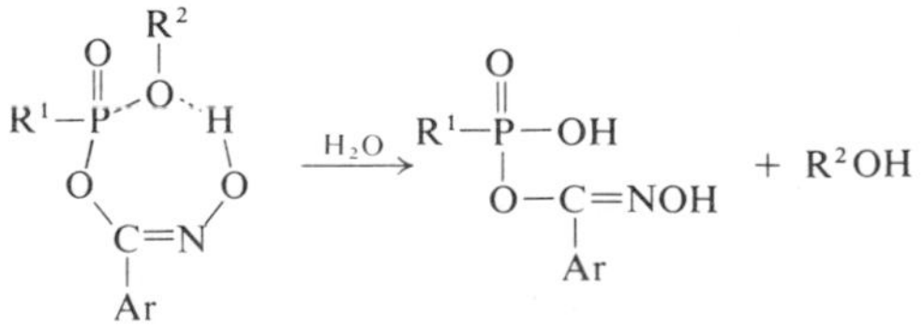

An important effect newly discovered is a dissociative mechanism (S_N1) for displacement at phosphorus in certain phosphinic chlorides[216]. Only chlorides with highly hindered P atoms (e.g. as in di-t-butylphosphinic chloride) follow this path, a process which is much slower than the highly preferred S_N2 process. The ion (phosphinylium) can probably be represented by the hybrid $R_2\overset{+}{P}{=}O \longleftrightarrow R_2P{\equiv}O^+$, with angles not necessarily those for a symmetrical trigonal structure.

Another mechanistic problem in phosphorus chemistry currently receiving attention is that of attack by trivalent phosphorus on halides offering alternatives to the normal Arbuzov pathway. Acetylenic and ketonic halides are among these structures. Acetylenic halides are reactive and are triphilic[217], offering three points of attack: on X (see Section 5.5.2), on C-1 in an addition–elimination process and on C-2. With phosphites, attack at C-1 gives the Arbuzov product and a detailed kinetic study[217] showed this to be the preferred event with chlorophenylacetylene. With the bromo-compound, attack occurred on the halogen as well as at C-1, for phenylacetylene could be obtained on protonation. Positive evidence for an addition at C-1 came from another source[218], attack of di-isopropyl phosphite on 1-bromo-ethynylcyclohexanol, which gave among other products the adduct below.

$$(Pr^iO)_2P(:O){\cdot}CBr{:}CH{\cdot}C(OH)(CH_2)_5$$

Phosphites can react with α-halogenoketones via the Arbuzov pathway or by an alternative known as the Perkow reaction, which gives enol phosphates. The mechanism of the Perkow reaction has been much discussed; the most recent report[219] supports a mechanism for reaction with p-substituted phenacyl halides in which attack occurs on carbonyl carbon, and the phosphorus moiety then migrates to oxygen.

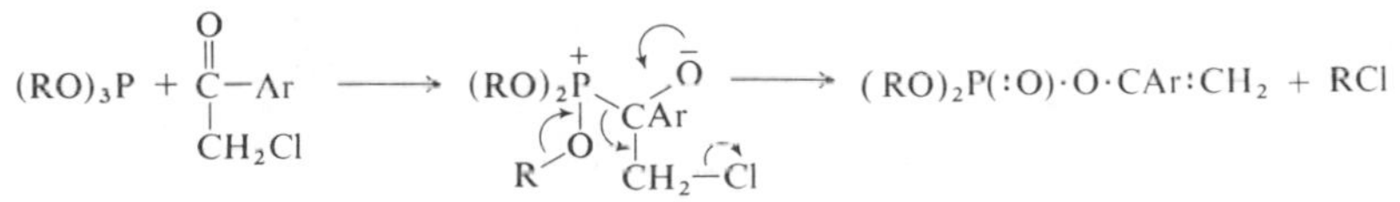

The kinetics are second order; electron withdrawing groups on phenyl cause faster reactions, correlated by Hammett's $\rho = +1.89$. This value is consistent with that for other reactions where nucleophilic attack on carbonyl carbon occurs.

A driving force in many reactions of organophosphorus chemistry is the formation of the highly stable phosphoryl group. Cases where phosphoryl is destroyed are therefore noteworthy and a new one is that of an apparent reversal of the Wittig reaction[220].

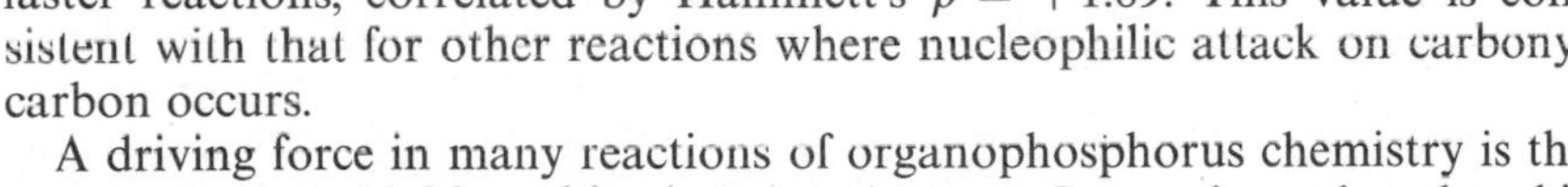

The phosphorane resulting is a resonance-stabilised ylid. Nucleophilicity in the phosphoryl group is not unknown; a recent example[221] is that of O-alkylation by $Et_3O^+BF_4^-$.

The mechanism of decarboxylation of phosphonoformic acid in acidic media has been studied in detail[222] and it is proposed that it involves an unprecedented protonated intermediate, $(HO)_3P^+{-}C(OH)_3$, which decomposes in the rate-controlling step to give the species $(HO)_3C^+$ and $(HO)_3P$. The latter tautomerises to its stable form, $HPO(OH)_2$. Presumably this mechanism might hold for decarboxylation of other α-phosphoryl carboxylic acids. Protonation of various phosphorus functions in very strongly acidic media has also been studied; ^{31}P and ^1H n.m.r. spectra suggest that phosphoryl groups are commonly protonated on oxygen, and that oxygen shares the positive charge with phosphorus through d_π–p_π bonding $(R_3 \overset{+}{P}\!-\!OH \longleftrightarrow R_3 P\!=\!\overset{+}{O}H)$[223].

5.10   N.M.R. SPECTRAL PROPERTIES

Natural-abundance ^{13}C spectroscopy has great promise in phosphorus chemistry, although a recent review of the literature reveals the undeveloped state of the field[224]. Phosphorus compounds possess a unique feature, coupling of ^{31}P with ^{13}C, and new instrumentation, particularly for utilising the pulse Fourier-transform technique with decoupling of ^1H from ^{13}C, makes possible the observation of ^{13}C–^{31}P coupling directly from the spectrum. Much of the current work is devoted to elucidation of factors influencing $J(CP)$ in various structures. Some generalities that are emerging are: (a) $^1J(CP)$ is greater for neutral four-coordinate species (50–150 Hz) than for trivalent (10–20 Hz); (b) $^2J(CP)$ is much smaller for both species, and $^3J(CP)$ smaller still or not observable. Chemical shifts also depend on the nature of the phosphorus substituent. Those containing phosphoryl groups cause deshielding relative to those with trivalent phosphorus; the magnitude of the difference is in the 5–10 p.p.m. range, but insufficient data of this type are available to permit a useful generalisation.

In a rather complete study[224], the effect of changes in X on ^{13}C parameters in over a dozen members of the series $XCH_2 \cdot P(\!:\!O)(OEt)_2$ was examined. The only important chemical shift effect was associated with the carbon bound to phosphorus, for which a spread of 79 p.p.m. was found. A similar spread is seen for CH_3X, and this revealed that the phosphono-group deshields this carbon by 13.2 p.p.m. by a direct electronic effect. The one-bond ^{13}C–^{31}P coupling constants varied over a range of c. 37 Hz, while again carbon atoms of the ethoxy groups were not appreciably affected by the nature of X. A correlation was observed with the C—P s-bond order obtained from approximate SCF–MO calculations.

Among applications of proton n.m.r. spectroscopy to phosphorus compounds, perhaps the most important recent one has been the use of lanthanide reagents to affect the chemical shifts through pseudo-contact or contact interactions. Phosphoryl[225] (but not thiophosphoryl[226]) groups are sites of attraction for interactions with the lanthanides; no interaction of useful magnitude has yet been observed for a phosphine. With triethylphosphine oxide[227], the CH_2 group resonance is shifted downfield by 7.1 p.p.m. by an equimolar amount of Eu(dipivaloylmethane)$_3$, (Eu(DPM)$_3$) while that of the Me group is shifted by 4.8 p.p.m. This provides excellent separation of the signals, and allows the coupling constants to be measured readily. With tri-n-butylphosphine oxide, similar shifts for the α- and the β- proton were observed; for the γ- and the δ-proton the shifts were 2.0 and 0.5 p.p.m. respectively. Pr(DPM)$_3$ also produced large shifts but in the upfield direction. For diethyl ethylphosphonate, the Me—CH_2—P signals were shifted upfield by Eu(NO$_3$)$_3 \cdot 6D_2O$, and downfield by the Pr analogue[228]. The ethoxy-group signals were similarly affected, but to a smaller degree. Ethoxy groups in organic phosphates and phosphites are strongly deshielded by Eu(DPM)$_3$; the protons of CH_2O

groups experience shifts of c. 8 p.p.m. at 1:1 molar ratio in both classes of compounds, and those of the Me group c. 3 p.p.m.[226].

^{31}P N.M.R. spectroscopy continues to receive attention, as attempts are made to develop understanding of the complex interplay of the various factors controlling the chemical shift. For purely aliphatic tertiary phosphines, a linear relationship between the chemical shift and the sum of the Taft polar substituent constants (σ^*), or of the half-neutralisation potentials, has been observed[229] (a separate line with slope of opposite sign correlated the results for some phenylphosphines). However, the significance of the relation is not clear, for its slope was not that expected for the operation of polar effects. Of the phosphines studied, that with greatest electron release to P ($Pr_3^i P$, $\Sigma\sigma^* = 0.57$) had a ^{31}P signal downfield from that of Me_3P ($\Sigma\sigma^* = 0$) by some 80 p.p.m.! The authors attributed this anomaly to the operation of overriding steric effects, a point considered by others and rejected as giving inconsistent correlations[230]. The most useful predictive device for ^{31}P shifts of phosphines remains that resting on the construction of empirical group contributions[230]; while these values are derived from experimental measurements, they seem to be closely associated with hyperconjugative trends among the substituents. In a series PH_3, RPH_2, R_2PH, R_3P, a correlation has been observed between the ^{31}P shift and the net charge on phosphorus calculated by SCF-methods[231]. These calculations give the correct sequence also to Me_3P and Et_3P; the smaller ^{31}P shift value of the latter ($+18.6$ p.p.m.) suggests greater positive charge on phosphorus, and this is what the calculations indicate (Me_3P, $+0.0341$; Et_3P, $+0.0359$). Correlations were also noted for phosphonium and phosphide ions.

Techniques for making ^{31}P n.m.r. measurements have improved to the point where quite dilute solutions may now be handled; with signal-averaging, solutions with a total P concentration of 10^{-3}–10^{-4} M have given strong signals[232]. One important feature of this is that direct observation of phosphonates in extracts from biological material may be made. In principle, phosphorus in phosphate form should be easily distinguished from that in phosphonate form (or in other forms with C—P bonds); phosphonates would have chemical shifts 15–20 p.p.m. downfield from phosphates. With the new instrumentation, it has been found possible[232] to observe signals for both types of phosphorus in some extracts from marine animals in which the concentrations are so low as to have provided no phosphonate signal by earlier techniques. The method is reasonably sensitive, is non-destructive to fragile forms, and may well become standard for studies in this field.

5.11 SOME BIOLOGICAL ASPECTS OF PHOSPHORUS CHEMISTRY

That compounds with C—P bonds could exist in nature was unknown until 1959, but now several such compounds have been found and their study constitutes an active area of research. Two reviews are available on the subject[233, 234]. Of more significant recent events, one has been the isolation of small amounts of the most common of these compounds, 2-aminoethyl-phosphonic acid (AEP), from human brain[235]; prior to this, the compound had been found intimately associated with lower forms of marine life and some fresh-water organisms. Techniques are now available for mass spectrometry of and separation by gas chromatography of aminophosphonic acids, as their N-acetyl dimethyl[235] and trimethylsilyl[236] derivatives. Identification by direct ^{31}P n.m.r. spectroscopy on extracts has already been mentioned[232].

Except for one substance to be noted, the natural C—P compounds are amino-derivatives of aliphatic phosphonic acids. A new versatile synthesis[237] consists of the addition of diethyl phosphonate to an α,β-unsaturated amide, followed by Hofmann hypobromite degradation of the amide function. Techniques are available for synthesis of various glyceryl derivatives of AEP and its N-methyl derivatives, which because of their similarity to phospholipids are named phosphonolipids. A typical sequence is that for the construction of a phosphonic analogue of a cephalin[238]; D-α,β-Distearin is phosphonylated with $PhCH_2O\cdot CO\cdot NMe\cdot(CH_2)_2\cdot P(:O)(OH)Cl$, followed by catalytic hydrogenolysis of the benzyloxycarbonyl group. Non-hydrolysable phosphinate analogues of lecithins,

$$ROCH_2\cdot CH(OR)\cdot (CH_2)_2\cdot P(:O)(O^-)\cdot (CH_2)_3 N^+ Me_3 (R = n\text{-}C_{18}H_{37}),$$

are also known[239].

One of the major events in this field was the announcement in 1969 of the isolation of a simple phosphonic acid from the fermentation broth of a streptomycetes[240]. The acid, named phosphonomycin, is of low toxicity and is an effective bactericidal agent for both Gram-positive and Gram-negative organisms; its activity compares favourably with tetracycline and chloroamphenicol, and it appears to have potential as a broad spectrum antibiotic. Along with this exciting announcement came the first[241] of several syntheses of phosphonomycin.

$$MeC\!:\!CMgBr + (Bu^nO)_2 P(:O)Cl \longrightarrow MeC\!:\!C\cdot P(:O)(OBu^n)_2$$

$$\xrightarrow[\text{Lindlar catalyst}]{H_2} \quad \begin{array}{c} Me \\ \diagup \\ H \end{array}\!\! C\!=\!C \!\!\begin{array}{c} P(:O)(OBu^n)_2 \\ \diagdown \\ H \end{array} \quad \xrightarrow{\text{conc. HCl}} \quad \begin{array}{c} Me \\ \diagup \\ H \end{array}\!\! C\!=\!C \!\!\begin{array}{c} P(:O)H_2 \\ \diagdown \\ H \end{array}$$

$$\xrightarrow{30\% \ H_2O_2} \quad \begin{array}{c} H \\ \diagup \\ Me \end{array}\!\! C\!\!\!\begin{array}{c} \\ O \end{array}\!\!\! C \!\!\begin{array}{c} H \\ \diagdown \\ PO_3H_2 \end{array}$$

The acid was resolved as the quinine salt; the benzylammonium salt of $[\alpha]_{405} -9.1°$ (H_2O) was identical to that from the natural product. Absolute configuration was determined by opening the epoxide ring with methanol to give the *threo*-1-hydroxy-2-methoxy derivative, and $KMnO_4$ oxidation to D-2-methoxypropionic acid. This related the configuration of the β-carbon to that of L-lactic acid, since inversion accompanies opening of the epoxide ring, and defined phosphonomycin as the $(-)$-$(1R, 2S)$ stereoisomer. A later synthesis[242] of the *cis*-propenylphosphonate system employed the following steps. Some other transformations of phosphonomycin have also been published[243], and the patent literature (to Merck and Co.) is voluminous.

$$HC\!:\!C\cdot CH_2OH \xrightarrow[\text{Et}_3N]{(\text{Bu}^t\text{O})_2PCl} HC\!:\!C\cdot CH_2O\cdot P(OBu^t)_2$$

$$\xrightarrow{45\,°C} H_2C\!:\!C\!:\!CH\cdot P(\!:\!O)(OBu^t)_2 \xrightarrow{H_2,\ Pd-C} \underset{H}{\overset{Me}{>}}C\!=\!C\underset{H}{\overset{P(\!:\!O)(OBu^t)_2}{<}}$$

Interest is also high at present in the synthesis of analogues of important natural phosphates where a C—P bond replaces C—O—P bonds. The application of Arbuzov or Michaelis–Becker reactions to 6-halogeno-hexoses or 5-halogenopentoses presents no difficulty, and for some years sugars with phosphonic groups appearing where phosphate groups are found naturally have been known. Attachment of phosphorus to the 1-position was accomplished[244] by reaction of the carbonyl group (see Section 5.3.2) of a blocked aldohexose with a phosphonate. 2-Ketohexoses reacted similarly and 2-deoxyhexose-2-phosphonates were prepared by Michael addition of a phosphonate to unsaturated sugar derivatives. Extension of the carbon chain of an aldose, with formation of a phosphorus derivative, has been accomplished[245] by using the Wittig reagent $Ph_3P\!:\!CH\cdot P(\!:\!O)(OPh)_2$, and mesylates and epoxides of sugars undergo reaction with Ph_2PLi to form phosphorus derivatives[246].

Phosphonate analogues of nucleoside 3'-phosphates may be obtained[247] by treating a blocked 3-keto-sugar with a phosphonate carbanion, reducing the resulting vinylphosphonate and then employing conventional means to establish a purine base at the 1-position. The linking of two nucleosides by a —$CH_2\cdot P(\!:\!O)(OH)O$— group has also been accomplished[248]. Yet another modification[249] of a natural structure is the synthesis of an ATP analogue in which the P—O—P bonds have been replaced by P—CH_2—P.

References

1. Rozinov, V. G., Mikhnevich, V. V. and Grechkin, E. F. (1970). *Zh. Obshch. Khim.*, **40**, 935
2. Rozinov, V. G., Mikhnevich, V. V. and Grechkin, E. F. (1970). *Izv. Nauch.-Issled. Inst. Nefte-Uglekhim. Sin. Irkutsk. Univ.*, **12**, 60; (*Chem. Abstr.* (1971). **75**, 534)
3. Kormachev, V. V., Tsivunin, V. S. and Koren, N. A. (1970). *Zh. Obshch. Khim.*, **40**, 1989
4. Moskva, V. V., Ismailov, V. M. and Razumov, A. I. (1970). *Zh. Obshch. Khim.*, **40**, 1489
5. Moskva, V. V., Nazvanova, G. F., Zykova, T. V. and Razumov, A. I. (1971). *Zh. Obshch. Khim.*, **41**, 1489
6. Moskva, V. V., Nazvanova, G. F., Zykova, T. V. and Razumov, A. I. (1971). *Zh. Obshch. Khim.*, **41**, 1493
7. Fridland, S. V., Tsivunin, V. S., Fridland, D. V. and Kamai, G. (1970). *Zh. Obshch. Khim.*, **40**, 1993
8. Pudovik, A. N., Khairullin, V. K. and Dmitrieva, G. V. (1970). *Zh. Obshch. Khim.*, **40**, 1034
9. Khairullin, V. K., Dmitrieva, G. V. and Pudovik, A. N. (1970). *Izv. Akad. Nauk SSSR, Ser. Khim.*, **871** (*Chem. Abstr.* (1971). **73**, 339)
10. Pudovik, A. N., Terent'eva, S. A. and Pudovik, M. A. (1970). *Zh. Obshch. Khim.*, **40**, 1707
11. Gazizov, T. K., Mareev, Y. M., Vinogradova, V. S., Pudovik, A. N. and Arbuzov, B. A. (1971). *Izv. Akad. Nauk SSSR, Ser. Khim.*, 1259 (*Chem. Abstr.* (1971). **75**, 378)
12. Pudovik, A. N., Batyeva, E. S., Shagidullin, R. R., Raevskii, O. A. and Pudovik, M. A. (1970). *Zh. Obshch. Khim.*, **40**, 1195
13. Fields, R., Haszeldine, R. N. and Wood, N. F. (1970). *J. Chem. Soc. C*, 1370

14. Fields, R., Haszeldine, R. N. and Kirman, J. (1970). *J. Chem. Soc. C*, 197
15. Kleiner, H. J. (1970). *Ger. Offen.*, 1, 902, 444 *(Chem. Abstr.* (1971). **73**, 479)
16. Aleksandrova, I. A. and Ufimteva, L. I. (1971). *Izv. Akad. Nauk SSSR, Ser. Khim.*, 1315
17. Ivanov, B. E., Kudryavtseva, L. A. and Bykova, T. G. (1970). *Izv. Akad. Nauk SSSR, Ser. Khim.*, 2063 *(Chem. Abstr.* (1971). **74**, 311)
18. Razumov, A. I., Yafarova, R. L. and Ismagilov, R. K. (1971). *Zh. Obshch. Khim.*, **41**, 1022
19. Konotopova, S. P., Chistokletov, V. N. and Petrov, A. A. (1971). *Zh. Obshch. Khim.*, **41**, 235
20. Draper, P. M., Chan, T. H. and Harpp, D. N. (1970). *Tetrahedron Letters*, 1687
21. Petrov, K. A., Raksha, M. A., Korotkova, V. P. and Shmidt, E. (1971). *Zh. Obshch. Khim.*, **41**, 324
22. Reichel, L. and Jahns, H. (1971). *Justus Liebigs Ann. Chem.*, **751**, 69
23. Petrov, K. A., Raksha, M. A. and Vinogradov, V. L. (1966). *Zh. Obshch. Khim.*, **36**, 729
24. Zolotova, M. V. and Konstantinova, T. V. (1970). *Zh. Obshch. Khim.*, **40**, 2131
25. Ogata, Y. and Tomioka, H. (1970). *J. Org. Chem.*, **35**, 597
26. Sander, M. (1960). *Chem. Ber.*, **93**, 1220
27. Pashinkin, A. P., Gazizov, T. K. and Pudovik, A. N. (1970). *Zh. Obshch. Khim.*, **40**, 28
28. Horner, L. and Röder, H. (1970). *Chem. Ber.*, **103**, 2984
29. Portnoy, N. A., Morrow, C. J., Chattha, M. S., Williams, J. C., Jr. and Aguiar, A. M. (1971). *Tetrahedron Letters*, 1397
30. Williams, J. C., Jr., Kuczkowski, J. A., Portnoy, N. A., Yong, K. S., Wander, J. D. and Aguiar, A. M. (1971). *Tetrahedron Letters*, 4749
31. Simalty, M. and Le Van Chau. (1970). *Tetrahedron Letters*, 4371
32. Fedorova, G. K., Shaturskii, Y. P., Moskalevskaya, L. S. and Kirsanov, A. V. (1970). *Zh. Obshch. Khim.*, **40**, 1167
33. Rizpolozhenskii, N. I. and Mukhametov, F. S. (1970). *Izv. Akad. Nauk. SSSR, Ser. Khim.*, 1087. *(Chem. Abstr.*, (1970). **73**, 355)
34. Frank, A. W. and Drake, G. L., Jr. (1971). *J. Org. Chem.*, **36**, 3461
35. Eichelberger, J. L. and Stille, J. K. (1971). *J. Org. Chem.*, **36**, 1840
36. Issleib, K. and Abicht, H. P. (1970). *J. Prakt. Chem.*, **312**, 456
37. Malevannaya, R. A., Tsvetkov, E. N. and Kabachnik, M. I. (1971). *Zh. Obshch. Khim.*, **41**, 1426
38. Khairullin, V. K., Dmitrieva, G. V. and Pudovik, A. N. (1970). *Izv. Akad. Nauk SSSR, Ser. Khim.*, 468 *(Chem. Abstr.* (1970). **73**, 342)
39. Bodnarchuk, N. S., Malovik, V. V., Derkach, G. I. and Kirsanov, A. V. (1971). *Zh. Obshch. Khim.*, **41**, 1464
40. Campbell, I. G. M. and Raza, S. M. (1971). *J. Chem. Soc. C*, 1836
41. Pudovik, A. N., Zimin, M. G. and Sobanov, A. A. (1970). *Zh. Obshch. Khim.*, **40**, 936
42. Komissarova, S. L., Valetdinov, R. K. and Kuznetsov, E. V. (1971). *Zh. Obshch. Khim.*, **41**, 322
43. Maier, L. (1971). *Helv. Chim. Acta*, **54**, 1434
44. Tsivunin, V. S., Krutskii, L. N., Ernazarov and Kamai, G. K. (1970). *Zh. Obshch. Khim.*, **40**, 2560
45. Padwa, A. and Eastman, D. (1971). *J. Org. Chem.*, **35**, 1173
46. Pudovik, A. N., Gareev, R. D., Aganov, A. V. and Stabrovskaya, L. A. (1971). *Zh. Obshch. Khim.*, **41**, 1232
47. Tyka, R. (1970). *Tetrahedron Letters*, 677
48. Fields, E. K. (1952). *J. Amer. Chem. Soc.*, **74**, 1528
49. Ivanov, B. E., Gorin, Y. A. and Krikhina, S. S. (1970). *Izv. Akad. Nauk. SSSR, Ser. Khim.*, **11**, 2627 *(Chem. Abstr.* (1971). **76**, 503)
50. Gross, H. and Costisella, B. (1971). *Justus Liebigs Ann. Chem.*, **750**, 44
51. Mastryukova, T. A., Lazareva, M. V. and Perekalin, V. V. (1971). *Izv. Akad. Nauk. SSSR, Ser. Khim.*, **6**, 1353 *(Chem. Abstr.* (1971). **75**, 980)
52. Schmidpeter, A. and Zeiss, W. (1971). *Chem. Ber.*, **104**, 1199
53. Schmidpeter, A. and Zeiss, W. (1971). *Angew. Chem., Int. Ed. Engl.*, **10**, 396
54. King, R. B. and Efraty, A. (1971). *J. Amer. Chem. Soc.*, **93**, 564
55. Seyferth, D. and Marmor, R. S. (1970). *Tetrahedron Letters*, 2493

56. Regitz, M. (1971). *Justus Liebigs Ann. Chem.*, **748**, 207
57. Seyferth, D., Marmor, R. S. and Hilbert, P. (1971). *J. Org. Chem.*, **36**, 1379
58. Marmor, R. S. and Seyferth, D. (1971). *J. Org. Chem.*, **36**, 128
59. Regitz, M., Scherer, H. and Anschütz, W. (1970). *Tetrahedron Letters*, 753
60. Maier, L. (1971). *Helv. Chim. Acta*, **54**, 275
61. Maier, L. (1970). *Helv. Chim. Acta*, **53**, 2069
62. Maier, L. (1971). *Helv. Chim. Acta*, **54**, 1651
63. Tsvetkov, E. N., Borisov, V., Sivriev, K., Malevannaya, R. A. and Kabachnik, M. I. (1970). *Zh. Obshch. Khim.*, **40**, 285
64. Gross, H. and Seibt, H. (1970). *J. Prakt. Chem.*, **312**, 475
65. Burg, A. B. and Kang, D. (1970). *J. Amer. Chem. Soc.*, **92**, 1901
66. Fields, R., Haszeldine, R. N. and Wood, N. F. (1970). *J. Chem. Soc. C*, 744
67. Janzen, A. F. and Pollitt, R. (1970). *Can. J. Chem.*, **48**, 1987
68. Henderson, W. A., Jr. and Streuli, C. A. (1960). *J. Amer. Chem. Soc.*, **82**, 5791
69. Stewart, A. P. and Trippett, S. (1970). *J. Chem. Soc. C*, 1263
70. Scherer, O. J. and Schieder, G. (1968). *Chem. Ber.*, **101**, 4184
71. Crofts, P. C. and Parker, D. M. (1970). *J. Chem. Soc. C*, 332
72. Corfield, J. R., De'ath, N. J. and Trippett, S. (1971). *J. Chem. Soc. C*, 1930
73. Kuchen, W. and Hägele, G. (1970). *Chem. Ber.*, **103**, 2114
74. Crofts, P. C. and Parker, D. M. (1970). *J. Chem. Soc. C*, 2342
75. Kuchen, W. and Hägele, G. (1970). *Chem. Ber.*, **103**, 2274
76. Hägele, G. and Kuchen, W. (1970). *Chem. Ber.*, **103**, 2885
77. Scherer, O. J. and Gick, W. (1970). *Z. Naturforsch. B*, **25**, 891
78. Scherer, O. J. and Gick, W. (1970). *Chem. Ber.*, **103**, 71
79. Stewart, A. P. and Trippett, S. (1970). *Chem. Commun.*, 1279
80. Crofts, P. C. and Parker, D. M. (1970). *J. Chem. Soc. C*, 2529
81. Field, M. and Schmutzler, R. (1970). *J. Chem. Soc. A*, 2359
82. Malisch, W., Rankin, D. and Schmidbaur, H. (1971). *Chem. Ber.*, **104**, 145
83. Zeliger, H. I. and Snyder, J. P. (1970). *Tetrahedron Letters*, 3313
84. Wilson, I. F. and Tebby, J. C. (1970). *Tetrahedron Letters*, 3769
85. Snyder, J. P. and Bestmann, H. J. (1970). *Tetrahedron Letters*, 3317
86. Zeliger, H. I., Snyder, J. P. and Bestmann, H. J. (1969). *Tetrahedron Letters*, 2199
87. Schweizer, E. E., Crouse, D. M., Minami, T. and Wehman, A. T. (1971). *Chem. Commun.*, 1000
88. Rakshys, J. W., Jr. and McKinley, S. V. (1971). *Chem. Commun.*, 1336
89. Richards, E. M. and Tebby, J. C. (1971). *J. Chem. Soc. C*, 1059
90. Lefèbvre, G. and Seyden-Penne, J. (1971). *Chem. Commun.*, 1308
91. Corey, E. J. and Shulman, J. I. (1970). *J. Org. Chem.*, **35**, 777
92. Bergmann, E. D. and Solomonovici, A. (1971). *Tetrahedron*, **27**, 2675
93. Lednicer, D. (1971). *J. Org. Chem.*, **36**, 3473
94. Bodnarchuk, N. D., Malovik, V. V. and Derkach, G. I. (1970). *Zh. Obshch. Khim.*, **40**, 1210
95. Pudovik, A. N., Nikitina, V. I. and Kurguzova, A. M. (1970). *Zh. Obshch Khim.*, **40**, 291
96. Kirilov, M. and Petrov, G. (1971). *Chem. Ber.*, **104**, 3073
97. Lavielle, G. and Reisdorf, D. (1971). *Compt. Rend. Acad. Sci., Ser. C*, **272**, 100
98. Kirilov, M. and Petrova, J. (1970). *Chem. Ber.*, **103**, 1047
99. Kirilov, M., Petrova, J. and Petkancin, K. (1971). *Chem. Ber.*, **104**, 173
100. Kirilov, M. and Petrova, J. (1970). *Tetrahedron Letters*, 2129
101. Pudovik, A. N., Yastrebova, G. E. and Pudovik, O. A. (1970). *Zh. Obshch. Khim.*, **40**, 499
102. Maier, L. (1970). *Helv. Chim. Acta*, **53**, 1948
103. Polikarpov, Y. M., Kulumbetova, K. Z., Medved, T. Y. and Kabachnik, M. I. (1970). *Izv. Akad. Nauk SSSR, Ser. Khim.*, 1326 (*Chem. Abstr.* (1970). **73**, 370)
104. Tavs, P. and Weitkamp, H. (1970). *Tetrahedron*, **26**, 5529
105. Moskva, V. V., Zykova, T. V., Ismailov, V. M. and Razumov, A. I. (1971). *Zh. Obshch. Khim.*, **41**, 93
106. Horner, L., Ertel, I., Ruprecht, H. and Belovský, O. (1970). *Chem. Ber.*, **103**, 1582
107. Horner, L., Hofer, W., Ertel, I. and Kunz, H. (1970). *Chem. Ber.*, **103**, 2718

108. Ecker, A., Boie, I. and Schmidt, U. (1971). *Angew. Chem., Int. Ed. Engl.*, **83**, 178
109. Schweizer, E. E. and Kopay, C. M. (1971). *J. Org. Chem.*, **36**, 1489
110. Swan, J. M. and Wright, S. H. B. (1971). *Aust. J. Chem.*, **24**, 777
111. Chatta, M. S. and Aguiar, A. M. (1971). *J. Org. Chem.*, **36**, 2719
112. Chattha, M. S. and Aguiar, A. M. (1971). *J. Org. Chem.*, **36**, 2720
113. Burt, D. W. and Simpson, P. (1971). *J. Chem. Soc. C*, 1971
114. Pudovik, A. N. and Khusainova, N. G. (1970). *Zh. Obshch. Khim.*, **40**, 1419
115. Carty, A. J., Hota, N. K., Ng, T. W., Patel, H. A. and O'Connor, T. J. (1971). *Can. J. Chem.*, **49**, 2706
116. King, R. B. and Efraty, A. (1970). *Inorg. Chim. Acta*, **4**, 123
117. Hewertson, W., Taylor, I. C. and Trippett, S. (1970). *J. Chem. Soc. C*, 1835
118. Kuchen, W. and Koch, K. (1970). *Z. Naturforsch B*, **25**, 1189
119. Borkent, G. and Drenth, W. (1970). *Recl. Trav. Chim. Pays-Bas*, **89**, 1057
120. Berlan, J., Capmau, M. L. and Chodkiewicz, W. (1971). *Compt. Rend. Acad. Sci., Ser. C*, **273**, 1107
121. Berlan, J., Capmau, M. L. and Chodkiewicz, W. (1971). *Compt. Rend. Acad. Sci., Ser. C*, **273**, 295
122. Pudovik, A. N., Khusainova, N. G., Tumoshina, T. V. and Raevskaya. (1971). *Zh. Obshch. Khim.*, **41**, 1476
123. Horner, L. and Binder, V. (1971). *Phosphorus*, **1**, 17
124. Callot, H. J. and Benezra, C. (1970). *Can. J. Chem.*, **48**, 3382
125. Griffin, C. E. and Daniewski, W. M. (1970). *J. Org. Chem.*, **35**, 1691
126. Schweizer, E. E. and Kim, C. S. (1971). *J. Org. Chem.*, **36**, 4033
127. Pudovik, A. N., Gareev, R. D., Aganov, A. V., Raevskaya, O. E. and Stabrovskaya, L. A. (1971). *Zh. Obshch. Khim.*, **41**, 1008
128. Pudovik, A. N., Gareev, R. D. and Aganov, A. V. (1971). *Zh. Obshch. Khim.*, **41**, 1017
129. Kolokol'tseva, I. G., Chistokletov, V. N. and Petrov, A. A. (1970). *Zh. Obshch. Khim.*, **40**, 2618
130. Kolokol'tseva, I. G., Chistokletov, V. N. and Petrov, A. A. (1970). *Zh. Obshch. Khim.*, **40**, 574
131. Kosovtsev, V. V., Chistokletov, V. N. and Pétrov, A. A. (1970). *Zh. Obshch. Khim.*, **40**, 2570
132. Pudovik, A. N. and Khusainova, N. G. (1970). *Zh. Obshch. Khim.*, **40**, 697
133. Portnoy, N. A., Morrow, C. J., Chattha, M. S., Williams, J. C., Jr. and Aguiar, A. M. (1971). *Tetrahedron Letters*, 1401
134. Portnoy, N. A., Yong, K. S. and Aguiar, A. M. (1971). *Tetrahedron Letters*, 2559
135. Chattha, M. S. and Aguiar, A. M. (1971). *Tetrahedron Letters*, 1419
136. Chattha, M. S. and Aguiar, A. M. (1971). *J. Org. Chem.*, **36**, 2892
137. Schindler, N. and Plöger, W. (1971). *Chem. Ber.*, **104**, 2021
138. King, R. B. and Kapoor, P. N. (1971). *J. Amer. Chem. Soc.*, **93**, 4158
139. King, R. B. and Kapoor, P. N. (1971). *Angew. Chem. Int. Ed. Engl.*, **10**, 734
140. Tsvetkov, E. N., Lobanov, D. I., Makhamatkhanov, M. M. and Kabachnik, M. I. (1971). *Tetrahedron*, **25**, 5623
141. Rosenberg, D. and Drenth, W. (1971). *Tetrahedron*, **27**, 3893
142. Quin, L. D., Breen, J. J. and Myers, D. K. (1971). *J. Org. Chem.*, **36**, 1297
143. Grim, S. O., Molenda, R. P. and Keiter, R. L. (1970). *Chem. and Ind. (London)*, 1378
144. Issleib, K. and Weichmann, H. (1971). *Z. Chem.*, **11**, 188
145. Maier, L. (1970). *U.S. Pat.*, 3, 518, 312. (*Chem. Abstr.* (1970). **73**, 370)
146. Grim, S. O., Yankowsky, A. W. and Briggs, W. L. (1971). *Chem. and Ind. (London)*, 575
147. Sommer, K. (1970). *Z. Anorg. Allgem. Chem.*, **376**, 37
148. Hewertson, W. and Taylor, I. C. (1970). *J. Chem. Soc. C*, 1990
149. Cooper, P., Fields, R. and Haszeldine, R. N. (1971). *J. Chem. Soc. C*, 3031
150. Issleib, K. and Lischewski, M. (1970). *J. Prakt. Chem.*, **312**, 135
151. Cilley, W. A., Nicholson, D. A. and Campbell, D. (1970). *J. Amer. Chem. Soc.*, **92**, 1685
152. Nicholson, D. A., Cilley, W. A. and Quimby, O. T. (1970). *J. Org. Chem.*, **35**, 3149
153. Maier, L. (1970). *Helv. Chim. Acta*, **53**, 1940
154. Maier, L. (1970). *Helv. Chim. Acta*, **53**, 1944
155. Knollmueller, K. O. (1970). *U.S. Pat.*, 3 534 125. (*Chem. Abstr.* (1971). **74**, 386)
156. Maier, L. (1971). *Phosphorus*, **1**, 105

157. Carter, R. P., Carroll, R. L. and Irani, R. R. (1967). *Inorg. Chem.*, **6**, 939
158. Maier, L. (1971). *Phosphorus*, **1**, 67
159. Motekaitis, R. J., Murase, I. and Martell, A. E. (1971). *Inorg. Nucl. Chem. Letters*, **7**, 1103
160. Nicholson, D. A. and Vaughn, H. (1971). *J. Org. Chem.*, **36**, 3843
161. Blaser, B., Worms, K. H., Germscheid, H. G. and Wollmann, K. (1971). *Z. Anorg. Allgem. Chem.*, **381**, 247
162. Nicholson, D. A. and Vaughn, H. (1971). *J. Org. Chem.*, **36**, 1835
163. Nicholson, D. A. (1970). *Ger. Offen.*, 1 948 475. (*Chem. Abstr.* (1970). **93**, 38)
164. Nicholson, D. A. and Campbell, D. (1971). *U.S. Pat.*, 3 579 570. (*Chem. Abstr.* (1971). **75**, 540
165. Bloom, S. M., Buckler, S. A., Lambert, R. F. and Merry, E. V. (1970). *Chem. Commun.*, 870
166. Kren, R. M. and Sisler, H. H. (1970). *Inorg. Chem.*, **9**, 836
167. Summers, J. C. and Sisler, H. H. (1970). *Inorg. Chem.*, **9**, 862
168. Ramirez, F. and Tsolis, E. A. (1970). *J. Amer. Chem. Soc.*, **92**, 7553
169. Quin, L. D. and Anderson, H. A. (1966). *J. Org. Chem.*, **31**, 1206
170. Seel, F. and Velleman, K. (1971). *Chem. Ber.*, **104**, 2967
171. Yamada, S. and Takeuchi, Y. (1971). *Tetrahedron Letters*, 3595
172. Horner, L., Winkler, H., Rapp, A., Mentrup, A., Hoffmann, H. and Beck, P. (1961). *Tetrahedron Letters*, 161
173. Munro, H. D. and Horner, L. (1970). *Tetrahedron*, **26**, 4621
174. Baechler, R. D. and Mislow, K. (1970). *J. Amer. Chem. Soc.*, **92**, 3090
175. Egan, W., Tang, R., Zon, G. and Mislow, K. (1971). *J. Amer. Chem. Soc.*, **93**, 6205
176. Egan, W. and Mislow, K. (1971). *J. Amer. Chem. Soc.*, **93**, 1805
177. Rauk, A., Andose, J. D., Frick, W. G., Tang, R. and Mislow, K. (1971). *J. Amer. Chem. Soc.*, **93**, 6507
178. Lambert, J. B., Jackson, G. F. III and Mueller, D. C. (1970). *J. Amer. Chem. Soc.*, **92**, 3093
179. McFarlane, H. C. E. and McFarlane, W. (1971). *Chem. Commun.*, 1589
180. Bissey, J. E., Goldwhite, H. and Rowell, D. G. (1970). *Org. Magn. Resonance*, **2**, 81
181. Marty, R., Houalla, D. and Wolf, R. (1970). *Org. Magn. Resonance*, **2**, 141
182. Powell, R. L., Posner, T. and Hall, C. D. (1971). *J. Chem. Soc. B*, 1246
183. Folli, U. and Iarossi, D. (1969). *Boll. Sci. Fac. Chim. Ind. Bologna*, **27**, 223. (*Chem. Abstr.* (1970). **73**, 342)
184. Luckenbach, R. (1971). *Tetrahedron Letters*, 2177
185. Stec, W., Okruszek, A. and Michalski, J. (1971). *Angew. Chem. Int. Ed. Engl.*, **10**, 494
186. Michalski, J., Okruszek, A. and Stec, W. (1970). *Chem. Commun.*, 1495
187. Horner, L. and Luckenbach, R. (1971). *Phosphorus*, **1**, 73
188. Van den Berg, G. R., Platenburg, D. H. J. and Benschop, H. P. (1971). *Chem. Commun.*, 606
189. Wadsworth, W. S., Jr. and Emmons, W. D. (1962). *J. Amer. Chem. Soc.*, **84**, 610
190. Farnham, W. B., Murray, R. K., Jr. and Mislow, K. (1971). *J. Amer. Chem. Soc.*, **93**, 5809
191. Korpiun, O. and Mislow, K. (1967). *J. Amer. Chem. Soc.*, **89**, 9784
192. Farnham, W. B., Murray, R. K., Jr. and Mislow, K. (1971). *Chem. Commun.*, 146
193. Benschop, H. P. and Platenburg, D. H. J. M. (1970). *Chem. Commun.*, 1098
194. Farnham, W. B., Murray, R. K., Jr. and Mislow, K. (1971). *Chem. Commun.*, 605
195. Meppelder, F. H., Benschop, H. P. and Kraay, G. W. (1970). *Chem. Commun.*, 431
196. Donohue, J., Mandel, N., Farnham, W. B., Murray, R. K., Jr., Mislow, K. and Benschop, H. P. (1971). *J. Amer. Chem. Soc.*, **93**, 3792
197. Nudelman, A. and Cram, D. J. (1971). *J. Org. Chem.*, **36**, 335
198. Reiff, L. P. and Aaron, H. S. (1970). *J. Amer. Chem. Soc.*, **92**, 5275
199. Szafraniec, L. J., Reiff, L. P. and Aaron, H. S. (1970). *J. Amer. Chem. Soc.*, **92**, 6391
200. Reiff, L. P., Szafraniec, L. J. and Aaron, H. S. (1971). *Chem. Commun.*, 366
201. Omelańczuk, J. and Mikolajczyk, M. (1971). *Tetrahedron*, **27**, 5587
202. Herriott, A. W. (1971). *J. Amer. Chem. Soc.*, **93**, 3304
203. Farnham, W. B., Lewis, R. A., Murray, R. K., Jr. and Mislow, K. (1971). *J. Amer. Chem. Soc.*, **92**, 5808
204. Emmick, T. L. and Letsinger, R. L. (1968). *J. Amer. Chem. Soc.*, **90**, 3459
205. De'ath, N. J., Ellis, K., Smith, D. J. H. and Trippett, S. (1971). *Chem. Commun.*, 714
206. Turnblom, E. W. and Katz, T. J. (1971). *J. Amer. Chem. Soc.*, **93**, 4066

207. Benschop, H. P. and Van den Berg, G. R. (1970). *Chem. Commun.*, 1431
208. Epstein, J., Cannon, P. L., Jr. and Sowa, J. R. (1970). *J. Amer. Chem. Soc.*, **92,** 7390
209. Weinberger, M. A., Greenhalgh, R. and Lutley, P. M. (1970). *Can. J. Chem.*, **48,** 1358
210. Epstein, J. (1970). *Science,* **170,** 1396
211. Blewett, F. and Watts, P. (1971). *J. Chem. Soc. B,* 881
212. Behrman, E. J., Biallas, M. J., Brass, H. J., Edwards, J. O. and Isaks, M. (1970). *J. Org. Chem.,* **35,** 3063
213. Behrman, E. J., Biallas, M. J., Brass, H. J., Edwards, J. O. and Isaks, M. (1970). *J. Org. Chem.,* **35,** 3069
214. Brass, H. J., Edwards, J. O. and Biallas, M. J. (1970). *J. Amer. Chem. Soc.,* **92,** 4675
215. Cadogan, J. I. G. and Eastlick, D. T. (1970). *Chem. Commun.,* 1546
216. Haake, P. and Ossip, P. S. (1971). *J. Amer. Chem. Soc.,* **93,** 6924
217. Fujii, A. and Miller, S. I. (1971). *J. Amer. Chem. Soc.,* **93,** 3694
218. Simpson, P. and Burt, D. W. (1970). *Tetrahedron Letters,* 4799
219. Arcoria, A. and Fisichella. (1971). *Tetrahedron Letters,* 3347
220. Ciganek, E. (1970). *J. Org. Chem.,* **35,** 1725
221. Nesterov, L. V., Kessel, A. Y., Samitov, Y. Y. and Musina, A. A. (1970). *Zh. Obshch. Khim.,* **40,** 1237
222. Warren, S. and Williams, M. R. (1971). *J. Chem. Soc. B,* 618
223. Olah, G. and McFarland, C. W. (1971). *J. Org. Chem.,* **36,** 1374
224. Gray, G. A. (1971). *J. Amer. Chem. Soc.,* **93,** 2132
225. Corfield, J. R. and Trippett, S. (1971). *Chem. Commun.,* 721
226. Ward, T. W., Allcox, I. L. and Wahl, G. H., Jr. (1971). *Tetrahedron Letters,* 4421
227. Cuddy, B. D., Treon, K. and Walker, B. J. (1971). *Tetrahedron Letters,* 4433
228. Sanders, J. K. M. and Williams, D. H. (1971). *Tetrahedron Letters,* 2813
229. Calderazzo, F., Losi, S. A. and Susz, B. P. (1971). *Helv. Chim. Acta,* **54,** 1156
230. Grim, S. O., McFarlane, W. and Davidoff, E. F. (1967). *J. Org. Chem.,* **32,** 781
231. Friedemann, R., Grundler, W. and Issleib, K. (1970). *Tetrahedron,* **26,** 2861
232. Glonek, T., Henderson, T. O., Hilderbrand, T. L. and Myers, T. C. (1970). *Science,* **169,** 192
233. Quin, L. D. (1967). *Topics Phosphorus Chem.,* **4,** 23
234. Kittredge, J. S. and Roberts, E. (1969). *Science,* **164,** 37
235. Alhadeff, J. A. and Daves, G. D., Jr. (1970). *Biochemistry,* **9,** 4866
236. Karlsson, K. A. (1970). *Biochem. Biophys. Res. Commun.,* **39,** 847
237. Barychi, J., Mastalerz, P. and Soroka, M. (1970). *Tetrahedron Letters,* 3147
238. Baer, E. and Pavanaram, S. K. (1970). *Can. J. Biochem.,* **48,** 979
239. Rosenthal, A. F. and Chodsky, S. V. (1971). *Biochim. Biophys. Acta,* **239,** 248
240. Hendlin, D., Stapley, E. O., Jackson, M., Wallick, H., Miller, A. K., Wolf, F. J., Miller, T. W., Chaiet, L., Kahan, F. M., Foltz, E. L., Woodruff, H. B., Mata, J. M., Hernandez, S. and Mochales, S. (1969). *Science,* **166,** 122
241. Christensen, B. G., Leanza, W. J., Beattie, T. R., Patchett, A. A., Arison, B. H., Ormond, R. E., Kuehl, F. A., Jr., Albers-Schonberg, G. and Jardetzky, O. (1969). *Science,* **166,** 123
242. Glamowski, E. J., Gal, G., Purick, R., Davidson, A. J. and Sletzinger, M. (1970). *J. Org. Chem.,* **35,** 3510
243. Girotra, N. N. and Wendler, N. L. (1969). *Tetrahedron Letters,* 4647
244. Paulsen, H., Grave, W. and Kuhne, H. (1971). *Tetrahedron Letters,* 2109
245. Paulsen, H., Bartsch, W. and Thiem, J. (1971). *Chem. Ber.,* **104,** 2545
246. Hall, L. D. and Steiner, P. R. (1971). *Chem. Commun.,* 84
247. Albrecht, H. P., Jones, G. H. and Moffatt, J. G. (1970). *J. Amer. Chem. Soc.,* **92,** 5511
248. Jones, G. H., Albrecht, H. P., Damodaran, N. P. and Moffatt, J. G. (1970). *J. Amer. Chem. Soc.,* **92,** 5510
249. Trowbridge, D. B. and Kenyon, G. L. (1970). *J. Amer. Chem. Soc.,* **92,** 2181
250. Issleib, K. and Malotki, P. V. (1970). *J. Prakt. Chem.,* **312,** 366

6
Sulphur Compounds

D. R. HOGG
University of Aberdeen

6.1 COMPOUNDS OF DIVALENT SULPHUR

6.1.1 Thiols

6.1.1.1 Preparation

Metal–ammonia systems continue to be used for the fission of C—S bonds. The ease of bond fission decreases in the order S-alkyl > S-vinyl > S-aryl, and thus $\alpha\beta$-unsaturated sulphides give the unsaturated thiol[1] with sodium or lithium in liquid ammonia. 1,3-Oxathiolanes[2], 1,3-oxathianes[2], 1,3-dithiolanes[3], 1,3-dithianes[3] and 3-benzylthio-2-hydroxypropanoic acid[4] have been similarly reduced by using calcium or sodium in liquid ammonia. The reaction is exploited in a recent method for the α-alkylation of ketones[5].

6.1.1.2 Nucleophilic substitutions

Thiols react[6] with o-carboxyphenyl o-carboxybenzenethiolsulphonate to give alkyl o-carboxyphenyl disulphides, which resist disproportionation except in alkaline solution. The thiols can be recovered by sodium boro-hydride reduction. This procedure is promising for latentiating pharmaco-logically active thiols. Thiol groups in proteins are selectively cyanylated with 2-nitro-5-thiocyanobenzoic acid, and the procedure is recommended[7] for blocking such groups and for radioactive labelling. The hydrolysis of N-benzyl-3-cyanopyridinium bromide is subject to nucleophilic catalysis by 2-mercaptoethanol[8], which is believed to add to the cyano-group to give a thioimidate. This reaction is of biochemical interest as various nitrilases are inhibited by reagents which bind thiol groups.

Methylthiolate ion reacts[9] at the ring carbon atoms of cyclopropane-carbaldehyde to give a complex mixture of products including the aldehydo-sulphides (1,2). In contrast n-butylthiolate ion gives methyl n-butyl sulphide with methyl cyclopropanecarboxylate. A cyclohexyne intermediate (3) is postulated[10] for the reaction of substituted 1-chlorocyclohexenes with thiols in the presence of a strong base, and normal and rearranged enethiol ethers are formed. 1-Halogenoalkynes undergo nucleophilic substitution at the halogen to give[11] the alkyne and a disulphide. Alkylthiolate ions similarly reduce[12] α-halogeno-βγ-acetylenic esters to the corresponding allenes and α-bromo-β-(alkylthio)crotonic esters to the β-(alkylthio)crotonic esters and a disulphide. An episulphonium ion intermediate (4) is postulated for the latter

$$MeS(CH_2)_2 \cdot CH{:}C(CHO) \cdot (CH_2)_2SMe$$

(1)

$$\begin{array}{c} CH_2 \\ | \\ CH \cdot CH{:}C(CHO) \cdot (CH_2)_2SMe \\ / \\ CH_2 \end{array}$$

(2)

(3)

(4)

reactions. Substitution reactions with 2-chloroadenosine[14], dichloroaceto-nitrile[15], and diaryl sulphinyl sulphones[16] are reported. In the last case, reaction occurred at the 'softer' sulphinyl centre, and as expected, the reaction was not acid-catalysed. The preparation and synthetic uses of alkynethiolates, $RC{:}CS^-$, have been reviewed[13].

6.1.1.3 Nucleophilic additions

α-Methylene groups conjugated with lactone carbonyl groups can[17] be protected by the addition of propane-1-thiol at pH 9.7. The group is removed by methylation and trituration of the salt so formed with sodium bicarbonate solution. Addition of thiols to $\Delta^{\alpha\beta}$-butenolides has been studied[18], as a model for similar additions which may play a key role in growth regulating pheno-mena. The rates of addition of thiols to αβ-unsaturated aldehydes have been measured[19] and comparisons drawn[20] with processes involving protein SH groups. Additions to ethyl ethynyl sulphone in the presence of Triton B under kinetic control ($-10\,°C$) gave[21] the cis-product by the rule of trans-nucleophilic addition, but additions to acetylenic ketones in protic and in aprotic solvents showed[22] substantial deviations from the rule. Such deviations have previously been observed only for aprotic solvents. Iron pentacarbonyl and hexacarbonyldi(ethylthio)di-iron, $[Fe(CO)_3SEt]_2$, catalyse[23] the addition of thiols to acrylic compounds and inhibit addition to alkenes and alkynes, presumably by radical trapping. The true catalyst, it is suggested, is an iron(II) dithiolate, $Fe(SR)_2$. Substituted mercaptoacetates add to 1-cyanoalka-1,2-dienes to give[24] 4-hydroxythiophen derivatives (5).

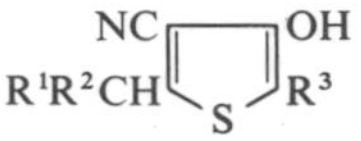

$$\text{R}^1\text{R}^2\text{CH}\overset{\displaystyle \text{NC} \quad \text{OH}}{\underset{\displaystyle \text{S}}{\quad}}\text{R}^3$$

(5)

6.1.1.4 Oxidation

Manganese(III) tris(acetylacetonate) quantitatively oxidises[25] thiols to disulphides in the absence of oxygen. The oxidation of 2-mercaptoethanol with o-iodosobenzoic acid[26] and of thiomalic acid with alkaline ferricyanide[27] have been studied kinetically and the variation of rate with pH explored. In addition to reducing unsaturated halides[11, 12], thiols also reduce α-halogenoketones and -esters[28] to the corresponding ketones or esters. The initially formed α-alkylthio-ester or -ketone, it is suggested, reacts with a thiolate ion to give the disulphide and a carbanion which can be methylated. The reaction of aluminium thiolates with carbonyl compounds does not resemble the Oppenauer oxidation. Methyl ketones undergo the aldol condensation but aldehydes and other ketones give[29] a mixture of the thioacetal and the enethiol ether. The intermediacy of sulphenyl iodides in the oxidation of thiols by iodine is further supported by studies of equilibrium in acetic acid[30], and by the formation[31] of disulphides and alkylthiopyrroles when the oxidation occurs in the presence of pyrroles, which compete with the thiol for the sulphenyl iodide. 2-Carboxypropane-2-sulphenyl iodide has been trapped[32] with p-chlorothiophenol as the unsymmetrical disulphide. A β- or a γ-carboxy group in the thiol assists[32] in the hydrolysis of the sulphenyl iodide to the sulphenic acid, which is then oxidised by iodine. Such thiols consequently require more than equimolar amounts of iodine for complete oxidation.

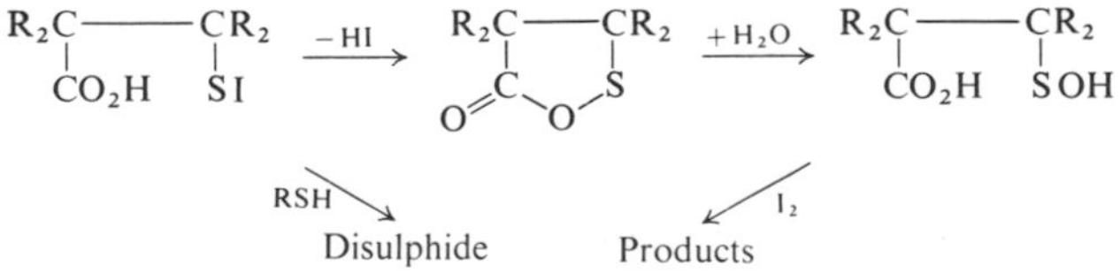

(6.1)

6.1.1.5 Unsaturated thiols

Ethenethiol readily polymerises[1], but the higher vinyl thiols tend to isomerise to the thione. Allyl thiols are also relatively unstable and polymerise to polysulphide thiols[33]. γ,γ-Dimethylallylthiol is photodimerised[33] to the *cis*- and *trans*-1,4-dithian (6), and the intermediate thiol (7) was also isolated.

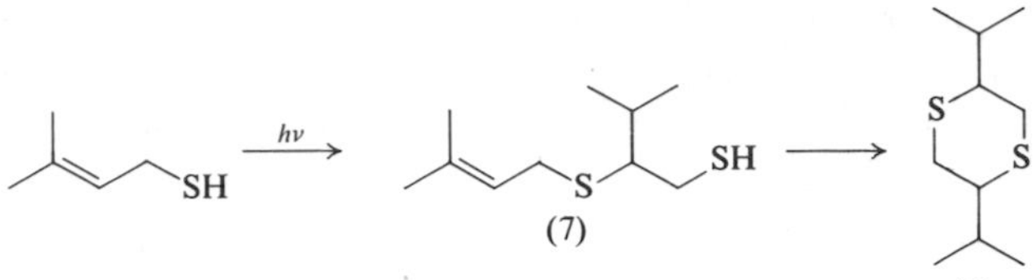

$$(6.2)$$

6.1.1.6 Dithiols

Unlike gem-diols, gem-dithiols are relatively stable with respect to the thione, thus cyclohexane-1,1-dithiol in acidified acetone[34] gives bis-(1-mercapto-cyclohexyl) sulphide, the dimer intermediate towards the cyclic trimer. gem-Dithiols form cyclic compounds with reactants having two electrophilic centres. Cyclohexane-1,1-dithiol and 1,2-dihalides give[34] the spiro-trithianes (8) with capture of a sulphur atom. Dipotassium 2-nitroethene-1,1-dithiolate and α-halogeno-aldehydes or -ketones give[35] substituted 3-nitrothiophen-2-thiols on acidification. Disodium 2,2-dicyanoethene-1,1-dithiolate gives[36] substituted benzo-1,3-dithioles (9) with picryl chloride or 1-chloro-2,4-dinitrobenzene with displacement of a nitro-group, and the diammonium salt on treatment with hydrogen sulphide followed by alkylation gives the mercaptothioamide, which reacts with[37] carbonyl compounds to give a 2,2-disubstituted 6-alkylthio-5-cyano-2,3-dihydro-4H-1,3-thiazine-4-thione (10), a new type of compound.

Ethane-1,2-dithiol with oxalyl chloride gives[38] the interesting dispirotri-cyclic compound (11) in which all twelve hydrogens are coincidentally magnetically equivalent. Furan acts as a diene with ethane-1,2-dithiol in the presence of BF_3 etherate giving[39] products resulting from 1,2- and 1,4-mono-addition. Benzalacetophenone forms[40] the 2:1 adduct. *cis*-Ethene-1,2-dithiol condenses with dichlorotrisulphane giving[41] pentathiepin (12) and with *cis*-3,4-dichlorocyclobutene to give[42] precursors (13) of the 10-electron 1,4-dithiocins, which were not aromatic.

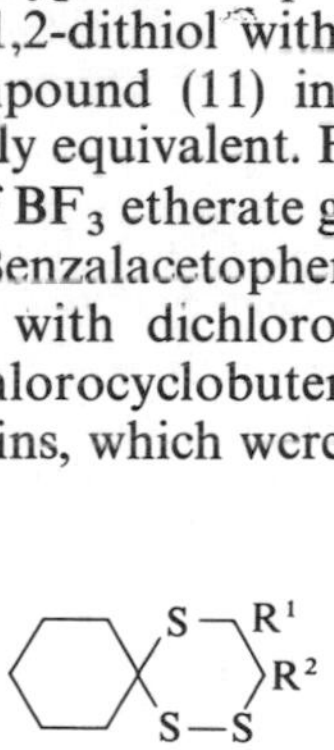

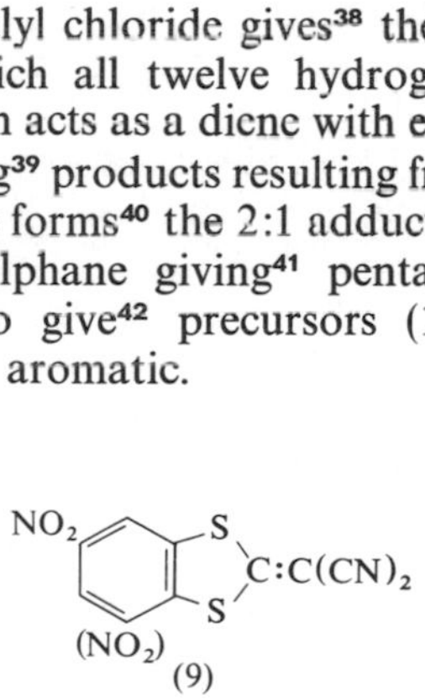

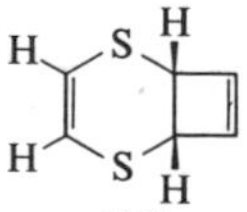

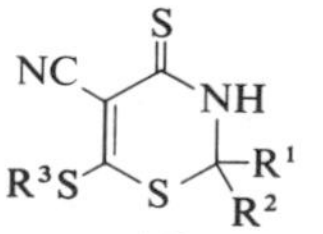

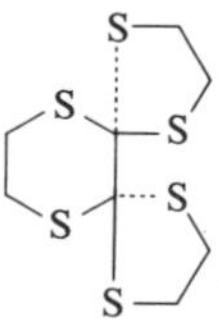

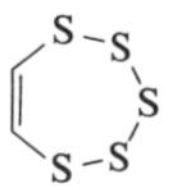

Trimethylene dithiotosylates react[43] with enamines, hydromethylene derivatives and active methylene groups to form acid-stable 1,3-dithianes of great synthetic utility. They are hydrolysed[44] to carbonyl compounds or reduced to the methylene compound[43] and serve as methylene blocking groups[43]. The 1,3-dithiotosylate is used in the interchange of carbonyl and methylene groups[45a] and in the degradation of a ketone to an aldehyde and an ester[45b].

6.1.1.7 Hydrodisulphides

Hydrodisulphides have been postulated as intermediates in enzyme reactions but only arylalkyl and alkyl derivatives are known. 2-Substituted ethyl hydrodisulphides may have been obtained[46] by the reaction of a thiol with acetylsulphenyl chloride, followed by alcoholysis of the product.

e.g. $MeCO_2 \cdot (CH_2)_2 SH \xrightarrow{MeCO \cdot SCl} MeCO_2 \cdot (CH_2)_2 S \cdot S \cdot COMe \xrightarrow{ROH-HCl}$

$$HO(CH_2)_2 S \cdot SH$$

The product was unstable at room temperature and unexpectedly insoluble in water. Basic amines react[47] with arylalkyl hydrodisulphides to give hydrogen sulphide (25 %), sulphur, disulphides, and polysulphides.

6.1.2 Derivatives of sulphenic acids

6.1.2.1 Sulphenyl chlorides

Dialkylaminosulphenyl chlorides, R_2NSCl, are prepared[48] from the corresponding aminosulphides $[(R_2N)_2S]$ and sulphur dichloride. They can be distilled under reduced pressure but decompose on being kept at room temperature. Their stability is dramatically increased by bulky alkyl substituents. α,α-Dichlorosulphenyl chlorides are obtained[49] by the reaction of sulphuryl chloride with S-benzyl sulphides, $PhCH_2 \cdot S \cdot CH_2X$, where X is an electron-withdrawing group.

Additions of sulphenyl compounds, mainly halides, to unsaturated systems have been reviewed[50]. Sulphenyl halides give a high ratio of the rate coefficient for exo-addition to norbornene to that for exo-addition to 7,7-dimethylnorbornene, as do other reagents which form a cyclic transition state or intermediate[51]. Additions to but-3-en-1-ynyltrimethylsilane[52] at the triple bond, and to ethyl but-2-ynoate[53] are reported. Treatment of ketones with thionyl chloride and pyridine gives[54] thietan-3-ones (14) by formation of the α-chlorosulphenyl chloride which reacts as the enol and subsequently loses more hydrogen chloride.

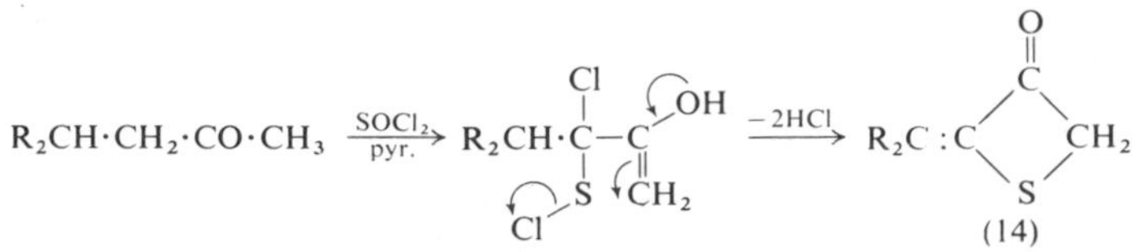

(6.3)

n-Alkanesulphenyl chlorides give[55] the relatively inaccessible 5-alkyl-thiopyrimidines with suitable pyrimidines, presumably by electrophilic substitution; s-alkanesulphenyl chlorides act[55] as chlorinating agents. Acetylsulphenyl chloride unexpectedly gives[56] disulphides with activated aromatic compounds. The expected monosulphide was shown not to be an intermediate.

The alkaline hydrolysis of trichloromethanesulphenyl chloride[57] gives 82% of trichloromethyl dichloromethanethiolsulphonate. It is suggested that this arises by loss of chloride ion from the trichloromethanesulphinate anion to give the sulphene, $Cl_2C{:}SO_2$, and subsequent addition of the anion to it. The exchange reaction between sulphenyl halides and disulphides[58] and the reaction with sulphides[59] may involve nucleophilic attack by the sulphur on the sulphenyl halide to give a sulphonium salt. In the latter case[59] the intermediate can undergo substitution, elimination or ylid formation.

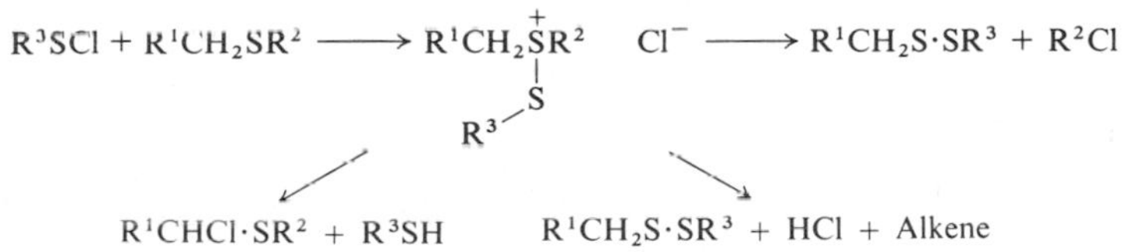

$$R^3SCl + R^1CH_2SR^2 \longrightarrow R^1CH_2\overset{+}{\underset{\underset{R^3}{\overset{|}{S}}}{S}}R^2 \quad Cl^- \longrightarrow R^1CH_2S{\cdot}SR^3 + R^2Cl$$

$$R^1CHCl{\cdot}SR^2 + R^3SH \qquad R^1CH_2S{\cdot}SR^3 + HCl + Alkene$$

$$(6.4)$$

6.1.2.2 Sulphenyl iodides

Although sulphenyl iodides have been little studied they are of considerable biological importance. t-Butylsulphenyl iodide is obtained[60] by iodination, in an inert medium, of the thiol, the silver thiolate, or the sulphenamide. It is stable at $-12\,°C$ for several days but is about 50% decomposed at $20\,°C$ after 3 days. Photolysis gives the disulphide and iodine. Sulphenyl iodides are implicated in thio oxidation (see p. 262) and in the formation[61] of benzo[b]-thiophens from β-aryl-α-mercaptoacrylic acids with iodine in an inert medium.

6.1.2.3 Sulphenates

N-Alkylthiophthalimides conveniently give[62] sulphenates with alkoxide ions in an alcohol. Methanesulphenyl chloride and lithium methoxide do not give[63] the ester, but give dimethoxymethane (via thioformaldehyde) and dimethyl disulphide. Alkyl halides are formed[64] when methanesulphenyl chloride reacts with alcohols, alkyl methanesulphinates or O-alkyl-S-methyl xanthates. Similar products occur in the chlorinolysis of alkyl methane-sulphinates and the reaction of methylsulphur trichloride with alcohols. An intermediate containing the MeS·O·C grouping appears to be involved and forms the alkyl halide by nucleophilic substitution. Tri-n-butylphosphine quantitatively abstracts sulphur from alkyl t-butylsulphenates to give[62] the ethers. The rearrangement of cis- and trans-but-2-enyl sulphenates to the sulphoxides involves[65a] a [2,3]-sigmatropic mechanism, but the rearrangement of benzyl and similar sulphenates may proceed[65b] by a radical-pair mechanism.

6.1.2.4 Sulphenamides

Methanesulphenyl chloride reacts[63] with secondary amines to give a low yield of the sulphenamide, together with the gem-diamine, and dimethyl di- and tri-sulphide. These reactions, and that with methoxide ion (see p. 265) may involve an E2 elimination giving thioformaldehyde, which reacts with an amine or methoxide ion to give the diamine or diether and hydrogen sulphide, which reacts further to give the di- and tri-sulphide. Such complications are avoided by preparing[66] sulphenamides from N-alkylthiophthalimides and primary or secondary amines, or by transamination[67]. Sulphenamides are also obtained[68] in good yield by the reaction of disulphides with amines in the presence of silver ion. This is an example of nucleophile–electrophile co-operation[69] in S—S bond fission. Cuprous salts catalyse the reaction of azides with thiols to give sulphenamides[70], presumably by nitrene insertion into the S—H or S—catalyst bond. Amine sulphides give[71] sulphenamides by successive alkylation and treatment with n-butyl-lithium. The preparation and reactions of sulphenamides have been reviewed[72].

The stability[73] of sulphenamides, $R^1S \cdot NR^2R^3$, decreases with increasing basicity of NHR^2R^3 and increases with increasing electron-withdrawal by R^1. In general, electrophilic attack occurs at nitrogen and this is followed by nucleophilic attack at sulphur with S—N bond fission, sometimes accom-panied by elimination if the reaction is intramolecular. Thus, carbon disul-phide undergoes[74] insertion (15) into the S—N bond to give aminocarbo-trithioates, and dimethylphenacylsulphonium bromide reacts with NN-diethylbenzenesulphenamide to give[75] ω-phenylthio-ω-(dimethylsulphur-anylidene)acetophenone (16) and the amine salt. Nucleophiles, e.g. acetyl-acetone[75], react at sulphur with S—N bond fission. N-Alkylthiophthalimide is desulphurised[76] by tris-(dimethylamino)phosphine, presumably by an initial nucleophilic displacement at sulphur to give the alkylphthalimide. This reaction complements the Gabriel method as secondary alkyl groups can be used and hence secondary alkylamines produced by hydrolysis. Trichloromethanesulphenyl chloride reacts[77] with 2-aminopyridines to form the new heterocyclic ring systems, 3-(2-pyridylimino)-3H-[1,2,4]thiadiazolo-[4,3,a]pyridines (17), presumably by formation of the sulphenamide and dis-placement of chloride ion.

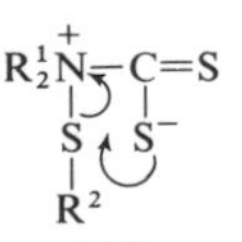

(15) 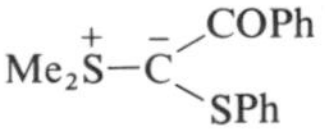   (16) 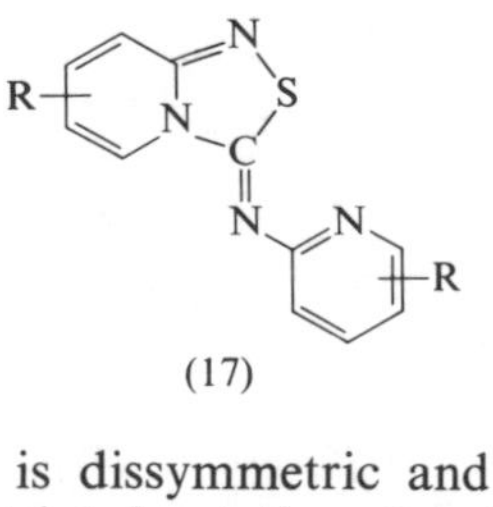   (17)

The sulphenamide groups, $R^1S \cdot NR^2R^3$ is dissymmetric and the axial chirality of this grouping derives[78] from restricted rotation about the N—S bond. The diastereoisomers (R,R)- and (R,S)-N-(1-α-naphthylethyl)-N-benzenesulphonyl trichloromethanesulphenamide have been prepared and their crystal structure[79], configuration, and spectra determined. The nitrogen has a nearly planar configuration, which is attributed to p_π–d_π bonding with the relatively electron-deficient sulphur.

6.1.2.5 Other derivatives

A new type of compound, an alkenyl thiosulphoxylate, $R^1 S \cdot S \cdot OR^2$, it is claimed[80], is formed by the thermal decomposition of *trans*-but-2-ene episulphoxide. The allylic sulphenic acid (18) initially formed, yields an equilibrium mixture of a thiosulphinate and the thiosulphoxylate (19). These isomers are interconverted by a [2,3]-sigmatropic process. The equilibrium mixture can be formed by peroxyacid oxidation of the disulphide. Sulphenic acid formation occurs only when the stereochemistry is favourable.

$$(6.5)$$

Sulphenyl thiocarbonates react[81] with thiosulphuric acid at $-40\,^\circ$C to give sulphenyl thiosulphates, $RS \cdot S \cdot SO_3^-$, a new type of sulphenyl derivative.

6.1.3 Sulphides

6.1.3.1 Preparation

The reaction of alkyl thiolsulphonates and the trichloromethide anion in 1,2-dimethoxyethane is the only general method[82] for the preparation of alkyl trichloromethyl sulphides. The phosphonate alkene synthesis has been adapted[83] to the preparation of methyl vinyl sulphides and hence, by hydrolysis with mercuric chloride in aqueous acetonitrile, to the preparation of unsymmetrical ketones. The relatively inaccessible sulphides, $(RS)_2C{:}CH \cdot Me$, are formed[84] by the elimination of an alkanethiol from 2-alkylthiopropionaldehyde mercaptals in basic solution. Cumulenic thioethers, $CH_2{:}C{:}C{:}CH \cdot SR$, most of which are pyrophoric and decompose explosively at room temperature, are obtained[85] together with the enyne thioether by treating suitable acetylenic bis-thioethers successively with n-butyl-lithium and water. The preparation and properties of tetra-alkylthioethylenes, $(RS)_2C{:}C(SR)_2$ have been described[86]. Chemically they are much less reactive than 'electron-rich alkenes' and should not be included in this category; furthermore electrophiles may react at sulphur rather than at carbon.

6.1.3.2 Halogenation

The kinetic isotope effect[87] and the lack of deuterium incorporation[49, 87] when dibenzyl sulphide is chlorinated in the presence of deuterium chloride, indicate that the slow step involves an E2 elimination from an initially formed halogenosulphonium salt.

$$R^2CH_2SR^1 + Cl_2 \; \rightleftharpoons \; R^2CH_2\overset{Cl}{\underset{|}{\overset{+}{S}}}R^1 \; Cl^- \; \xrightarrow{\; E2 \;} \; R^2CH\overset{Cl^-}{\overset{+}{\cdots}}SR^1 \; + \; HCl$$

$$(20) \hspace{12em} \Big\downarrow \hspace{6em} (6.6)$$

$$R^2CHCl\cdot SR^1$$

Chlorination[88] with N-chlorosuccinimide in methylene chloride at 0 °C gave the 1:1 adduct analogous to (20). A similar intermediate is postulated[89] for chlorination with iodobenzene dichloride. With bromine and dibenzyl sulphide at low temperatures a crystalline adduct was isolated[87]. On being warmed it gave benzyl α-bromobenzyl sulphide, benzyl bromide, benzyl-sulphenyl bromide and dibenzyl disulphide. These products and the isotope effects are consistent with a rate-limiting reaction between bromide ion and a single intermediate analogous to (20), which can undergo substitution at carbon or elimination. The reaction of bromine with methyl phenyl sulphide in aqueous methanol gives the corresponding sulphoxide. The rate of oxidation is sensitive[90] to polar and steric factors, and is retarded by bromide ion. Acetate ion, which is a stronger nucleophile than the solvent, increases the rate. The rate-limiting step is considered to be solvolysis of the bromo-sulphonium ion analogous to (20).

6.1.3.3 Oxidation and reduction

Sulphides are oxidised[91] in high yields to sulphoxides, uncontaminated by sulphones, with 2,4,4,6-tetrabromocyclohexadienone in aqueous dioxan, or by organic hydroperoxides in the absence of alkali[92]. 2-Alkylthiophenyl aryl sulphoxides in concentrated sulphuric acid undergo[93] internal oxidation–reduction to give 2-alkylsulphinyl phenyl aryl sulphides, with oxygen transfer usually from lower to higher sulphoxides. The reaction may involve hydration of the dipositive intermediate formed by loss of water from the protonated sulphoxide. The reductive fission of sulphides with lithium aluminium hydride–cupric chloride[94] and of 1-ethylcyclo-alkyl ethyl sulphides with various reagents[95], of which Raney nickel or sodium and ethanol gave the highest yield, have been described.

6.1.3.4 Miscellaneous reactions

Chloromethyl methyl sulphide and antimony pentachloride react with potassium t-butoxide to give[96] the 3,5,5-trimethyl-1,3-oxathiolanium cation, (21), rather than a product derived from the carbene. Carbethoxycarbene, from ethyl diazoacetate, gives[97] alkylmercaptoacetates and alkenes with dialkyl sulphides. Methyl esters exchange alkyl groups with sulphides on being heated[98], by formation of the sulphonium carboxylate, $R^1R^2R^3\overset{+}{S}\overline{O}, CR^4$.

6.1.3.5 Unsaturated sulphides

Optically active 1-phenylethyl vinyl sulphide is obtained[99] without racemisation by addition of the corresponding thiolate ion to acetylene. Electron diffraction studies show[100] that methyl vinyl sulphide, unlike the ether, occurs only in the *syn*-form. Diallyl and allyl benzyl sulphide, on lithiation[101], undergo a [2,3]-sigmatropic rearrangement,

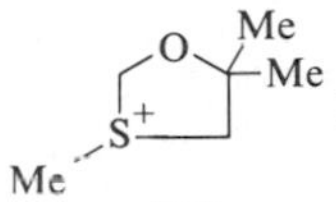

$$\text{ICH}_2\text{S·CH}_2\text{·CH:CH}_2 \xrightarrow[\text{(ii) MeI}]{\text{(i) Bu}^n\text{Li}(-30\,^\circ\text{C})} \text{PhCH(SMe)·CH}_2\text{·CH:CH}_2 \quad (6.7)$$

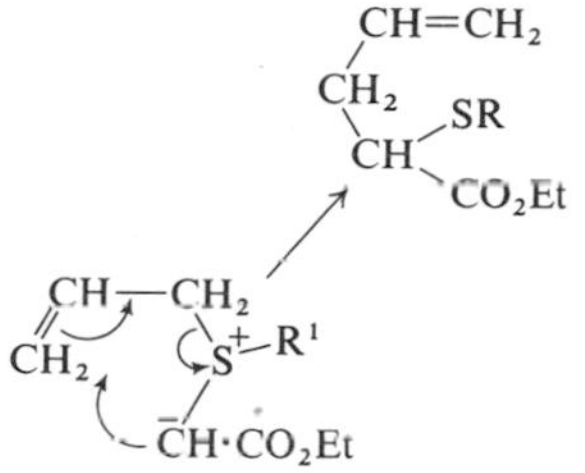

and the addition[102] of allylthiol to chloro- and to diethyl amino-propiolonitrile occurs with a thio-Claisen rearrangement to give the thiono-acid derivative, e.g. (22). Carbethoxycarbene undergoes insertion[97] into the double bond and into the C–S bond of allyl sulphides, in the latter case with rearrangement (23). Synthetic uses of unsaturated sulphides have been discussed[103].

6.1.3.6 β-Ketosulphides

Sulphenyl chlorides react with the readily available 1,3,2-dioxaphospholenes (24) to give[104] α-chloro-β-ketosulphides which react with potassium thiocyanate to give the substituted 3-oxazoline-2(1*H*)-2-thiones with phenyl migration, and on reduction yield the synthetically important β-ketosulphides equation (6.8).

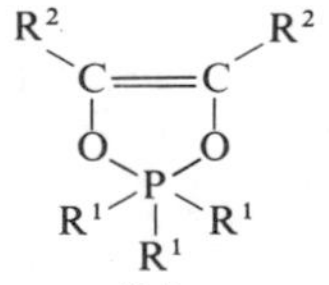

$$R^2\text{CO·CR}^2\text{Cl·SR}^3 \xrightarrow[\text{(ii) NaOH}]{\text{(i) HSiCl}_3\text{–Bu}^n_3\text{N}} R^2\text{CO·CHR}^2\text{·SR}^3 \quad (6.8)$$

The synthetic uses of β-ketosulphides and related compounds, including sulphoxides and sulphones, have been summarised[105]. Further reactions giving α-diketones[106], have been described. The photolysis of β-ketosulphides having γ-hydrogens proceeds[107] by a type II photo-elimination giving the methyl ketone and a sulphur-containing polymer.

6.1.4 Disulphides

6.1.4.1 Preparation

Unsymmetrical disulphides are obtained in good yields by the reaction of thiols with N-alkylthiophthalimides[108] or with sulphenyl thiocarbonates[81]. Alkylthioboranes react[109] with methyl benzenesulphenate to give alkyl phenyl sulphides. Sulphonyl[110] and acyl[111] disulphides are prepared from sulphenyl halides and pyridinium alkanethiolsulphonates or thioacids respectively. Alternatively the acylsulphenyl halide and thiol may[112] be used in the latter preparation. Dialkyl disulphides can be synthesised from alkenes[113] by hydroboration, reaction of the trialkylborane with sulphur, and alkaline hydrolysis. Sulphonyl, sulphinyl, and sulphenyl derivatives are reduced[114] to symmetrical disulphides with trichlorosilane in the presence of a tertiary amine. Thiols react[115] with chloroamines to give disulphides in good yields. The sulphenamide, it is suggested, is the initial product and forms a disulphide with the thiol. Hydrogen sulphide reacts with primary and secondary aminoalkanethiosulphuric acids, $RNH(CH_2)_n \cdot S \cdot SO_3H$ (R = H or alkyl), to give[116] the bis-aminoalkyl disulphides as salts.

6.1.4.2 Sulphur–sulphur bond fission

Interest in S—S bond fission has been stimulated by the possibility that radiation damage to biologically important compounds might occur in this manner. Methyl and pivaloyl acetyl disulphide are resistant to disproportionation[111] and have half-lives of over a month at 100 °C or on irradiation. Disproportionation is catalysed by thiolate ion. The products of thermolysis appear to arise from an initial S—S bond heterolysis, whereas photolysis apparently involves homolytic fission. In contrast the half-life for hydrolysis or ethanolysis at 100 °C is only a few hours. Disproportionation of an acetyl 2-substituted-ethyl disulphide in dioxan at 100 °C is anchimerically assisted[112] by amino- and acetamido-groups, and o-carboxy groups similarly assist[117] in the disproportionation of diaryl disulphides in alkaline solution. The sulphide and the disulphide group in 1,2,5-trithiepane react[118] independently.

A new mechanism[119] has been proposed for the alkaline hydrolysis of disulphides having acidic α-hydrogens, e.g. α,α-dithiosuccinic acid and meso-1,2-dithiane-3,6-dicarboxylic acid (25). Direct α-elimination, which hitherto was considered to be the most likely mechanism, is currently the most difficult to authenticate[119b].

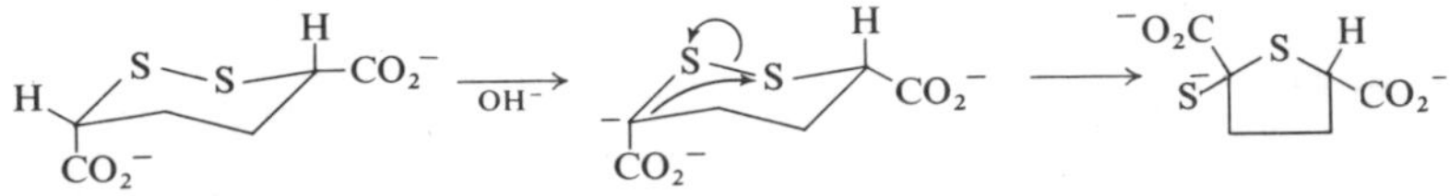

(25)

(6.9)

Tris-(diethylamino)phosphine selectively desulphurises certain cystine[120] derivatives to L-lanthionine compounds, and cyclic disulphides[121] to sulphides. The reaction proceeds with inversion at one of the α-carbon atoms, and kinetic studies[122] are consistent with a nucleophilic displacement at the most negative sulphur atom followed by a rapid S_N2 reaction of the liberated thiolate ion, giving the sulphide and the phosphothioic triamide, $(Et_2N)_3PS$. Disulphide groups in peptides and proteins are completely and rapidly reduced[123] to give the thiols by dithiothreitol in liquid ammonia, without apparent side-reactions.

6.1.4.3 Oxidation

Certain biologically important disulphides are susceptible to sensitised photo-oxidations which may involve singlet oxygen. Accordingly the reactions of dialkyl disulphides with triphenylphosphite ozonide, which can give singlet oxygen, have been studied[124]. Thiolsulphinate was initially the major product, but induced decomposition of the ozonide and direct oxidation cannot be discounted.

6.1.4.4 Thiosulphoxide formation

The ready rearrangement and desulphurisation of α-substituted diallyl disulphides compared with those of alkyl allyl disulphides is attributed[125] to the formation of a branched disulphide intermediate (26), namely a thiosulphoxide. The rearrangement is energetically similar to the allyl sulphoxide–sulphenate rearrangement, which proceeds through a [2,3]-sigmatropic mechanism. The disulphide rearrangement is considered to involve two such processes consecutively.

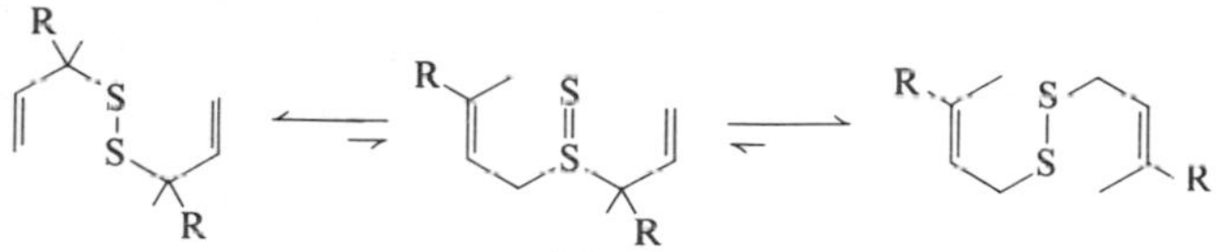

(26)   (6.10)

6.1.5 Trisulphides

Trisulphides are obtained[108] by the action of hydrodisulphides on N-alkyl-thiophthalimides. Desulphurisation with phosphines depends[126] on the reagent. Triphenylphosphine extrudes the central sulphur atom from dibenzyl trisulphide and thiodehydrogliotoxin[127], but tris-(diethylamino)phosphine removes a terminal sulphur. Tri-n-butylphosphine exhibits intermediate behaviour.

The desulphurisation and thermal isomerisation of alkenyl tri- and polysulphides have been studied[128]. The biosynthesis of bis-(3-oxoundecyl)di-, tri-, and tetra-sulphide, isolated[129] from the Hawaiian algae, *Dictyopteris plagiogramma* and *D. australis*, may involve the reaction of sulphur or hydrogen disulphide with the thiol or the disulphide. Hydrogen disulphide does not [130], however, react similarly with cysteine or dibenzyl disulphide, although diaryl and cyclic disulphides undergo insertion into the S—S bond.

6.1.6   Substitution at divalent sulphur

Evidence of an addition–elimination mechanism for nucleophilic substitution at divalent sulphur has been obtained from kinetic studies on triphenylmethylsulphenyl derivatives. The high value of the Brønsted coefficient, β, for amine nucleophiles[131] indicates considerable bond formation in the transition state, while the 'element effect' for sulphenyl halides[132] suggests that bond fission is unimportant. Paradoxically values of β for the corresponding reactions of sulphenate esters[133, 132b] are consistent with an S_N2-type mechanism, showing the doubtful reliability of these mechanistic criteria. The kinetics of the reactions of a dimethylmethylthiosulphonium salt, $[Me_2S^+SMe]BF_4^-$, with dimethyl sulphide and disulphide have been interpreted[134] in terms of an S_N2 fission of the S—S bond, although intermediate formation cannot be excluded. The rate of S—S bond fission is substantially increased by methylation, thiomethylation or protonation at sulphur. Equilibrium is rapidly attained and lies far to the left in the following system.

$$Me_2S^+SMe + MeS \cdot SMe \rightleftharpoons Me_2S + MeS^+(SMe)_2$$

$$\left(\begin{array}{c} PhN\!\!-\!\!N \\ R\,\underset{\underset{S}{\diagdown}}{\overset{\diagup}{C}H}\,\overset{\parallel}{C}\!\!-\!\!S \end{array} \right)_2$$

(27)

This explains the catalysis of disulphide interchange by $Me_2\overset{+}{S}\cdot SMe$. An apparently clear case of unimolecular S—S bond fission has been found[135] for the disulphide (27). Nucleophilic and radical substitution reactions at sulphur have been reviewed[119b, 136].

6.1.7   Thioketones and related compounds

Thione-enethiol tautomerism has been studied[137] for 2,4-dimethylpentane-3-thione and 2,6-dimethylheptane-4-thione. At equilibrium at 40 °C *c.* 45% of thione was observed. Tetramethyl-3-thiocyclobutane-1,3-dione, unlike most aliphatic thioketones, has no tendency to dimerise or polymerise and thus can be used to study the chemistry of thioketones without such complications. It reacts[138] with diazomethane to form a thiadiazoline (28), which loses nitrogen at room temperature to give the episulphide. Bis-(trifluoromethyl)thioketene is sufficiently stable to be stored[139] at room temperature for several months with little decomposition. This compound[139], and methyl cyanodithioformate[140], MeS·CS·CN, like thioketones, readily undergo the Diels–Alder reaction with dienes. A new method for converting thiocarbonyl into carbonyl compounds by heating them with propylene oxide at 170–180 °C in the presence of triethylamine or boron trifluoride etherate has been

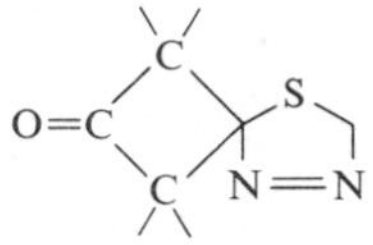

(28)

developed[141]. The reaction gives propylene sulphide and the carbonyl compound, presumably through a cyclic intermediate. Trialkyl phosphites react with α-chlorothioketones to give[142] not only the reported vinyl sulphide, e.g. $(RO)_2P(:O)\cdot SCMe:CH_2$, but also the episulphide.

6.2 COMPOUNDS OF TETRAVALENT SULPHUR

6.2.1 Sulphonium salts

Optically active sulphonium salts have been prepared[143] with partial racemisation, from optically active sulphoxides by o-ethylation with triethyloxonium tetrafluoroborate followed by alkylation with a dialkylcadmium involving displacement of ethoxide ion. Rates of racemisation of methyl cyclic sulphonium salts are[144] several orders of magnitude smaller than those of the acyclic analogues, thus suggesting that racemisation occurs by pyramidal inversion at sulphur. The protons *trans* to the lone pair on sulphur in these 5-membered ring sulphonium salts undergo exchange[145] much faster than the other methylene protons, i.e. re-protonation occurs with retention, as postulated on theoretical grounds[146]. Results from 6-membered ring and acyclic compounds are not[145] in such close agreement with theory. The proportion of Hofmann elimination, and the *cis/trans* ratio, from 2- and 3-pentyldimethylsulphonium bromide increase[147] when n-butoxide is replaced by s-butoxide ion, in keeping with base strength and size, but the trend is reversed with t-butoxide ion. This is tentatively attributed to an α',β-*syn*-elimination as in quaternary ammonium salts. Dimethylprop-2-ynylsulphonium bromide isomerises[148] in ethanol to the allene, which reacts with β-ketoesters, β-ketosulphones or β-diketones in the presence of sodium ethoxide to give furans. Vinylsulphonium salts similarly[148] give cyclopropane derivatives, the dimethyl sulphide being displaced in this case by the carbon of the enolate ion. Allyl sulphonium salts and sulphones are cleaved[149] by potassium cyanide to give methacrylonitrile and the sulphide or sulphinic acid respectively. The reaction is thought to involve double bond migration via the ylid, addition of hydrogen cyanide, and elimination. 2,3-Di-(methoxycarbonyl)prop-2-enyl dimethylsulphonium bromide undergoes self-condensation[150] in a basic medium to give the vinyl sulphide (29). Biphenyl-2,2′-bis-

$$CH_2:C(CO_2Me)\cdot[CH(CO_2Me)]_2\cdot C(CO_2Me):CHSMe$$
(29)

(dialkylsulphonium) salts give[151] dibenzothiophen on being heated, by an unusual internal nucleophilic aromatic substitution.

6.2.2 Sulphoxides and related compounds

6.2.2.1 Stereochemistry

Alkyl aryl sulphoxides without acidic or basic groups have been partially resolved[152] by inclusion into cyclodextrins, and decomposition of the adducts. The photo-induced racemisation of the cyclic sulphoxide (30) has been shown[153] to involve simultaneous pyramidal inversion and sulphenate ester formation. Methyl p-tolyl sulphoxide reacts[154] with N-sulphinyl-toluene-p-sulphonamide, TsN:SO, or with NN'-bis(toluene-p-sulphonyl)-sulphur di-imide, (TsN)$_2$S, in pyridine to give the sulphilimine, TsN:SMe·C$_6$H$_4$Me-p, with inversion, but in benzene the latter reagent gives[155] retention. The reaction with diastereoisomeric cyclic sulphoxides was claimed[156] to give inversion in both solvents although the yields were somewhat low. A linear displacement is precluded by kinetic studies[154] and a cyclic transition state or intermediate is postulated for both inversion and retention, [e.g. the intermediate (31) for retention involving the (+)-R-sulphoxide]. Alkaline hydrolysis of the sulphilimine gives[154] the sulphoxide

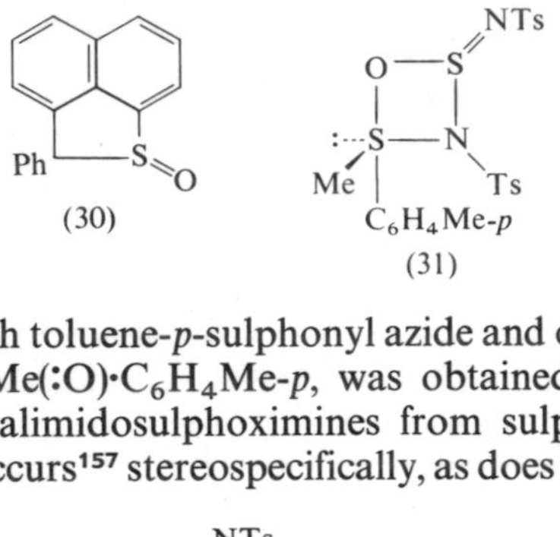

(30)

(31)

with inversion. With toluene-p-sulphonyl azide and copper the sulphoximine derivative, TsN:SMe(:O)·C$_6$H$_4$Me-p, was obtained[154] with retention. The formation of phthalimidosulphoximines from sulphoxides and N-amino-phthalimide also occurs[157] stereospecifically, as does the reverse reaction with

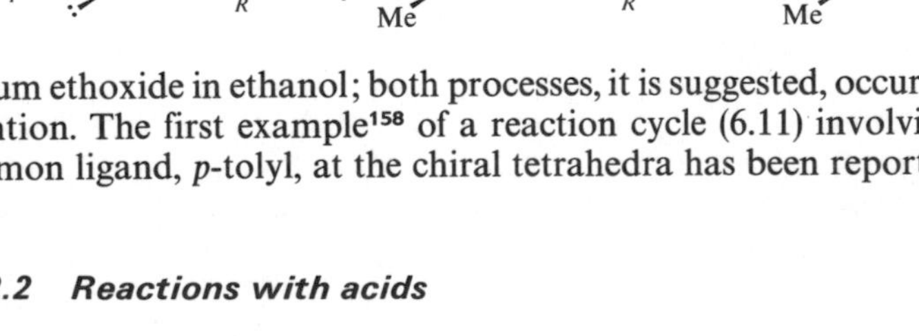

$$(6.11)$$

sodium ethoxide in ethanol; both processes, it is suggested, occur with overall retention. The first example[158] of a reaction cycle (6.11) involving only one common ligand, p-tolyl, at the chiral tetrahedra has been reported.

6.2.2.2 Reactions with acids

Although sulphoxides and sulphones are reported[159] to be protonated on sulphur in 'superacid' solutions, the acid-catalysed racemisation and oxygen-exchange of sulphoxides is considered to proceed through an O-protonated

intermediate. Kinetic studies of the acid-catalysed racemisation of methyl
p-tolyl sulphoxide with chloride ion, and of the acid-catalysed reduction
with iodide ion, show that in perchloric and in sulphuric acid both processes
follow the same mechanism to the rate-limiting step[160], and suggest that the
reaction is specific acid catalysed[161]. A possible shift towards general acid
catalysis with the isopropyl sulphoxide[161] and a ρ-value[162] of -1.05 are
consistent with the following scheme in which (iii) and (iv) may merge with
increasing α-alkylation.

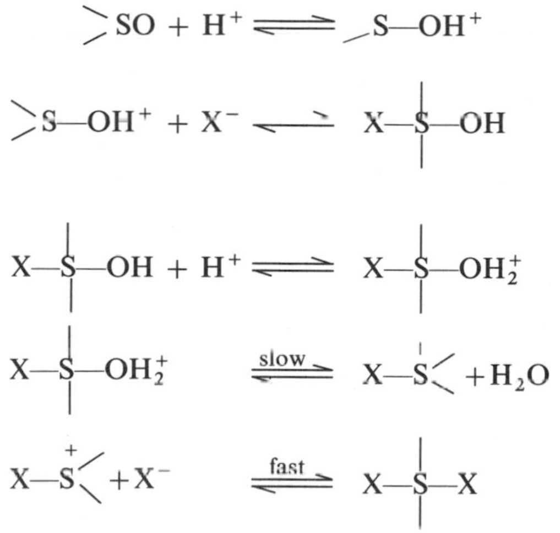

$$\text{\Large $>$}SO + H^+ \rightleftharpoons \text{\Large $>$}S\text{---}OH^+ \tag{i}$$

$$\text{\Large $>$}S\text{---}OH^+ + X^- \rightleftharpoons X\text{---}\overset{|}{\underset{|}{S}}\text{---}OH \tag{ii}$$

$$X\text{---}\overset{|}{\underset{|}{S}}\text{---}OH + H^+ \rightleftharpoons X\text{---}\overset{|}{\underset{|}{S}}\text{---}OH_2^+ \tag{iii}$$

$$X\text{---}\overset{|}{\underset{|}{S}}\text{---}OH_2^+ \overset{\text{slow}}{\rightleftharpoons} X\text{---}S\text{\Large $<$} + H_2O \tag{iv}$$

$$X\text{---}\overset{+}{S}\text{\Large $<$} + X^- \overset{\text{fast}}{\rightleftharpoons} X\text{---}\overset{|}{\underset{|}{S}}\text{---}X \tag{v}$$

Formation of the sulphodichloride, R_2SCl_2, leads to loss of chirality due to
a plane of symmetry, or to dissociation of the intermediate, $R_2\overset{\delta+}{S}\text{---}Cl\text{---}\overset{\delta-}{Cl}$, into
the sulphide and halogen[163]. The structure of the 1:1 adduct of thiophane and
bromine supports[163] the latter hypothesis. Chlorination of dimethyl sulphide
gives[164] the same product as the reaction of dimethyl sulphoxide with hydrogen chloride in the presence of molecular sieves, and the reaction of an aryl
benzyl sulphoxide with hydrogen chloride in acetic acid–acetic anhydride
gives[165] cleavage products as does chlorination of an aryl benzyl sulphide.
Both of these observations support a common intermediate for chlorination
of sulphides and chloride-ion catalysed racemisation of sulphoxides. The use
of aqueous acetic acid, however, inhibited[165] cleavage but not racemisation.
It has been suggested[165] as an alternative that (ii) is the slow step and that
racemisation in the presence of water may proceed by (iii) and ligand exchange
to give the 1,1-diol. Reduction with iodide ion may involve[161] a nucleophilic
displacement by iodide ion on iodine.

6.2.2.3   Oxygen exchange

Oxygen exchange reactions of sulphoxides have been reviewed[166]. Benzyl
p-tolyl sulphoxide is considered to undergo oxygen exchange in acetic
anhydride by acylation followed by an S_N2 displacement by acetate ion.
Racemisation occurs by exchange and another process which, it is suggested[167]

involves C—S heterolysis following acylation. The Pummerer reaction of methyl p-tolyl sulphoxide occurs simultaneously with oxygen exchange and has $\rho = -1.6$ and $k_H/k_D = 2.9$, which is consistent with the slow step being ylid formation from the acylated sulphoxide, followed by intermolecular migration of acetate ion.

6.2.2.4 Halogenation

Chlorination of sulphoxides with sulphuryl chloride gives the α-chlorosul-phoxide, and the sulphone, which is considered[168] to be formed by the hydrolysis of the chloro-oxosulphonium ion (32). In agreement with this, the relative yield of sulphone is increased by the addition of water, and β-, γ- and δ-hydroxysulphoxides give β-, γ- and δ-chlorosulphones. The latter reactions involve cyclic intermediates.

$$R^1SO\cdot CH_2R^2 \xrightarrow{SO_2Cl_2} R^1\overset{+}{S}(:O)Cl\cdot CH_2R^2\ \bar{C}l \xrightarrow{H_2O} R^1SO_2\cdot CH_2R^2$$

$$(32)$$

$$\downarrow -HCl$$

$$R^1\overset{+}{S}(:O)Cl\cdot\bar{C}HR^2 \longrightarrow R^1SO\cdot CHClR^2$$

α-Halogenation of (+)-methyl p-tolyl sulphoxide gives the (+)-halogeno-sulphoxide, whereas the enantiomer is surprisingly obtained[169] in the presence of silver ion, although no bonds are broken at the chiral centre nor is there diastereoisomeric interaction in the transition state.

6.2.2.5 Carbanion reactions

The diastereoisomer formed in the deuteriation of the lithio-derivative of benzyl methyl sulphoxide depends[170] on the dielectric constant of the aprotic solvent. With benzyl methyl or benzyl t-butyl sulphoxide in THF, deuteri-ation and reaction with acetone occur[171] with preferential retention (>94%) of carbanion configuration whereas reaction with methyl iodide occurs with inversion. These conclusions have been contested[172]. Optically active epoxides have been obtained[171] from these sulphoxides. Methyl methylthiomethyl sulphoxide reacts with sodium hydride and alkyl halides to give[173] the aldehyde dimethylmercaptal S-oxides, RCH(SMe)·SOMe, which are readily hydrolysed to the aldehydes. This is a useful alternative to the 1,3-dithiane aldehyde synthesis[44].

6.2.2.6 Reduction

Sulphoxides are reduced to sulphides under mild conditions with iron pentacarbonyl[174], sodium borohydride–cobalt chloride[175], and with O,O-dialkyl dithiophosphates[176], which also reduce sulphilimines.

6.2.2.7 Dimethyl sulphoxide

Dimethyl sulphoxide continues to be exploited as an oxidising agent and, in the presence of other reagents, converts alkynes into α-diketones[177], alkyl halides into carbonyl compounds[178], methylpyridines into formyl-pyridines[179], and cyclohexa-1,4-diene dioxide into catechol[180]. The reactions of hydroxy compounds and amides with dimethyl sulphoxide, dicyclohexyl-carbodi-imide and phosphoric acid (Moffatt oxidation) have been studied[181]. The nucleophilic substrate is thought to react with the DCC–DMSO adduct to form a sulphonium ylid which rearranges or collapses to give the product. The nucleophilic reactivity of the methylsulphinylmethide anion has been used[182] to convert cyclo-alkylcarboxylic esters into cycloalkyl ketosulphoxides, $RCO \cdot CH_2 \cdot SO \cdot Me$, and hence into diketones. This anion also gives the products of 1,2-addition, and 1,4-addition followed by deproto-nation and cyclisation with benzylideneacetophenone[183], methylates 2-methylstilbene[184], gives 3-methoxybenzyne and hence 3-methoxy-2-methyl-thiophenol with 3-bromoanisole[185] but reacts with p-chlorobenzotri-fluoride at the trifluoromethyl group[186]. Phenols are cleanly ortho-thiomethyl-ated[187] with dimethyl sulphoxide in the presence of the pyridine–sulphur trioxide complex and triethylamine, by a mechanism similar to that for the Moffatt oxidation[181]. Dimethyl sulphoxide reacts[188] as a nucleophile with α-chloro-oximes, which are also biphilic, to give the carboxylic acid, dimethyl sulphide and oxides of nitrogen.

6.2.2.8 Unsaturated sulphoxides

Treatment of (+)-R-but-3-yn-2-ol with toluene-p-sulphenyl chloride in pyridine at −70 °C gave[189] optically active buta-1,2-dienyl p-tolyl sulphoxide by a cyclic irreversible rearrangement. But-2-yn-1,4-diol similarly gave the disulphoxide. A complete description of the energy surface for the [2,3]-sigmatropic rearrangement of allyl p-trifluoromethylbenzenesulphenate to the sulphoxide, and the thermal racemisation of the latter is now possible[190]. Although the equilibrium lies towards the sulphoxide the energy difference between the two forms is only c. 3 kcal mol^{-1}. Electron-withdrawing groups and non-polar solvents influence the equilibrium by their destabilising effect on the ground-state of the sulphoxide. Addition of piperidine[191] and of bromine[192] to aryl vinyl sulphoxides occurs initially on the side of the double bond remote from the aryl substituent.

6.2.3 Sulphilimines

Methyl N-chlorobenzimidate reacts[193] exothermically with dimethyl sulphide to give the dimethylsulphilimine, $PhCO \cdot N{:}SMe_2$, but with dipropyl sulphide the sulphenamide was formed. Primary arylamines condense[194] with sul-phoxides in the presence of triethylamine and phosphorus pentoxide to give the compounds, $ArN{:}SR_2$. The S—N bond in N-sulphonylsulphilimines is more[195] dipolar than the S—O bond in sulphoxides, but both classes of compounds undergo *cis*-pyrolytic eliminations at 100–130 °C, and the *N*-acyl derivatives undergo[196] Pummerer-type rearrangements with acid chlorides or anhydrides. Aryl methyl and aryl benzyl *N*-sulphonylsulphilimines react[197] with cyanide ion and other nucleophiles in dimethyl sulphoxide to give the sulphide.

6.2.4 Sulphinic acids and their derivatives

Sodium in liquid ammonia quantitatively removes the benzyl group from benzyl sulphones giving[198] the sulphinate ion. The hitherto unknown perfluoroalkanesulphinic acids can be synthesised[199] by reduction of the sulphonyl chloride with hydrazine. Dipole moment studies indicate[200] that sulphinic esters have the *gauche*-conformation as is assumed for the acid. Sulphinates undergo pyrolytic *syn*-eliminations involving a five-membered cyclic transition state similar to that for sulphoxides, and a comparison of the rate of this reaction for chiral sulphinates has been used[201] to establish configuration at sulphur. Acetylenic sulphinates undergo[189] a cyclic irreversible rearrangement to give the dienyl sulphone in contrast to the acid-catalysed rearrangement of alkyl and alkenyl sulphinates, which proceeds by a carbonium-ion mechanism. The dienyl sulphone rearranges further.

$$HC\text{:}C\cdot CHRO\cdot SOAr \xrightarrow{130\,°C} RCH\text{:}C\text{:}CH\cdot SO_2Ar \rightarrow RC\text{:}C\cdot CH_2\cdot SO_2Ar$$

Sodium methanesulphinate reacts with thiophosgene by addition to the thiocarbonyl group and by displacement of halogen to give[202] bis-(methylsulphonyl)methyl methanethiosulphonate, $(MeSO_2)_2CH\cdot S\cdot SO_2Me$.

6.3 COMPOUNDS OF HEXAVALENT SULPHUR

6.3.1 Sulphones

6.3.1.1 Preparation

Structures of the type $(R^1SO_2)_2C\text{:}C(SR^2)_2$ and $R^1SO_2\cdot CR^2\text{:}C(SR^3)_2$ are obtained[203] by treating mono- or di-sulphones with potassium t-butoxide and carbon disulphide in dimethyl sulphoxide, followed by alkylation. Oxidation of one of the former gave di(methylsulphonyl)methane, not tetra(methylsulphonyl)ethene, presumably by hydration and fission of the tetrasulphone. The sulphone obtained by treating sulphur, in the presence of formic acid and a tertiary amine, with monosubstituted ethylenes having an electron-withdrawing substituent may arise[204] from an intermediate sulphinic acid. Recent interest in negatively substituted epoxides has led[205] to the synthesis of $\alpha\beta$-epoxy-sulphones, -sulphoxides, and -sulphonamides[206] from the α-chloromethyl derivatives and carbonyl compounds by the Darzens reaction[205b], or metallation and addition[205a, 206].

6.3.1.2 Carbanion reactions

gem-Dilithio derivatives in which the dianions are capable of being stabilised by $d_\pi - p_\pi$ interaction are given[207] by sulphones or sulphoxides, with an excess of n-butyl-lithium in THF–hexane. An example of this new class of compound with a good leaving group, chlorine, at the dianionic carbon atom is reported[208, 209] and its reaction with α-chloroamines is described[209]. The kinetics of formation of a trans-stilbene by the Ramberg–Bäcklund reaction from threo-α-bromo-α-methylbenzyl α-methylbenzyl sulphone are consistent[210] with rapid formation of the episulphone, which undergoes a rate-limiting decomposition, although in other cases loss of halide ion has been rate limiting. It is suggested that the proton is removed from a conformation in which it is flanked by the two sulphonyl oxygen atoms, the resultant asymmetric carbanion undergoes inversion, and reaction then occurs with inversion at the other carbon atom. α-Sulphonyl carbanions were shown to be asymmetric and protonated with retention. α-Chlorodicyclopropyl sulphone gave[211] β-t-butoxydicyclopropyl sulphone with potassium t-butoxide instead of bicyclopropylidene, the Ramberg–Bäcklund product. α,α-Dichlorodibenzyl sulphone and triethylenediamine in dimethyl sulphoxide yielded[212] 2,3-diphenylthi-iren-1,1-dioxide (>90%), which gave diphenylacetylene when heated.

6.3.1.3 Other reactions

Di-n-propyl sulphone is chlorinated in the β- and the γ-position[213] by sulphuryl chloride in the dark, although other common polar halogenating agents give α-chlorination. Addition of peroxide gave a different product pattern. The acid dissociation constants and rates of bromination for certain disulphones have been measured[214]. Aryl pent-2-yl sulphones undergo elimination reactions with strong bases which have a high degree of ElcB character[215].

6.3.2 Sulphonic acids and their derivatives

6.3.2.1 Sulphonic acids

s- and t-Alkanesulphonic acids are obtained[216] from n-alkanesulphonic esters by metalation, alkylation and hydrolysis. βγ-Unsaturated sulphonic acids, unlike related carboxylic acids, do not readily rearrange to αβ-unsaturated acids, indicating[217] a decreased conjugative interaction. Sulphonic acids react[159] in 'superacid' solutions at $20\,^{\circ}C$ to give the methane-sulphonylium ion, or with higher homologues, the carbonium ion.

6.3.2.2 Sulphonates and sultones

A reaction similar to the Claisen ester condensation[218] has been reported.

$$\text{e.g. } RCH_2 \cdot SO_2 \cdot OAr \xrightarrow[5\,^{\circ}C]{Bu^tO^- - THF} RCH \cdot SO_2 \cdot OAr$$

$$\underset{SO_2 \cdot CH_2R}{\big|}$$

The mechanism is thought to involve a sulphene intermediate or an S_N2 displacement by a carbanion. A mechanism similar to that of the Claisen condensation is considered unlikely. The relative reactivity of sulphonates as leaving groups[219, 220] shows the following order: $CF_3 \cdot SO_2 \cdot O-(120\,000) > CF_3 \cdot CH_2 \cdot SO_2 \cdot O-(300) > $ tosyl (3) $> CH_3 \cdot SO_2 \cdot O-(1)$; the wide variation in rate reflects the high charge development at sulphur in the S_N1 transition state[219]. Vinyl sulphonates are obtained from aldehydes or ketones with trifluoromethanesulphonic anhydride[221a] and from ethyl perfluoroethyl ketone with methanesulphonyl chloride[221b]. 1,2- and 1,3-Alkanedisulphonates react[222] with n-butyl-lithium below $-10\,^{\circ}C$ to give 5- and 6-membered cyclic sultones by displacement of sulphonate anion from the carbanion, e.g. (33). Similar compounds are also obtained[223] by sulphonation of terminal

$$\begin{array}{c} (CH_2)_n \cdot CH_2 \overset{\frown}{-} O\,SO_2Me \\ MeCH \\ \diagdown \\ \overset{|}{O} \cdot SO_2 \overset{(-)}{CH_2} \\ (33) \quad [n = 0, 1] \end{array}$$

alkenes with sulphur trioxide. Kinetic and isotopic studies are consistent with hydrolysis by a substantially $B_{AL}1$-E1 mechanism, but 14% of S—O bond fission by a bimolecular reaction was found with propane-1,3-sultone. The isomerisation of dimethyl sulphite to methyl methanesulphonate involves S-methylation of the methyl sulphite ion[224].

6.3.2.3 Thiolsulphonates

Methyl methanethiolsulphonate reacts[225] with bis-(phenylsulphonyl)methane in the presence of base to give products resulting from nucleophilic substitution at S^{II}, $(PhSO_2)_2CH \cdot SMe$ and $(PhSO_2)_2C(SMe)_2$. Rate studies are consistent with an S_N2 or an addition–elimination mechanism. The second substitution is much slower than the first. With aromatic compounds and aluminium chloride, aryl methyl sulphides are formed[226]. Alkyl alkanethiolsulphonates are partially desulphurised by tris(triethylamino)phosphine to give[227] the sulphone or a mixture of a sulphone and a sulphinate ester.

6.3.2.4 Sulphonyl chlorides

α-Cyanosulphonyl chlorides, $RCH(CN) \cdot SO_2Cl$, a new class of compound, are obtained[228] from aldehyde cyanohydrins by chlorination and reaction with sodium sulphite. They react with aldehyde enamines giving thietane 1,1-dioxides and with enamines from cyclic ketones to give α-cyanosulphones. Methanesulphonyl chloride forms a mixture of the primary and the secondary chloride with 1-phenylthiopropan-2-ol in the presence of pyridine[229]. The esters formed initially readily undergo substitution by the liberated chloride ion. Contrary to earlier reports, sulphonyl halides react[230] with the sodium salt of ethyl acetoacetate to give ethyl sulphonylacetate, $RSO_2 \cdot CH_2 \cdot CO_2Et$. The kinetics of the hydrolysis and alcoholysis of sulphonyl chlorides have been studied[231].

6.3.2.5 Sulphenes

Methylsulphene has been obtained by an 'abnormal' route[232] from the 1-chloroethanesulphinate anion or from the acid and triethylamine. This reaction shows that attack of the nucleophile at carbon is possible in the collapse of sulphenes to product. Sulphenes have also been identified in the thermal rearrangement of allyl vinyl sulphone[233] and in the thermolysis of thiete 1,1-dioxide[234]. Methanedisulphonyl dichloride gives[235] products that could arise from the disulphene. The i.r. spectrum of sulphene has been recorded[236] and evidence has been adduced that sulphene formation involves an E2 mechanism[237]. The cyclo-addition of sulphene and optically active enamines proceeds by a product-like transition state to give[238] an optically active product by asymmetric induction. A new ring system, pyrido[1,2-d]-[1,3,4]dithiazine as in (34), has been obtained[239] by the addition of sulphene to pyridines.

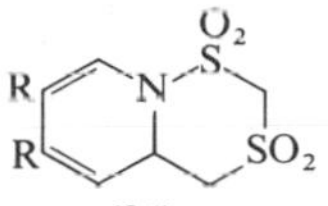

(34)

6.3.2.6 Anhydrides

Mixed sulphonic–carboxylic anhydrides are best prepared[240] from the sulphonic acid and the acyl chloride. On thermolysis below 130 °C the sulphonic anhydride is obtained in good yield. Mixed anhydrides readily cleave ethers to give the carboxylic esters, but with amines the sulphonate salt is the major product. The mixed anhydride is the acylating species[241] in solutions of methanesulphonic acid in acetic anhydride.

6.3.2.7 Derivatives of iminosulphonic acids and sulphoximines

Sulphonimidoyl chlorides, $R^1SO:NR^2) \cdot Cl$, are a comparatively new class of compounds with chiral sulphur. They are obtained[242] by the chlorination of sulphinamides or from sulphinyl chlorides and dichloroamines[243]. They react with water and alcohols to give sulphonamides by a base-catalysed reaction which involves an intermediate iminosulphene, e.g. $CH_2:SO(:NR)$. Nucleophilic substitutions at the tetracoordinated sulphur proceed[244] with inversion.

Sulphoximines, $R^1SO(:NR^2)R^3$, are reduced[245] by aluminium amalgam to the sulphinamide with retention of configuration at sulphur. Dialkyl sulphone di-imides, $R_2^1S(:NH)_2$, are prepared[246] from the sulphides by reaction with t-butyl hypochlorite and ammonia; they undergo[247] base-catalysed condensations with malonic esters to give the novel 3,5-dioxo-3H-4,5-dihydro-1,2,6-thia(IV)diazines (35).

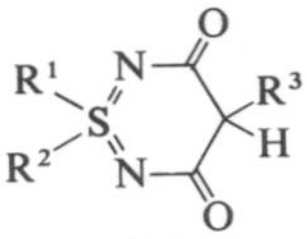

(35)

6.4 SULPHUR YLIDS

An insight into the complementary behaviour of oxysulphonium and sulphonium ylids has been provided[248] by the synthesis of the betaine inter-mediates expected for methylene transfer reactions to benzaldehyde (36, 37) and benzalacetophenone (38). These compounds decompose to give the expected epoxides and cyclopropane derivatives. Betaine (36) was in equili-brium with the ylid and benzaldehyde. It was concluded that attack of

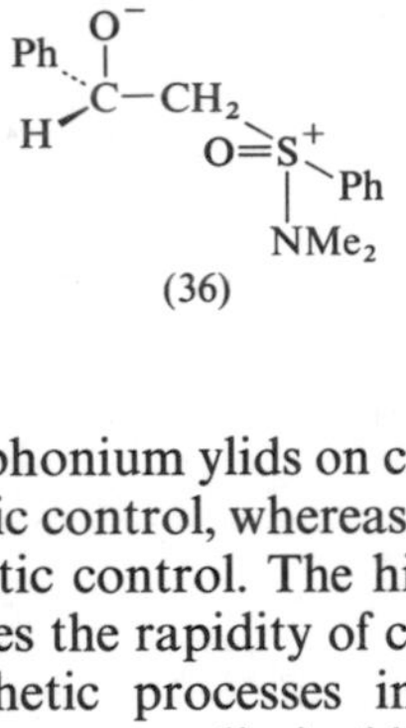

(36) 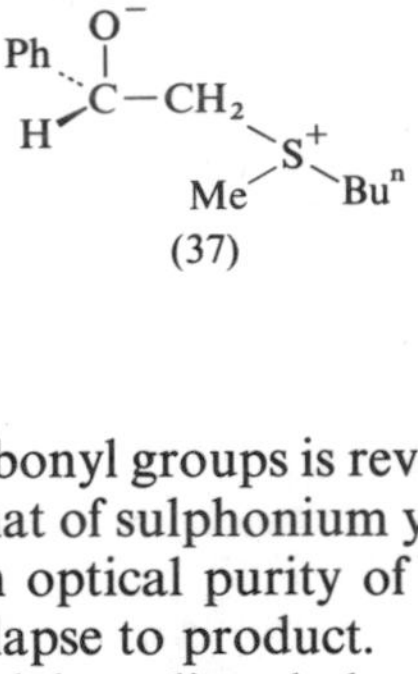   (37) 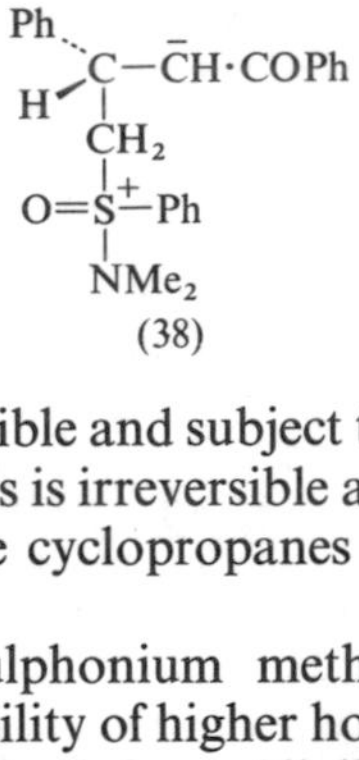   (38)

oxysulphonium ylids on carbonyl groups is reversible and subject to thermo-dynamic control, whereas that of sulphonium ylids is irreversible and subject to kinetic control. The high optical purity of the cyclopropanes produced, indicates the rapidity of collapse to product.

Synthetic processes involving dimethyloxysulphonium methylide and its congeners are limited by the frequent inavailability of higher homologues. This can be overcome by the use of ylids derived from (dialkylamino)-oxysulphonium salts, $R^2SO(NMe_2) \cdot CH_2R^2\bar{B}F_4$ which have similar re-activity[249] and are readily available from the sulphide by oxidation, sulphoxime formation, dimethylation and proton removal. The ylid derived from 3-chloropropylphenyl sulphide gives[250] relatively inaccessible spiropentane derivatives (e.g. (39) from benzalacetophenone). N-Toluene-p-sulphonyl sulphoximines give stable anions[251], which offer advantages over existing reagents for methylene transfer reactions. Limitations of space prevent further discussion of syntheses involving ylids.

Spectral and chemical evidence indicates[252] restricted rotation about the C-carbonyl bond in dimethylsulphonium phenacylide with the cis-form predominating. Similar results have been obtained[253] for stabilised oxy-sulphonium ylids.

Allyldimethylsulphonium salts react with n-butyl-lithium giving[254] methyl but-3-enyl sulphide by a [2,3]-sigmatropic rearrangement of the ylid. The ylid derived from the phenol (40) rearranged[255] similarly, but the Stevens rearrangement of benzylmethylsulphonium phenacylide proceeds[256] through

a radical pair. Products from the photolysis of diphenylsulphonium allylide are best explained[257] by an α-elimination to give a carbene, and hence cyclopropene, accompanied by homolysis of the S—Ph bond. The thermal or the metal-catalysed decompositions did not give evidence of an α-elimination. The reaction of methyl sulphoxides with Grignard reagents also appears[258] to involve decomposition of an ylid to a carbene. Carbenes could not be detected[259] in the decomposition of bis(dialkylsulphonium) methylides, which furthermore do not display nucleophilic reactivity. The chemistry of sulphur ylids has been reviewed[260].

References

1. Brandsma, L. (1970). *Recl. Trav. chim. Pays-Bas.*, **89**, 593
2. Eliel, E. L. and Doyle, T. W. (1970). *J. Org. Chem.*, **35**, 2716
3. Newman, B. C. and Eliel, E. L. (1970). *J. Org. Chem.*, **35**, 3641
4. Hope, D. B. and Wälti, M. (1970). *J. Chem. Soc. C*, 2475
5. Coates, R. M. and Sowerby, R. L. (1971). *J. Amer. Chem. Soc.*, **93**, 1027
6. Field, L. and Giles, P. M. (1971). *J. Org. Chem.*, **36**, 309
7. Degani, Y., Neumann, H. and Patchornik, A. (1970). *J. Amer. Chem. Soc.*, **92**, 6969
8. Zervos, C. and Cordes, E. H. (1971). *J. Org. Chem.*, **36**, 1661
9. van der Maeden, F. P. B., Steinberg, H. and de Boer, Th. J. (1971). *Recl. Trav. chim. Pays-Bas.*, **90**, 423
10. Caubere, P. and Brunet, J. J. (1971). *Tetrahedron*, **27**, 3515
11. Verplocgh, M. C., Donk, L., Bos, H. J. T. and Drenth, W. (1971). *Recl. Trav. chim. Pays-Bas.*, **90**, 765
12. Verny, M. (1970). *Bull. Soc. Chim. Fr.*, 1942, 1947
13. Arens, J. F., Brandsma, L., Schuijl, P. J. W. and Wijers, H. E. (1970). *Quart. Reports Sulfur Chem.*, **5**, 1
14. Maguire, M. H., Nobbs, D. N., Einstein, R. and Middleton, J. C. (1971). *J. Medicin. Chem.*, **14**, 415
15. Hashimoto, N., Kawano, Y. and Moreta, K. (1970). *J. Org. Chem.*, **35**, 828
16. Kice, J. L. and Campbell, J. D. (1971). *J. Org. Chem.*, **36**, 2289
17. Kupchan, S. M., Giacobbe, T. J. and Krull, I. S. (1970). *Tetrahedron Letters*, 2859
18. Kupchan, S. M., Giacobbe, T. J., Krull, I. S., Thomas, A. M., Eakin, M. A. and Fessler, D. C. (1970). *J. Org. Chem.*, **35**, 3539
19. Esterbauer, H. (1970). *Monatsh.*, **101**, 782
20. Schauenstein, H., Taufer, M., Esterbauer, H., Kylianek, K. and Seelich, Th. (1971). *Monatsh*, **102**, 517
21. Prilezhaeva, E. N., Laba, V. I., Snegotskii, V. I. and Shekhtnan, R. I. (1970). *Izv. Akad. Nauk. S.S.S.R. Ser. Khim.*, 1602
22. Prilezhaeva, E. N., Vasil'ev, G. S., Mikhelashvili, I. L. and Bogdanov, V. S. (1970). *Izv. Akad. Nauk. S.S.S.R. Ser. Khim.*, 1922
23. Kandror, I. I., Petrovskii, P. V., Petrova, R. G. and Freidlina, R. Kh. (1970). *Izvest. Akad. Nauk. S.S.S.R. Ser. Khim.*, 2222
24. Kay, I. T. and Punja, N. (1970). *J. Chem. Soc. C*, 2409

25. Nakaya, T., Arabori, H. and Imoto, M. (1970). *Bull. Chem. Soc. Jap.*, **43**, 1888
26. Leslie, J. (1970). *Can. J. Chem.*, **48**, 3104
27. Kachwaha, O. P., Sinha, B. P. and Kapoor, R. C. (1970). *Indian J. Chem.*, **8**, 806
28. Oki, M., Funakoshi, W. and Nakamura, A. (1971). *Bull. Chem. Soc. Jap.*, **44**, 828, 832
29. Lalancette, J. M., Beauregard, Y. and Bhereur, M. (1970). *Can. J. Chem.*, **48**, 1093, 2983
30. Danehy, J. P., Doherty, B. T. and Egan, C. P. (1971). *J. Org. Chem.*, **36**, 2525
31. Beveridge, S. and Harris, R. L. N. (1971). *Aust. J. Chem.*, **24**, 1229
32. Danehy, J. P., Egan, C. P. and Switalski, J. (1971). *J. Org. Chem.*, **36**, 2530
33. Takabe, K., Katagiri, T. and Tanaka, J. (1970). *Tetrahedron Letters*, 4805
34. Burakevich, J. V., Lore, A. M. and Volpp, G. P. (1970). *J. Org. Chem.*, **35**, 2102
35. Henriksen, L. and Autrup, H. (1970). *Acta. Chem. Scand.*, **24**, 2629
36. Yokoyama, M. (1970). *J. Org. Chem.*, **35**, 283
37. Yokoyama, M. (1971). *J. Org. Chem.*, **36**, 2009
38. Coffen, D. L., Chambers, J. Q., Williams, D. R., Garrett, P. E. and Canfield, N. D. (1971). *J. Amer. Chem. Soc.*, **93**, 2258
39. Rindone, B. and Scolastico, C. (1971). *J. Chem. Soc. C*, 3339
40. Hideg, K. and Lloyd, D. (1971). *J. Chem. Soc. C*, 3441
41. Fehér, F. and Langer, M. (1971). *Tetrahedron Letters*, 2125
42. Coffen, D. L., Poon, Y. C. and Lee, M. L. (1971). *J. Amer. Chem. Soc.*, **93**, 4628
43. Woodward, R. B., Pachter, I. J. and Scheinbaum, M. L. (1971). *J. Org. Chem.*, **36**, 1137
44. Corey, E. J. and Erikson, B. W. (1971). *J. Org. Chem.*, **36**, 3553; Vedejs, E. and Fuchs, P. L. (1971). *J. Org. Chem.*, **36**, 366
45a. Marshall, J. A. and Roebke, H. (1969). *J. Org. Chem.*, **34**, 4188
45b. (1970). *Tetrahedron Letters*, 1555
46. Tsurugi, J., Abe, Y. and Kawamura, S. (1970). *Bull. Chem. Soc. Jap.*, **43**, 1890
47. Tsurugi, J., Abe, Y., Nakabayashi, T., Kawamura, S., Kitao, T. and Niwa, M. (1970). *J. Org. Chem.*, **35**, 3263
48. Armitage, D. A. and Tso, C. C. (1971). *Chem. Commun.*, 1413
49. Phillips, W. G. and Ratts, K. W. (1971). *J. Org. Chem.*, **36**, 3145
50a. Hogg, D. R. (1970). *Mech. Reaction Sulphur Chem.*, **5**, 87
50b. Kühle, E. (1971). *Synthesis*, 563, 617
51. Brown, H. C. and Liu, K.-T. (1971). *J. Amer. Chem. Soc.*, **93**, 7335
52. Kochetkov, B. B., Stadnichuk, M. D. and Petrov, A. A. (1971). *Zhur. Obshch. Khim.*, **41**, 715
53. Verny, M. and Vessière, R. (1970). *Bull. Soc. Chim. Fr.*, 746
54. Krubsack, A. J., Higa, T. and Slack, W. E. (1970). *J. Amer. Chem. Soc.*, **92**, 5258
55. Gray, E. A., Hulley, R. M. and Snell, B. K. (1970). *J. Chem. Soc. C*, 986
56. Fujisawa, T. and Kobayashi, N. (1971). *J. Org. Chem.*, **36**, 3547
57. Dykman, E. (1971). *Chem. Commun.*, 1400
58. Pietra, F. and Vitali, D. (1970). *J. Chem. Soc. B*, 623
59. Oki, M. and Kobayashi, K. (1970). *Bull. Chem. Soc. Jap.*, **43**, 1223, 1229
60. Field, L., Vanhorne, J. L. and Cunningham, L. W. (1970). *J. Org. Chem.*, **35**, 3267
61. Chakrabarti, P. M. and Chapman, N. B. (1970). *J. Chem. Soc. C*, 914
62. Barton, D. H. R., Page, G. and Widdowson, D. A. (1970). *Chem. Commun.*, 1466
63. Armitage, D. A. and Clark, M. J. (1971). *J. Chem. Soc. C*, 2840; (1970). *Chem. Commun.*, 104
64. Douglass, I. B., Norton, R. V., Cocanour, P. M., Koop, D. A. and Kee, M.-L. (1970). *J. Org. Chem.*, **35**, 2131
65a. Rautenstrauch, V. (1970). *Chem. Commun.*, 526
65b. Baldwin, J. E., Erikson, W. F., Hackler, R. E. and Scott, R. M. (1970). *Chem. Commun.*, 576
66. Harpp, D. N. and Back, T. G. (1971). *Tetrahedron Letters*, 4953
67. Armitage, D. A., Clark, M. J. and White, A. M. (1971). *J. Chem. Soc. C*, 3141
68. Bentley, M. D., Douglass, I. B., Lacadie, J. A., Weaver, D. C., Davis, F. A. and Eitelman, S. J. (1971). *Chem. Commun.*, 1625
69. Kice, J. L. (1968). *Accounts Chem. Res.*, **1**, 58
70. Saegusa, T., Ito, Y. and Shimizu, T. (1970). *J. Org. Chem.*, **35**, 2974
71. Richards, J. L. and Tarbell, D. S. (1970). *J. Org. Chem.*, **35**, 2079
72. Brown, C. and Grayson, B. T. (1970). *Mech. Reaction Sulphur Chem.*, **5**, 93
73. Heimer, N. E. and Field, L. (1970). *J. Org. Chem.*, **35**, 3012

74. Dunbar, J. E. and Rogers, J. H. (1970). *J. Org. Chem.*, **35**, 279
75. Mukaiyama, T., Hosoi, K., Inokuma, S. and Kumamoto, T. (1971). *Bull. Chem. Soc. Jap.*, **44**, 2453
76. Harpp, D. N. and Orwig, B. A. (1970). *Tetrahedron Letters*, 2691
77. Potts, K. T. and Armbruster, R. (1970). *J. Org. Chem.*, **35**, 1965
78. Raban, M. and Lauderback, S. K. (1971). *J. Amer. Chem. Soc.*, **93**, 2781
79. Kay, J., Glick, M. D. and Raban, M. (1971). *J. Amer. Chem. Soc.*, **93**, 5224
80. Baldwin, J. E., Höfle, G. and Se Chun Choi. (1971). *J. Amer. Chem. Soc.*, **93**, 2811
81. Brois, S. J., Pilot, J. F. and Barnum, H. W. (1970). *J. Amer. Chem. Soc.*, **92**, 7629
82. Kloosterziel, H. and van der Ven, S. (1970). *Recl. Trav. chim. Pays-Bas.*, **89**, 1017
83a. Corey, E. J. and Shulman, J. I. (1970). *J. Org. Chem.*, **35**, 777
83b. Mukaiyama, T., Kumamoto, T., Fukuyama, S. and Taguchi, T. (1970). *Bull. Chem. Soc. Jap.*, **43**, 2870
84. Atavin, A. S., Mikhaleva, A. I., Trofimov, B. A., Kalabin, G. A. and Vasil'ev, N. P. (1971). *Izvest. Akad. Nauk S.S.S.R. Ser. Khim.*, 614
85. Mantione, R., Alves, A., Montijn, P. P., Wildschut, G. A., Bos, H. J. T. and Brandsma, L. (1970). *Recl. Trav. chim. Pays-Bas.*, **89**, 97
86. Coffen, D. L., Chambers, J. Q., Williams, D. R., Garrett, P. E. and Canfield, N. D. (1971). *J. Amer. Chem. Soc.*, **93**, 2258
87. Wilson, G. E. and Huang, M. G. (1970). *J. Org. Chem.*, **35**, 3002
88. Vilsmaier, E. and Sprügel, W. (1971). *Annalen*, **747**, 151
89. Vilsmaier, E. and Sprügel, W. (1971). *Annalen*, **749**, 62
90. Miotti, U., Modena, G. and Sedea, L. (1970). *J. Chem. Soc. B*, 802
91. Caló, V., Ciminale, F., Lopez, G. and Todesco, P. E. (1971). *Int. J. Sulphur Chem. A.*, **1**, 130
92. Ogata, Y. and Suyama, S. (1971). *Chem. Ind. (London)*, 707
93. Numata, T. and Oae, S. (1971). *Int. J. Sulphur Chem.*, **A, 1**, 6
94. Mukaiyama, T., Narasaka, K., Maekawa, K. and Furusato, M. (1971). *Bull. Chem. Soc. Jap.*, **44**, 2285
95. Gelli, G., Marongiu, E. and Solinas, V. (1970). *Gazzeta.*, **100**, 796
96. Olofson, R. A. and Hansen, D. W. Jr. (1971). *Tetrahedron*, **27**, 4209
97. Ando, W., Yagihara, T., Kondo, S., Nakayama, K., Yamato, H., Nakaido, S. and Nigita, T. (1971). *J. Org. Chem.*, **36**, 1732
98. Nigita, T., Matsuyama, H. and Ando, W. (1971). *Int. J. Sulphur Chem.*, **A, 1**, 47
99. Chicllini, E., Marchetti, M. and Ceccarelli, G. (1971). *Int. J. Sulphur Chem.*, **A, 1**, 73
100. Samdal, S. and Seip, H. M. (1971). *Acta. Chem. Scand.*, **25**, 1903
101a. Biellman, J. F. and Ducep, J. B. (1971). *Tetrahedron Letters*, 33
101b. Rautenstrauch, V. (1971). *Helv. Chim. Acta.*, **54**, 739
102. Sasaki, T., Kojima, A. and Ohta, M. (1971). *J. Chem. Soc. C*, 196
103. Brandsma, L., Schiujl, P. J. W., Schiujl-Laros, D., Meijer, J. and Wijers, H. E. (1971). *Int. J. Sulphur Chem. B*, **6**, 85
104. Harpp, D. N. and Mathiaparanam, P. (1970). *Tetrahedron Letters*, 2089; (1971). *J. Org. Chem.*, **36**, 2540, 2886
105. Russell, G. A. and Ochrymowycz, L. O. (1970). *J. Org. Chem.*, **35**, 764, 2106
106. Otsuji, Y., Tsuji, Y., Yoshida, A. and Imoto, E. (1971). *Bull. Chem. Soc. Jap.*, **44**, 219
107. Caserio, M. C., Lauer, W. and Novinson, J. (1970). *J. Amer. Chem. Soc.*, **92**, 6082
108. Boustany, K. S. and Sullivan, A. B. (1970). *Tetrahedron Letters*, 3547; Harpp, D. N. and Back, T. G. (1971). *J. Org. Chem.*, **36**, 3828; Harpp, D. N., Ash, D. K., Back, T. G., Gleason, J. G., Orwig, B. A., VanHorn, W. F. and Snyder, J. P. (1970). *Tetrahedron Letters*, 3551
109. Cragg, R. H., Husband, J. P. N. and Weston, A. F. (1970). *Chem. Commun.*, 1701
110. Abe, Y., Nakabayashi, T. and Tsurugi, J. (1971). *Bull. Chem. Soc. Jap.*, **44**, 2744
111. Field, L., Hanley, W. S., McVeigh, I. and Evans, Z. (1971). *J. Medicin. Chem.*, **14**, 202
112. Field, L., Hanley, W. S. and McVeigh, I. (1971). *J. Org. Chem.*, **36**, 2735
113. Yoshida, Z-i., Okushi, T. and Manabe, O. (1970). *Tetrahedron Letters*, 1641
114. Chan, T. H., Montillier, J. P., VanHorn, W. F. and Harpp, D. N. (1970). *J. Amer. Chem. Soc.*, **92**, 7224
115. Sisler, H. H., Kotia, N. K. and Highsmith, R. E. (1970). *J. Org. Chem.*, **35**, 1742
116. Klayman, D. L., Kenny, D., Silverman, R. B., Tomaszewski, J. E. and Shine, R. J. (1971). *J. Org. Chem.*, **36**, 3681

117. Field, L., Giles, P. M. and Tuleen, D. L. (1971). *J. Org. Chem.*, **36**, 623
118. Field, L. and Foster, C. H. (1970). *J. Org. Chem.*, **35**, 749
119a. Danehy, J. P. and Elia, V. J. (1971). *J. Org. Chem.*, **36**, 1394
119b. Danehy, J. P. (1971). *Int. J. Sulphur Chem. B.*, **6**, 103
120. Harpp, D. N. and Gleason, J. G. (1971). *J. Org. Chem.*, **36**, 73
121. Harpp, D. N. and Gleason, J. G. (1970). *J. Org. Chem.*, **35**, 3259
122. Harpp, D. N. and Gleason, J. G. (1971). *J. Amer. Chem. Soc.*, **93**, 2437
123. Meienhofer, J., Czombos, J. and Maeda, H. (1971). *J. Amer. Chem. Soc.*, **93**, 3081
124. Murray, R. W., Smetana, R. D. and Block, E. (1971). *Tetrahedron Letters*, 229
125. Höfle, G. and Baldwin, J. E. (1971). *J. Amer. Chem. Soc.*, **93**, 6308
126. Harpp, D. N. and Ash, D. K. (1970). *Chem. Commun.*, 811
127. Safe, S. and Taylor, A. (1969). *Chem. Commun.*, 1466
128. Tidd, B. K. (1971). *Int. J. Sulphur Chem. C.*, **6**, 101
129. Moore, R. E. (1971). *Chem. Commun.*, 1168
130. Safe, S. and Taylor, A. (1970). *J. Chem. Soc. C*, 432
131. Cuiffarin, E., Senatore, L. and Isola, M. (1971). *J. Chem. Soc. B*, 2187; Cuiffarin, E. and Senatore, L. (1970). *J. Chem. Soc. B*, 1680
132a. Cuiffarin, E. and Guaraldi, G. (1970). *J. Org. Chem.*, **35**, 2006
132b. Senatore, L., Cuiffarin, E. and Sagramora, L. (1971). *J. Chem. Soc. B*, 2191
133. Senatore, L., Cuiffarin, E. and Fava, A. (1970). *J. Amer. Chem. Soc.*, **92**, 3035
134. Smallcombe, S. H. and Caserio, M. C. (1971). *J. Amer. Chem. Soc.*, **93**, 5826
135. Kiwan, A. M. and Irving, H. M. N. H. (1970). *Chem. Commun.*, 928; (1971). *J. Chem. Soc. B*, 901
136. Pryor, W. A. and Smith, K. (1970). *J. Amer. Chem. Soc.*, **92**, 2731
137. Paquer, D. and Vialle, J. (1971). *Bull. Soc. Chim. Fr.*, 4407
138. Diebert, C. E. (1970). *J. Org. Chem.*, **35**, 1501
139. Raasch, M. S. (1970). *J. Org. Chem.*, **35**, 3470
140. Vyas, D. M. and Hay, G. W. (1971). *Can. J. Chem.*, **49**, 3755; (1971). *Chem. Commun.*, 1411
141. Veno, Y., Nakai, T. and Okawara, M. (1970). *Bull. Chem. Soc. Jap.*, **43**, 168
142. Gaydou, E., Peiffer, G. and Guillemonat, A. (1971). *Tetrahedron Letters*, 239
143. Anderson, K. K. (1971). *Chem. Commun.*, 1051
144. Garbesi, A., Corsi, N. and Fava, A. (1970). *Helv. Chim. Acta.*, **53**, 1499
145. Barbarella, G., Garbesi, A. and Fava, A. (1971). *Helv. Chim. Acta.*, **54**, 341, 2297
146. Wolfe, S., Rauk, A. and Csizmadia, I. G. (1969). *J. Amer. Chem. Soc.*, **91**, 1567; Wolfe, S Rauk, A., Tel, L. M. and Csizmadia, I. G. (1970). *Chem. Commun.*, 96
147. Feit, I. N., Schadt, F., Lubinkowski, J. and Saunders, W. H. Jr. (1971). *J. Amer. Chem. Soc.*, **93**, 6606
148. Batty, J. W., Howes, P. D. and Stirling, C. J. M. (1971). *Chem. Commun.*, 534
149. Horner, L., Hofer, W., Ertel, I. and Kunz, H. (1970). *Chem. Ber.*, **103**, 2718
150. Baldwin, J. E., Walker, J. A., Labuschagne, A. J. H. and Schneider, D. F. (1971). *Chem. Commun.*, 1382
151. Allen, D. W., Braunton, P. N., Millar, I. T. and Tebby, J. C. (1971). *J. Chem. Soc. C.*, 3454
152. Mikolajczyk, M., Drabowicz, J. and Cramer, F. (1971). *Chem. Commun.*, 317
153. Schultz, A. G. and Schlessinger, R. H. (1970). *Chem. Commun.*, 1294
154. Cram, D. J., Day, J., Rayner, D. R., v. Schriltz, D. M., Duchamp, D. J. and Garwood, D. C. (1970). *J. Amer. Chem. Soc.*, **92**, 7369
155. Christensen, B. W. (1971). *Chem. Commun.*, 597
156. Cook, R. E., Glick, M. D., Rigau, J. J. and Johnson, C. R. (1971). *J. Amer. Chem. Soc.*, **93**, 924
157. Colonna, S. and Stirling, C. J. M. (1971). *Chem. Commun.*, 1591
158. Williams, T. R., Booms, R. E. and Cram, D. J. (1971). *J. Amer. Chem. Soc.*, **93**, 7338
159. Olah, G. A., Ku, A. T. and Olah, J. A. (1970). *J. Org. Chem.*, **35**, 3904, 3908, 3925, 3929
160. Landini, D., Modena, G., Montanari, F. and Scorrano, G. (1970). *J. Amer. Chem. Soc.*, **92**, 7168
161. Landini, D., Modena, G., Quintily, U. and Scorrano, G. (1971). *J. Chem. Soc. B*, 2041
162. Ockuni, I. and Fry, A. (1971). *J. Org. Chem.*, **36**, 4097
163. Allegra, G., Wilson, G. E., Benedetti, E., Pedone, C. and Albert, R. (1970). *J. Amer. Chem. Soc.*, **92**, 4002

164. Rynbrandt, R. H. (1971). *Tetrahedron Letters*, 3553
165. Kwart, H. and Omura, H. (1971). *J. Amer. Chem. Soc.*, **93**, 7250
166. Oae, S. (1970). *Quart. Rept. Sulphur Chem.*, **5**, 53
167. Kise, M. and Oae, E. (1970). *Bull. Chem. Soc. Jap.*, **43**, 1421, 1426
168. Durst, T. and Tin, K.-C. (1971). *Can. J. Chem.*, **49**, 2374
169. Cinquini, M., Colonna, S. and Montanari, F. (1970). *Chem. Commun.*, 1441
170. Durst, T., Frazer, R. R., McClory, M. R., Swingle, R. B., Viau, R. and Wigfield, Y. Y. (1970). *Can. J. Chem.*, **48**, 2148
171. Durst, T., Viau, R. and McClory, M. R. (1971). *J. Amer. Chem. Soc.*, **93**, 3077
172. Nishihata, K. and Nishio, M. (1971). *Chem. Commun.*, 958
173. Ogura, K. and Tsuchihasi, G. T. (1971). *Tetrahedron Letters,* 3151
174. Alper, H. and Keung, E. C. H. (1970). *Tetrahedron Letters*, 53
175. Chasar, D. W. (1971). *J. Org. Chem.*, **36**, 613
176. Nakanishi, A. and Oae, S. (1971). *Chem. Ind. (London)*, 960
177. Wolfe, S., Pilgrim, W. R., Garrard, T. F. and Chamberlain, P. (1971). *Can. J. Chem.*, **49**, 1098
178. Epstein, W. W. and Ollinger, J. (1970). *Chem. Commun.*, 1338
179. Markovac, A., Stevens, C. L., Ash, A. B. and Hackley, B. E. (1970). *J. Org. Chem.*, **35**, 841
180. McKague, B. (1971). *Can. J. Chem.*, **49**, 2447
181. Lerch, U. and Moffatt, J. G. (1971). *J. Org. Chem.*, **36**, 3391, 3861, 3686; Marino, J. P., Pfitzner, K. E. and Olofson, R. A. (1971). *Tetrahedron*, **27**, 4181, 4195; Kerr, D. A. and Wilson, D. A. (1970). *J. Chem. Soc. C*, 1718
182. Russell, G. A. and Hamprecht, G. (1970). *J. Org. Chem.*, **35**, 3007
183. Gautier, J. A., Miocque, M., Plat, M., Moskowitz, H. and Blanc-Guenee, J. (1970). *Tetrahedron Letters*, 895
184. James, B. G. and Pattenden, G. (1971). *Chem. Commun.*, 1015
185. Birch, A. J., Chamberlain, K. B. and Oloyede, S. S. (1971). *Aust. J. Chem.*, **24**, 2178
186. Meschino, J. A. and Plampin, J. N. (1971). *J. Org. Chem.*, **36**, 3636
187. Claus, P. (1971). *Monatsch*, **102**, 913
188. Biffin, M. E. C. and Paul, D. B. *Tetrahedron Letters*. (1971). 3849
189. Smith, G. and Stirling, C. J. M. (1971). *J. Chem. Soc. C*, 1530
190. Tang, R. and Mislow, K. (1970). *J. Amer. Chem. Soc.*, **92**, 2100
191. Abbott, D. J., Colonna, S. and Stirling, C. J. M. (1971). *Chem. Commun.*, 471
192. Abbott, D. J. and Stirling, C. J. M. (1971). *Chem. Commun.*, 472
193. Papa, A. J. (1970). *J. Org. Chem.*, **35**, 2837
194. Claus, P. and Vycudilik, W. (1970). *Monatsch*, **101**, 396, 405
195. Oae, S., Tsujihara, K. and Furukawa, N. (1970). *Tetrahedron Letters*, 2663
196. Kise, H., Whitfield, G. F. and Swern, D. (1971). *Tetrahedron Letters*, 4839
197. Oae, S., Aida, T., Tsujihara, K. and Furukawa, N. (1971). *Tetrahedron Letters*, 1145
198. Hope, D. B., Morgan, C. D. and Wälti, M. (1970). *J. Chem. Soc. C.*, 270
199. Roesky, H. W. (1971). *Angew. Chem. Int. Ed. Eng.*, **10**, 810
200. Exner, O., Dembech, P. and Vivarelli, P. (1970). *J. Chem. Soc. B.*, 278
201. Jones, D. N. and Higgins, W. (1970). *J. Chem. Soc. C*, 81
202. Nilsson, N. H., Jacobsen, C. and Senning, A. (1970). *Chem. Commun.*, 658
203. Truce, W. E., Tracy, J. E. and Gorbaty, M. L. (1971). *J. Org. Chem.*, **36**, 237
204. Gibson, H. W. and McKenzie, D. A. (1970). *J. Org. Chem.*, **35**, 2994
205a. Durst, T. and Tin, K.-C. (1970). *Tetrahedron Letters*, 2369
205b. Tavares, D. F., Estep, R. E. and Blezard, M. (1970). *Tetrahedron Letters*, 2373
206. Truce, W. E. and Christensen, L. W. (1971). *J. Org. Chem.*, **36**, 2538
207. Kaiser, E. M., Solter, L. E., Schwarz, R. A., Beard, R. D. and Hauser, C. R. (1971). *J. Amer. Chem. Soc.*, **93**, 4237
208. Truce, W. E. and Christensen, L. W. (1971). *Chem. Commun.*, 588
209. Böhme, H. and Stammberger, W. (1971). *Annalen*, **754**, 56
210. Bordwell, F. G., Doomes, E. and Corfield, P. W. R. (1970). *J. Amer. Chem. Soc.*, **92**, 2581
211. Paquette, L. A. and Houser, R. W. (1971). *J. Org. Chem.*, **36**, 1015
212. Phillips, J. C., Swisher, J. V., Haidukewych, D. and Morales, O. (1971). *Chem. Commun.*, 22
213. Tabushi, I., Tamara, Y. and Yoshida, Z. (1971). *Tetrahedron Letters*, 3893

214. Bell, R. P. and Cox, B. G. (1971). *J. Chem. Soc. B,* 652
215. Colter, A. K. and Miller, R. E. (1971). *J. Org. Chem.,* **36,** 1898
216. Truce, W. E. and Vrencur, D. J. (1970). *J. Org. Chem.,* **35,** 1226
217. Petrova, R. G., Kandror, I. I. and Freidlina, R. Kh. (1970). *Izvest. Akad. Nauk. S.S.S.R. Ser. Khim.,* 1083
218. Truce, W. E. and Christensen, L. W. (1970). *J. Org. Chem.,* **35,** 3969
219. Crossland, R. K., Wells, W. E. and Shiner, V. J. Jr. (1971). *J. Amer. Chem. Soc.,* **93,** 4217
220. Crossland, R. K. and Servis, K. L. (1970). *J. Org. Chem.,* **35,** 3195
221a. Dueber, T. E., Stang, P. J., Pfeifer, W. D., Summerville, R. H., Inhoff, M. A., Schleyer, P.v.R., Hummel, K., Bocher, S., Harding, C. E. and Hanack, M. (1970). *Angew. Chem. Int. Ed. Eng.,* **9,** 521
221b. Truce, W. E. and Liu, L. K. (1970). *Tetrahedron Letters,* 517
222. Durst, T. and Tin, K.-C. (1970). *Can. J. Chem.,* **48,** 845
223. Mori, A., Nagayama, M. and Mandai, H. (1971). *Kogyo Kagaku Zasshi.,* **74,** 715. (1971). *Bull. Chem. Soc. Jap.,* **44,** 1669
224. Brook, A. J. W. and Robertson, R. K. (1971). *J. Chem. Soc. B,* 1161
225. Bosscher, J. K. and Kloosterziel, H. (1970). *Recl. Trav. chim. Pays-Bas.,* **89,** 402
226. Bosscher, J. K., Kraak, E. W. A. and Kloosterziel, H. (1971). *Chem. Commun.,* 1365
227. Harpp, D. N., Gleason, J. C. and Ash, D. K. (1971). *J. Org. Chem.,* 322
228. Sammers, M. P., Wylie, C. M. and Hoggett, J. G. (1971). *J. Chem. Soc. C,* 2151
229. Khan, M. S. and Owen, L. N. (1971). *J. Chem. Soc. C,* 1448
230. Huppatz, J. L. (1971). *Aust. J. Chem.,* **24,** 653
231. Tonnet, M. L. and Hambly, A. N. (1970). *Aust. J. Chem.,* **23,** 2427, 2435; Foon, R. and Hambly, A. N. (1971). *Aust. J. Chem.,* **24,** 713
232. King, J. F. and Beatson, R. P. (1970). *Chem. Commun.,* 663
233. King, J. F. and Harding, D. R. K. (1971). *Chem. Commun.,* 959
234. King, J. F., de Mayo, P., McIntosh, C. L., Piers, K. and Smith, D. L. (1970). *Can. J. Chem.,* **48,** 3704
235. Hirai, K. and Tokura, N. (1970). *Bull. Chem. Soc. Jap.,* **43,** 488
236. King, J. F., Marty, R. A., de Mayo, P. and Verdun, D. L. (1971). *J. Amer. Chem. Soc.,* **93,** 6304
237. King, J. F. and Lee, T. W. S. (1971). *Can. J. Chem.,* **49,** 3724
238. Paquette, L. A., Freeman, J. P. and Maiorana, S. (1971). *Tetrahedron,* **27,** 2599
239. Grossert, J. S. (1970). *Chem. Commun.,* 305
240. Karger, M. H. and Mazur, Y. (1971). *J. Org. Chem.,* **36,** 528, 532, 540
241. Casadevall, A. and Commeyras, A. (1970). *Bull. Soc. Chim. Fr.,* 1856
242. Jonsson, E. U., Bacon, C. C. and Johnson, C. R. (1971). *J. Amer. Chem. Soc.,* **93,** 5306
243. Johnson, C. R. and Jonsson, E. U. (1970). *J. Amer. Chem. Soc.,* **92,** 3815
244. Jonsson, E. U. and Johnson, C. R. (1971). *J. Amer. Chem. Soc.,* **93,** 5308
245. Schroeck, C. W. and Johnson, C. R. (1971). *J. Amer. Chem. Soc.,* **93,** 5305
246. Haake, M. (1970). *Tetrahedron Letters,* 4449
247. Haake, M. (1970). *Angew. Chem. Int. Edn. Eng.,* **9,** 373
248. Johnson, C. R. and Schroeck, C. W. (1971). *J. Amer. Chem. Soc.,* **93,** 5303
249. Johnson, C. R., Haake, M. and Schroeck, C. W. (1970). *J. Amer. Chem. Soc.,* **92,** 6594
250. Johnson, C. R., Katekar, G. F., Huxol, R. F. and Janiga, E. R. (1971). *J. Amer. Chem. Soc.,* **93,** 3772
251. Johnson, C. R. and Katekar, G. F. (1970). *J. Amer. Chem. Soc.,* **92,** 5753
252. Dale, A. J. and Frøyen, P. (1970). *Acta. Chem. Scand.,* **24,** 3772
253. Konda, K. and Tunemoto, D. (1970). *Chem. Commun.,* 1361
254. LaRochelle, R. W., Trost, B. M. and Krepski, L. (1971). *J. Org. Chem.,* **36,** 1126
255. Baldwin, J. E. and Erikson, W. F. (1971). *Chem. Commun.,* 359
256. Baldwin, J. E., Erikson, W. F., Hackler, R. E. and Scott, R. M. (1970). *Chem. Commun.,* 576
257. Trost, B. M. and LaRochelle, R. W. (1970). *J. Amer. Chem. Soc.,* **92,** 5804
258. Manya, P., Sekera, A. and Rumpf, P. (1970). *Tetrahedron,* **26,** 467
259. Lillya, C. P., Miller, E. F. and Miller, P. (1971). *Int. J. Sulphur Chem. A,* **1,** 89
260. Johnson, A. W. (1970). *Organic Compounds of Sulphur Selenium and Tellurium,* Vol. 1, 248, Senior Reporter D. H. Reid (London: The Chemical Society) and refs. therein.

7
Carbonyl Compounds

R. BRETTLE
University of Sheffield

7.1   MONOCARBONYL COMPOUNDS: ALDEHYDES AND KETONES

7.1.1   Synthesis

7.1.1.1   *Syntheses based on nucleophilic acylation*

Several recent aldehyde and ketone syntheses are based on the idea of nucleophilic acylation[1]. A beautifully simple and now obvious example of this is the use of the anion from an *O*-protected aldehyde cyanohydrin as an acyl carbanion equivalent. The hydroxy group is protected by reaction with ethyl vinyl ether, and the anion can then be prepared by using lithium di-isopropylamide and alkylated with a primary or a secondary alkyl halide. Successive treatment with dilute aqueous acid and base then generates the ketone.

One of the best known of these methods is the one introduced by Corey and Seebach in 1965 based on the use of sulphur-stabilised carbanions from 1,3-dithianes. The preparation of 1,3-dithiane itself has now been described in *Organic Syntheses*[3] and its uses as a source not only of aldehydes and ketones but of acyloins, α-diketones and α-keto-acids have been reviewed[4]. A major difficulty until recently has been the final conversion of the intermediate 2-substituted 1,3-dithianes into the carbonyl compounds. However, oxidative cleavage by, for example, *N*-chlorosuccinimide in the presence of silver(I) ions in aqueous solvent systems has now been shown to be effective for a wide variety of dithianes and has made possible the synthesis of α-dicarbonyl compounds from 2-acyl-1,3-dithianes[5]. The original method[4], based on mercury(II) chloride fails with 2-(acetoxy-3,3-diethoxypropyl)-1,3-dithiane, but the aldehyde is obtained by the use of mercury(II) oxide and boron trifluoride in aqueous tetrahydrofuran[6], and can be converted into the α,β-alkenic aldehyde by treatment with 1,5-diazabicyclo[5.4.0]undec-5-ene. Since β-acetoxyalkyl-1,3-dithianes are available from epoxides[4] this result opens up a route to α,β-alkenic aldehydes.

Saturated aldehydes can now be obtained by the action of either of the foregoing reagents[5, 6] on 2-alkyl-1,3-dithianes, which can be prepared either by the alkylation of the 1,3-dithiane anion or by the ionic reduction of 2-alkylidene-1,3-dithianes by using trifluoracetic acid as a proton donor and triethylsilane as a hydride-ion donor.

One route to alkylidenedithianes is from an aldehyde (Figure 7.1) and thus the sequence shown in Figure 7.1 provides a method for aldehyde homologation[7]. The thioacetal monoxides ((1); $R^1 = R^2 = H$ or alkyl) are converted into the corresponding aldehydes or ketones by the action of catalytic amounts of a mineral acid in anhydrous organic solvents[8]; they can be prepared by the alkylation of methyl methylthiomethyl sulphoxide ((1); $R^1 = R^2 = H$) or by the oxidation of the corresponding thioacetals, which are available from formaldehyde dimethylthioacetal[8]. 1,3-Dithiane 1-oxide cannot be substituted for the sulphoxide ((1); $R^1 = R^2 = H$) in this modification[9], but the use of

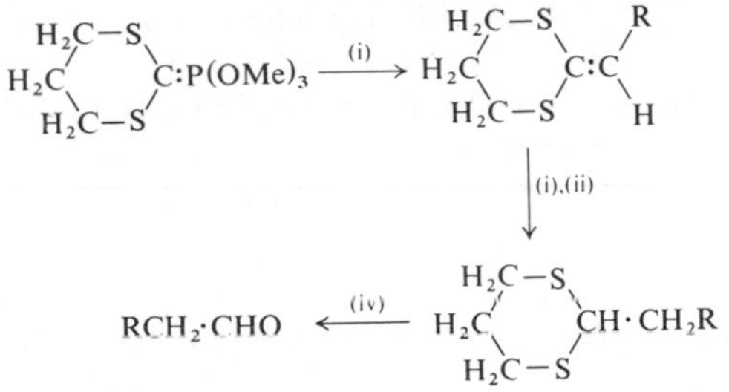

Reagents: (i) RCHO, (ii) $CF_3 \cdot CO_2H$, (iii) Et_3SiH, (iv) $Hg(II), H_2O$

Figure 7.1

thioacetal monoxides to prepare other classes of carbonyl compounds will clearly be investigated. Another ingenious and versatile synthesis, shown in Figure 7.2 uses diethyl methylthiomethylphosphonate to prepare enethiol

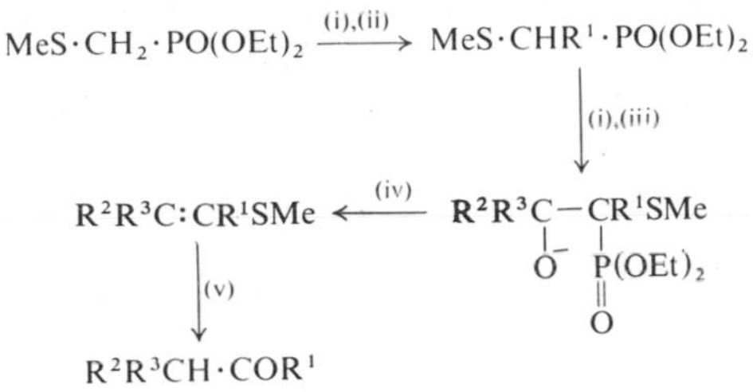

Reagents: (i) Bu^nLi, (ii) R^1X, (iii) R^2R^3CO, (iv) heat (v) $Hg(II), H_2O$

Figure 7.2

ethers by the phosphonate modification of the Wittig reaction, and a ketone can then be obtained by mercury(II) ion-catalysed hydrolysis[10].

Vinylogous nucleophilic acylation can be effected through the anion (2), easily preparable from epichlorohydrin, which is chemically equivalent to the 2-formylvinyl anion (3). For example, condensation with 1-bromopentane

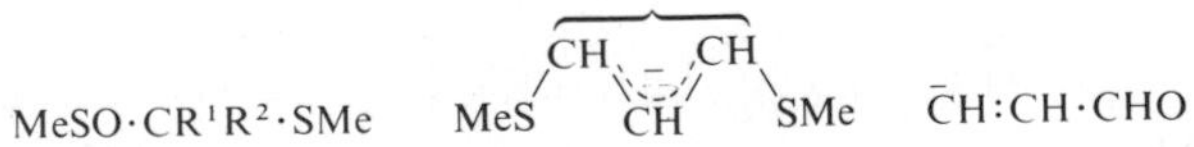

gives a product which is hydrolysed in the presence of mercury(II) ions to trans-oct-2-enal, which is best prepared by this route, and condensation with aldehydes and ketones gives products similarly transformed into *trans*-4-hydroxyalk-2-enals[11].

7.1.1.2   Syntheses based on tetrahydro-oxazines

Developments of the versatile aldehyde and ketone syntheses of Meyers continue to appear. Thus alkylation of the lithium salt of 2,4,6-tetramethyl-5,6-dihydro-1,3(4H)oxazine (4) with αω-dihalides, followed by reduction to the tetrahydro-derivative by sodium borohydride, and acidic hydrolysis leads to ω-halogenoaldehydes in which the α- and the β-carbon atoms are derived

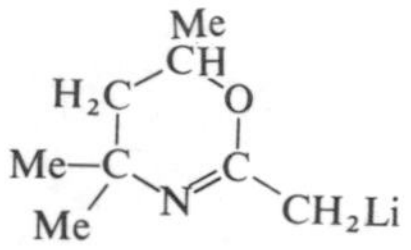

(4) 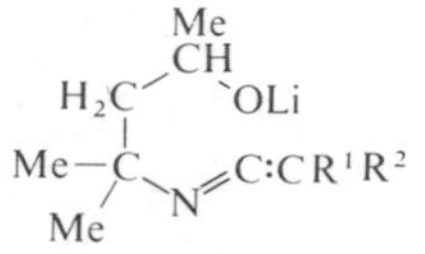   (5)

from (4)[12]. Standard transformations of the ω-halogenoalkyl group at the dihydro-oxazine-derivative stage provide routes to many other ω-substituted aldehydes[12]. Salts analogous to (4) from dihydro-oxazines in which the 2-alkyl-substituent is larger than methyl can be prepared at −78 °C and used to prepare α-deuterated aldehydes, but near room temperature they are largely present as the lithium salts of hydroxyketenimines (5). Routes to the parent dihydro-oxazines had been established earlier. Reaction of (5) with an alkyl-lithium introduces a new alkyl group and hydrolysis of the resultant dilithium

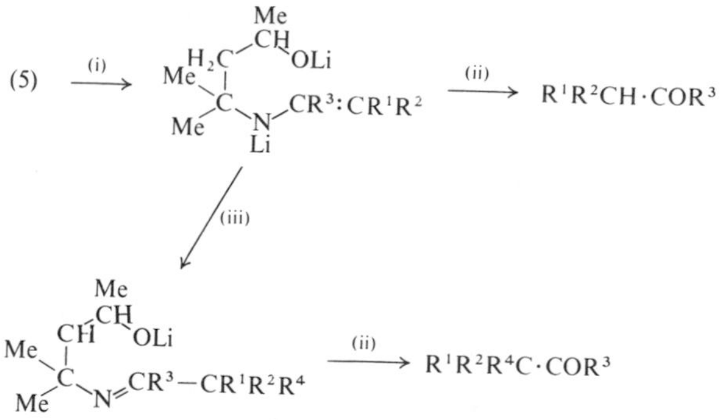

Reagents: (i) R³Li, (ii) H₃O⁺, (iii) R⁴I

Figure 7.3

salt gives a ketone. Alternatively, the dilithium salt can be further alkylated with an alkyl halide to give a lithio-imine which on hydrolysis gives a more highly alkylated ketone. These reactions[13], which provide one of the best routes to ketones containing a quaternary carbon atom at the α-position are illustrated in Figure 7.3. An aldehyde synthesis based on N,4,4-trimethyl-2-oxazolinium salts is *not* applicable in the aliphatic series[14], but one based on

2,4-dimethylthiazole ought to be applicable, although this has not yet been demonstrated[15]. This is also the present position with a new ketone synthesis based on 2-alkyl-N,4,4-trimethyldihydro-oxazolium salts[16].

7.1.1.3 Syntheses based on organoboranes

Ketone syntheses based on hydroboration continue to be developed. One very efficient synthesis[17] involves the reaction of a trialkylborane with cyanide ion to give a trialkylcyanoborate. Treatment of this with just 1 mol of trifluoroacetic anhydride causes two of the alkyl groups to migrate to carbon, and oxidative hydrolysis of the intermediate with alkaline hydrogen peroxide then gives a ketone. Tri-n-octylborane gives heptadecan-9-one in 95% yield. By using a trialkylborane in which one group is the thenyl group, which has a low aptitude for B → C migration and in which the other two groups can be introduced one at a time by the hydroboration of two different alkenes, the synthesis can be adapted to the preparation of unsymmetrical ketones[18]. The conditions are so mild that 8-iodo-oct-1-ene, and no doubt other functionalised alkenes, can be used in this synthesis.

A few years ago the Purdue school showed that some but not all α,β-alkenic aldehydes and ketones undergo β-alkylation on reaction with trialkylboranes. It has now been shown that this is a radical reaction[19]. Hydrolysis of the intermediate vinyloxyboranes ((6); X = BR_2^5) gives ketones, and in this way, by using the hydroboration products from pent-1-ene, isopropenyl methyl ketone is converted nearly quantitatively into a mixture of 3 parts of 3-methylnonan-2-one and 1 part of 3,5-dimethyloctan-2-one[19] (see also p. 294).

$$R^1R^2CH \cdot CR^3 {:} CR^4OX$$

(6)

Alternatively, treatment of the distilled vinyloxyboranes (e.g. (6); R^1 = Bu^n; $R^2 = R^3$ = H; R^4 = Me; X = BR_2^5) with an alkyl-lithium gives a specific lithium enolate (e.g. (6); R^1 – Bun, $R^2 = R^3$ = H; R^4 = Me; X = Li) which can then be alkylated in the conventional manner to give an α-alkylated ketone[20]. In those cases in which the extent of the initial addition of the trialkylborane is minimal, such as with crotonaldehyde, ethylideneacetone, and mesityl oxide, the reaction is catalysed by oxygen, by diacetyl peroxide, or by photochemical irradiation in propan-2-ol[21]. Some of the α,β-alkenic ketones can themselves be made by a similar process involving the catalysed addition of a trialkylborane to butynone, thus allowing the stepwise introduction of two different alkyl groups in the β-position of the final saturated methyl ketone[22]. Highly hindered ketones can be prepared by the simple, or if necessary catalysed, addition of a B-alkylborocyclane such as a B-alkyl-borinane, a B-alkyl-3,5-dimethylborinane or a B-alkyl-3,6-dimethylborepane, to an α,β-alkenic ketone[23]. All these reagents, and especially the last one, can readily be prepared by the hydroboration of the appropriate highly branched alkenes with bis-borinane, bis-(3,5-dimethyl)borinane or bis-(3,6-dimethyl)-borepane respectively, and B-t-butyl-3,5-dimethylborinane, a convenient source of the t-butyl group can be prepared by the action of t-butyl-lithium on B-methoxy-3,5-dimethylborinane[23]. By converting a terminal alkene into a B-alkylboracyclane rather than a trialkylethylene, the product with an unbranched β-alkyl group can be obtained to the virtual exclusion of the isomeric ketone (see p. 293)[23]. Another route to highly hindered ketones is based on the monohydroboration of alkynes. Monohydroboration of alk-1-ynes by mono- or di-alkylboranes gives almost exclusively a bond from the terminal carbon atom to boron; this is also true of alk-3-en-1-ynes. The regiospecificity in the addition of a dialkylborane to an internal alkyne depends on the two alkyl groups, but where they differ in ramification there is a strong tendency for the new boron–carbon bond to be formed at the less hindered position. For example, when the groups are primary alkyl and t-butyl the product after oxidative hydrolysis is almost exclusively the alkyl neopentyl ketone[24]. Monohydroboration of symmetrical conjugated dialkynes, readily available by Glaser coupling, can be achieved and places the boron atom at the propargylic position; oxidative hydrolysis then gives an α,β-acetylenic ketone[25]. This process appears to be preferable to the acid-catalysed monohydration of the dialkyne.

7.1.1.4 Syntheses from acids and acid derivatives

The preparation of ketones by the reaction of carboxylic acids with organo-lithium reagents has been reviewed in *Organic Reactions*[26]. The method has been used mainly for methyl ketones, but is applicable to other alkyl and to vinyl ketones. The review includes a detailed account of the recent procedure due to House in which a preformed lithium carboxylate is used in the reaction with the alkyl-lithium to prevent tertiary alcohol formation. Acid chlorides react with organocopper reagents, e.g. a lithium dialkylcopper, to give ketones[27]. The conditions are very mild and groups known not to react with organocopper reagents such as carboxy, alkoxycarbonyl, carboxamide, halogenoalkyl, and carbonyl are likely to be compatible with the method. α,β-Acetylenic ketones can be prepared by the action of copper(I) acetylides on acid chlorides, including α,β-alkenic acid chlorides, in the presence of lithium iodide in aprotic solvents[28]. Methyl ketones can be obtained by the reaction of acid chlorides with trimethylsilylmethylmagnesium chloride, which gives an oxosilane readily hydrolysed to the ketone[29]. The best route[30] to hindered aliphatic ketones from acid chlorides is by their reaction with organometallic reagents in the presence of copper(I) iodide, which prevents the reduction of the carbonyl group. Alkyl-lithium compounds are the preferred reagents, but Grignard reagents are only a little less satisfactory and are much more generally available. The scope is limited by the availability of the requisite highly hindered acid chlorides. Careful attention to the conditions ensures yields of c. 80%. Ketones can also be obtained from Grignard reagents and esters, provided that the reaction is conducted in N^1,N^2,N^3-hexamethyl phosphoric triamide[31]. In this solvent the ketonic product is extensively enolised and so protected from further reaction. α-Fluoro-esters react with Grignard reagents at $-60\,°C$ to give α-fluoroketones[32] and α-chloro- or α-bromo-esters react with propargylmagnesium bromide to give a mixture of acetylenic and allenic ketones, which on alkaline isomerisation leads exclusively to the α-halogenoallenic ketones[33]. α-Chloroglycidic esters ((7); $X = CO_2R^3$) react with Grignard reagents at $-78\,^{\circ}C$ to give α-chloro-α,β-epoxyketones ((7); $X = COR^4$); at $-50\,^{\circ}C$ further reaction occurs to give the corresponding tertiary alcohols ((7); $X = C(OH)R_2^4$)[34]. The starting esters ((7); $X = CO_2R^3$) can be made by the

$$R^1R^2C\overset{\displaystyle O}{\overset{\diagup\diagdown}{-}}CClX \qquad CCl_2{:}C(OR)OMCl \qquad Na_2Fe(CO)_4$$

(7)   (8)   (9)

Darzens reaction from α,α-dichloroesters or by ring closure of α,α-dichloro-β-hydroxy-esters, which can be obtained by the reaction of aldehydes or ketones with reagents of the type (8), available by the action of Grignard reagents (M = Mg) or zinc metal (M = Zn) on alkyl trichloracetates[35]. The glycidic esters ($X = CO_2R^3$) rearrange at $80\,^{\circ}C$ in the presence of pyridine to give β-chloro-α-ketoesters[36]. Exceptionally ethyl 2,2-diethoxybutanoate reacts with lithium dimethylcopper to give 3,3-diethoxypentan-2-one; the method is not generally applicable to esters[37]. Symmetrical ketones can be prepared by the reaction of a Grignard reagent, which provides both the alkyl groups, with an N,N-disubstituted carbamate[38].

Acid chlorides can be converted into aldehydes by treatment with disodium tetracarbonylferrate(II) (9), prepared by the reduction of iron pentacarbonyl with sodium amalgam, and quenching the product with acetic acid[39]. This is an acceptable alternative to the Rosenmund reduction, for which the use of ethyldi-isopropylamine as the hydrogen chloride acceptor has been recommended[40]; the hydrochloride is soluble in acetone, which can be used as the solvent. The reduction proceeds at room temperature and atmospheric pressure with a palladium catalyst[40].

The reduction of carboxylic acids to aldehydes with lithium in methylamine has recently been developed into a useful synthetic method[41]. Isolated alkenic bonds are unaffected but conjugated and skipped diene systems are reduced. This reduction, and that of N-substituted amides[42] can also be effected electrochemically in methylamine containing lithium chloride, with ethanol as a proton source.

A route from acids to aldehydes via β-formyl acids is discussed in Section 7.3.1.2.

7.1.1.5   Syntheses from alkenes

The oxidation of alkenes by mercury(II) salts has been summarised[43] for the first time since the initial results appeared in 1898. Terminal alkenes give α,β-alkenic carbonyl compounds; but-1-ene gives butenone and prop-1-ene gives acrolein. Buta-1,3-diene gives a mixture of butenone and 2,5-dihydrofuran. In the presence of iron(III) ions terminal alkenes give saturated ketones; propene gives acetone and but-1-ene gives butanone. Electrochemical oxidation has been used to regenerate mercury(II) ions from the mercury(I) ions produced in the oxidation of propene, thus leading to a continuous process for the oxidation of propene[44] to acrolein by mercury(II) nitrate in nitric acid at 60 °C. The oxidation of alkenes in aqueous media by palladium(II) salts has been reviewed[45]. Both alk-1- and alk-2-enes give high yields of methyl ketones. α,β-Alkenic acids, ketones, nitriles and nitro-compounds give the β-oxo-derivatives, and so α,β-alkenic acids give methyl ketones by decarboxylation of the intermediate β-keto-acids. A convenient method for oxidising terminal alkenes to methyl ketones involves methoxymercuration, followed by transmetallation with a palladium(II) salt to give an intermediate which breaks down to give the ketone[46]. The process can be made catalytic in palladium by adding copper(II) chloride. In this way hexan-2-one can be obtained quantitatively from hex-1-ene under very mild conditions. A promising aldehyde/ketone synthesis based on alkene oxythallation is *not* satisfactory in the aliphatic series[47].

Some recent results from the long-known oxidation of alkenes by selenium dioxide are interesting[48]. The specific oxidation of the methyl groups in either of a pair of stereoisomeric 2-alkylalk-2-enes can be achieved and leads predominantly in each case to the *trans*-aldehyde (see Figure 7.4). For 2-methylalk-2-enes (R = Me) the proportion of the *trans*-isomer is

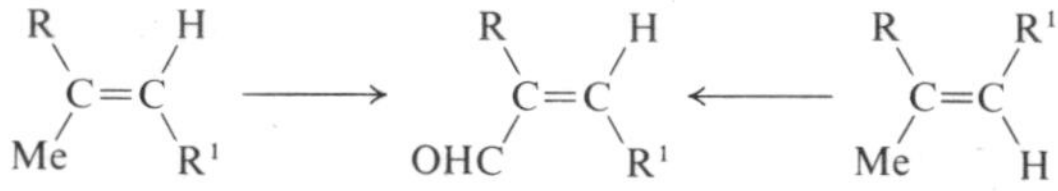

Figure 7.4

c. 99%, but this drops to 55% for 2,2,3-trimethylalk-3-enes (R = t-butyl). Oxidation of a 2-methylalk-2-en-1-ol, whether *cis* or *trans* in configuration, similarly gives exclusively the *trans*-2-methylalk-2-enal.

The oxidation of alkenes to ketones in one step by using sulphur in the presence of steam as the oxidising agent has been discussed[49]; this is an important technical method.

An interesting set of ketonic products is formed by the oxidation of an enolisable ketone with manganese(III) or ceric(IV) salts in the presence of an

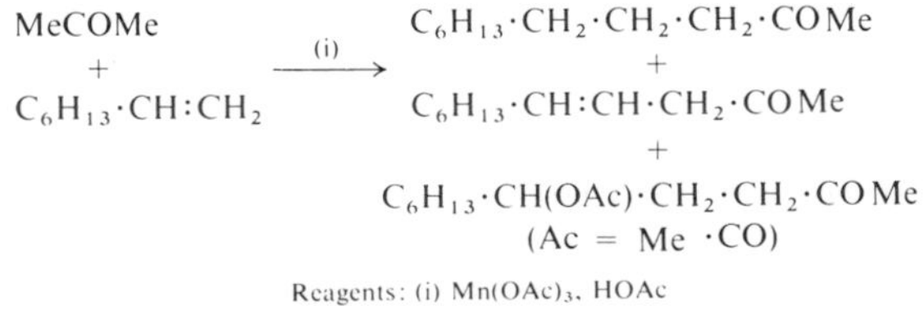

Reagents: (i) Mn(OAc)₃, HOAc

Figure 7.5

alk-1-ene[50]. A typical case is shown in Figure 7.5. The proportions of the products depend upon the conditions. The reaction involves the production of an enolate radical, which adds to the alkene to give a radical which is more easily oxidised to a cation than the enolate radical. The products then arise by hydrogen abstraction by the intermediate radical, or by elimination from or solvolysis of the cation. With 2-methylbutan-4-one the initial hydrogen abstraction is from the methyl not the methine group.

7.1.1.6 Syntheses from alcohols

Dipyridinechromium(VI) oxide in dichloromethane—the Collins reagent[51]—is the perfect oxidant for primary and secondary alcohols. The acetonide of butane-1,2,4-triol gives the corresponding aldehyde in 99% yield[52] and octan-2-ol gives octan-2-one is 97% yield[53]. The reagent prepared in situ[51] seems to give somewhat lower yields than when the dipyridinechromium oxide is prepared separately[54] and then dissolved in dichloromethane, but the in situ method avoids both the considerable, though well recognised, fire hazard in the preparation of the oxide and problems due to its hygroscopic nature. The reagent oxidises internal alkynes to conjugated alkynones in moderate yields; dec-2-yne gives dec-2-yn-4-one, but dec-1-yne is unaffected[55]. It has been used earlier for a similar allylic oxidation of alkenes[54].

7.1.1.7 Syntheses from halides

The oxidation of halides to carbonyl compounds by dimethyl sulphoxide under anhydrous conditions is catalysed by silver(I) ions; the addition of silver perchlorate allows the reaction to be performed at room temperature. Bromides can be converted into aldehydes with one more carbon atom by the action of disodium tetracarbonylferrate(II) (9) in the presence of triphenylphosphine, followed by quenching the product with acetic acid; hexanal was obtained in 99% yield from 1-bromopentane, but yields are poorer with secondary bromides[57].

7.1.1.8 Syntheses based on the alkylation of aldehydes and ketones

A new high-yield route to tertiary alkyl ketones allows the introduction at an α-methylenic position of a primary alkyl group together with a methyl group. Thus 2,2-dimethylpentan-4-one can easily be prepared from pentan-3-one. The α-methylene group is converted by standard methods into a butylthiomethylene derivative, reduction of which with lithium in liquid ammonia generates an α-methyl group and a lithium enolate group in which the double bond is directed towards the newly methylated position, which can then be alkylated in the usual way[58]. Hydrolysis without alkylation provides a simple method of α-methylation. Specific metal enolates have, of course, been prepared in the past, for example through enol trimethylsilyl ethers. A recent method[59] starts from an α-chloroketone, both isomers of which are frequently accessible. Reaction with ethyl diphenylphosphinate gives an enol diphenylphosphinate, cleavage of which with an alkyl-lithium then gives an alkylatable lithium enolate of specific constitution. It has also been shown that aldehyde enol O-tributyltin derivatives (10), which can be obtained from the enol acetates with tributylmethoxytin, are alkylated by halides[60]. The reaction, which had previously been reported for ketones[61], is particularly useful for generating a tertiary α-carbon atom; for example pivalaldehyde is easily prepared from isobutyraldehyde through (10) with R = Me.

$$R_2C:CH \cdot OSnBu_3^n \qquad Bu^tCO \cdot CH_2Pr^i \qquad Bu^tCO \cdot CHPr_2^i$$

$$\qquad (10) \qquad\qquad\qquad (11) \qquad\qquad\qquad (12)$$

The lithium divinylcopper–tributylphosphine complex is a much better reagent for the introduction of a vinyl group by 1,4-addition to an α,β-alkenic ketone than the vinylmagnesium halides; butenone gives hex-5-en-2-one in good yield[62]. Highly substituted ketones can be prepared by the reaction of α-bromoketones with a lithium dialkylcopper; in this way an isopropyl group can be introduced into (11) to give (12). The yields are poor, but acceptable in view of the directness of the method[63].

7.1.1.9 Syntheses based on β-keto-sulphoxides and -sulphones

The Corey–Chaykovsky ketone synthesis can be extended by alkylating a β-ketosulphoxide with an electrophilic alkene before the reductive removal of the sulphur[64]. In this way electron-withdrawing groups, e.g. cyano, can be introduced, and the method provides a route to 1,5-diketones and δ-keto-esters. β-Ketosulphones, like other moderately acidic compounds can be alkylated by Brandstrom's ion-pair extraction technique, in which the tetra-butylammonium salt of the anion is extracted into an organic solvent and then alkylated in a separate step[65]. More vigorous conditions (sodium hydride) had been used previously. Efficient conversion of β-ketosulphones into ketones can be simply achieved by electrochemical reduction in an undivided cell[66].

7.1.1.10 Miscellaneous syntheses

Further indications of the variety of carbonyl compounds which can be prepared by using metalloaldimines ((13); M = Li or MgBr) derived from 1,1,3,3-tetramethylbutyl isocyanide have been given[67]. Routes to aldehydes,

$$Bu^t CH_2 \cdot CMe_2 N:CRM \qquad R_3 SiCH \overset{O}{\overbrace{}} CHX \qquad XCH_2 \cdot CHO$$

$$\qquad (13) \qquad\qquad\qquad (14) \qquad\qquad\qquad (15)$$

ketones with at least one α-methylene group, α-keto-acids and α- and β-hydroxyketones are being developed.

A promising new synthesis is based on the acid-catalysed rearrangement and cleavage of trialkylsilylepoxides, which can be prepared by the epoxidation of the corresponding trialkylsilylalkenes. So far only disubstituted epoxides of type (14) have been studied; these give to the aldehydes (15)[68]. Trialkylsilylalkenes can be prepared by the addition of trialkylsilanes to alk-1-ynes or by standard synthetic procedures from trimethylsilylacetyl-ene. Hydroxyepoxides of the type (14) with $X = CR^1R^2OH$, obtained from 1-trimethylsilylalk-1-yn-3-ols are dehydrated under the conditions for the rearrangement and cleavage so that the products in these cases are α,β-alkenic aldehydes. cis-4-Methylhex-3-enal and cis-5-methylhept-4-enal have both been prepared from a common precursor with great elegance by the routes shown in Figure 7.6, in which the crucial steps are both thermal sigmatropic rearrangements[69].

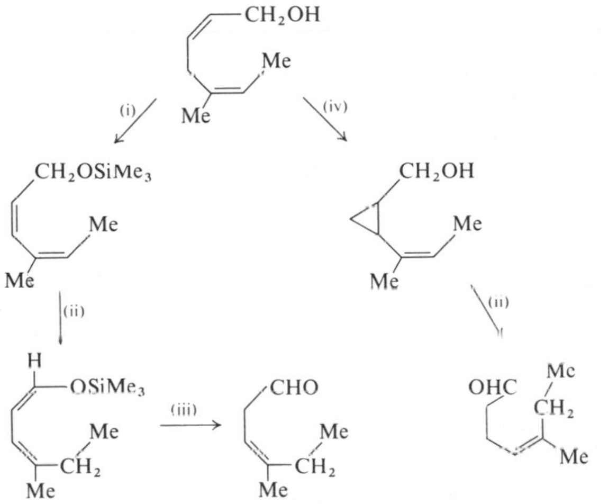

Reagents: (i) Me₃Si·N:N·SiMe₃, (ii) heat, (iii) MeOH, pH7, (iv) CH₂I₂, ZnCu

Figure 7.6

The best route to hexamethylacetone is by the reductive condensation of pivalonitrile with sodium to give the imine, which can then be hydrolysed to the ketone; the production and fragmentation of the radical anion from pivalonitrile is probably involved[70].

Some allenic aldehydes are now readily available by the base-catalysed elimination of hydrogen chloride from 3-chloro-alk-2-enals, which are them-selves easily synthesised from alkanones by a Vilsmeier reaction. For example, 2-methylbuta-2,3-dienal can be made from butanone by this method[71].

7.1.2 Regeneration from derivatives

The oxidation of ketone acetals (2,2-dialkyldioxolans) with triphenylmethyl tetrafluoroborate discovered by Barton[72] can be thought of either as a useful way to remove the protecting group or alternatively as an extremely good route from 1,2-diols to acyloins. It is illustrated in Figure 7.7. Acetals derived from primary-secondary diols (Figure 7.7; $R^3 = H$) give hydroxymethylene ketones. The reagent has been used to cleave a hemithioacetal (1,3-oxathiolan)

but failed to cleave a dithioacetal (1,3-dithiolan). However, both these protecting groups can be removed under very mild conditions by oxidative
hydrolysis with chloramine-T (cf. Section 7.1.1.1)[73].

Several reagents will effect the deoximation of oximes by reduction to the
imine, which is then rapidly hydrolysed. The best of these seems to be titan-
ium(III) chloride used in an acetate buffer at room temperature, with which

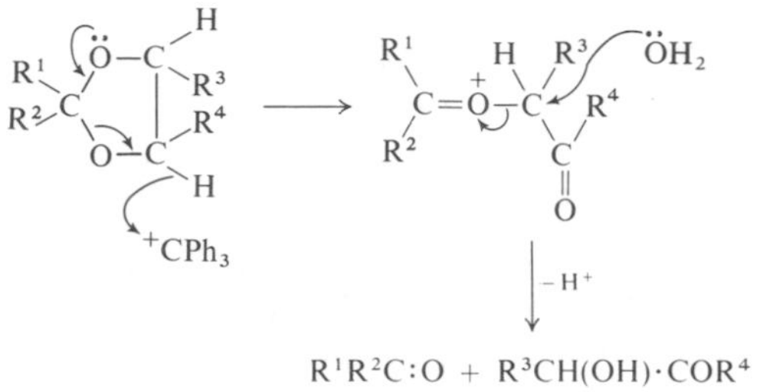

Figure 7.7

the reduction can be followed by the accompanying colour change[74]. It
works for both aldoximes and ketoximes, including conjugated alkenic
oximes, and is the only satisfactory reagent for the deoximation of α-diketone
monoximes. Thallium(III) nitrate is another useful reagent[75], although other
functional groups may be attacked; e.g. isolated double bonds are oxidised.
This reagent also regenerates carbonyl compounds from semicarbazones and
phenylhydrazones, but not from 2,4-dinitrophenylhydrazones. Another
method, which should be applicable to aliphatic compounds, the use of which
may permit selective deoximations, is the reduction of oxime O-acetates with
chromium(II) acetate[76]. Acid-sensitive groups are unaffected, but α-chloro-
and α-acetoxy-substituents suffer reductive cleavage.

As an alternative to exchange with keto-acids, non-conjugated alkenic
and saturated ketones can be regenerated from their 2,4-dinitrophenyl-
hydrazones by boiling them with one mol equiv^{-1} of toluene-p-sulphonic
acid hydrate in chloroform[77].

7.1.3   Reactions

7.1.3.1   Oxidation

Saturated ketones can be dehydrogenated to the corresponding α,β-alkenic
ketones by heating them with hydrogen peroxide or t-butyl hydroperoxide
in the presence of catalytic amounts of various transition-metal salts[78]. The
autoxidation of ketones in aprotic solvents in the presence of strong bases
such as potassium t-butoxide at low temperatures allows the α-hydro-
peroxyketones to be isolated in almost 100% yield[79]. Aldehydes are converted
into α-hydroxy-acids with the same number of carbon atoms[80] by oxidation
with oxygen in aqueous pyridine containing copper(II) chloride at 70 °C.
Aldehydes can be converted into the corresponding carboxylic acids through

acetals, which on ozonolysis give esters; 1,1-dimethoxyheptane from heptanal gives methyl heptanoate on ozonolysis at room temperature[81].

7.1.3.2 Reduction

Amalgamated aluminium in dichloromethane is a useful system for the pinacolic reduction of ketones, because the resultant aluminium salts are soluble in the medium[82]. Reduction of α,β-alkenic aldehydes and ketones with sodium borohydride in propan-2-ol gives not only the allylic and saturated alcohols but also the 1,3-ether-alcohol arising by reduction of the β-alkoxy-substituted aldehyde or ketone formed by Michael addition of propan-2-ol to the substrate[83]. This side-reaction is due to the production of a very strong base on the addition of the borohydride to the anhydrous alcohol, and can be completely suppressed by the addition of a few per cent of water. The possibility of base-catalysed epimerisation at the α-position to a carbonyl group during borohydride reductions can likewise be abolished by the addition of water[84]. Of several new reagents for the reduction of chiral α,β-alkenic ketones exclusively to the allylic alcohols, di-isopinocamphyl-t-butyl borohydride, prepared from di-isopinocamphenylborane and t-butyl-lithium in N^1,N^2,N^3-hexamethyl phosphoric triamide at $-120\,°C$, gave the highest diastereoisomeric differentiation[85].

Aldehydes and ketones can be deoxygenated by reduction of their toluene-p-sulphonylhydrazones. Sodium borohydride has been used, but it has now been shown that sodium cyanotrihydridoborate (16) is a superior reagent; fewer alkenic by-products are formed, and very few functional groups are

$$Na^+\bar{B}H_3CN$$

(16)

incompatible with the new reagent[86]. This also makes (16) a most important reagent for the reductive amination of many carbonyl compounds including aldehydes, unhindered ketones, α-keto-acids and β-keto-esters. In nearly all cases conditions leading to yields higher than those obtainable by other reductive amination procedures can be found[87].

7.1.3.3 Reactions at the α-position

An up-to-date review of the Favorskii rearrangement, the skeletal rearrangement of α-halogenated ketones under the influence of bases, which was last reviewed in 1960, has appeared[88]. A new method for the controlled hydroxymethylation of ketones[89] should be applicable in the aliphatic field. The ketone is converted into its 2-hydroxymethylene derivative with ethyl formate, and the sodium enolate of this β-dicarbonyl compound is reduced with aluminium hydride to the corresponding 2-hydroxymethyl ketone.

1-Bromo-1-chloro-1-fluoroacetone has been prepared by sequential halogenation[90] and resolved through its 'menthydrazide'[91]. Strong bases bring about the haloform reaction to produce the chiral bromochlorofluoromethanes, which might prove suitable chiral n.m.r. solvents for the investigation of enantiotopic ligands in a chiral environment.

7.2 DICARBONYL COMPOUNDS

7.2.1 α-Dicarbonyl compounds

Internal alkynes are oxidised to α-diketones by ruthenium tetroxide in much better yields than can be achieved by ozonolysis[92]. All alkynes can be oxidised to the corresponding α-diketones or α-keto-aldehydes by N-bromosuccinimide in anhydrous dimethyl sulphoxide, though full details have not yet been published[93]. Glyoxal can easily be obtained from chloracetaldehyde by a Kornblum oxidation with dimethyl sulphoxide[94]. A combination of two well-established specific oxidising agents allows the conversion of diethylenic 1,2-diols to the corresponding diethylenic 1,2-diketones[95]. The diol is oxidised to an acyloin with Fetizon's reagent (silver carbonate) and the acyloin is further oxidised to the α,β-α',β'-dialkenic diketone by bismuth oxide. Acyloins are also obtainable as their enediol bis-trimethylsilyl ethers by the condensation of esters with sodium in the presence of trimethylsilyl chloride; this reaction has recently been reviewed[96]. Treatment of the ene diol bis-ethers with bromine gives the corresponding α-diketones in high yield[97].

7.2.2 β-Dicarbonyl compounds

A very general and efficient synthesis of enolisable β-dicarbonyl systems based on alkylative coupling followed by sulphur extrusion has been developed by Eschenmoser[98]. It is illustrated schematically in Figure 7.8. For example, condensation of thiobutyric acid with 1-bromobutan-2-one

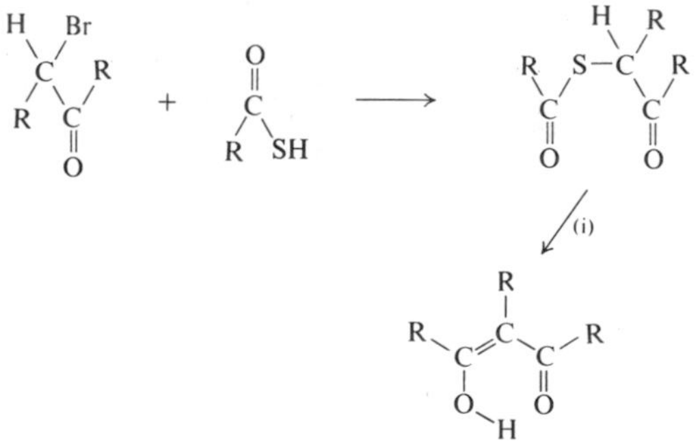

Reagent: (i) R₃P

Figure 7.8

gives the thioester and treatment of this with bis-(3-dimethylaminopropyl)-phenylphosphine (17) in the presence of lithium bromide gives octane-3,5-dione in c. 80 % overall yield.

$$PhP[(CH_2)_3NMe_2]_2$$

(17)

Other trivalent-phosphorus reagents are less efficient than (17). The reaction has also been applied by using two components each containing a ketone group protected as a cyclic acetal, thereby leading to the protected form of a tetraketone. The use of α-bromoesters in place of bromoketones leads to β-keto-esters.

7.2.3 γ-Dicarbonyl compounds

Two new syntheses of γ-diketones involve the use of a nitro-group as a protected form of a keto-group. In one method[99] a nitroalkane anion is used as the equivalent of an acyl carbanion (cf. Section 7.1.1.1) and is added to an α,β-alkenic ketone in a Michael reaction to give a γ-nitroketone. This is then converted into the γ-diketone by reduction to the imine with titanium(III) chloride followed by hydrolysis (cf. deoximation in Section 7.1.2). In this way heptane-2,5-dione can easily be prepared from butenone and 1-nitropropane. The method can be applied in the presence of an acetylenic linkage, which is unaffected by the reducing agent. In the other method[100] a ketone with an α-methyl or α-methylene group is condensed with 1-dimethylamino-2-nitropropene (18) in the presence of potassium ethoxide to give an α,β-alkenic γ-nitroketone as its aci-potassium salt. This is then methylated on oxygen in dimethyl sulphoxide to give the methyl nitronic ester, which on being heated in ethanol loses formaldehyde and gives the corresponding γ-keto-oxime, from which the α,β-alkenic γ-diketone is then regenerated by acidic hydrolysis. In this case the conversion of the nitro

$$Me_2NCH:CMeNO_2 \qquad CHO\cdot CMe_2\cdot CMe_2\cdot CHO \qquad Me_2C(CHO)\cdot O\cdot CH:CMe_2$$

$$(18) \qquad\qquad (19 \qquad\qquad (20)$$

$$MeCO\cdot CHEt\cdot CHEt\cdot COMe$$

$$(21)$$

group into a ketonic function is based on earlier methods developed by Kornblum. By this method acetone can be converted into trans-hex-3-ene-2,5-dione.

γ-Enol acetates react with ynamines to give enaminolactones which can be hydrolysed and decarboxylated to give γ-diketones[101]. The enol lactone from laevulinic acid reacts with 1-diethylaminoprop-1-yne to give the precursor of heptane-2,5-dione.

Various dimerisation reactions lead to γ-dicarbonyl compounds. For

$$Ph_2C:N\cdot N:CMeEt \xrightarrow[\text{(iii)}]{\text{(i),(ii)}} EtCO\cdot (CH_2)_2\cdot COEt$$

Reagents: (i) $Li\overset{+}{N}Pr^i_2$, (11) Cu^ICl, (iii) H_3O^+

Figure 7.9

example, certain ketazines, capable of α-metallation, can be dimerised to give bis-azines, which on hydrolysis give γ-diketones[102]. Details of the reagents are included with the example shown in Figure 7.9. The oxidative dimerisation of aldehydes can be accomplished with lead (IV) dioxide, activated manganese (IV) dioxide or nickel (IV) peroxide[103]. A mixture of

C → C and C → O coupled products is formed. For example isobutyraldehyde gives the γ-dialdehyde (19) in 34% yield, together with 45% of the C → O coupled product (20). The oxidation of ketones with lead dioxide had been reported earlier, but it has recently been shown that this route to γ-diketones is improved by carrying out the oxidation in toluene[104]. The yield from pinacolone is low and 4,4-dimethyl-1-phenylpentan-3-one, incorporating a solvent residue, is also formed together with bibenzyl. The reaction has also been applied to ethyl α-methylacetoacetate, (which unlike the parent β-keto-ester cannot be 'dimerised' with iodine under basic conditions) when diethyl α,α'-diacetyl-α,α'-dimethylsuccinate is amongst the products[105]. This γ-diketone is also amongst the products of the electrochemical oxidation of ethyl sodio-α-methylacetoacetate[106]. A practicable electrochemical route to simple γ-dicarbonyl compounds involves the electrolysis of carboxylic acid salts in the presence of α,β-alkenic aldehydes or ketones[107]. Electrochemically generated alkyl radicals add to the β-position of the conjugated systems to give radicals which then dimerise. Anodic oxidation of acetate ions in the presence of butenone for example, gives equal amounts of the (±)- and the meso-form of 3,4-diethylhexane-2,5-dione (21). If two different α,β-alkenic carbonyl compounds are present then crossed coupling occurs; anodic oxidation of acetate ion in the presence of a mixture of butenone and methacrolein gives the γ-keto-aldehyde (22) in 23% yield.

7.2.4 Other dicarbonyl compounds

The dimerisation of α,β-alkenic β-alkyl ketones can be effected by heating them with cyclohexyl isocyanide in the presence of copper(I) oxide[108]. In this way ethylidene acetone gives the 1,5-diketone (23). Co-dimerisation with an

MeCO·CHEt·CMeEt·CHC   MeCH:C(COMe)·CHMe·CH$_2$·COMe

(22)   (23)

α,β-alkenic ketone lacking a β-substituent is possible, but these ketones cannot themselves be dimerised.

Certain β-chloroketones undergo a Wurtz-type reaction with sodium amalgam, to give 1,6-diketones[109].

7.3 KETO-ACIDS AND THEIR DERIVATIVES

7.3.1 Synthesis

7.3.1.1 α-Keto-acids and their derivatives

β,β-Dialkyl-α-keto-esters (24) can be synthesised in excellent yields by the autoxidation of the corresponding α-cyano-esters (25) in the presence of potassium t-butoxide[110]. The α-cyano-esters (25) are easily obtained by the copper(I) iodide-catalysed addition of Grignard reagents to alkyl alkylidenecyanoacetates. The reaction of t-butyl alcohol with a cyanohydrin in the presence of sulphuric acid gives the N-t-butylamide of the related α-hydroxy-acid, which can be oxidised with chromium(VI) oxide to the amide of the

$$R^1R^2R^3C\cdot CO\cdot CO_2R^4 \qquad\qquad R^1R^2R^3C\cdot CH(CN)\cdot CO_2R^4$$

$$(24) \qquad\qquad\qquad\qquad (25)$$

corresponding α-keto-acid, hydrolysis of which gives the α-keto-acid itself[111].

7.3.1.2 β-Keto-acids and their derivatives

Several new β-keto-ester or β-keto-acid syntheses have appeared. The important general synthesis of β-keto-esters developed by Eschenmoser[98], which is based on the extrusion of sulphur from thioesters obtained by condensing α-bromo-esters with thio-acids, has been discussed in Section 7.2.2.

It has been known for 30 years that the sodium enolates of α-branched esters can be acylated by acid chlorides but that diacylation occurs with simple esters. Recently it has been shown that the use of the lithium enolate, prepared by the use of lithium cyclohexylisopropylamide in tetrahydrofuran at $-78\,^\circ$C permits the acylation of both simple and α-branched esters by an acid chloride[112]. The method is equivalent to a crossed-Claisen condensation, but without the production of any symmetrical by-products. The di-anions of carboxylic acids are now available by the action of lithium di-isopropyl-amide on acids with an α-hydrogen atom, and these, too, can be acylated at the α-position. For example, such a dianion reacts with ethyl formate to give an α-formylcarboxylate, which on acidification is decarboxylated to give an aldehyde[113]. The procedure provides a method for converting an acid into the corresponding aldehyde (see Section 7.1.1.4). Acylation of the dianions of carboxylic acids with esters other than formates similarly leads to a β-ketocarboxylate ion, which can be trapped with trimethylchlorosilane[114]. Solvolysis with neutral methanol then gives the β-keto-acid and, of course, these undergo very ready thermal decarboxylation to give ketones. Details are still awaited of another synthesis of β-keto-acids[115], based on the action of heat on alkylidene bis-(1,1-trimethylsilyl ethers) (26) which leads to the fully protected enol form of a β-keto-acid (27) from which the free acid can be obtained by methanolysis. This method is clearly related to the observation made some years ago that the action of heat on the 1-ethoxyvinyl esters of certain strong carboxylic acids such as trichloracetic acid, leads to β-keto-esters. It has now been shown[116] that the 1-ethoxyvinyl esters of weak carboxylic acids such as acetic acid (28), on treatment with zinc chloride behave analogously and give the enol-carboxylates of β-keto-esters (29) which can be hydrolysed to the β-keto-esters by dilute aqueous acid. The conversion of (28) into (29) is not brought about by heat, and the original reactions are not promoted by zinc chloride. Compounds of the type (28) are easily prepared by the addition of carboxylic acids to ethoxyacetylene in the presence of mercury(II) ions.

Full details have appeared of Bestmann's routes to β-keto-esters based on phosphorus ylids[117]. The reaction of an acid chloride with a Wittig reagent derived from an α-halogeno-ester and triphenylphosphine gives an α-triphenylphosphonio-β-keto-ester (30), which can be reduced electrolytically to the β-keto-ester. Alternatively, if the acid chloride possesses an α-hydrogen

$$R_2C:C(OSiMe_3)_2$$

(26)

$$R_2C:C(OSiMe_3) \cdot CR_2 \cdot CO_2 \cdot SiMe_3$$

(27)

$$CH_2:C(OEt)O \cdot COR$$

(28)

$$RC(O \cdot COR):CH_2CO_2Et$$

(29)

$$R^1CH_2 \cdot CO \cdot R^2(\overset{+}{P}Ph_3) \cdot CO_2Et$$

(30)

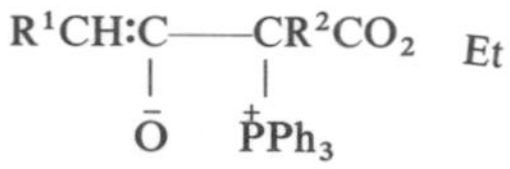

$$R^1CH:C\overset{|}{\underset{\bar{O}}{\rule{0pt}{0pt}}}\overset{|}{\underset{\overset{+}{P}Ph_3}{\rule{0pt}{0pt}}}CR^2CO_2 \quad Et$$

(31)

atom the addition of a second molar proportion of the Wittig reagent leads to the production of a betaine (31) which loses triphenylphosphine oxide to give an α,β,γ-allenic ester. Such esters add secondary amines at the β-position to give enamino-esters which can be hydrolysed to the same β-keto-esters as are obtained by the first route. The overall yields by the two routes are similar.

2-Diazo-1,3-diketones, available from the parent β-diketones by diazo-group exchange by using toluene-p-sulphonyl azide, on photolysis in water undergo the Wolff rearrangement and give β-keto-acids[118]. In this way (32) gives (33). A big improvement on an earlier method is the use of triethyl-oxonium tetrafluoroborate to catalyse the insertion of a CH·CO$_2$Et group derived from ethyl diazoacetate into the α-C—C bond of a ketone[119]. Ethyl α-ethylpropionylacetate can be prepared from diethyl ketone in 86% yield by this modified route.

$$Bu^t \cdot CO \cdot CN_2 \cdot CO \cdot Me$$

(32)

$$Bu^t \cdot CO \cdot CHMe \cdot CO_2H$$

(33)

The useful intermediate di-t-butyl acetonedicarboxylate can best be synthesised from acetonedicarboxylic acid by reaction with 2-methylpropene in sulphuric acid[120], a method employed earlier for malonic and other acids. Several recently reported routes are distinctly less favourable.

7.3.1.3 Polyketo-acids and their derivatives

Ethyl acetoacetate and dimethyl malonate react with keten in the presence of a little sodium chloroacetate to give the polyketo-esters (34) and (35) respectively, in a reaction which seems to be generally applicable to β-dicarbonyl systems. Reaction of ethyl orthocarbonate with propargyl-

$$(CH_3CO)_3C \cdot CO_2Et$$

(34)

$$(CH_3CO)_2C(CO_2Me)_2$$

(35)

magnesium bromide gives triethyl orthobut-3-ynoate, the lithium salt of which can be acylated with an acid anhydride. Michael addition of ethanol to the resultant α,β-acetylenic ketone then gives the enol ether of a 3,5-diketo-ortho-ester[122].

7.4 REACTIONS OF β-DI- AND POLY-CARBONYL SYSTEMS

7.4.1 Alkylation and acylation

Much detailed work continues to be reported concerning the alkylation of β-dicarbonyl systems through enolate anions. It has been shown, for example, that the C-alkylation of methyl acetoacetate and of pentane-2,4-dione by using optically active s-butyl bromide proceeds with partial retention of configuration in the s-butyl group, but that the parallel O-alkylation involves almost complete inversion of configuration[123]. The stereochemistry of the O-alkylation of the alkali metal salts of ethyl acetoacetate by ethyl toluene-p-sulphonate in $N^1N^2N^3$-hexamethyl phosphoric triamide has been studied in detail[124]. In the concentration range 0.01 to 0.1 M the reactive particle for all the alkali metals is the free enolate ion[125] and only the (E)-form of the enol ether, ethyl β-ethoxycrotonate, is formed. This result is attributed to the fact that the free enolate ion adopts the W-shaped conformation (36). On the other hand, the bidentate chelate form of the sodium enolate adopts the

(36)   (37)   (38)

U-shaped conformation (37) and to this is attributed[124] the fact that the O-alkylation of ethyl sodioacetoacetate by triethyloxonium tetrafluoroborate in ether gives the (Z)-form of ethyl β-ethoxycrotonate[126]. At a concentration as high as 1 M of the sodium derivatives in N^1,N^2,N^3-hexamethyl phosphoric triamide alkylation with ethyl toluene-p-sulphonate still gives almost exclusively ethyl (E)-β-ethoxycrotonate, but this is believed to arise under these conditions through a solvent-separated ion-pair in the W-shaped conformation (36)[124]. An intimate ion-pair would be expected to adopt the U-shaped conformation (37). It is not surprising that the results with the lithium salt often differ from those with the sodium, potassium and caesium salts[124]. Amongst other work in this area recently has been a comparison of the rate constants for the O- and C-alkylation of ethyl α-ethylsodioacetoacetate by ethyl toluene-p-sulphonate and by ethyl iodide in NN-dimethylformamide, which shows that the rates for ethyl iodide are higher at both nucleophilic centres, so that the ambident enolate ion is a relatively 'soft' base[127].

The monoalkylation products of heptane-2,4,6-trione, obtained by the action of alkyl halides in dimethyl sulphoxide in the presence of potassium carbonate, exist predominantly in the acyclic form at ambient temperatures, but the 3,5-dialkylated products, obtained similarly, exist as the cyclic tautomers (38)[128].

The γ-alkylation and γ-acylation of dianions of β-dicarbonyl compounds has recently been covered by *Organic Reactions*[129]. Acetoacetic esters cannot be alkylated in the γ-position under the usual Hauser–Harris conditions, by using two equivalents of sodamide in liquid ammonia, either because the di-anions are not formed or because if they are formed they do not react with alkyl halides at $-33\,°C$. However the di-anion from methyl acetoacetate can certainly be prepared by treating the monosodium derivative, prepared by means of sodium hydride, with butyl-lithium, although it cannot be made by treating methyl acetoacetate with two equivalents of butyl-lithium owing to the formation of carbonyl addition products. The di-anion undergoes the expected γ-alkylation on treatment with an alkyl halide at ambient temperatures, but the reaction is sluggish at $-23\,°C$ and does not take place at $-78\,°C$. The sequence can be repeated, leading to the γ,γ-dialkylated derivative[130]. The acetoacetate ester di-anions prepared either in this way or by using two equivalents of lithium di-isopropylamide, also undergo aldol type condensations with aldehydes and ketones at the γ-position[131]. The reactions are again very dependent on temperature, but good yields can be obtained with ketones; the yields are less satisfactory with aldehydes and no condensation was observed with formaldehyde. It had been reported earlier that heptane-2,4,6-trione could be carboxylated at the terminal position through the trianion prepared by the action of four equivalents of lithium di-isopropylamide and it has now been shown that the anion can also be acylated at the terminal position by using methyl benzoate[132]. The product is the first example of a tetraketone of the poly-β-carbonyl type with one aliphatic terminal group. It, too, can be converted into a polyanion with an excess of lithium di-isopropylamide and can thereby be benzoylated and carboxylated at the terminal position.

$$PdCl_2(PPh_3)_2 \qquad Pd[Ph_2P(CH_2)_2PPh_2]_2 \qquad Pd(MeCO\cdot CH\cdot COMe)_2$$

$$(39) \qquad\qquad (40) \qquad\qquad\qquad (41)$$

Pentane-2,4-dione and ethyl acetoacetate react with alka-1,3-dienes in the presence of palladium catalysts to give $1:2$- and $1:4$-adducts. For example the major product from ethyl acetoacetate and buta-1,3-diene in the presence of the complex (39) and sodium phenate is the α-octa-2,7-dienyl derivative. Isoprene similarly gives the 2,7-dimethylocta-2,7-dienyl derivative (corresponding to tail-to-tail coupling of the two isoprene units)[133]. When the catalyst is the related complex (40), ethyl acetoacetate and butadiene can be caused to react to give comparable amounts of the two $1:1$-adducts, the α-but-2-enyl and the α-1-methylprop-2-enyl derivative[134]. In the presence of palladium catalysts allylic exchange between allylic ethers, esters, alcohols and amines, and the active methylene positions of pentane-2,4-dione and of methyl acetoacetate can occur. Methyl acetoacetate and octa-2,7-dienyl phenyl ether in the presence of the catalyst (39) and sodium phenolate give a high yield of α-octa-2,7-dienylacetoacetate together with a small amount of the α,α-disubstituted keto-ester[135]. Similarly, in the presence of triphenylphosphine and the palladium complex (41) either of the allylic alcohols, octa-2,7-dien-1-ol or octa-1,7-dien-3-ol reacts with pentane-2,4-

dione to give 3-(octa-2,7-dienyl)pentane-2,4-dione together with a little of the disubstituted diketone[136].

7.4.2 Reduction

The lithium aluminium hydride reduction of β-diketones under forcing conditions gives predominantly *trans*-allylic alcohols, with small amounts of saturated mono- and di-hydric alcohols. For example hexane-2,4-dione gives (E)-hex-3-en-2-ol and (E)-hex-2-en-4-ol in 52 and 21% yield respectively[137].

The products in the electrochemical reduction of β-dicarbonyl compounds depend very much on the conditions. A careful examination of the products from the reduction of pentane-2,4-dione in hydrochloric acid at a mercury cathode has recently been completed[138]. The most important product is the rearranged acyloin 3-hydroxy-3-methylbutan-2-one ((42); R = OH), which is

$$\text{RCMe}_2 \cdot \text{COMe} \qquad \qquad \overset{\displaystyle \text{CH}_2}{\underset{}{\text{Me(OH)C}\!\!-\!\!\text{C(OH)Me}}}$$

$$(42) \qquad\qquad\qquad\qquad (43)$$

accompanied by the simple ketone ((42); R = H). The latter had previously been reported to be the only product of the Clemmensen reduction of pentane-2,4-dione with amalgamated zinc in hydrochloric acid; in both cases these rearranged products probably arise through 1,2-dimethylcyclopropane-1,2-diol (43), the product of intramolecular pinacol formation. The electrochemical reduction also leads to (E)-4,5-dimethyloct-4-ene-2,7-dione, which is derived from an intermolecular pinacol-forming process, together with its cyclisation product 3-acetyl-4-hydroxy-1,2,4-trimethylcyclopent-1-ene.

7.5 PHOTOCHEMISTRY

A recent paper[139] contains an extensive compilation of spectroscopic data, including results from i.r., u.v. and n.m.r. spectroscopy, which serves to establish the conformations of many conjugated dienones. Except for some highly substituted compounds which are non-planar or tautomerise to α-pyrones, the evidence is in favour of conformations which are sufficiently close to planarity to be labelled meaningfully *s-cis* or *s-trans*. The diene parts of the chromophores uniformly adopt an *s-trans* conformation, but the conformation of the enone part of the chromophores depends on the configuration of the α,β-alkenic bond and on the substitution pattern at the α- and the β-position. Irradiation of any of the configurations of a conjugated dienone in ether, at concentrations in the range 10^{-2} to 10^{-4} M, rapidly produces a photostationary state in which generally only the *trans,trans-*, *cis,trans-* and *trans,cis-*configurations are present[140]. A detailed study has been made of the photoisomerisation of hepta-3,5-dien-2-one, in which none of the all-*cis*-isomer is formed, although the photostationary state involves the other three configurations. The all-*cis*-isomer is probably converted into the *cis,trans*-isomer by a dark reaction involving valence tautomerism to an α-pyrone[141].

The photochemistry of saturated aliphatic ketones continues to be explored, although most of the chemical and photochemical problems in this area have now been solved. A summary has appeared recently[142] covering the photoelimination reactions of ketones possessing at least one hydrogen atom in the γ-position, and the accompanying photocyclisation to give cyclobutanols. These reactions are illustrated in Figure 7.10; the photofragmentation is analogous to the McLafferty rearrangement observed on

$$R^1CO \cdot CR_2^2 \cdot CHR_2^3 + CHR_2^4 \xrightarrow{\;hv\;} R^1COCHR_2^2 \; + \; CR_2^3 = CR_2^4$$

$$+$$

$$\begin{array}{cc}
OH & R^2 \\
| & | \\
R^1 - C - C - R^2 \\
| & | \\
R^4 - C - C - R^3 \\
| & | \\
R^4 & R^3
\end{array}$$

Figure 7.10

electron impact. Recent papers have reported experiments which are consistent with the involvement of both singlet and triplet 1,4-diradicals as intermediates in these processes. The pathways from the excited singlet and triplet states were separated in the usual way by carrying out the photochemical reactions in the presence and in the absence of dienes, which are efficient triplet quenchers. In experiments with (S)-(+)-5-methylheptan-2-one it was shown[143] by recovering 5-methylheptan-2-one before the photodecomposition had gone to completion that extensive racemisation occurs in both polar and non-polar solvents when the reaction is from the triplet state, but that no racemisation occurs when the reaction is from the singlet state. From this it was concluded that the excited carbonyl $n\pi^*$ triplet states decay to the triplet ground state, which then undergoes cyclic γ-hydrogen atom abstraction to give a long-lived triplet 1,4-diradical, in which β,γ-bond rotation occurs more rapidly than spin inversion and the inverse hydrogen atom transfer which leads to the observed racemisation, and that the singlet $n\pi^*$ state decays by a different route. Evidence from the measurement of quantum yields for the various processes involved in the photodecomposition of (S)-(+)-5-methylheptan-2-one and a series of related ketones having varying degrees of alkyl-substitution at the γ-position shows that non-radiative relaxation from the lowest excited $n\pi^*$ singlet state is important and increases as the nature of the γ-hydrogen atom changes from primary to secondary to tertiary[144]. It is concluded that the singlet 1,4-diradical generated by hydrogen abstraction from the first excited singlet state undergoes chemical transformations without spin inversion and before β,γ-bond rotation can occur, so that racemisation is not observed along this pathway owing to the rapidity of the reverse hydrogen atom transfer. The incursion of a chemical process in the non-radiative decay to the ground state is in line with theoretical predictions. The conclusions are confirmed by experiments on the photochemical decomposition of methyl *erythro*- and *threo*-3,4-dimethyl-6-oxoheptanoate which give both the (*E*)- and the (*Z*)-form of methyl 3-methyl pent-3-enoate by a Norrish Type II cleavage[145]. The reactions from the excited singlet state via the short-lived singlet 1,4-diradical are highly stereospecific, but the reactions from the triplet state via the long-lived triplet 1,4-diradical show very little selectivity. Much less work has been done on the photochemical reactions of aliphatic aldehydes in solution, but recent work has shown that simple aldehydes with at least one hydrogen atom at the γ-position undergo the same Norrish Type II photoelimination and photocyclisation reactions as the corresponding ketones[146]. Both the singlet and the triplet $n\pi^*$ states are direct precursors of the products.

During the photoreduction of simple ketones in very concentrated solutions in methylcyclopentane at $-70\,^\circ$C radicals have been detected by passing the irradiated solutions through the cavity of an electron-spin resonance spectrometer[147]. Among these radicals are ones such as (44) derived from acetone, and radicals such as isopropyl and t-butyl arising by

$$Me_2\dot{C}OH \qquad\qquad Me\dot{C}(OH)\cdot COMe$$

$$(44) \qquad\qquad\qquad (45)$$

the cleavage of a bond to an α-carbon atom, the so-called Norrish Type I cleavage, in the excited state of the corresponding methyl ketone; the other fragment from these cleavages gives rise to the radical (45).

Several photochemical reactions of α-diketones have been known for a long time, including 1,2- and 1,4-cyclo-additions and the competing hydrogen abstraction, α-cleavage, and photoenolisation processes. A new type of product has now been isolated from the photochemical reaction of butanedione (biacetyl) with certain methyl-substituted alkenes: an acyclic alkenic

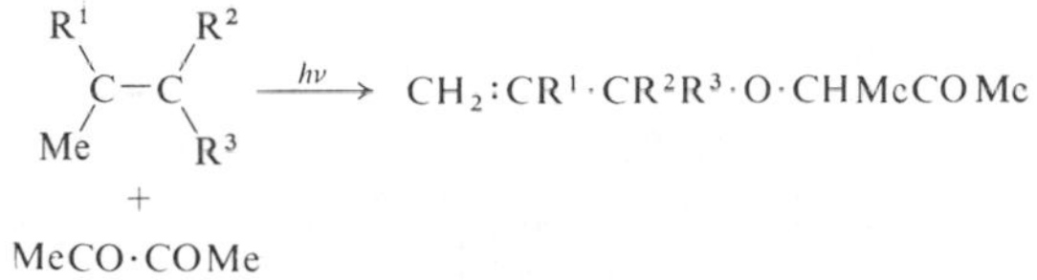

$$R^1 \quad R^2$$
$$\underset{Me}{\overset{}{\diagup}}C - C\underset{R^3}{\overset{}{\diagdown}} \xrightarrow{h\nu} CH_2\!:\!CR^1\cdot CR^2R^3\cdot O\cdot CHMeCOMe$$
$$+$$
$$MeCO\cdot COMe$$

Figure 7.11

keto-ether is amongst the primary photochemical products[148]. The reaction is shown in Figure 7.11; for 2,3-dimethylbut-2-ene this product ($R^1 = R^2 = R^3 = Me$) is isolated in 70% yield and none of the expected oxetan is formed. The genesis of these alkenic keto-ethers involves an attack on the alkene by the $n\pi^*$ triplet state of the diketone, followed by an intramolecular disproportionation of the resultant 1,4-diradical.

The photochemical cleavage of certain β-keto-acid anilides to give a ketone and phenyl isocyanate, shown in Figure 7.12, is analogous to a Norrish Type II cleavage; the process is also observed in the mass spectrometer[149]. The photolysis of ethereal solutions of β-keto-esters leads to

$$RCO\cdot CR^1R^2\cdot CO\cdot NHPh \xrightarrow{h\nu} RCO\cdot CR^1R^2H + PhNCO$$

Figure 7.12

homolytic cleavage of the α-bond, the Norrish type I cleavage[150]. In the presence of cyclohexane, the acyl group is replaced by a cyclohex-2-en-1-yl residue. For example ethyl α-methylpropionylacetate gives ethyl propionate in the absence of cyclohexane and ethyl α-(cyclohex-2-en-1-yl)propionate in its presence.

References

1. Seebach, D. (1969). *Angew. Chem. Internat. Edn.*, **8**, 639
2. Stork, G. and Maldonado, L. (1971). *J. Amer. Chem. Soc.*, **93**, 5286
3. Breslow, R. (editor). (1970). *Org. Synth.*, **50**, 72
4. Seebach, D. (1969). *Synthesis*, **1**, 17
5. Corey, E. J. and Erickson, B. W. (1971). *J. Org. Chem.*, **36**, 3553
6. Vedejs, E. and Fuchs, P. L. (1971). *J. Org. Chem.*, **36**, 366
7. Carey, F. A. and Neergaard, J. R. (1971). *J. Org. Chem.*, **36**, 2731
8. Ogura, K. and Tsuchihashi, G. (1971). *Tetrahedron Letters*, 3151; Nieuwenhuyse, M. and Louw, R. (1971). *ibid.*, 4142
9. Carlson, R. M. and Helquist, P. M. (1966). *J. Org. Chem.*, **33**, 2596
10. Corey, E. J. and Shulman, J. I. (1970). *J. Org. Chem.*, **35**, 777
11. Corey, E. J., Erickson, B. W. and Noyari, R. (1971). *J. Amer. Chem. Soc.*, **93**, 1724
12. Meyers, A. I., Malone, G. R. and Adickes, H. W. (1970). *Tetrahedron Letters*, 3715
13. Meyers, A. I., Smith, E. M. and Jurjevich, A. F. (1971). *J. Amer. Chem. Soc.*, **93**, 2314
14. Meyers, A. I. and Collington, E. W. (1970). *J. Amer. Chem. Soc.*, **92**, 6676
15. Altman, L. J. and Richheimer, S. L. (1971). *Tetrahedron Letters*, 4709
16. Meyers, A. I. and Smith, E. M. (1970). *J. Amer. Chem. Soc.*, **92**, 1084
17. Pelter, A., Hutchings, M. G. and Smith, K. (1970). *Chem. Commun.*, 1528
18. Pelter, A., Hutchings, M. G. and Smith, K. (1971). *Chem. Commun.*, 1049
19. Kabalka, G. W., Brown, H. C., Suzuki, A., Honma, S., Arase, A. and Itoh, M. (1970). *J. Amer. Chem. Soc.*, **92**, 710
20. Pasto, D. J. and Wojtkowski, P. W. (1971). *J. Org. Chem.*, **36**, 1790
21. Brown, H. C. and Kabalka, G. W. (1970). *J. Amer. Chem. Soc.*, **92**, 712, 714
22. Suzuki, A., Nozawa, S., Itoh, M., Brown, H. C., Kabalka, G. W. and Holland, G. W. (1970). *J. Amer. Chem. Soc.*, **92**, 3503
23. Brown, H. C. and Negishi, E. (1971). *J. Amer. Chem. Soc.*, **93**, 3777 and refs. therein
24. Zweifel, G., Clark, G. M. and Polston, N. L. (1971). *J. Amer. Chem. Soc.*, **93**, 3395
25. Zweifel, G. and Polston, N. L. (1970). *J. Amer. Chem. Soc.*, **92**, 4068
26. Jorgenson, M. J. (1970). *Organic Reactions*, **18**, 1
27. Posner, G. H. (1970). *Tetrahedron Letters*, 4647
28. Normant, J. F. and Bourgain, M. (1970). *Tetrahedron Letters*, 2659
29. Chan, T. H., Chang, E. and Vinokur, E. (1970). *Tetrahedron Letters*, 1137
30. Dubois, J. E., Boussu, M. and Lion, C. (1971). *Tetrahedron Letters*, 829
31. Emptoz, G., Huet, F. and Jubier, A. (1971). *Compt. Rend.*, **273C**, 1543
32. Elkik, E. and Assadi-Far, H. (1970). *Bull. Soc. Chim. France*, 991
33. Couffignal, R. and Gaudemar, M. (1970). *Bull. Soc. Chim. France*, 3157
34. Coutrot, P., Coubret, J. C. and Villieras, J. (1971). *Tetrahedron Letters*, 1553
35. Villieras, J., Castro, B. and Ferracutti, N. (1971). *Bull. Soc. Chim. France*, 1450
36. Villieras, J. and Ferracutti, N. (1970). *Bull. Soc. Chim. France*, 2699
37. Humphrey, S. A., Herrmann, J. L. and Schlesinger, R. M. (1971). *Chem. Commun.*, 1244
38. Michael, U. and Hornfield, A. B. (1970). *Tetrahedron Letters*, 5219
39. Watanabe, Y., Mitsudo, T., Tanaka, M., Yamamoto, K., Okajima, T. and Takegami, Y. (1971). *Bull. Chem. Soc. Japan*, **44**, 2569
40. Peters, J. A. and van Bekkum, H. (1971). *Rec. Trav. Chim.*, **90**, 1323
41. Bendenbaugh, A. O., Bendenbaugh, J. H., Bergin, W. A. and Adkins, J. D. (1970). *J. Amer. Chem. Soc.*, **92**, 5774
42. Benkeser, R. A., Watanabe, H., Mels, S. J. and Sabol, M. A. (1970). *J. Org. Chem.*, **35**, 1210
43. Arzoumanian, H. and Metzger, J. (1971). *Synthesis*, 527
44. Coulson, J. M., Fleishmann, M., Goodridge, F. and Robb, I. D. (1971). *British Pat.*, 1 232 011. *(Chem. Abstr.)*, 1971, **75**, 44323

45. Hüttel, R. (1970). *Synthesis*, 225
46. Rodeheaver, G. T. and Hunt, D. F. (1971). *Chem. Commun.*, 818
47. McKillop, A., Hunt, J. D., Taylor, E. C. and Kienzle, F. (1970). *Tetrahedron Letters*, 5275
48. Bhalerao, U. T. and Rapoport, H. (1971). *J. Amer. Chem. Soc.*, **93**, 4835, (1971). *ibid.*, **93**, 5311
49. Suzuki, S. and Ransley, D. L. (1971). *Ind. Eng. Chem., Prod. Res. Dev.*, **10**, 179
50. Heiba, E. I. and Desau, R. M. (1971). *J. Amer. Chem. Soc.*, **93**, 524
51. Collins, J. C., Hess, W. W. and Frank, F. J. (1968). *Tetrahedron Letters*, 3363
52. Corey, E. J., Shirahama, H., Yamamoto, H., Terashima, S., Venkateswaren, A. and Schaaf, T. K. (1971). *J. Amer. Chem. Soc.*, **93**, 1490
53. Ratcliffe, R. and Rodehorst, R. (1970). *J. Org. Chem.*, **35**, 4000
54. Dauben, W. G., Lorber, M. and Fullerton, D. S. (1969). *J. Org. Chem.*, **34**, 3587
55. Shaw, J. E. and Sherry, J. J. (1971). *Tetrahedron Letters*, 4379
56. Epstein, W. P. and Allinger, J. (1970). *Chem. Commun.*, 1338
57. Cooke, M. P., Jr. (1970). *J. Amer. Chem. Soc.*, **92**, 6080
58. Coates, R. M. and Sowerby, R. L. (1971). *J. Amer. Chem. Soc.*, **93**, 1027
59. Borowitz, I. J., Casper, E. W. R. and Crouch, R. K. (1971). *Tetrahedron Letters*, 105
60. Odic, Y. and Pereyne, M. (1970). *Compt. Rend.*, **270C**, 100
61. Odic, Y. and Pereyne, M. (1969). *Compt. Rend.*, **269C**, 469
62. Hooz, J. and Layton, R. B. (1970). *Canad. J. Chem.*, **48**, 1626
63. Dubois, J. R. and Lion, C. (1971). *Compt. Rend.*, **272C**, 1377; Dubois, J. E., Lion, C. and Moulineau, C. (1971). *Tetrahedron Letters*, 177
64. Nozaki, H., Mori, T. and Kawanisi, M. (1968). *Canad. J. Chem.*, **46**, 3767; Vig, O. P., Matta, K. L., Schgal, J. M. and Sharma, S. D. (1970). *J. Indian Chem. Soc.*, **47**, 894
65. Samuelsson, B. and Lamm, B. (1971). *Acta Chem. Scand.*, **25**, 1555
66. Lamm, B. and Samuelsson, B. (1970). *Acta Chem. Scand.*, **24**, 561
67. Walborsky, H. M., Morrison, W. H. and Niznik, G. E. (1970). *J. Amer. Chem. Soc.*, **92**, 6675. cf. Walborsky, H. M. and Niznik, G. E. (1969). *J. Amer. Chem. Soc.*, **91**, 7778
68. Stork, G. and Colvin, E. (1971). *J. Amer. Chem. Soc.*, **93**, 2080
69. Corey, E. J., Yamamoto, H., Herron, D. K. and Achiura, K. (1970). *J. Amer. Chem. Soc.*, **92**, 6635; Corey, E. J. and Herron, D. K. (1971). *Tetrahedron Letters*, 1641
70. Hartzler, H. D. (1971). *J. Amer. Chem. Soc.*, **93**, 4527
71. Schelhorn, H., Frischleder, H. and Hauptmann, S. (1970). *Tetrahedron Letters*, 4315
72. Barton, D. H. R., Magnus, P. D., Smith, G. and Zurr, D. (1971). *Chem. Commun.*, 861
73. Emerson, D. W. and Hynberg, H. (1971). *Tetrahedron Letters*, 3445; Huurdeman, W. J. F., Wynberg, H. and Emerson, D. W. (1971). *Tetrahedron Letters*, 3449
74. Timmis, G. H. and Wildsmith, E. (1971). *Tetrahedron Letters*, 195
75. McKillop, A., Hunt, J. D., Naylor, R. D. and Taylor, E. C. (1971). *J. Amer. Chem. Soc.*, **93**, 4918
76. Corey, E. J. and Richman, J. E. (1970). *J. Amer. Chem. Soc.*, **92**, 5276
77. Nambudiry, M. E. N. and Krishna Rao, G. S. (1971). *Aust. J. Chem.*, **24**, 2183
78. Baltz, H., Bierling, B., Kirschke, K., Oberender, H. and Schulz, M. (1971). *German Pat.*, 2 050 566 additional to 2 050 565 (*Chem. Abstr.* (1971). **75**, 19706)
79. Gersmann, H. R. and Bickel, A. F. (1971). *J. Chem. Soc. B*, 2230
80. Charman, H. B. (1970). *British Pat.* 1 215 012 (*Chem. Abstr.* (1971), **74**, 87410)
81. Deslongchamps, P. and Moreau, C. (1971). *Canad. J. Chem.*, **49**, 2465
82. Schreibmann, A. A. P. (1970). *Tetrahedron Letters*, 4271
83. Johnson, M. R. and Rickborn, B. (1970). *J. Org. Chem.*, **35**, 1041
84. Hack, V., Fryberg, E. C. and McDonald, E. (1971). *Tetrahedron Letters*, 2629
85. Corey, E. J., Albonico, S. M., Koelliker, U., Shaaf, T. K. and Varma, R. K. (1971). *J. Amer. Chem. Soc.*, **93**, 1491
86. Hutchins, R. O., Maryanoff, B. E. and Milewski, C. A. (1971). *J. Amer. Chem. Soc.*, **93**, 1793
87. Borch, R. F., Bernstein, M. D. and Durst, H. D. (1971). *J. Amer. Chem. Soc.*, **93**, 2897
88. Akhren, A. A., Ustynyuk, T. K. and Titov, Yu. A. (1970). *Russian Chem. Revs.*, **39**, 732
89. Corey, E. J. and Cane, D. E. (1971). *J. Org. Chem.*, **36**, 3070
90. Barrett, G. C., Hall, D. M., Hargreaves, M. K. and Modari, B. (1971). *J. Chem. Soc. C*, 279
91. Hargreaves, M. K. and Modari, B. (1971). *J. Chem. Soc. C*, 1013
92. Gopal, H. and Gordon, A. J. (1971). *Tetrahedron Letters*, 2941
93. Wolfe, S., Pilgrim, W. R., Garrard, T. F. and Chamberlain, P. (1971). *Canad. J. Chem.*, **49**, 1099

94. German Patent. (1970). 2 022 567. (*Chem. Abstr.* (1971). **74**, 41917)
95. Sa Le Thi Thuan and Wiemann, J. (1971). *Compt. Rend.,* **272C**, 233
96. Rühlmann, K. (1971). *Synthesis,* 236
97. Strating, J., Reiffers, S. and Wynberg, H. (1971). *Synthesis,* 209, 211
98. Roth, M., Dubs, P., Gotschi, E. and Eschenmoser, A. (1971). *Helv. Chim. Acta,* **54**, 710
99. McMurry, J. E. and Melton, J. (1971). *J. Amer. Chem. Soc.,* **93**, 5309
100. Severin, T. and Kullmer, H. (1971). *Chem. Ber.,* **104**, 440
101. Ficini, J. and Genet, J. P. (1971). *Tetrahedron Letters,* 1565
102. Kauffmann, T. and Schoenfelder, M. (1970). *Annalen,* **731**, 37
103. Leffingwell, J. C. (1970). *Chem. Commun.,* 357; French Pat. 2 035 537 (*Chem. Abstr.,* (1971). **75**, 88111)
104. Brettle, R. (1970). *Chem. Commun.,* 342
105. Brettle, R. and Seddon, D. (1970). *J. Chem. Soc. C,* 1320
106. Brettle, R. and Seddon, D. (1970). *J. Chem. Soc. C,* 1317
107. Chkir, M. and Lelandais, D. (1971). *Chem. Commun.,* 1369
108. Saegusa, T., Ito, Y., Tomita, S. and Kinoshita, H. (1970). *J. Org. Chem.,* **35**, 670
109. Mogto, J. K., Wiemann, J. and Kossanyi, J. (1970). *Ann. Chim. (Paris),* **5**, 471
110. Rabjohn, N. and Harbert, C. A. (1970). *J. Org. Chem.,* **35**, 3240
111. Anatol, J. and Medete, A. (1971). *Synthesis,* 538; (1971). *Compt. Rend.,* **272C**, 1157
112. Rathke, M. W. and Deitch, J. (1971). *Tetrahedron Letters,* 2953
113. Pfeffer, P. E. and Silbert, L. S. (1970). *Tetrahedron Letters,* 699
114. Kuo, Y. N., Yahner, J. A. and Ainsworth, C. (1971). *J. Amer. Chem. Soc.,* **93**, 6321
115. Kuo, Y. N. and Ainsworth, C. (unpublished work) quoted in ref. 114 . .
116. Wassermann, H. H. and Wentland, S. H. (1970). *Chem. Commun.,* 1
117. Bestmann, H. J., Graf, G., Hartung, H., Kolewa, S. and Vilsmeier, E. (1970). *Chem. Ber.,* **103**, 2794
118. Karobitsyna, I. K. and Nikolaev, V. A. (1971). *J. Org. Chem. USSR,* **7**, 411
119. Mock, W. L. and Hartmann, M. E. (1970). *J. Amer. Chem. Soc.,* **92**, 5767
120. Danishefsky, S., Crawley, L. S., Solomon, D. M. and Heggs, P. (1971). *J. Amer. Chem. Soc.,* **93**, 2356
121. Eck, H. and Prigge, H. (1970). *Annalen,* **731**, 12
122. Finding, R. and Schmidt, U. (1970). *Angew. Chem. Internat. Ed.,* **9**, 456
123. Suama, M., Sugita, T. and Ichikawa, K. (1971). *Bull. Soc. Chem. Japan,* **44**, 1999
124. Kurts, A. L., Macias, A., Beletskaya, I. P. and Reutov, O. A. (1971). *Dokl. Akad. Nauk. S.S.S.R.,* **197**, 1088; (1971). *Tetrahedron Letters,* 3037
125. Kurts, A. L., Dem'yanov, P. I., Macias, A., Beletskaya, I. P. and Reutov, O. A. (1970). *Dokl. Akad. Nauk. S.S.S.R.,* **195**, 1117
126. Mastryukova, T. A., Shipov, A. E., Abalyaeva, V. V., Kugutcheva, E. E. and Kabachnik, M. I. (1965). *Dokl. Akad. Nauk, S.S.S.R.,* **164**, 340
127. Kurts, A. L., Macias, A., Beletskaya, I. P. and Reutov, O. A. (1971). *Tetrahedron,* **27**, 4759
128. Gelin, S. and Rouet, J. (1971). *Bull. Soc. Chim. France,* 1874
129. Harris, T. M. and Harris, C. M. (1969). *Organic Reactions,* **17**, 155
130. Weiler, L. (1970). *J. Amer. Chem. Soc.,* **92**, 6702
131. Huckin, S. N. and Weiler, L. (1971). *Tetrahedron Letters,* 4835
132. Harris, T. M. and Murphy, G. P. (1971). *J. Amer. Chem. Soc.,* **93**, 6708
133. Hata, G., Takahashi, K. and Miyake, A. (1969). *Chem. and Ind.,* 1836
134. Takahashi, K., Hata, G. and Miyake, A. (1970). *German Pat.* 1 955 664 (*Chem. Abstr.,* (1970), **73**, 34822)
135. Hata, G., Takahashi, K. and Miyake, A' (1970). *Chem. Commun,* 1392
136. Atkins, K. E., Walker, W. E. and Manyik, R. M. (1970). *Tetrahedron Letters,* 3821
137. Frankenfeld, J. W. and Tyler, W. E. (1971). *J. Org. Chem.,* **36**, 2110
138. Thomsen, A. D. and Lund, H. (1971). *Acta Chem. Scand.,* **25**, 1576
139. Kluge, A. F. and Lillya, C. P. (1971). *J. Org. Chem.,* **36**, 1977
140. Kluge, A. F. and Lillya, C. P. (1971). *J. Org. Chem.,* **36**, 1988
141. Kluge, A. F. and Lillya, C. P. (1971). *J. Amer. Chem. Soc.,* **93**, 4458
142. Wagner, P. J. (1971). *Accounts Chem. Res.,* **4**, 168
143. Yang, N. C. and Elliott, S. P. (1969). *J. Amer. Chem. Soc.,* **91**, 7550
144. Yang, N. C. and Elliott, S. P. (1969). *J. Amer. Chem. Soc.,* **91**, 7551
145. Stephenson, L. M., Cavigli, P. R. and Parlett, J. L. (1971). *J. Amer. Chem. Soc.,* **93**, 1984
146. Coyle, J. D. (1971). *J. Chem. Soc. B,* 2254

147. Paul, H. and Fischer, H. (1971). *Chem. Commun.*, 1038
148. Ryang, S.-H., Shima, K. and Sakurai, H. (1970). *Tetrahedron Letters*, 1091; (1971). *J. Amer. Chem. Soc.*, **93,** 5270
149. Reisch, J. and Niemeyer, D. H. (1971). *Tetrahedron Letters*, 4637
150. Tokuda, M., Hataya, M., Imai, J., Itoh, M. and Suzuki, A. (1971). *Tetrahedron Letters*, 3133

8
Carboxylic Acids

N. POLGAR
University of Oxford

8.1   GENERAL TECHNIQUES

The following account concerns some recent publications on the application of general techniques which are not limited to the groups of compounds included in certain individual sections of this chapter.

The application of gas-liquid chromatography (g.l.c.) to the determination of the structure of esters of fatty acids has been reviewed[1].

A review[2] on the mass spectrometry of fatty acid derivatives is devoted mainly to the interpretation of the mass spectra of fatty acids and the structural factors bearing upon the spectra.

Deuterated natural products which have become available by the cultivation of algae in D_2O represent valuable tools for various biological studies. With reference to the structural problems involving these compounds, the g.l.c. and mass spectrometric properties of methyl esters derived from perdeuteriated fatty acids, including mono-, di-, and tri-enoic acids, as well as methyl-branched acids, have been investigated[3]; these acids were obtained from Scenedesmus obliquus which had been grown in D_2O.

A recent review[4] on ozonolysis deals with unsaturated fatty acids in particular; apart from theoretical aspects, analytical applications as well as preparative uses are discussed.

A useful procedure[186] for the esterification of sterically hindered carboxylic acids should also be mentioned here. It involves the use of triethyloxonium tetrafluoroborate in dichloromethane in the presence of ethyldi-iso-propylamine under mild conditions.

8.2 MONOCARBOXYLIC ACIDS

8.2.1 Normal-chain acids

8.2.1.1 Saturated acids

The photobromination of alkanoic acids containing 6–18 carbon atoms yielded complex mixtures[5] including notable quantities of *threo*- and *erythro*-dibromides, believed to be formed through a two-step radical mechanism. The isomer (1) was predominant, and no compound was formed with the ·CHBr·CHBr· system in a terminal position, or adjoining the carboxy group. The following order of reactivity of carbon units was established: $CH_2 > CH_3 > CH_2 \cdot CO_2 H > CHBr$.

$$MeCHBr \cdot CHBr \cdot (CH_2)_n \cdot CO_2 H$$

(1)

In connection with studies designed to localise deuterium atoms distributed in the carbon chain of deuterated long chain fatty acids, the oxidative degradation of n-alkyl chains by means of potassium permanganate has been investigated[6]. Apparently oxidative cleavage of the carbon chain by potassium permanganate does not occur haphazardly. In oxidative reactions of docosane, docosan-1-ol, docosanoic acid, and methyl docosanoate, the C—C bonds in the middle of the molecule were more easily cleaved under the conditions employed, and certain differences were noted in the sensitivity of C—C linkages in the neighbourhood of polar groups.

The terminal chain-elongation of fatty acids by reaction with methyl or ethyl radicals has been studied[7] with regard to a hypothesis concerning the formation of normal-chain fatty acids and hydrocarbons found in some meteorites. Whereas random addition of the radicals would be expected to yield complex mixtures including branched-chain compounds, reactions of the radicals and monolayers of potassium palmitate or n-heptadecanoate on an aqueous surface gave, under the conditions employed, continuous-chain acids with longer carbon chains than those of the reactant acids. Thus, it is found that an extension of alkyl chains by alkyl radicals can be achieved if the alkyl chains are suitably packed in a monolayer surface film, i.e. the reaction takes place only at the terminal methyl groups of the alkyl chains, the latter projecting out of the aqueous surface.

For studies of the mechanism of certain enzyme reactions the *R*- and *S*-forms of acetic acid have been prepared with the methyl groups consisting of one hydrogen, one deuterium, and one tritium atom attached to carbon[187]; the two specimens were obtained by chemical synthesis starting from phenylacetylene, and by utilising stereospecific reactions.

8.2.1.2 Monoenoic acids

(a) *Natural acids*—A series of 5-enoic acids with 22, 24 and 26 carbon atoms have been isolated[8] from *Mycobacterium phlei*; these acids have been suggested as precursors for the mycolic acids (see Section 8.4.5) synthesised by the same organism.

The occurrence of *cis*-hexadec-5-enoic acid and its homologues in several species of *Bacillus* in unusually large proportions for this genus has also been reported[9].

There is continued interest in establishing the location of the double bonds in the various naturally occurring fatty acids, with a view to understanding the biogenesis of these compounds. A detailed study on the position of the double bond in the mono-unsaturated C_{14}–C_{26} acids of *Mycobacterium smegmatis* and *M. bovis* BCG has been published[10].

cis-Tetracos-15-enoic and *cis*-hexacos-17-enoic acids are present in large amounts in the fat of the commercially available *Tropaeolum speciosum* seeds[11]. A rich source is thus available for the laboratory preparation of these acids.

A valuable procedure[12] for isolating unsaturated fatty acids involves fractionation of the corresponding methyl esters as their mercury(II) acetate addition products on a column of partially methylated dextran (Sephadex LH-20). This procedure permits the separation of the components of mixtures of unsaturated fatty acids, as well as of intact lecithin species, according to their degree of unsaturation.

(b) *Structural studies* — Valuable information for structural studies of unsaturated fatty acids is provided by the examination of the physical and spectroscopic properties of various series of fatty acids.

The g.l.c. equivalent chain-lengths have been determined for a number of isomeric methyl octadecenoates and octadecynoates[13]. The equivalent chain-length values generally increase with the distance of the double bond from the ester group, but show little change from the 6- to the 10-position. Smaller differences were found among the isomeric *trans*-enoic esters than among the *cis*-unsaturated compounds. For the methyl octadecynoates there was a greater and more regular increase in the equivalent chain-length values as the triple bond moved away from the ester group.

A study of the thin-layer chromatographic (t.l.c.) and g.l.c. properties of various *cis*-octadecenyl compounds, derived from methyl *cis*-2- to *cis*-17-octadecenoates, has also been published[14].

The location of alkenic linkages in long-chain esters may conveniently be established by methoxymercuration–demercuration, followed by combined g.l.c. and mass spectrometry (ms) of the resulting methoxy-derivatives[15]. Another recent procedure[16] for determining the position of alkenic linkages involves oxidation of the corresponding epoxides with ethereal meta-periodic acid, followed by identification of the resulting aldehydes by n.m.r. spectroscopy, m.s., or g.l.c.; a convenient procedure involves the combination of n.m.r. with m.s., or g.l.c. with m.s.

It has been suggested[17] that the derivatives obtainable by stereospecific adduct formation between mercury(II) acetate and unsaturated fatty acids are suitable for n.m.r. determination of the *cis/trans* ratio.

(c) *Syntheses* — In recent years much attention has been directed towards introducing carbon–carbon double bonds with known configuration at a certain point. The various stereospecific and highly stereoselective procedures which are now available have been reviewed[18].

The isomeric *trans*-octadecenoic acids have been synthesised[19] by standard procedures.

Long-chain fatty acids with a terminal double bond are useful intermediates for the synthesis of various long-chain compounds. Ethyl nonadec-18-enoate has been prepared[20] by an anodic synthesis involving the mixed coupling of undec-10-enoic acid and ethyl hydrogen sebacate by using a platinum anode and a mercury cathode.

(d) *Reactions* – The hydrogenation of unsaturated fatty acids and related compounds with homogeneous and heterogeneous catalysts has been recently reviewed[21]. In a study[22] of the competitive hydrogenation of isomeric methyl octadecenoates, mixtures containing a radioactively labelled and an unlabelled isomer were subjected to catalytic hydrogenation; the isomers were thus compared with each other under conditions which required competition for catalyst surface and hydrogen.

The reduction of simple $\alpha\beta$-unsaturated acids with an excess of lithium in liquid ammonia[23] gives high yields of saturated acids.

The stereospecific hydration of the Δ^9 double bond of oleic acid to give the naturally occurring stereoisomers of related hydroxy-acids has been the subject of further studies. It has been shown[24] that a soluble enzyme preparation from a *Pseudomonad* can catalyse the interconversion of oleic and D-10-hydroxystearic acid.

In continuation of studies concerning the mechanism of the Varrentrapp reaction, various alkenoic, alkynoic, and alkadienoic acids have been subjected to fusion with potassium deuteroxide[25]. It has been suggested earlier[26] that the Varrentrapp reaction occurs by reversible migration of the double bond along the carbon chain (both towards and away from the carboxy group), and the $\alpha\beta$-ethylenic acid anion (2), when formed, undergoes fission of the retro-aldol or retro-Claisen type via the intermediate hydroxy acid (3). Examination of the deuteriated degradation products by n.m.r. and mass spectroscopy showed that their deuteriation patterns are consistent with a stepwise reversible migration of the double bond during the fusion with alkali.

$$RCH{:}CH{\cdot}CO_2^- \qquad\qquad RCH(OH){\cdot}CH_2{\cdot}CO_2^-$$

$$(2) \qquad\qquad\qquad\qquad (3)$$

Among the reactions studied, the addition of nitrones to unsaturated esters is to be noted. Esters of undec-10-enoic, oleic and linoleic acid, as well as hexadec-1-ene, have been converted by 1,3-addition of various diaryl nitrones into isoxazolidines[27]; the latter, on reduction, gave amino-alcohols and alcohols.

The addition of NN-dichlorourethane to alkenoic acids, resulting in the formation of aziridines, has been compared[28] with the preparation of these compounds by the iodine–isocyanate method. The latter is stereospecific: *cis*-alkenes give rise to *cis*-aziridines, and *trans*-alkenes to *trans*-aziridines. The addition of NN-dichlorourethane results in mixtures of *cis*- and *trans*-aziridines in which the *trans*-isomers predominate.

Topics reviewed include alkene reactions catalysed by transition-metal compounds, with particular consideration of unsaturated fatty acids[29], and the allylic halogenation and oxidation of unsaturated esters[30].

8.2.1.3 Di- and poly-enoic acids

(a) *Natural acids*—The capacity of bacteria to synthesise dienoic fatty acids has been demonstrated[31] for *Bacillus licheniformis* by using labelled palmitic acid. This organism can convert palmitic acid into hexadeca-5,10-, as well as into hexadeca-7,10-dienoic acid, the latter containing a methylene-interrupted unsaturated system. These biosyntheses are shown to be temperature-dependent, involving two distinct desaturation systems.

Two highly unsaturated fatty acids to occur as constituents of aje, the body fat of the Mexican and Central American insect *Llaveia axin*[32]. From studies of these acids, including u.v. spectroscopy, and m.s. of the deuteriated esters, it is concluded that both acids are conjugated penta-unsaturated, with the conjugated system in the terminal position. They have been assigned the structures: dodecapenta-3,5,7,9,11-enoic acid (4), and tetradecapenta-5,7,9,11-13-enoic acid (5) respectively.

$$CH_2{:}CH{\cdot}(CH{:}CH)_4{\cdot}CH_2{\cdot}CO_2H \qquad\qquad CH_2{:}CH{\cdot}(CH{:}CH)_4{\cdot}(CH_2)_3{\cdot}CO_2H$$

(4)   (5)

The odd-numbered polyunsaturated fatty acids have been reviewed[33].

(b) *Structural studies*—The alkenic proton signals of the methyl esters derived from certain conjugated fatty acids, namely octadeca-9c,11t- and octadeca-9t,11t-dienoic acid, as well as α- and β-eleostearic acid have been studied[34] by using the double resonance method.

In another series of investigations the effect of varying the number of methylene groups between the two unsaturated centres of various methyl *cis,cis*- and *trans,trans*-octadeca-dienoates, and the corresponding -diynoates, upon the n.m.r. spectra[35], as well as the g.l.c. and silver-ion t.l.c. properties of these esters[36] have been studied.

(c) *Syntheses*—*cis*-Buta-1,3-diene-1-carboxylic acid (7) has been synthesised by reduction of α-pyrone (6) with sodium borohydride in acetonitrile[37].

(6)   (7)

In the course of comparative studies of the properties of closely related unsaturated acids (cf. Refs. 35 and 36) the 5,12-, 6,12-, 7,12-, 8,12-, 9,12-, 10,12-, 6,8-, 6,9-, 6,10-, and 6,11-octadecadiynoic acids were prepared by various known procedures; these diynoic acids were then converted by reduction into the corresponding *cis,cis*- and *trans,trans*-octadecadienoic acids[38]. The *cis,cis*-dienoic acids were prepared by hydrogenation of the diynoic acids over Lindlar's catalyst, and most of the *trans,trans*-dienoic acids by reduction of the diynoic acids with lithium in liquid ammonia. The *trans,trans*-6,9- and *trans,trans*-9,12-dienoic acids could not be obtained by the foregoing procedure; these acids were prepared by stereomutation of the *cis,cis*-isomers with nitrogen dioxide.

Methyl octadeca-5,8,11-trienoate, eicosa-10,13-dienoate, and eicosa-7,10,-13-trienoate were synthesised[39] by standard procedures involving the reaction of propargyl bromides with the di-Grignard reagents derived from ω-acetylenic acids, followed by stereospecific reduction of the resulting poly-acetylenic acids. The preparation of labelled acids corresponding to these esters, as well as of some related acids, by similar procedures has also been described[40].

A new general procedure[41] for the synthesis of polyenoic acids via the corresponding acetylenic acids (10) is based upon the reaction of alkadiynoic acids (8), with propargyl halides[42] ((9); X = halogen), the latter being obtainable by *O*-alkyl cleavage of the corresponding methyl ethers with acetyl bromide or iodide in the presence of the appropriate zinc halide[43]. The alkadiynoic acids (8), having the triple bonds in the $n-4$ and $n-1$ positions (*n* indicates the number of carbon atoms in the chain in the sense recommended by the IUPAC–IUB Commission on Biochemical Nomenclature), are synthesised[42] from $(n-1)$-alkyn-1-ols (11), which by elongation (via their tetrahydropyranyl-derivatives) with propargyl toluene-*p*-sulphonate give the alkadiynols (12); the latter are convertible, by oxidation with chromic acid, into the diynoic acids (8). The intermediate diynols (12) can be purified by distillation, thus removing unreacted starting material. Moreover, the use of diynoic in place of monoynoic acids in the condensation with the propargyl halides (9) has the advantage that for the synthesis of the desired poly-ynoic acid (10), a halide (9) with one less triple bond can be employed. This procedure has been used for syntheses of poly-ynoic acids (10) with four, five, and six triple bonds. Stereospecific reduction of the triple bonds gave the corresponding polyenoic acids.

$$H(C\vdots C\cdot CH_2)_2\cdot(CH_2)_z\cdot CO_2H + Me(CH_2)_x\cdot(C\vdots C\cdot CH_2)_yX \longrightarrow$$

$$(8) \qquad\qquad\qquad (9)$$

$$Me(CH_2)_x\cdot(C\vdots C\cdot CH_2)_{y+2}\cdot(CH_2)_z\cdot CO_2H$$

$$(10)$$

$$HC\vdots C\cdot(CH_2)_nOH \xrightarrow{HC\vdots C\cdot CH_2OTs} HC\vdots C\cdot CH_2\cdot C\vdots C\cdot(CH_2)_nOH$$

$$(11) \qquad\qquad\qquad (12)$$

$$[\text{Ts = toluene-}p\text{-sulphonyl}]$$

The hydrogenation of skipped poly-ynoic acids and esters over Lindlar's catalyst has been the subject of a recent study[44]. Optimum results for hydrogenation to all-*cis*-polyenoic acids are obtained by use of light petroleum, ethyl acetate, or acetone as solvent at room temperature, in the presence of a small amount of quinoline. From consideration of the mechanism and kinetics involved, it appears that *trans*-double bonds cannot be directly formed from triple bonds, and their formation, in those cases where some *trans*-isomers are found, occurs only by isomerisation of *cis*-double bonds.

The principal sex attractant of the black carpet beetle, *Attagenus megatoma*, identified as tetradeca-3t,5c-dienoic acid (13), has been synthesised starting from nonan-1-ol[45], or dec-1-yne[46].

$$\text{Me(CH}_2)_7 \cdot \overset{c}{\text{CH}} \overset{}{:} \text{CH} \cdot \overset{t}{\text{CH}} \overset{}{:} \text{CH} \cdot \text{CH}_2 \cdot \text{CO}_2\text{H}$$

(13)

Octadeca-2t,9c,12c-trienoic acid (18), an attractant for the honey bee, has been synthesised from linoleic acid[47]. The latter was converted into the tetrabromo-acid (14). Bromination of this acid in the 2-position by the thionyl chloride–bromine method, followed by reduction of the resulting bromo-acid chloride, gave the pentabromo-octadecan-1-ol (15). This was debrominated with activated zinc dust to give material containing the 2-bromo-9,12-dien-1-ol (16); double bond migration occurred to an appreciable extent. Oxidation of this material to a product containing the acid (17), followed by dehydrobromination of the corresponding methyl ester with 1,5-diazabicyclo[5.4.0]undec-5-ene, yielded two major products, one of which was shown to be essentially the desired acid (18).

$$\text{Me(CH}_2)_4 \cdot (\text{CHBr} \cdot \text{CHBr} \cdot \text{CH}_2)_2 \cdot (\text{CH}_2)_6 \cdot \text{CO}_2\text{H}$$

(14)

$$\text{Me(CH}_2)_4 \cdot (\text{CHBr} \cdot \text{CHBr} \cdot \text{CH}_2)_2 \cdot (\text{CH}_2)_5 \cdot \text{CHBr} \cdot \text{CH}_2\text{OH}$$

(15)

$$\text{Me(CH}_2)_4 \cdot (\text{CH} \overset{}{:} \text{CH} \cdot \text{CH}_2)_2 \cdot (\text{CH}_2)_5 \cdot \text{CHBr} \cdot \text{CH}_2\text{OH}$$

(16)

$$\text{Me(CH}_2)_4 \cdot (\text{CH} \overset{}{:} \text{CH} \cdot \text{CH}_2)_2 \cdot (\text{CH}_2)_5 \cdot \text{CHBr} \cdot \text{CO}_2\text{H}$$

(17)

$$\text{Me(CH}_2)_4 \cdot (\overset{c}{\text{CH}} \overset{}{:} \text{CH} \cdot \text{CH}_2)_2 \cdot (\text{CH}_2)_4 \cdot \overset{t}{\text{CH}} \overset{}{:} \text{CH} \cdot \text{CO}_2\text{H}$$

(18)

(d) *Reactions* — Studies on the biohydrogenation of linoleic acid in *Butyrivibrio fibrisolvens* indicate that the initial reaction is the conversion of the *cis*-12-double bond of linoleic acid into a *trans*-11-bond, thus yielding octadeca-9c,11t-dienoic acid which is then hydrogenated to give octadec-11t-enoic acid[48]. The latter acid also results from the biohydrogenation of octadeca-9c,11t,13c-trienoic acid[49]; the reduction was found to occur by *cis*-addition to the *D*-side of the intermediate octadeca-9c,11t-dienoic acid.

Hydrogenations with copper chromite are known to cause extensive rearrangement of the double bond system. In a study[50] involving the reduction of β-eleostearates and other unsaturated esters, including *trans,trans*-conjugated dienoates, with deuterium and copper chromite, the positional and geometric rearrangement of the double bonds was investigated. In the course of these studies it was also found that monoenoates were formed from conjugated *trans,trans*-dienoates by 1,2- and 1,4-addition.

In view of the importance of alkaline isomerisation in procedures involving the spectrophotometric estimation of various polyenoic acids in fats and oils, the isomerisation of methyl linoleate and methyl linolenate with potassium t-butoxide has been studied[51], and a detailed account is given of the products formed.

A soluble enzyme preparation from a *Pseudomonad* which catalyses the stereospecific hydration of the double bond of oleic acid (cf. Ref. 24) also catalyses[52] the stereospecific hydration of the Δ^9-double bond of linoleic acid to yield D-10-hydroxyoctadec-12c-enoic acid.

The formation of 1,4-epoxides from methyl linoleate and related esters by means of toluene-*p*-sulphonic acid in the presence of an appropriate solvent has been studied[53].

Conjugated octadecadienoic acids are known to react with numerous dienophiles. Thus, ethylene adds to octadeca-9t,11t-dienoic acid with the formation of the C_{20} cyclohexene fatty acid (19). Various peroxy-acid oxidations[54] of this cyclohexene acid, as well as the ozonolysis products[55] of the corresponding methyl ester are reported.

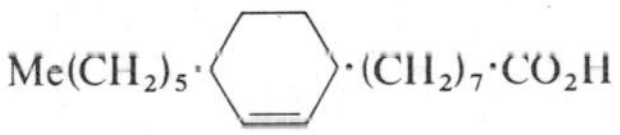

$$Me(CH_2)_5 \cdot \langle \rangle \cdot (CH_2)_7 \cdot CO_2H$$

(19)

Another series of studies deals with addition reactions of methyl octadeca-9t,11t-dienoate with methyl methacrylate and with methyl crotonate[56], as well as with methyl but-3-enoate[57].

The partial cyclisation of trienoic long-chain carboxylic acids at 180 °C with potassium hydroxide in ethylene glycol has been studied comparatively[58]. The acids investigated include trienoic acids with isolated double bonds, namely octadeca-9c,12c,15c- and octadeca-9t,12t,15t-trienoic acid, as well as octadeca-9c,11t,13t- and octadeca-9t,11t,13t-trienoic acid, both containing a conjugated system. Cyclohexa-1,3-diene derivatives ((20); $n+m = 11$) are formed, and it seems that the ring closure, involving a polycentric mechanism, is preceded by an ionic isomerisation. The resulting conjugated trienic system, with a central *cis*-double bond, can undergo cyclisation. Octadeca-9c,12c,15c-trienoic acid yields primarily the acid (20) ($n = 3$,

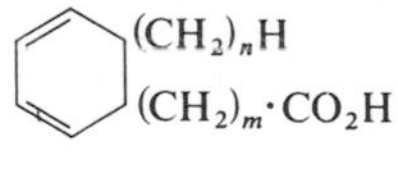

$$\langle \rangle \begin{array}{l} (CH_2)_n H \\ (CH_2)_m \cdot CO_2H \end{array}$$

(20)

$m = 8$), whereas those trienoic acids which do not possess a central *cis*-double bond give rise to mixtures containing a wide range of isomers, as a result of prototropic migration of a conjugated triene system; with octadeca-9t,12t,15t-trienoic acid the conjugated triene system is expected to be formed during isomerisation.

8.2.1.4 *Alkynoic acids*

Some of the papers on alkenoic acids already discussed also include studies of the corresponding alkynoic acids. Thus g.l.c. equivalent chain-lengths have been determined for isomeric methyl octadecynoates[13], and the n.m.r. spectra, as well as the g.l.c. and silver-ion t.l.c. properties of methyl octade-cadiynoates have been investigated[35, 36]. The isomeric octadecynoic acids have been synthesised[19], as well as various di- and poly-ynoic acids, as inter-mediates for syntheses of di- and poly-enoic acids (cf. Refs. 38, 39, 41, 42, 44).

The study[25] on the mechanism of the Varrentrapp reaction also includes alkynoic acids.

In the course of studies of the constituents of seed oils, the presence of 9,10-epoxyoctadec-12-ynoic as well as 9,10-epoxyoctadeca-3t,12c-dienoic acid in these oils has been reported[59, 60].

Crepenynic acid ((24); R = H) represents a key intermediate in Bu'lock's scheme for the biosynthesis of polyacetylenes from fatty acids. The synthesis of labelled crepenynic acid is, therefore, of interest for biosynthetic studies. A successful route[61] starts from hept-1-yne (21) which is converted, by means of ethylene oxide, into non-3-yn-1-ol (22). Conversion of the corresponding bromide into non-3-yn-1-yltriphenylphosphonium bromide (23), followed by its reaction with methyl 8-formyloctanoate (obtained by ozonolysis of methyl oleate), leads to methyl crepenynate ((24); R = Me); the corresponding acid ((24); R = H) is obtained by alkaline hydrolysis.

$$Me(CH_2)_4 \cdot C \vdots CH \xrightarrow{\quad \overset{CH_2}{\underset{CH_2}{|}}O \quad} Me(CH_2)_4 \cdot C \vdots C \cdot CH_2 \cdot CH_2OH$$

$$(21) \hspace{5cm} (22)$$

$$Me(CH_2)_4 \cdot C \vdots C \cdot CH_2 \cdot CH \vdots CH \cdot (CH_2)_7 \cdot CO_2R \leftarrow Me(CH_2)_4 \cdot C \vdots C \cdot CH_2 \cdot CH_2 \overset{+}{P}Ph_3 \ \bar{B}r$$

$$(24) \hspace{5cm} (23)$$

Crepenynic acid ((24); R = H) has been used as the starting point for a synthesis[60] of racemic helenynolic acid (26). This synthesis involves epoxyda-tion of the acid (24) (R = H) to give the cis-epoxy-acid (25). The latter is converted by base-catalysed rearrangement into racemic helenynolic acid (26).

$$(24) \ \ (R = H) \longrightarrow Me(CH_2)_4 \cdot C \vdots C \cdot CH_2 \cdot \overset{O}{\overset{/\backslash}{CH-CH}} \cdot (CH_2)_7 \cdot CO_2H$$

$$(25)$$

$$Me(CH_2)_4 \cdot C \vdots C \cdot CH \vdots CH \cdot CH(OH) \cdot (CH_2)_7 \cdot CO_2H$$

$$(26)$$

It is suggested[60] that the above route, involving the stages epoxidation and base-catalysed rearrangement, might be the pathway by which helen-ynolic acid, occurring in Helichrysum bracteatum seed oil, is produced from crepenynic acid which is also present; the same seed oil also contains the intermediate cis-epoxy-acid (25).

ω-Iodoacetylenic fatty acids (27) ($n = 1$–10) have been prepared[62] from alkynoic acids by means of iodine and sodium hydroxide; those acids with $n = 5$ or 6 are reported to show anti-fungal activity against certain organisms.

$$IC\vdots C \cdot (CH_2)_n \cdot CO_2H$$

(27)

8.2.2 Cycloalkanoic and cycloalkenoic long-chain acids

8.2.2.1 Natural acids

Three cyclopropane fatty acids containing respectively, 17, 18, and 19 carbon atoms (including the methylene groups of the cyclopropane ring) have been isolated[63] from the millipede *Graphidostreptus tumuliporus*, by means of column chromatography, urea fractionation, and preparative g.l.c. of their methyl esters. An interesting feature is the presence, as the most abundant component, of the C_{18}, i.e. an even-numbered, cyclopropane fatty acid; the monocyclopropane fatty acids occurring as natural products are usually odd-numbered.

An unsaturated fatty acid found to be 12,13-methylenetetradec-9-enoic acid (28) has been isolated[64] from a *Clostridium* species.

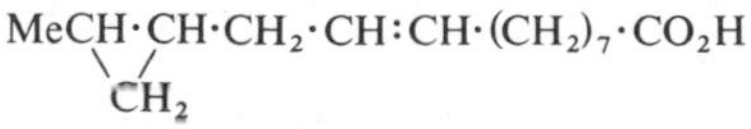

MeCH·CH·CH$_2$·CH⋮CH·(CH$_2$)$_7$·CO$_2$H $\langle\ \rangle$—(CH$_2$)$_n$·CO$_2$H

     \\/
     CH$_2$

(28) 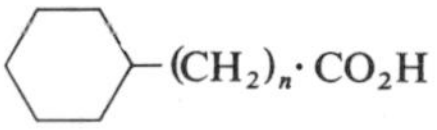   (29)

In studies of the cell lipids of the thermophilic bacterium *Bacillus acido-caldarius*, the principal components of the saponifiable lipids have been identified[65] as 11-cyclohexylundecanoic acid ((29); $n = 10$) and 13-cyclohexyl tridecanoic acid ((29); $n = 12$). It is of interest that these organisms have been exposed to an environment of very high temperatures and low pH.

8.2.2.2 Syntheses

For studies of their g.l.c. and spectroscopic properties a series of isomeric methyl *cis,cis*-dimethyleneoctadecanoates have been prepared[66] by the Simmons–Smith reaction, from the following methyl *cis,cis*-octadecadi-enoates: 5,12-, 6,12-, 7,12-, 6,11-, 8,12-, 6,10-, 9,12-, and 6,9-dienoates.

Methyl sterculate, the methyl ester of the principal fatty acid from *Sterculia foetida* seed fat, has been synthesised[67] from methyl stearolate (30). This

Me(CH$_2$)$_7$·C⋮C·(CH$_2$)$_7$·CO$_2$Me   CH·CO$_2$Et

                                  / \\
             Me(CH$_2$)$_7$·C=C·(CH$_2$)$_7$·CO$_2$Me

(30)   (31)

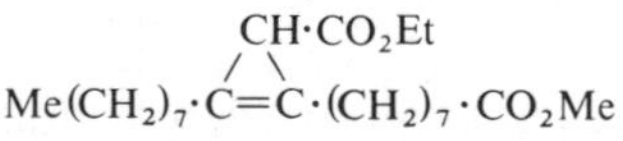

acetylenic ester, on reaction with ethyl diazoacetate in the presence of copper bronze gave the di-ester (31) which was hydrolysed to give the di-acid (32). The corresponding di-acid chloride (33), obtained by using oxalyl chloride, was selectively decarbonylated by treatment with anhydrous zinc chloride to give the cyclopropenium-ion acid chloride (34). This was converted into the corresponding methyl ester (35), which on reduction with sodium borohydride gave methyl sterculate (36).

$$\overset{\displaystyle CH \cdot CO_2H}{\overset{\diagup\diagdown}{Me(CH_2)_7 \cdot C=C \cdot (CH_2)_7 \cdot CO_2H}}$$

(32)

$$\overset{\displaystyle CH \cdot COCl}{\overset{\diagup\diagdown}{Me(CH_2)_7 \cdot C=C \cdot (CH_2)_7 \cdot CO \cdot Cl}}$$

(33)

$$\overset{\displaystyle {}^+CH}{\overset{\diagup\diagdown}{Me(CH_2)_7 \cdot C=C \cdot (CH_2)_7 \cdot CO \cdot Cl}}$$

(34)

$$\overset{\displaystyle {}^+CH}{\overset{\diagup\diagdown}{Me(CH_2)_7 \cdot C=C \cdot (CH_2)_7 \cdot CO_2Me}}$$

(35)

$$\overset{\displaystyle CH_2}{\overset{\diagup\diagdown}{Me(CH_2)_7 \cdot C=C \cdot (CH_2)_7 \cdot CO_2Me}}$$

(36)

A procedure[68] recently developed involves the direct decarbonylation of the di-ester (31) by means of fluorosulphonic or chlorosulphonic acid. Thus, by employing fluorosulphonic acid in methylene chloride, the cyclopropenium-cation ester (35) is directly obtained, and the preparation of the cyclopropene ester (36) from methyl stearolate (30) can be achieved in a three-step synthesis.

Methyl sterculate (36) has also been obtained[69] by photolysis of diazomethane in the presence of methyl stearolate; the ester (36) has been isolated from the irradiation mixture by column chromatography on silicic acid. In this one-step synthesis carbene, which adds to the triple bond, is generated by photolysis of the diazo-compound.

Malvalic acid, a major constituent of cotton-seed oil, and, together with sterculic acid, widely distributed in various species of the order *Malvales*, is the C_{18} homologue of sterculic acid. A synthesis[70] started from dec-1-yne

$$Me(CH_2)_7 \cdot C\vdots CH$$

(37)

$$Me(CH_2)_7 \cdot C\vdots C \cdot (CH_2)_6 Cl$$

(38)

$$\overset{\displaystyle CH \cdot CO_2H}{\overset{\diagup\diagdown}{Me(CH_2)_7 \cdot C=C \cdot (CH_2)_6 Cl}}$$

(39)

$$\overset{\displaystyle CH \cdot CO \cdot Cl}{\overset{\diagup\diagdown}{Me(CH_2)_7 \cdot C=C \cdot (CH_2)_6 Cl}}$$

(40)

$$\overset{\displaystyle {}^+CH}{\overset{\diagup\diagdown}{Me(CH_2)_7 \cdot C=C \cdot (CH_2)_6 Cl}}$$

(41)

$$\overset{\displaystyle CH_2}{\overset{\diagup\diagdown}{Me(CH_2)_7 \cdot C=C \cdot (CH_2)_6 X}}$$

(42)

(37) which was converted, by coupling its lithium derivative with 1,6-dichlorohexane, into 1-chlorohexadec-7-yne (38). Reaction of this with ethyl diazoacetate in the presence of copper bronze, followed by hydrolysis of the resulting cyclopropene ester furnished the cyclopropene acid (39). The corresponding acid chloride (40), under the action of anhydrous zinc chloride,

lost carbon monoxide, with the formation of the cyclopropenium ion (41). Reduction of the latter with sodium borohydride or lithium aluminium hydride gave the 1-chlorocyclopropene derivative ((42); X = Cl) which by standard procedures, via the nitrile ((42); X = CN), was converted into methyl malvalate ((42); X = CO_2Me). Two alternative syntheses are also described in the same paper.

In the course of this work 1-chlorodec-4-yne (43) was converted by an analogous sequence of stages into methyl 5,6-methanoundec-5-enoate (44).

$$Me(CH_2)_4 \cdot C \vdots C \cdot (CH_2)_3 Cl$$

(43)

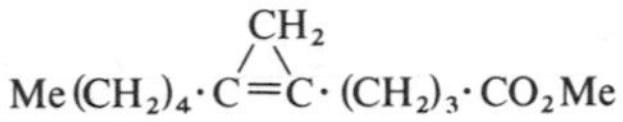

$$Me(CH_2)_4 \cdot \overset{\displaystyle CH_2}{\overset{\displaystyle /\backslash}{C}} = C \cdot (CH_2)_3 \cdot CO_2 Me$$

(44)

Attention was then turned to synthesis of labelled methyl malvalate containing ^{14}C at specific positions[71, 72]. For the preparation of the ester labelled in the methylene group of the cyclopropene ring a route analogous to that for the synthesis of methyl sterculate (36) was followed. The starting point was methyl hexadec-8-ynoate, and for the reaction with the latter ethyl 2-^{14}C-diazoacetate was employed. Methyl 1-^{14}C-malvalate was obtained according to the scheme described above for the synthesis of methyl malvalate, with the modification that the chlorine in the chlorocyclopropene ((42); X = Cl) was displaced by ^{14}CN. 10-^{14}C-Malvalate resulted on employing 3-^{14}C-dec-1-yne for the synthesis (37)–(42). A route for the synthesis of methyl 9-^{14}C-malvalate is also shown[72] to be well defined.

8.2.3 Alkyl-branched acids

8.2.3.1 General techniques

A method for the quantitative separation of the components and analysis of mixtures of saturated branched-chain carboxylic acids and their esters by programmed-temperature g.l.c. has been described[73]. Comparative data are given for the retention times and temperatures of various C_7- and C_8-carboxylic acids with α-quaternary and α-tertiary carbon atoms and their esters.

8.2.3.2 Monoalkyl-branched acids

(a) Syntheses — Methyl 3-methyltridecanoate (48), required as an intermediate for another synthesis, has been obtained[74] by a route based upon Stetter's chain-lengthening procedure, involving 5-methylcyclohexane-1,3-dione ((45); X = H) as a chain-extender. Alkylation of this dione with the allylic halide, 1-bromohept-2-ene, in the presence of aqueous ethyldi-

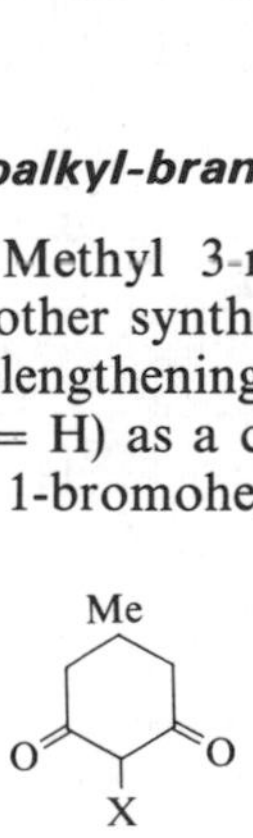

Me

O   O

X

(45)

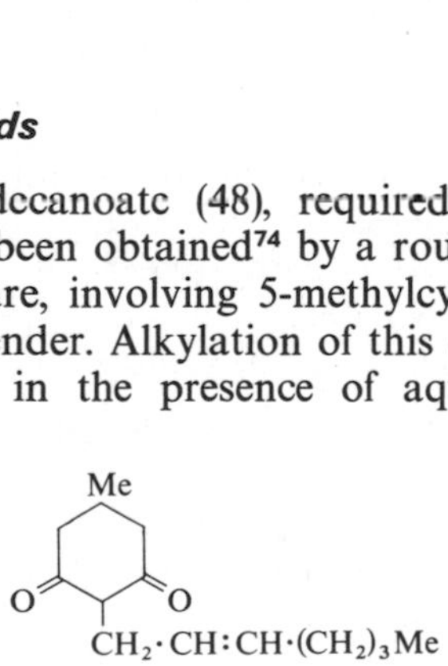

Me

O   O

$CH_2 \cdot CH \vdots CH \cdot (CH_2)_3 Me$

(46)

isopropylamine gave 2-(hept-2-enyl)-5-methylcyclohexane-1,3-dione (46) which was converted, on reductive cleavage by the action of sodium hydroxide and hydrazine, into the acid (47). Catalytic hydrogenation of the corresponding methyl ester gave the ester (48). 2,5-Dimethylcyclohexane-1,3-dione ((45); X = Me) by analogous procedures gave methyl 3,6-dimethyltridecanoate.

$$Me(CH_2)_3 \cdot CH\!:\!CH \cdot (CH_2)_4 \cdot CHMe \cdot CH_2 \cdot CO_2H$$

(47)

$$Me(CH_2)_9 \cdot CHMe \cdot CH_2 \cdot CO_2Me$$

(48)

For studies of the metabolism of iso-acids, tritium-labelled 14-methylpenta-1, 15-methylhexa- and 16-methylhepta-decanoic acid have been prepared by anodic synthesis[75].

Methyl 14-methyl-*cis*-hexadec-8-enoate (53), as well as the corresponding alcohol, which are sex attractants of *Trogoderma* species, have been synthesised[76] from 3-methylpent-1-yne (49). Bromination of the latter gave the 1-bromo-derivative (50) which was converted, by coupling with propargyl alcohol, into the diynol (51), and thence, by hydrogenation, into 6-methyloctan-1-ol ((51); X = CH$_2$OH). Oxidation of this alcohol gave the aldehyde (52)(X = CHO). Wittig reaction of the latter with the triphenylphosphinylide of methyl 8-bromo-octanoate led to the ester (53).

$$MeCH_2 \cdot CHMe \cdot C\!:\!CH$$

(49)

$$MeCH_2 \cdot CHMe \cdot C\!:\!CBr$$

(50)

$$MeCH_2 \cdot CHMe \cdot C\!:\!C \cdot C\!:\!C \cdot CH_2OH$$

(51)

$$MeCH_2 \cdot CHMe \cdot (CH_2)_4 \cdot X$$

(52)

$$MeCH_2 \cdot CHMe \cdot (CH_2)_4 \cdot CH\!:\!CH \cdot (CH_2)_6 \cdot CO_2Me$$

(53)

Methyl esters of various alkenyl- and alkyl-substituted long-chain acids have been synthesised by the Wittig reaction between alkylidene triphenylphosphoranes, RCH:PPh$_3$ (where R varied from H to n-C$_{17}$H$_{35}$) and methyl 12-oxo-octadecanoate or 10-oxohexadecanoate[77]. Detailed studies of the mass spectral fragmentation processes of these esters are described.

$$R^1R^2C(OH) \cdot C\!:\!CH \qquad R^1R^2C\!:\!CH \cdot CHO \qquad R^1R^2C\!:\!CH \cdot CH\!:\!CH \cdot CO_2H$$

(54)   (55)   (56)

An attempted preparation of 2,4-dienoic acids (56) is based upon the reaction of malonic acid with tertiary α-acetylenic carbinols (54), which are expected to rearrange to the aldehydes (55) under the reaction conditions employed[78].

(b) *Reactions* — The Wohl–Ziegler bromination of methyl tiglate (57) and methyl angelate (58) has been studied[79] in connection with the possible synthetic use of the resulting γ-bromo-esters for introducing isoprene units.

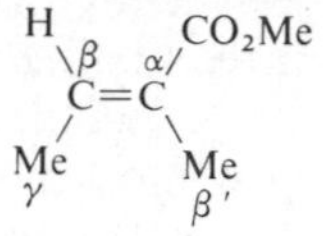

$$
\begin{array}{cc}
\underset{\gamma}{\overset{H}{\underset{\text{Me}}{\diagdown}}}\underset{\beta}{C}=\underset{\beta'}{\overset{\alpha}{\underset{\text{Me}}{C}}}\overset{CO_2Me}{\diagup} &
\overset{Me}{\diagdown}C=C\overset{CO_2Me}{\diagup}\\
(57) & (58)
\end{array}
$$

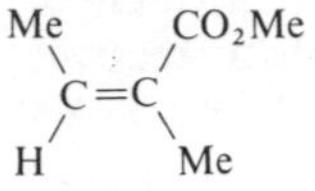

It is shown that *N*-bromosuccinimide can attack not only the γ-methyl group of the tiglic ester, but also the methyl group in the β'-position. Thus, the bromination of methyl tiglate results in a 2:1 mixture of methyl γ-bromo-tiglate and methyl β'-bromotiglate; methyl angelate shows similar behaviour.

8.2.3.3 *Isoprenoid acids*

The stereochemistry of the phytanic (3,7,11,15-tetramethylhexadecanoic) acid (59) occurring in the tissue lipids of humans suffering from Refsum's syndrome has been investigated[80]. In these studies the optical rotatory dispersion (o.r.d) curve of the ketone, 6,10,14-trimethylpentadecan-2-one (60), resulting from oxidative degradation of the acid was compared with that of the ketone, of known configuration, which is formed on ozonolysis of phytol. The close correspondence of these curves indicated that both ketones have essentially the same (6-D, 10-D) configuration. From this comparison, together with the optical rotation data for the isolated acid (59) it was concluded that the latter has the 3-DL,7-D,11-D-configuration, with the 3-L-diastereoisomer predominating.

$$Me_2CH\cdot(CH_2)_3\cdot CHMe\cdot(CH_2)_3\cdot CHMe\cdot(CH_2)_3\cdot CHMe\cdot CH_2\cdot CO_2H$$

(59)

$$Me_2CH\cdot(CH_2)_3\cdot CHMe\cdot(CH_2)_3\cdot CHMe\cdot(CH_2)_3\cdot COMe$$

(60)

A recent paper[81] describes experiments towards the synthesis of the C_{19} retinoic acid analogues (61) and (62) from isopropyl t-butyl ketone.

$$Me_2C\!:\!CBu^t\!\cdot\!C\!:\!C\cdot CMe\!:\!CH\cdot CH\!:\!CH\cdot CMe\!:\!CH\cdot CO_2H$$

(61)

$$Me_2C\!:\!CBu^t\!\cdot\!CH\!:\!CH\cdot CMe\!:\!CH\cdot CH\!:\!CH\cdot CMe\!:\!CH\cdot CO_2H$$

(62)

8.2.3.4 *Acids with repeating ·CHMe·CH$_2$ units*

(a) *Natural acids* — Polymethyl substituted acids with methylene-interrupted branching occur as constituents of mycobacterial lipids, as well as component acids of the preen-gland wax of various birds. The acids from mycobacterial lipids include $(+)-\alpha\beta$-unsaturated acids having the general structure (63) with the L-configuration of the asymmetric centres (named mycolipenic of phthienoic acid), and $(-)$-saturated acids (64) having the D-configuration in respect of the asymmetric centres (named mycoceranic or mycocerosic acids). A discussion of these acids is included in a recent review[82]. The acids from preen-gland wax were, in most of the cases so far studied, lower homologues of the $(-)$-acids (64), with the D-configuration of their asymmetric centres, but some representatives with the L-configuration of the C-2 centre have also been described. While the main constituents of mycoceranic acids have the structure (64) with $m = 2$ or 3, respectively, and $n = 19$, the most abundant polymethyl substituted acids of the preen-gland waxes have these structures with $n = 1$ or 2. Some recent reports deal with a variety of these acids, isolated from the preen-gland waxes of the flamingo[83], rook[84], and tufted duck[85]. A review of the chemistry of preen-gland waxes of waterfowl has been published[86].

$$\overset{L}{Me(CH_2)_n \cdot CHMe \cdot CH_2} \cdot \overset{L}{CHMe \cdot CH:CMe \cdot CO_2H}$$

(63) $n = 14\text{–}19$

$$Me(CH_2)_n \cdot (CHMe \cdot CH_2)_m \cdot CHMe \cdot CO_2H$$

(64)

A series of $(+)$-polymethyl substituted acids having the general structural features of (64) have now been shown to occur in a sulpholipid isolated from *Mycobacterium tuberculosis*[87]. The structure of these acids has been established mainly by mass spectrometric studies, which indicate that a main representative is a pentamethyl substituted C_{31}-acid (64; $n = 15$, $m = 4$). Their *dextro*rotation suggests that they are configurationally related to the $(+)-\alpha\beta$-unsaturated acids (63) with the L-configuration of their asymmetric centres. These acids occur together with a series of hydroxylated derivatives (65); the principal member of this series is the octamethyl-substituted C_{40}-acid ((65); $m = 7$)[87]. The hydroxy group of these acids was always located adjacent to the methyl branch which is farthest away from the carboxy group.

$$Me(CH_2)_{14} \cdot CH(OH) \cdot (CHMe \cdot CH_2)_m \cdot CHMe \cdot CO_2H$$

(65)

(b) *Syntheses* — Various optically active methylene-interrupted polymethyl substituted acids and their esters have been synthesised[88], including methyl 2(D), 4(D), 6(D), 8(D)-tetramethyloctacosanoate (70), the methyl ester of a major component of the mycoceranic (mycocerosic) acids (see above). The synthetic route employed represents a general method for the synthesis of the individual steroisomers of methylene-interrupted polymethyl substituted long-chain acids. It is based upon a step-wise introduction of the asymmetric centres with the formation of diastereoisomeric 2-methyl-substituted esters, which are separated by preparative g.l.c.[89]. The synthesis of the ester (70) involved methyl 2(D), 4(D), 6(D)-trimethyldodec-11-enoate ((68); $n = 2$) as an intermediate. The latter has been prepared[90] from L-$(+)$-5-acetoxy-4-

methylpentanoic acid (66), an industrial by-product, which by anodic coupling with pente-4-noic acid, followed by hydrolysis, gave the alcohol ((67); X = OH). The corresponding iodide ((67); X = I) was condensed with diethyl methylmalonate; hydrolysis of the product, followed by decarboxylation of the liberated acid gave a mixture of diastereoisomers which were separated by preparative g.l.c. of their methyl esters. The resulting ester ((68); $n = 1$) was converted by another chain-elongation employing the methylmalonic ester method, into the trimethyl-substituted ester ((68); $n = 2$). A further repetition of this procedure yielded methyl 2(D),4(D),6(D), 8(D)-tetramethyltetradec-13-enoate ((68); $n = 3$). This ester, on oxidation with potassium permanganate in acetic acid, gave methyl 2(D),4(D),6(D), 8(D)-tetramethyl-12-carboxydodecanoate (69). By anodic coupling of the latter with n-heptadecanoic acid the ester (70) was obtained.

$$AcOCH_2 \cdot \overset{L}{C}HMe \cdot (CH_2)_2 \cdot CO_2H$$

(66)

$$CH_2{:}CH \cdot (CH_2)_4 \cdot \overset{D}{C}HMe \cdot CH_2X$$

(67)

$$CH_2{:}CH \cdot (CH_2)_4 \cdot (\overset{D}{C}HMe \cdot CH_2)_n \cdot \overset{D}{C}HMe \cdot CO_2Me$$

(68)

$$HO_2C \cdot (CH_2)_4 \cdot (\overset{D}{C}HMe \cdot CH_2)_3 \cdot \overset{D}{C}HMe \cdot CO_2Me$$

(69)

$$Me(CH_2)_{19} \cdot (\overset{D}{C}HMe \cdot CH_2)_3 \cdot \overset{D}{C}HMe \cdot CO_2Me$$

(70)

($\pm$)-4,6-Dimethylocta-2,4-dienoic acid (72), a major degradation product of both (+)- and (−)-sclerotiorin, may be regarded as structurally related to the acids discussed in this section, as far as the relative positions of the methyl branches are concerned. This acid has been synthesised[91] by a Wittig condensation of 2,4-dimethylhex-2t-enal (71) with ethoxycarbonylmethylene triphenylphosphorane, prepared from triphenylphosphine and ethyl bromo-

$$MeCH_2 \cdot CHMe \cdot CH{:}CMe \cdot CHO$$

(71)

$$MeCH_2 \cdot CHMe \cdot CH{:}CMe \cdot CH{:}CH \cdot CO_2H$$

(72)

acetate. The aldehyde (71) was obtained by base-catalysed condensation of 2-methylbutyraldehyde and propionaldehyde.

8.2.3.5 *The* Cecropia *juvenile hormones*

The juvenile hormones (73) and (74) have received considerable attention. Their chemistry, as well as that of their analogues, has been reviewed[92, 93], and numerous syntheses including various stereoselective routes leading to these compounds have been published. Only the more recent developments will now be considered.

$$\underset{O}{EtC Me} \overset{c}{-} CH \cdot (CH_2)_2 \cdot CR \text{:} CH \cdot (CH_2)_2 \cdot CMe \text{:} CH \cdot CO_2 Me$$

$$\begin{aligned}&(73)\ R = Et\\&(74)\ R = Me\end{aligned}$$

It has been shown[94] that the material isolated from the *Cecropia* moth, consisting of the C_{18} (73) and C_{17} (74) hormone, exhibits *dextro*rotation. Therefore, attention has been turned to syntheses of the enantiomeric forms of these compounds.

A synthesis[95] of both enantiomeric forms of the natural C_{18} hormone, methyl *cis*-10,11-epoxy-7-ethyl-3,11-dimethyl-*trans*-2,*trans*-6-tridecadieno-ate, as well as its *trans,trans,trans*-isomer, was carried out according to a scheme earlier described[96] for producing racemic juvenile hormone. Methyl 6-hydroxy-3-methyl-7-methylenenon-2*t*-enoate (75), on reaction with the dimethyl acetal (76) of 3-chloro-3-methylpentan-2-one, prepared in both enantiomeric forms as described below, gave, via a Claisen rearrangement, the enantiomeric forms of the chloro-oxo-ester (77). Each of the enantiomers was reduced with sodium borohydride to give pairs of diastereoisomeric chlorohydrins (78) which were separated by t.l.c. Treatment of each of the four resulting chlorohydrins with methanolic potassium carbonate furnished the corresponding epoxy-compounds; one of the (+)-isomers, with $[\alpha]_D$ +12.2 °C, was the juvenile hormone, corresponding to the (+)-form of the ester (73).

The enantiomeric forms of the ketal (76) were prepared from 2-methyl-butyraldehyde (2-formylbutane) which was converted, by the action of cupric chloride in *NN*-dimethylformamide, into 2-chloro-2-formylbutane and thence, by oxidation with potassium permanganate, into 2-chloro-2-methyl butyric acid. Reaction of the corresponding acid chloride with (−)-α-(1-

$$\underset{(75)}{H_2C \text{:} CEt \cdot CH(OH) \cdot (CH_2)_2 \cdot CMe \overset{t}{\text{:}} CH \cdot CO_2 Me} + \underset{(76)}{EtCMeCl \cdot CMe(OMe)_2}$$

↓

$$\underset{(77)}{EtCMeCl \cdot CO \cdot (CH_2)_2 \cdot CEt \overset{t}{\text{:}} CH \cdot (CH_2)_2 \cdot CMe \overset{t}{\text{:}} CH \cdot CO_2 Me}$$

↓

$$\underset{(78)}{EtCMeCl \cdot CHOH \cdot (CH_2)_2 \cdot CEt \overset{t}{\text{:}} CH \cdot (CH_2)_2 \cdot CMe \overset{t}{\text{:}} CH \cdot CO_2 Me}$$

↓

C_{18} juvenile hormone

naphthyl)-ethylamine gave a mixture of diastereoisomeric amides which were separated by preparative t.l.c. The enantiomeric forms of 2-chloro-2-methylbutyric acid, obtained by hydrolysis of the amides, were then converted, via the corresponding methyl ketones, into the acetals (76).

Both enantiomeric forms of the C_{18} juvenile hormone have been synthesised[97] from starting materials of known absolute configuration by the following sequence of reactions.

The enantiomeric forms of the acetal 2,2-dimethoxy-3-methylpentan-3-ol (79) were obtained from 3-methylpent-1-yn-3-ol, the hydrogen phthalate of which was resolved by fractional crystallisation of the brucine salt. The hydrogen phthalates were hydrolysed, and each enantiomer was converted, by reaction with methanol with the aid of a mercuric oxide–boron trifluoride etherate–trifluoroacetic acid catalyst, into the respective enantiomer of the acetal (79). In another series of experiments the (−)-acetal was shown, by conversion into the corresponding methyl ketone, and oxidation of the latter to (−)-2-hydroxy-2-methylbutyric acid, to have the R-configuration.

Treatment of the enantiomeric forms of the acetal (79) with the hydroxy-ester (75) gave, via a Claisen rearrangement, the ketols (80), which were reduced with sodium borohydride to the diols (81). The latter were converted via the monotosylates into the respective stereoisomers of the epoxide (73). The dextrorotatory juvenile hormone was found to result from the S-(+)-enantiomer of the acetal (79) via the threo-diol (81), thus indicating that the dextrorotatory natural hormone has the 10(R), 11(S)-configuration.

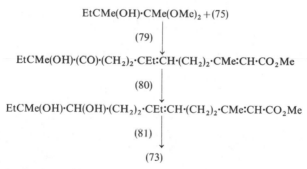

A review on insect juvenile hormone analogues has been published recently[98].

8.3 DI- AND POLY-CARBOXYLIC ACIDS

8.3.1 Normal-chain acids

A new synthesis[99] of trans-dodec-2-ene-1,12-dioic acid (traumatic acid) (86), yielding this acid free from positional and geometric isomers, involves conversion of methyl undec-10-enoate (82), by the action of N-bromo-

$$H_2C:CH \cdot (CH_2)_8 \cdot CO_2Me$$

(82)

succinimide, into the bromo-ester (83), from which, via the nitrile (84), the dioic acid (85) is obtained. The acid (85), on being heated *in vacuo*, is converted into the acid (86).

$$XCH_2 \cdot CH(OH) \cdot (CH_2)_8 \cdot CO_2R$$

(83; X = Br, R = Me)
(84; X = CN, R = Me)
(85; X = CO$_2$H, R = H)

$$HO_2C \cdot CH:CH \cdot (CH_2)_8 \cdot CO_2H$$

(86)

8.3.2   Alkyl-branched acids

In connection with studies of the antibiotic borrelidin, the two *meso*-forms of 2,4,6-trimethylpimelic acid (89) have been synthesised[100]. The *meso*-compounds, m.p. 97 and 130 °C respectively, were isolated from a mixture of stereoisomers obtained from α,α'-dimethylglutaric anhydride via the lactone (87) and the bromo-ester (88), and their configurations were determined by x-ray crystallography[101]. The compound of m.p. 130 °C which yields a dimethyl ester identical with a degradation product of the antibiotic, had the 2(R),4(R),6(S)-configuration.

$$CH_2 \cdot CHMe \cdot CH_2 \cdot CHMe \cdot CO$$
$$| \qquad\qquad\qquad\qquad |$$
$$O \underline{\qquad\qquad\qquad\qquad\qquad}$$

(87)

$$BrCH_2 \cdot CHMe \cdot CH_2 \cdot CHMe \cdot CO_2Me$$

(88)

$$HO_2C \cdot CHMe \cdot CH_2 \cdot CHMe \cdot CH_2 \cdot CHMe \cdot CO_2H$$

(89)

Syntheses of various dialkylated pimelic acids by using appropriately substituted glutaric acids as intermediates have also been described[102].

During studies involving the controlled synthesis of carotenoids containing *cis*-double bonds, methyl natural bixin (95) has been synthesised[103]. The synthesis was achieved by a condensation between 5-methoxycarbonyl-3-

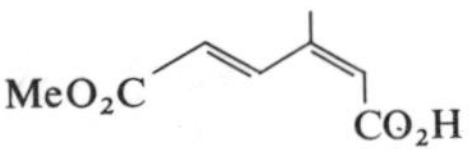

(90) 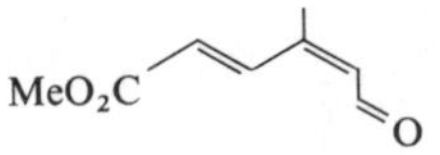   (91) 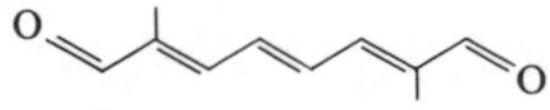   (92)

methylpenta-*cis*-(Z)-2-*trans*-(E)-4-dien-1-al (91) and the phosphorane (94). The former was obtained via the *cis*-*trans*-β-methylmuconic half-ester (90), by conversion of the latter into its acid chloride, followed by reduction with lithium tri-t-butoxyaluminium hydride. The phosphorane (94) was prepared by way of the triene-dial (92), which, by the appropriate Wittig reaction and reduction gave the hydroxy-ester (93).

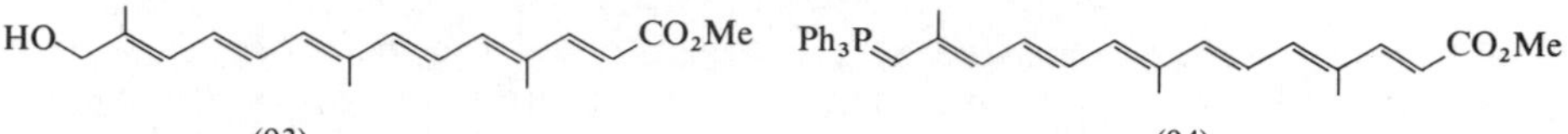

(93)

(91) + (94) ⟶

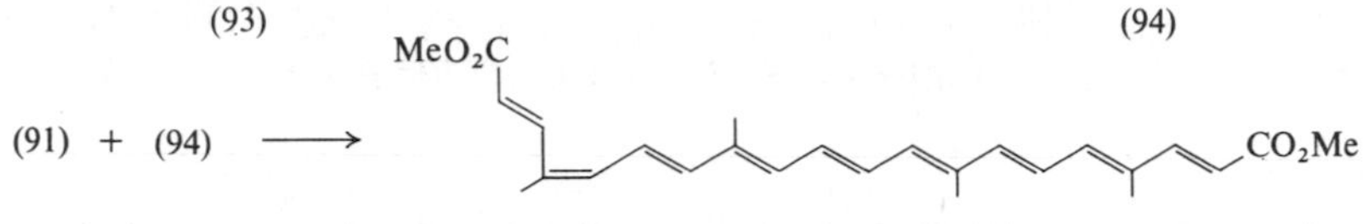

(95)

The above synthesis confirms the *cis*-(Z)-4-structure previously assigned to methyl natural bixin on the basis of n.m.r. studies. Moreover, it represents the first stereochemically controlled direct total synthesis of a carotenoid containing a methylated *cis*-double bond.

A toxic antibiotic, bongkrekic acid, produced by the action of the bacterium *Pseudomonas cocovenenans* on partially defatted coconut, has been shown[104] to be a tricarboxylic acid containing two pairs of conjugated double bonds, both conjugated with a carboxy group, as well as two isolated double bonds, three methyl groups, one methoxy group, and a ring system. From chemical and spectroscopic evidence obtained for the acid and its reduction product, the structure (96) has now been proposed[105] for bongkrekic acid; for one of the isolated double bonds the two possible positions are shown by dotted lines. Preliminary investigations involving ozonolysis of the acid are stated to indicate that the isolated double bond is at C-3.

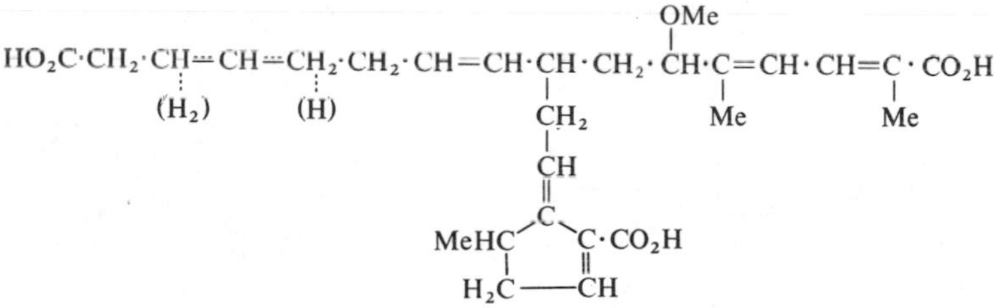

(96)

The dicarboxylic acids belonging to the mycolic acid group are included in Section 8.4.5.

8.4 HYDROXY ACIDS

8.4.1 General synthetic procedures

A convenient procedure[106] for the preparation of aliphatic α-hydroxy-acids involves reaction of thallium(III) acetate with a carboxylic acid. On hydrolysis, the resulting α-acyloxy-derivative gives the α-hydroxy-acid.

For the synthesis of β-hydroxy-acids a procedure using α-lithiated carboxylic acid salts has recently been described[107]. These lithiated derivatives react with carbonyl compounds according to the following scheme.

$$R^1R^2CO + LiCXY\cdot CO_2Li \longrightarrow R^1R^2C(OH)\cdot CXY\cdot CO_2H$$

Although several acids and ketones failed to undergo the above reaction, possibly because of steric hindrance, the procedure is shown to give satisfactory results for the preparation of β-hydroxy-acids, usually prepared by hydrolysis of the esters resulting from a Reformatsky reaction.

8.4.2 Normal-chain acids

8.4.2.1 Natural acids

A number of papers deal with α-hydroxy-acids as constituents of yeast[108] and bacterial phospholipids[109, 110].

Myrmicacin, isolated from the metathoracic gland of leaf-cutting ants, has been shown[111] to be the (−)-form of 3-hydroxydecanoic acid (97). It occurs together with its lower homologues, and appears to act as a herbicide by preventing the germination of undesirable spores.

$$Me(CH_2)_6\cdot CH(OH)\cdot CH_2\cdot CO_2H$$

(97)

The acids isolated from the seed oil of *Monnina emarginata* include 13(L)-hydroxyoctadeca-9c, 11t-dienoic acid (98)[112]. This conjugated dienolic acid is the enantiomer of coriolic acid.

$$Me(CH_2)_4\cdot \overset{L}{C}H(OH)\cdot CH\overset{t}{:}CH\cdot CH\overset{c}{:}CH\cdot (CH_2)_7\cdot CO_2H$$

(98)

An unsaturated trihydroxy-acid isolated from wheat bran has been suggested[113] to have the structure (99).

$$Me(CH_2)_5\cdot CH(OH)\cdot CH_2\cdot CH\overset{t}{:}CH\cdot CH(OH)\cdot (CH_2)_2\cdot CH(OH)\cdot (CH_2)_3\cdot CO_2H$$

(99)

8.4.2.2 Syntheses

The hydroxy-acid (101) has been synthesised[114] from the acetoxy-compound (100) by condensation with a halogenoacetate, followed by Clemmensen reduction of the crude condensation product.

$$AcO(CH_2)_{11}\cdot CO\cdot CH_2S\cdot OMe$$

(100)

halogenoacetate

$$HO(CH_2)_{14}\cdot CO_2H$$

(101)

A synthesis[115] of the *trans,trans*-dienoic hydroxy-acid (107), reported to arise as a by-product of prostaglandin biogenesis, starts from the chloronitrile (102). Reduction of the latter with lithium aluminium triethoxyhydride gave the chloroaldehyde (103) which reacted with the Grignard reagent from 4-methoxybut-3-en-1-yne to afford the methoxyenynol ((104), X = Cl). The latter was reduced to the dienol (105; X = Cl), and thence converted, via the nitrile ((105); X = CN), into the aldehydo-acid (106). This compound reacted with pentylmagnesium bromide to give the acid (107).

$$CN \cdot (CH_2)_6Cl \longrightarrow OHC \cdot (CH_2)_6Cl \ + \ MeOCH{:}CH \cdot C{:}CMgBr \longrightarrow$$

(102)   (103)

$$MeOCH{:}CH \cdot C{:}C \cdot CH(OH) \cdot (CH_2)_6X \longrightarrow MeOCH{:}CH \cdot CH{:}CH \cdot CH(OH) \cdot (CH_2)_6X$$

(104)   (105)

$$OHC \cdot CH{:}CH \cdot CH{:}CH \cdot (CH_2)_6 \cdot CO_2H \longrightarrow Me(CH_2)_4 \cdot CH(OH) \cdot CH{:}CH \cdot CH{:}CH \cdot (CH_2)_6 \cdot CO_2H$$

(106)   (107)

A partial synthesis[116] of hydroxylated conjugated dienoic acids involved allylic bromination of methyl linoleate, followed by acetylation and hydrolysis of the products. Several isomeric mono- and di-hydroxylated dienoic acids were obtained.

The *threo*-isomer of 9,10,16-trihydroxyhexadecanoic acid (108), which has been recently[117] synthesised, has now been resolved[118] by fractional crystallisation of the (−)-ephedrine salt; the enantiomeric acids were obtained with $[\alpha]_D + 24.6$ and $-24.45\,°C$, respectively. This resolution was of interest, because the synthetic acid had a somewhat lower melting point than that of natural aleuritic acid obtained from shellac. Since the natural acid was stated to have a very small rotation, it was suggested[117] that it may be one enantiomorph of the *threo*-form of the acid (108), or contain an excess of one enantiomorph.

$$HO(CH_2)_6 \cdot CH(OH) \cdot CH(OH) \cdot (CH_2)_7 \cdot CO_2H$$

(108)

Syntheses of hydroxy-acids containing acetylenic linkages are discussed in Section 8.2.1.4.

8.4.2.3 *Reactions*

(a) *Saturated acids* — Further studies of the reactions occurring in concentrated alkali metal hydroxides, by examining the behaviour of various functional groups present in long-chain fatty acids, have been published. They include studies involving the alkali fusion of epoxy-acids, as well as of a series of 11-alkoxyundecanoic acids[119]. While for the epoxy-acids several reaction pathways, including β-elimination, occur, with the alkoxy-acids β-elimination takes place almost exclusively. Initially undec-10-enoic and

11-hydroxyundecanoic acid are formed which are converted into nonanoic and undecanedioic acid, respectively.

The alkali fusion of 2-hydroxyoctadecanoic acid at 300 °C gives[120] a 90% yield of heptadecanoic acid. A simple method is thus provided for converting a fatty acid by α-hydroxylation and subsequent alkali fusion into the nor-acid.

In the course of the above studies[120] the alkali fusion of aleuritic acid (108) was re-investigated. The cyclohexane carboxylic acid (111) has been found to be the major product, and its formation was explained by considering the known reactions of hydroxy-acids. The intermediate compound (110) could be isolated, and its formation indicated that the reaction involves dehydrogenation of the 9,10-glycol function of aleuritic acid to give the diketo-acid (109); the latter would be expected to undergo a benzilic rearrangement with the formation of the compound (110).

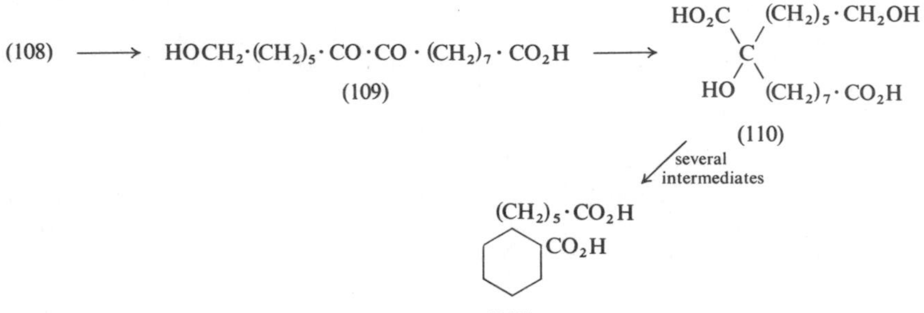

$$(108) \longrightarrow HOCH_2 \cdot (CH_2)_5 \cdot CO \cdot CO \cdot (CH_2)_7 \cdot CO_2H \longrightarrow$$

$$\underset{(109)}{}$$

$$\begin{array}{c} HO_2C \quad (CH_2)_5 \cdot CH_2OH \\ \diagdown \diagup \\ C \\ \diagup \diagdown \\ HO \quad (CH_2)_7 \cdot CO_2H \end{array}$$

(110)

several intermediates

$$\begin{array}{c} (CH_2)_5 \cdot CO_2H \\ \text{⬡} \quad CO_2H \end{array}$$

(111)

In another study[121] the previously reported alkaline fission of 9(10)-hydroxy-10(9)-oxo-octadecanoic acid (112) to give nonanedioic and nonanoic acid has been confirmed. It is shown that this fission involves autoxidation.

$$\left. \begin{array}{l} Me(CH_2)_7 \cdot CH(OH) \cdot CO \cdot (CH_2)_7 \cdot CO_2H \\ Me(CH_2)_7 \cdot CO \cdot CH(OH) \cdot (CH_2)_7 \cdot CO_2H \end{array} \right\} \quad (112)$$

In connection with polymerisation studies a series of anhydrosulphites (114) have been prepared[122] by reaction of thionyl chloride with α-hydroxy-acids (113).

$$\begin{array}{cc} R^1R^2C(OH) \cdot CO_2H & \qquad R^1R^2C\!-\!CO \\ & \qquad \underset{\diagdown}{O} \quad \underset{\diagup}{O} \\ (113) & \qquad SO \end{array}$$

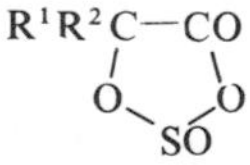

(114)

(b) *Unsaturated acids* — The reactions of methyl ricinoleate and ricinelaidate have been studied further. In particular, the reactivity of the homoallylic system of these compounds has been investigated by studies of the O-toluene-p-sulphonate of methyl ricinelaidate[123] and the O-methyl sulphonate of methyl ricinoleate[124]. In a suitably buffered solution the O-methylsulphonate (mesyl ester) of methyl ricinoleate ((115), X = mesyl) gives with methanol the 9-methoxy-10,11-methylene-compound (116); with acetic acid the corresponding acetoxy-derivative is formed, and with water the hydroxy-derivative. Under certain conditions elimination predominates, and similar results were obtained with the toluene-*p*-sulphonate of methyl ricinelaidate.

$$Me(CH_2)_5 \cdot CH(OX) \cdot CH_2 \cdot CH{:}CH \cdot (CH_2)_7 \cdot CO_2Me$$

(115)

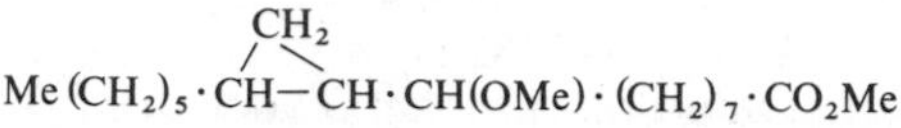

$$Me(CH_2)_5 \cdot CH{-}CH \cdot CH(OMe) \cdot (CH_2)_7 \cdot CO_2Me$$

(116)

In studies of the oxymercuration–demercuration of long-chain unsaturated esters the influence of substituent hydroxy-, acetoxy-, and methoxy-groups on the nature of the resulting products has been investigated[125]. Some of the unsaturated hydroxy-esters gave high yields of 1,4-epoxides (2,5-disubstituted tetrahydrofurans). Other reports include the formation of 1,4-epoxides from γ-hydroxy-alkenes by epoxidation[126], and by free-radical cyclisation of certain hydroxy-esters[127]. The mechanism and stereochemistry of such cyclisations have been studied[128] by employing certain trihydroxy-esters obtainable from hydroxyalkenoic esters.

8.4.3 Mevalonic acid

The chemistry of mevalonic acid has been reviewed[129].

The optically active forms of mevaldic acid were obtained[130] by resolution of (±)-3-hydroxy-3-methyl-5,5-dimethoxypentanoic acid (117) The resolution was achieved by fractionation of the quinine salt, and the resulting (*R*)- and (*S*)-enantiomer of mevaldic acid (118) were converted, by reduction with sodium borohydride, into the respective stereoisomers of mevalonolactone (119).

$$(MeO)_2CH \cdot CH_2 \cdot C(OH)Me \cdot CH_2 \cdot CO_2H \qquad OCH \cdot CH_2 \cdot C(OH)Me \cdot CH_2 \cdot CO_2H$$

(117)   (118)

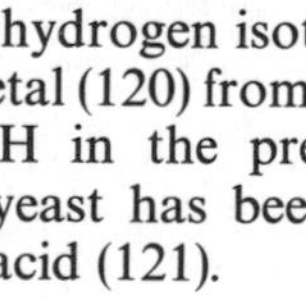

(119)

This synthesis also provides a route to the 3(*R*)- as well as the 3(*S*)-mevalonolactone labelled with hydrogen isotopes at C-5.

Reduction of the hemithioacetal (120) from coenzyme A and (±)-mevaldic acid with 4(*R*)-4-[³H]-NADPH in the presence of 3-hydroxy-3-methyl-glutaryl CoA reductase from yeast has been shown[131] to give the 5(*S*)-5-[³H₁]-form of 3(*R*)-mevalonic acid (121).

$$CoA \cdot SCH(OH) \cdot CH_2 \cdot C(OH)Me \cdot CH_2 \cdot CO_2H$$

(120)

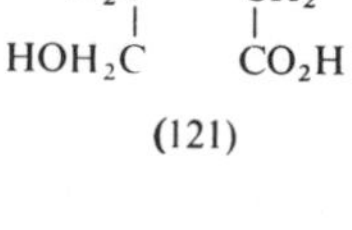

(121)

The above work, together with the syntheses[132, 133] which were announced simultaneously and are discussed below, completes the syntheses of mevalono-lactones tritiated in the six possible methylene positions; the other five stereospecifically labelled mevalonolactones have been previously described[134].

The synthesis[132] of 5(S)-5-[^{3}H$_1$]-mevalonic acid starts from 1,1-diethoxy-3-methylbut-3-ene (122), obtainable from methallylmagnesium chloride and ethyl orthoformate. Hydrolysis of this acetal gave 3-methylbut-3-enal (123) which was reduced by adding it to a tetrahydrofuran solution of lithium borohydride previously treated with tritiated water. The product, 1-[^{3}H]-3-methylbut-3-enol (124), was oxidised with dimethyl sulphoxide–dicyclo-hexylcarbodi-imide–phosphoric acid to give the aldehyde (125). This was reduced with liver alcohol-dehydrogenase and an excess of reduced nicotina-mide–adenine dinucleotide to the S-1-[^{3}H$_1$]-isopentenol (126) which was converted via the bromide (127) and the cyanide (128) into mevalonic acid (129); the latter was isolated as the lactone.

$$CH_2{:}CMe{\cdot}CH_2{\cdot}CH(OEt)_2 \longrightarrow CH_2{:}CMe{\cdot}CH_2{\cdot}CHO$$

(122)   (123)

$$CH_2{:}CMe{\cdot}CH_2{\cdot}CTO \longleftarrow CH_2{:}CMe{\cdot}CH_2{\cdot}CHTOH$$

(125)   (124)

$$CH_2{:}CMe{\cdot}CH_2CH(OH){\cdots}T$$

(126)

$$CH_2X{\cdot}C(OH){\cdot}CH_2{\cdot}\overset{Me}{\underset{OH}{C}}{\cdots}T$$

(127; X = Br)
(128; X = CN)
(129; X = CO$_2$H)

Another synthesis[133] proceeded via 5-[^{3}H]-mevaldic acid. The latter was obtained by way of 1-[^{3}H]-3-oxobutyraldehyde dimethylacetal (130) and the hydroxy-ester (131), prepared from the the acetal (130) by a Reformatsky reaction.

$$MeCO{\cdot}CH_2{\cdot}CHT(OMe)_2 \longrightarrow MeO_2C{\cdot}CH_2{\cdot}CMe(OH){\cdot}CH_2{\cdot}CT(OMe)_2$$

(130)   (131)

The preparation of 3(R)-mevalonolactone acetate has been described[135]. The latter has the useful property of being much less soluble in ether than the racemate. This difference in solubility presented a convenient method for determining the chirality of the enzymically produced[131] mevalonate.

8.4.4   The prostaglandins

An outstanding achievement is the synthesis of all the known primary prostaglandins.

The first total synthesis of the prostaglandins F$_{1\alpha}$ (132) and E$_1$ (133) was published in 1969[136].

An adaptation of the stereocontrolled syntheses[137] of the racemic forms of the prostaglandins $F_{2\alpha}$ and E_2 resulted in the first total syntheses[138] of the naturally-occurring optically-active forms of these prostaglandins.

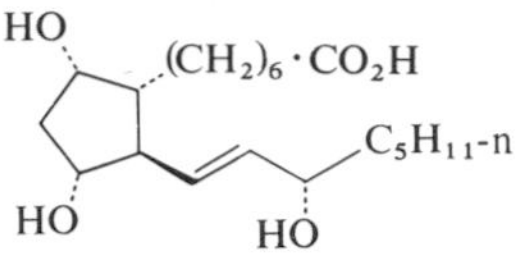

$$(132) \qquad\qquad (133)$$

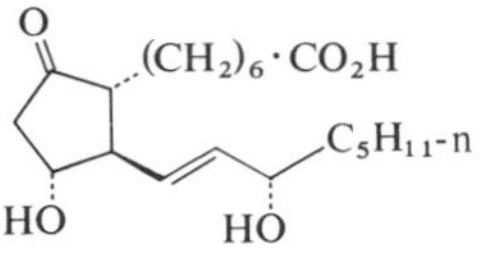

Hydrolysis of the readily obtainable[137] lactone (134) gave the hydroxy-acid (135). The (+)-form of this acid, obtained by crystallisation of its (+)-ephedrine salt, gave on treatment of its sodium salt with potassium tri-iodide, the iodolactone (136). The latter was converted according to the procedure described earlier[137] via the optically active forms of the intermediates (137), (138) and (139) into the aldehyde (140) which, on reaction with the

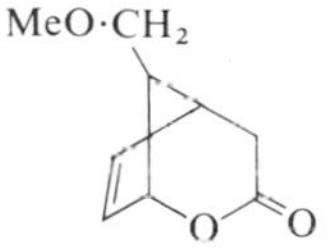

(134)

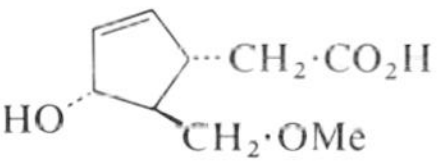

(135)

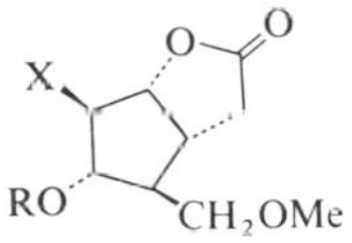

(136) X = I; R = H
(137) X = I; R = Ac
(138) X = H; R = Ac

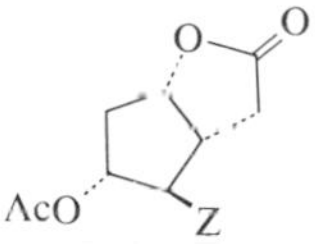

(139) Z = CH₂OH
(140) Z = CHO
(141) Z = CH:CH·CO·C₅H₁₁-n

sodio-derivative of dimethyl 2-oxoheptylphosphonate formed stereospecifically the trans-enone lactone (141). Reduction of the keto-group gave the hydroxy-compound (142). The bis-tetrahydropyranyl-derivative (143) of the latter, on reduction to the corresponding lactol, followed by a Wittig reaction, was converted into the acid (144). Removal of the tetrahydropyranyl groups afforded prostaglandin $F_{2\alpha}$ (145); oxidation of the tetrahydropyranyl-derivative (144), followed by hydrolysis gave prostaglandin E_2 (146).

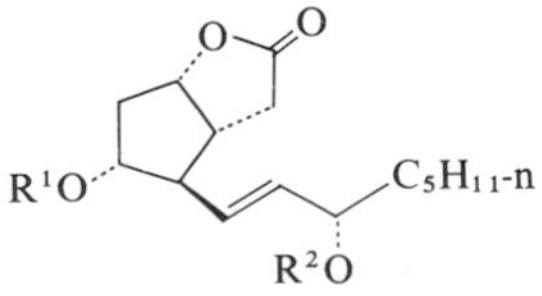

(142) R¹ = Ac; R² = H
(143) R¹ = R² = tetrahydropyranyl (THP)

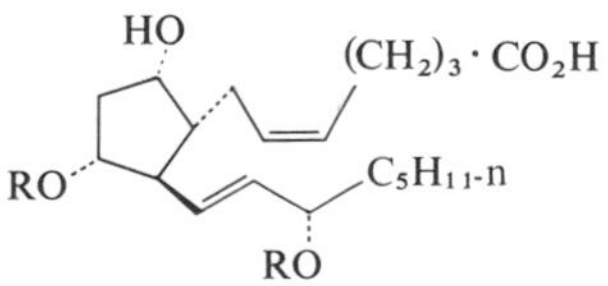

(144) R = THP
(145) R = H

The bistetrahydropyranyl ether (144) was also employed[139] for a second total synthesis of prostaglandin $F_{1\alpha}$ and E_1; the four primary prostaglandins were thus obtained from a common intermediate.

Further information on the literature concerning syntheses of the above prostaglandins is available from a recent review[140].

The first syntheses of prostaglandin $F_{3\alpha}$ (154) and E_3 (155) in the optically-active, naturally-occurring form, were accomplished[141] by way of the optically-active hydroxy-lactone (147) as an intermediate. This lactone was converted,

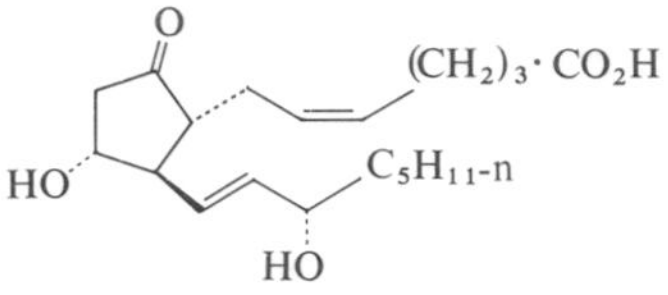

(CH$_2$)$_3$·CO$_2$H

C$_5$H$_{11}$-n

HO

HO

(146)

via the alcohol (148), into the aldehyde (149). Condensation of the latter with the S-(+)-phosphonium salt (150), synthesised from S-(−)-malic acid, gave the compound (151). The bis-tetrahydropyranyl-derivative (152) was then reduced to the corresponding lactol which, on reaction with the Wittig reagent derived from 5-triphenylphosphoniovaleric acid, produced

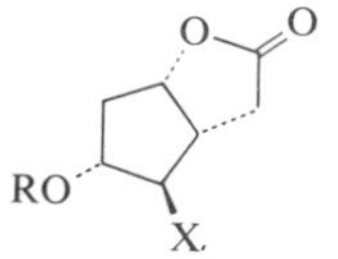

RO

X.

(147) R = H; X = CH$_2$·O·CH$_2$Ph
(148) R = THP; X = CH$_2$OH
(149) R = THP; X = CHO

(149) + Ph$_3$P$^+$H$_2$C—CH$_2$—CH(OH)—CH$_2$—CH:CH—CH$_2$—CHI$^-$

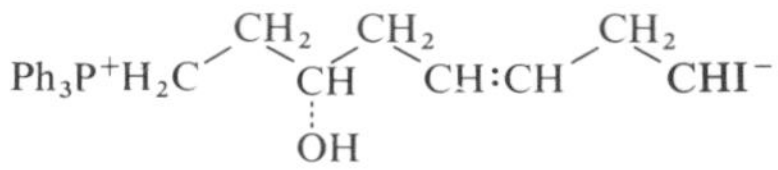

(150)

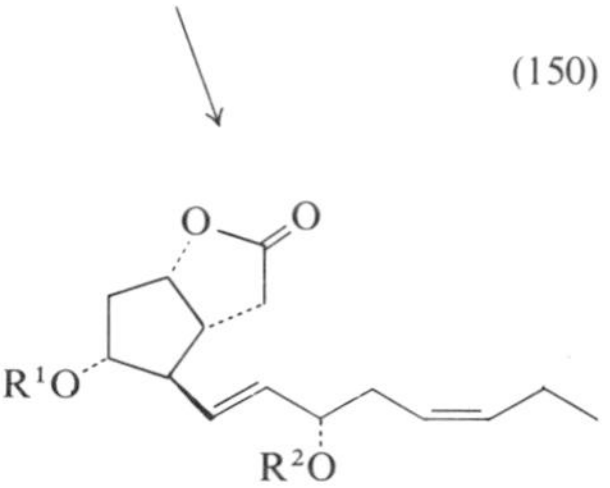

R^{1}O

R^{2}O

(151) R^1 = THP, R^2 = H
(152) R^1 = R^2 = THP

the bis-tetrahydropyranyl-derivative (153); removal of the tetrahydro-pyranyl groups gave prostaglandin $F_{3\alpha}$ (154). Oxidation of the bis-tetrahydropyranyl-derivative (153), followed by removal of the tetrahydropyranyl groups afforded prostaglandin E_3 (155).

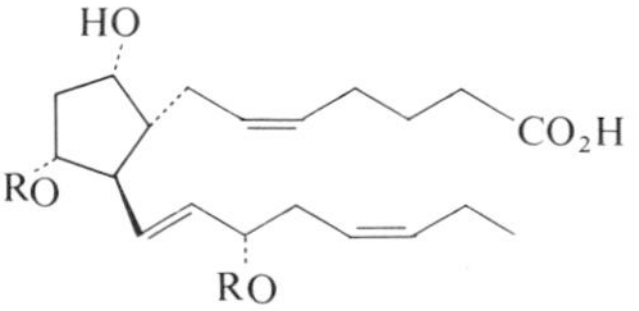

HO

RO

RO

CO$_2$H

(153) R = THP
(154) R = H

HO

HO

CO$_2$H

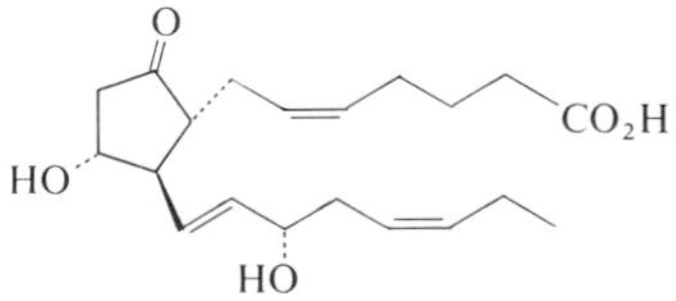

(155)

A novel prostaglandin derivative was isolated[142] during the biosynthetic conversion of arachidonic acid into prostaglandin by rat stomach homogenates. The structure (156) proposed for this compound is based upon spectroscopic evidence, and the products obtained from oxidative ozonolysis.

Two biologically-active cyclohexane analogues of the prostaglandins have been prepared[143] in order to study the effect of ring size on the biological properties.

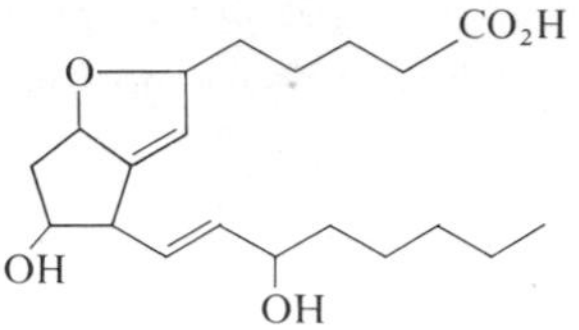

(156)

8.4.5   The mycolic acids

The mycolic acids (β-hydroxy-acids with a long alkyl chain in the α-position) from *Corynebacterium hofmanii* (the mycolic acids occurring in *Corynebacteria* have been named corynomycolic acids) have been shown[144] to include monoenoic and dienoic acids which were separated by argentation chromatography. The monoenoic acids to include a mixture of the C_{34}-acids (157) and (158), containing the double bond in the main chain, and side chain, respectively.

$$Me(CH_2)_7 \cdot CH{:}CH \cdot (CH_2)_7 \cdot CH(OH) \cdot CH(CO_2H) \cdot (CH_2)_{13}Me$$

(157)

$$Me(CH_2)_{14} \cdot CH(OH) \cdot CH(CO_2H) \cdot (CH_2)_6 \cdot CH{:}CH \cdot (CH_2)_7 Me$$

(158)

The dienoic acid fraction consisted of a C_{36}-acid (159) with one double bond in the main chain, and one double bond in the side chain.

$$Me(CH_2)_7 \cdot CH{:}CH \cdot (CH_2)_7 \cdot CH(OH) \cdot CH(CO_2H) \cdot (CH_2)_6 \cdot CH{:}CH \cdot (CH_2)_7 Me$$

(159)

Studies of the mycolic acids from several strains of *Nocardia* (nocardic acids) demonstrated[145] the presence of a large variety of acids ranging from C_{34} to C_{66}, including tri- and tetra-enoic acids.

The lower mycolic acids, occurring mostly in *Corynebacteria*, have now also been found as constituents of lipid fractions from *Mycobacteria*[146] and *Nocardia*[147].

The mycolic acids isolated from a strain of *Mycobacterium paratuberculosis* were shown[148] to include, in addition to dicyclopropane acids, mixtures of keto- and di-carboxylic acids. For these acids the general structures (160) and (161) were proposed.

$$Me(CH_2)_{17} \cdot CHMe \cdot CO \cdot (C_nH_{2n-2}) \cdot CH(OH) \cdot CH(CO_2H) \cdot C_{22}H_{45}$$

(160; $n = 34$–39)

$$HO_2C \cdot (C_nH_{2n-2}) \cdot CH(OH) \cdot CH(CO_2H) \cdot C_{22}H_{45}$$

(161; $n = 32$–39)

In the course of this investigation an ester was isolated which, on hydrolysis, gave eicosan-2-ol and dicarboxylic mycolic acids; the structure (162) has been proposed for the ester. It is of interest that according to a biogenetic scheme proposed earlier[149] such esters would arise as intermediates in the formation of dicarboxylic mycolic acids from higher ketomycolic acids by a Baeyer–Villiger oxidation of the latter.

$$Me(CH_2)_{17}{\cdot}CHMeO{\cdot}CO{\cdot}(C_nH_{2n-2}){\cdot}CH(OH){\cdot}CH(CO_2H){\cdot}C_{22}H_{45}$$

$$(162; n = 32\text{--}39)$$

In further studies on the biogenesis of the mycolic acids, the 5-enoic acids from *Mycobacterium phlei* already mentioned (Section 8.2.1.2) were compared[8] with a series of dienoic mycolic acids occurring in the same organism. This comparison showed that the terminal portions RCH:CH· of the mycolic acids corresponded to those of the 5-enoic acids, thus suggesting that the latter might serve as precursors in the biogenesis of these mycolic acids.

A further report[150] concerning the stereochemistry of the mycolic acids at C-2 and C-3 has also been published. By applying the method of molecular rotation differences to various mycolic acids from *Nocardia* and *Mycobacteria* it has been shown that all the acids studied had the 2R,3R-configuration earlier assigned to certain representatives of the mycolic acids.

In a series of investigations[151, 152] involving the lipids of *Mycobacterium bovis* isolated from the lungs of infected mice it has been demonstrated that mycolic acid, as well as phthiocerol dimycocerosate and tuberculostearic acid are present in the bacterial cells grown *in vivo*.

Recent work on mycobacterial cell wall fractions containing mycolic acids has been reviewed[153].

8.5   LACTONES

8.5.1   α-Lactones

The occurrence of α-lactóne intermediates in various reactions has received

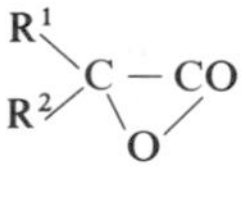

further attention. The reactions proposed to involve such intermediates include certain nucleophilic substitution reactions[188], free-radical substitution reactions[189], the ozonolysis of ketenes[190], as well as the photolysis of a monomeric malonyl peroxide[191].

8.5.2   β-Lactones

The structure (163) has been proposed[154] for an antibiotic isolated from an unidentified fungus, believed to be a *Cephalosporium*. This methyl-branched

2,4-dienoic acid with a β-lactone function at C-12 and C-13 appears to be the first-known naturally occurring β-lactone.

$$HOCH_2 \cdot CH \cdot CH \cdot (CH_2)_4 \cdot CHMe \cdot CH_2 \cdot CMe:CH \cdot CMe:CH \cdot CO_2H$$
$$\underset{OC-O}{|\quad|}$$

(163)

A new general β-lactone synthesis[155] involves the bromolactonisation of βγ-unsaturated acids. Thus, bromination of the sodium salt of 2,2-dimethyl-but-3-enoic acid (164) produced the bromolactone (165). This behaviour is shown to be in contrast to the known formation of β-iodo-γ-lactones by iodolactonisation of βγ-unsaturated acids.

$$CH_2:CH \cdot CMe_2 \cdot CO_2H \qquad\qquad BrCH_2 \cdot CH \cdot CMe_2 \cdot CO$$
$$\qquad\qquad\qquad\qquad\qquad\qquad\qquad \underset{O}{|}\underline{\qquad\qquad}|$$

(164)

(165)

A laboratory synthesis of β-propiolactone by condensation of formaldehyde and ketene in the presence of a zinc chloride catalyst has been described[156].

8.5.3 γ-Lactones

Two diastereoisomeric γ-lactones (166) and (167), derived from 3-methyl-4-hydroxycaprylic acid, have been isolated from the wood of three *Quercus* species, together with γ-nonalactone[157]. Their configurations were established by spectroscopic studies, in particular by comparative n.m.r. spectroscopy.

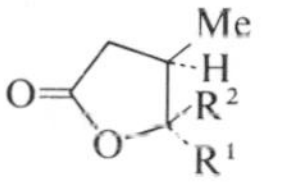

(166) R¹ = Buⁿ, R² = H
(167) R¹ = H, R² = Buⁿ

The stereochemistry formerly assigned to canadensolide, a mould metabolite produced by *Penicillium canadense*, showing antigerminative activity against fungi, has been shown[158] to be incorrect by synthesis of (±)-canadensolide (172) and its C-5 epimer (173). Reaction of the Grignard reagent from hex-1-yne (168) with trimethyl ethanetricarboxylate (169) in the presence of cuprous chloride in tetrahydrofuran gave the acetylenic ester

$$Bu^nC:CH + MeO_2C \cdot CH_2 \cdot CH(CO_2Me)_2$$

(168)   (169)

$$Bu^nC:C \cdot CH(CO_2Me) \cdot CH(CO_2Me)_2$$

(170)

$$Bu^nCH:CH \cdot CH(CO_2Me) \cdot CH(CO_2Me)_2$$

(171)

(172   (173)

(170), which was reduced to the ethylenic ester (171). The latter was converted into the compounds (172) and (173) by procedures involving hydroxylation of the ethylenic linkage, removal of one carboxy group, and treatment with formaldehyde.

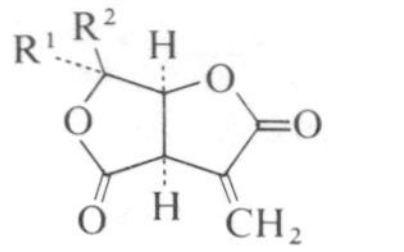

(172) R^1 = H, R^2 = Bu^n
(173) R^1 = Bu^n, R^2 = H

A general method[159] for the synthesis of α-methylenebutyrolactones involves carboxylation of a butyrolactone (174) with methyl methoxymagnesium carbonate, and reaction of the resulting acid (175) with a mixture of aqueous formaldehyde and diethylamine, followed by treatment of the product with sodium acetate in acetic acid. The α-methylenebutyrolactones (176) are thus obtained in two simple steps.

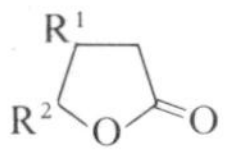

(174) 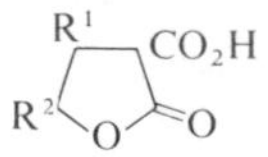   (175) 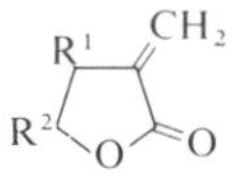   (176)

Another simple route[160] to various α-methylene-γ-lactones is based upon the reaction of organozinc compounds derived from α-(bromomethyl)acrylic esters (177) with aldehydes or ketones (178). The methylenelactones (179) are stated to result from this reaction in very good yields.

$$CH_2:C(CH_2Br)\cdot CO_2R \quad + \quad R^1R^2CO \longrightarrow$$

(177)   (178)

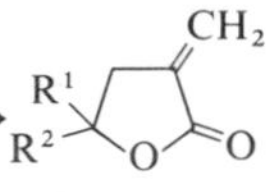

(179)

β-Alkyl-substituted γ-butyrolactones have been prepared[161] by the reaction of some alkylacetylenes with carbon monoxide in the presence of iron pentacarbonyl.

Substituted γ-lactones (180) and (181) have been distinguished[162] from each other by means of differences in equivalent chain-lengths and retention indices on two stationary phases (one of them polar and the other non-polar). The apparent polarity of the lactones, detected in this way, is found to depend on the steric α-hindrance of the carbonyl group.

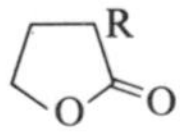

(180) 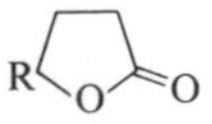   (181)

There have been further reports on investigations involving butenolides. Thus, four butenolide derivatives isolated from *Umbelliferae* have been shown[163] to have the structures (182)–(185). These compounds, differing from each other by the number of double bonds in the side-chains, are probably formed from the respective unsaturated C_{18}-acids as precursors.

Syntheses of butenolides published include the formation of but-2-enolides by heating methyl γ-bromo-$\alpha\beta$-unsaturated carboxylates with iron powder at 120–150 °C[164]; lactonisation takes place with elimination of methyl bromide. This procedure is based upon the observation that butenolides are sometimes obtained as by-products on brominating $\alpha\beta$-unsaturated carboxylic esters with N-bromosuccinimide.

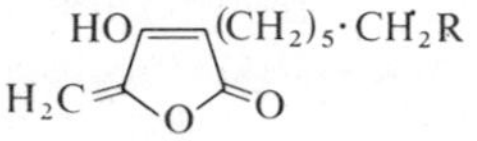

$$HO-(CH_2)_5 \cdot CH_2R$$

H$_2$C, O, O structure

(182) R = $(CH:CH)_3 \cdot (CH_2)_3Me$
(183) R = $CH:CH \cdot CH_2 \cdot CH:CH \cdot (CH_2)_4Me$
(184) R = $CH:CH \cdot (CH_2)_7Me$
(185) R = $(CH:CH)_3 \cdot CH_2 \cdot CH_2 \cdot CH:CH_2$

4-Alkyl-2-methylbut-2-en-4-olides (188) have been prepared[165] from methyl ketones (186) by reaction with sodium pyruvate, followed by reduction of the resulting 4-oxo-acids (187) with sodium borohydride and lactonisation of the products.

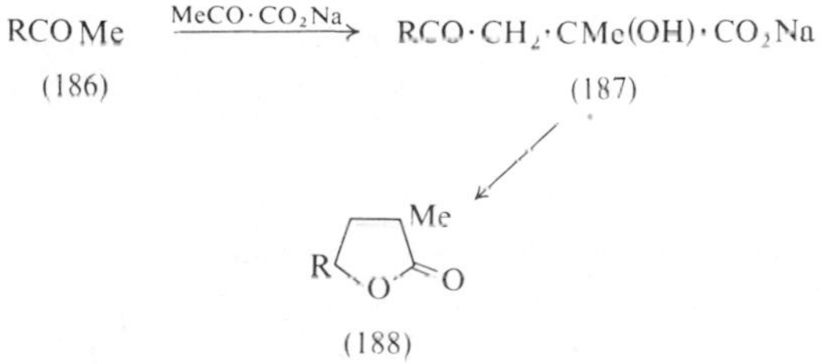

$$RCOMe \xrightarrow{MeCO \cdot CO_2Na} RCO \cdot CH_2 \cdot CMe(OH) \cdot CO_2Na$$

(186) (187)

(188) structure with Me, R, O, O

Syntheses of α-isocyanato- (189) and α-isocyano-γ-butyrolactones (190) have been described and their reactions discussed[166]. With alcohols, amines, and phenylhydrazines the isocyanatolactone (189) forms urethanes, ureas, and semicarbazides respectively, which undergo rearrangements to form as-triazines and hydantoins. The isocyanolactone (190), after basic opening of the lactone ring, rearranges to 4-carbethoxy-5,6-dihydro-4H-1,3-oxazine (191).

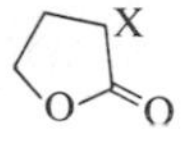

(189) X = NCO
(190) X = NC

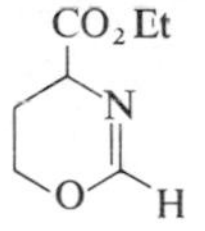

(191) structure with CO$_2$Et, N, O, H

Investigations of various reactions of unsaturated γ-lactones have also been reported. Thus, in connection with studies relating to the carcinogenicity of γ-lactones the reactions of 4-hydroxypent-2-enoic acid lactone (192) with several model compounds have been investigated[167]. The Michael addition

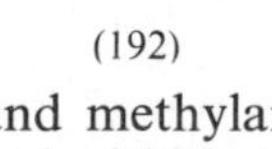

(192) structure with Me, O, O

of benzyl thiol, benzylamine, and methylamine to this lactone was found to take place readily, as compared with that of imidazole and glycine amide. No stable addition products were obtained with guanidine.

The alcoholysis of but-2-enolides in the presence of acid catalysts, has been shown[168] to involve isomerisation to but-3-enolides as the preponderant reaction. The respective 4-oxo-esters thus result, in addition to secondary products.

In studies[169] on the reactivity of 2,2-dibromo-γ-butanolide (193), obtained[170] by bromination of γ-butanolide with phosphorus tribromide, this dibromo-butanolide was found to yield, by the action of hot alkali, 2-bromo-4-hydroxy-*trans*-but-2-enoic acid (194) together with 2-bromoacrylic acid (195). The resulting products are also found to include acetaldehyde, presumed to arise by the action of alkali on the acid (195).

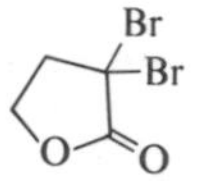

$$\text{HOCH}_2\cdot\text{CH:CBr}\cdot\text{CO}_2\text{H} \quad + \quad \text{H}_2\text{C:CBr}\cdot\text{CO}_2\text{H}$$

(194)   (195)

(193)

Mevalonic acid lactone has already been considered in Section 8.4.3.

8.5.4 δ-Lactones

An improved procedure, for industrial application, involving the preparation of δ-lactones from dihydroresorcinol (196) has been described[171]. The alkyl substituents are introduced by reactions in aqueous dioxan and the derivatives (197) are converted in a known manner, via the keto-acid (198) and hydroxy-acid (199) into the lactone (200).

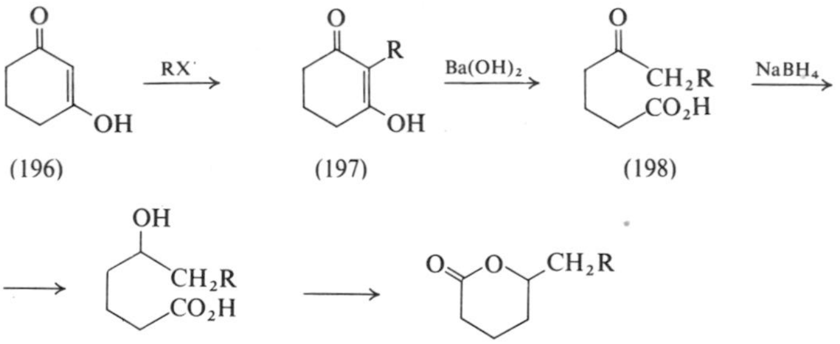

(196)   (197)   (198)

(199)   (200)

Another synthesis[172] is based upon the reduction of 4-alkyl-4,4-dicar-bethoxybutanals (201), followed by heating the resulting hydroxy-esters with metaphosphoric acid.

$$(\text{EtO}_2\text{C})_2\text{CR}\cdot\text{CH}_2\cdot\text{CH}_2\cdot\text{CHO}$$

(201)

8.5.5   Higher lactones

A 14-membered-ring lactone isolated from the seed oil of *Monnina emarginata* has been shown[173] by spectroscopic and chemical evidence to be (S)-13-

hydroxyoctadeca-9c,11t-dienoic acid lactone. The corresponding hydroxy-acid has been discussed in Section 8.4.2.1.

The 12-membered-ring-lactone (202) has been isolated[174] from a fungus.

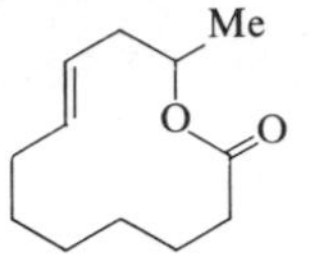

(202)

8.6 LACTAMS

8.6.1 α-Lactams

α-Lactams (203) have been shown[175] to undergo O-alkylation by reaction with a trialkyloxonium tetrafluoroborate. Thus, 1,3-di-t-butylaziridinone ((203); $R^1 = R^2 = Bu^t$) is converted by this reaction into the ester (204). A

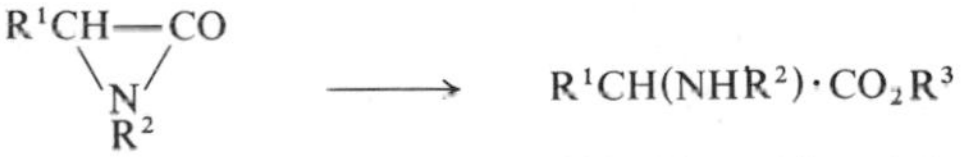

$$R^1CH(NHR^2)\cdot CO_2R^3$$

(204) $R^1 = R^2 = Bu^t$

(203)

new method is, therefore, provided for synthesising esters of N-substituted amino acids.

8.6.2 β-Lactams

A method for the isolation of the β-lactam function (205) of penicillins has been described[176]. This provides a convenient intermediate for synthesis of various β-lactam antibiotics.

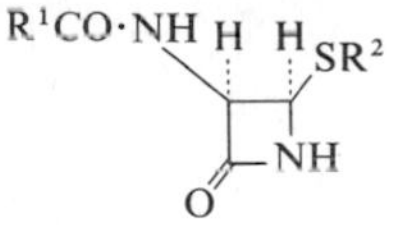

(205) R^2 – protecting group

The stereochemistry of the formation of β-lactams by the reaction of an acid chloride with an imine has been studied[177].

N-Bromo- and N-chloro-β-lactams have been shown[178] to rearrange in the presence of alkenes or alkynes to give, respectively, β-bromo- and β-chloro-alkyl isocyanates.

8.6.3 γ-Lactams

Corydalactam, identified as trans-3-ethylidene-2-pyrrolidone (211), has been synthesised[179] as follows. The acetal (207), obtained from α-acetyl-γbutyrolactone (206) and ethylene glycol, was converted by ammonolysis into the lactam (208), and thence into the ketone (209). Reduction of the latter to a mixture of epimeric carbinols (210), followed by dehydration, gave the *trans*-ethylidene compound (211).

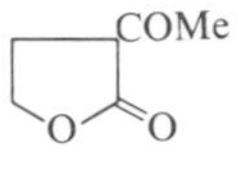

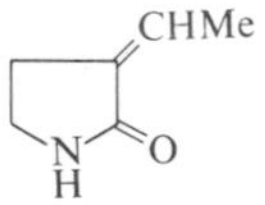

Other syntheses of γ-lactams include the catalytic carbonylation of cyclo-propylamine[180], and a procedure involving hydrolysis of β-cyanoketones[181].

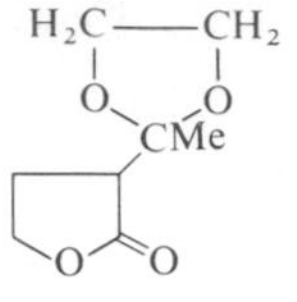

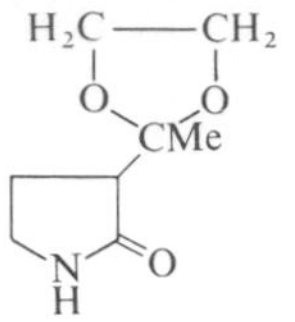

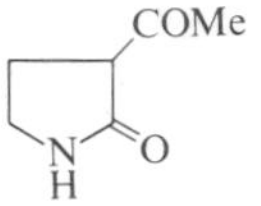

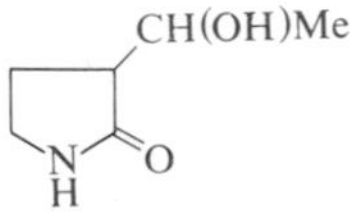

8.6.4 Higher lactams

Comparative studies on the hydrolysis of the lactam linkage of 6- and 5-membered rings[182, 183] as well as that of a 7-membered ring[184] have been reported.

The conversion of hexanolactam into a cyclic hydroxamic acid in only two steps, via an *O*-alkyl imino-ester, has been described[185].

References

1. Jamieson, G. R. (1970). *Topics in Lipid Chemistry* (F. D. Gunstone, editor), Vol. 1, 107 (London: Logos Press)
2. McCloskey, J. A. (1970). *Topics in Lipid Chemistry* (F. D. Gunstone, editor), Vol. 1, 369. (London: Logos Press)
3. Wendt, G. and McCloskey, J. A. (1970). *Biochemistry*, **9**, 4854
4. Pryde, E. H. and Cowan, J. C. (1971). *Topics in Lipid Chemistry* (F. D. Gunstone, éditor), Vol. 2, 1. (London: Logos Press)
5. Ucciani, E., Pierri, F. and Naudet, M. (1970). *Bull. Soc. Chim. Fr.,* 791
6. Nguyen Dinh-Nguyen and Raal, A. (1970). *Acta Chem. Scand.,* **24,** 3416
7. Johnson, C. B. and Wilson, A. T. (1971). *Lipids,* **6,** 181
8. Asselineau, C. P., Lacave, C. S., Montrozier, H. L. and Promé, J.-C. (1970). *European J. Biochem.,* **14,** 406
9. Kaneda, T. (1971). *Biochem. Biophys. Res. Commun.,* **43,** 298
10. Hung, J. G. C. and Walker, R. W. (1970). *Lipids,* **5,** 720
11. Litchfield, C. (1970). *Lipids,* **5,** 144
12. King, R. J. and Clements, J. A. (1970). *J. Lipid Res.,* **11,** 381
13. Scholfield, C. R. and Dutton, H. J. (1970). *J. Amer. Oil Chem. Soc.,* **47,** 1
14. Gunstone, F. D. and Lie Ken Jie, M. (1970). *Chem. Phys. Lipids,* **4,** 139
15. Abley, P., McQuillin, F. J., Minnikin, D. E., Kusamran, K., Maskens, K. and Polgar, N. (1970). *Chem. Commun.,* 348
16. Kusamran, K. and Polgar, N. (1971). *Lipids,* **6,** 961

17. Schaumburg, K. (1970). *Lipids, 5,* 505
18. Reucroft, J. and Sammes, P. G. (1971). *Quart. Rev. Chem. Soc., 25,* 135
19. Barve, J. A. and Gunstone, F. D. (1971). *Chem. Phys. Lipids, 7,* 311
20. Suhara, Y. and Miyazaki, S. (1970). *Bull. Chem. Soc. Jap., 43,* 3924
21. Frankel, E. N. and Dutton, H. J. (1970). *Topics in Lipid Chemistry.* (F. D. Gunstone, editor), Vol. 1, 161. (London: Logos Press)
22. Scholfield, C. R., Mounts, T. L., Butterfield, R. O. and Dutton, H. J. (1971). *J. Amer. Oil Chem. Soc., 48,* 237
23. Shaw, J. E. and Knutson, K. K. (1971). *J. Org. Chem., 36,* 1151
24. Niehaus, W. G., Jr., Kisic, A., Bednarczyk, D. J. and Schroepfer, G. J., Jr. (1970). *J. Biol. Chem., 245,* 3790
25. Ansell, M. F., Radziwill, A. N. and Weedon, B. C. L. (1971). *J. Chem. Soc. C,* 1851
26. Ackman, R. G., Linstead, Sir P., Wakefield, B. J. and Weedon, B. C. L. (1960). *Tetrahedron, 8,* 221
27. Basu, H. and Schlenk, H. (1971). *Chem. Phys. Lipids, 6,* 266
28. Foglia, T. A., Maerker, G. and Smith, G. R. (1970). *J. Amer. Oil Chem. Soc., 47,* 384
29. Bird, C. W. (1971). *Topics in Lipid Chemistry* (F. D. Gunstone, editor), Vol. 2, 247. (London: Logos Press)
30. Naudet, M. and Ucciani, E. (1971). *Topics in Lipid Chemistry* (F. D. Gunstone, editor), Vol. 2, 99. (London: Logos Press)
31. Fulco, A. J. (1970). *J. Biol. Chem., 245,* 2985
32. Cason, J., Davis, R. and Sheehan, M. H. (1971). *J. Org. Chem., 36,* 2621
33. Schlenk, H. (1970). *Progr. Chem. Fats and Lipids, 9,* 587. (Oxford: Pergamon Press)
34. Suzuki, O., Hashimoto, T., Hayamizu, K. and Yamamoto, O. (1970). *Lipids, 5,* 457
35. Gunstone, F. D., Lie Ken Jie, M. and Wall, R. T. (1969). *Chem. Phys. Lipids, 3,* 297
36. Gunstone, F. D. and Lie Ken Jie, M. (1970). *Chem. Phys. Lipids, 4,* 131
37. Kirmse, W. and Lechte, H. (1970). *Justus Liebig's Ann. Chem., 739,* 235
38. Gunstone, F. D. and Lie Ken Jie, M. (1970). *Chem. Phys. Lipids, 4,* 1
39. Sprecher, H. W. (1971). *Biochim. Biophys. Acta, 231,* 122
40. Budny, J. and Sprecher, H. (1971). *Biochim. Biophys. Acta, 239,* 190
41. Kunau, W.-H., Lehmann, H. and Gross, R. (1971). *Hoppe-Seyler's Z. Physiol. Chem., 352,* 542
42. Kunau, W.-H. (1971). *Chem. Phys. Lipids, 7,* 101
43. Kunau, W.-H. (1971). *Chem. Phys. Lipids, 7,* 108
44. Steenhoek, A., Van Wijngaarden, B. H. and Pabon, H. J. J. (1971). *Rec. Trav. Chim., 90,* 961
45. Burkholder, W. E., Silverstein, R. M., Rodin, J. and Gorman, J. E. (1970). *Chem. Abstr., 72,* 131 440c; *U.S. Pat.* 3 501 566
46. Rodin, J. O., Leaffer, M. A. and Silverstein, R. M. (1970). *J. Org. Chem., 35,* 3152
47. Starratt, A. N. and Boch, R. (1971). *Can. J. Biochem., 49,* 251
48. Kepler, C. R., Tucker, W. P. and Tove, S. B. (1970). *J. Biol. Chem., 245,* 3612
49. Rosenfeld, I. S. and Tove, S. B. (1971). *J. Biol. Chem., 246,* 5025
50. Koritala, S. and Selke, E. (1971). *J. Amer. Oil Chem. Soc., 48,* 222
51. Mounts, T. L. and Dutton, H. J. (1970). *Lipids, 5,* 997
52. Schroepfer, G. J., Jr. and Niehaus, W. G., Jr. (1970). *J. Biol. Chem., 245,* 3798
53. Abbot, G. G., Gunstone, F. D. and (in part) Hoyes, S. D. (1970). *Chem. Phys. Lipids, 4,* 351
54. Dufek, E. J., Gast, L. E. and Friedrich, J. P. (1970). *J. Amer. Oil Chem. Soc., 47,* 47
55. Dufek, E. J., Cowan, J. C. and Friedrich, J. P. (1970). *J. Amer. Oil Chem. Soc., 47,* 51
56. Suzuki, O. (1970). *Yukagaku, 19,* 1075 (*Chem. Abstr., 74,* 41983u)
57. Suzuki, O. (1970). *Yukagaku, 19,* 1081 (*Chem. Abstr., 74,* 41987y)
58. Sagredos, A. N., von Mikusch, J. D. and Wolf, V. (1971). *Justus Liebig's Ann. Chem., 745,* 169
59. Earle, F. R. (1970). *J. Amer. Oil Chem. Soc., 47,* 510
60. Conacher, H. B. S. and Gunstone, F. D. (1970). *Lipids, 5,* 137
61. Bradshaw, R. W., Day, A. C., Jones, Sir E. R. H., Page, C. B., Thaller, V. and (in part) Vere Hodge, R. A. (1971). *J. Chem. Soc. C,* 1156
62. Ueno, A. and Maeda, T. (1970). *Yakugaku Zasshi, 90,* 1578 (*Chem. Abstr., 74,* 75 979n)
63. Oudejans, R. C. H. M., van der Horst, D. J. and van Dongen, J. P. C. M. (1971). *Biochemistry, 10,* 4938

64. Chan, M., Himes, R. H. and Akagi, J. M. (1971). *J. Bacteriol.*, **106**, 876
65. de Rosa, M., Gambacorta, A. and Minale, L. (1971). *Chem. Commun.*, 1334
66. Gunstone, F. D., Lie Ken Jie, M. and Wall, R. T. (1971). *Chem. Phys. Lipids*, **6**, 147
67. Gensler, W. J., Floyd, M. B., Yanase, R. and Pober, K. W. (1970). *J. Amer. Chem. Soc.*, **92**, 2472
68. Williams, J. L. and Sgoutas, D. S. (1971). *J. Org. Chem.*, **36**, 3064
69. Altenburger, E. R., Berry, J. W. and Deutschman, A. J. (1970). *J. Amer. Oil Chem. Soc.*, **47**, 77
70. Gensler, W. J., Pober, K. W., Solomon, D. M. and Floyd, M. B. (1970). *J. Org. Chem.*, **35**, 2301
71. Gensler, W. J., Pober, K. W., Solomon, D. W., Yanase, R. and Floyd, M. B. (1970). *Chem. Commun.*, 287
72. Gensler, W. J., Solomon, D. M., Yanase, R. and Pober, K. W. (1971). *Chem. Phys. Lipids*, **6**, 280
73. Eidus, Ya. T., Puzitskii, K. V. and Yang, Yung-Ping. (1970). *Zh. Anal. Khim.*, **25**, 1413 (*Chem. Abstr.*, **73**, 137139d)
74. Grimwood, P. D., Minnikin, D. E., Polgar, N. and Walker, J. E. (1971). *J. Chem. Soc. C*, 870
75. Björkhem, I. and Danielsson, H. (1970). *European J. Biochem.*, **14**, 473
76. De Graw, J. I. and Rodin, J. O. (1971). *J. Org. Chem.*, **36**, 2902
77. Chasin, D. G. and Perkins, E. G. (1971). *Chem. Phys. Lipids*, **6**, 8
78. Ploquin, J. and Sparfel, L. (1970). *Ann. Chim. (Paris)*, **5**, 143
79. Löffler, A., Pratt, R. J., Rüesch, H. P. and Dreiding, A. S. (1970). *Helv. Chim. Acta*, **53**, 383
80. Lough, A. K. (1970). *Lipids*, **5**, 201
81. Augustyn, O. P. H., de Wet, P., Garbers, C. F., Lourens, L. C. F., Neuland, E., Schneider, D. F. and Steyn, K. (1971). *J. Chem. Soc. C*, 1878
82. Polgar, N. (1971). *Topics in Lipid Chemistry* (F. D. Gunstone, editor), Vol. 2, 207. (London: Logos Press)
83. Bertelsen, O. (1970). *Arkiv Kemi*, **32**, 17
84. Jacob, J. and Glaser, A. (1970). *Z. Naturforsch.*, **25B**, 1435
85. Jacob, J. and Zeman, A. (1970). *Z. Naturforsch.*, **25B**, 1438
86. Odham, G. and Stenhagen, E. (1971). *Accounts Chem. Res.*, **4**, 121
87. Goren, M. B., Brokl, O., Das, B. C. and Lederer, E. (1971). *Biochemistry*, **10**, 72
88. Odham, G., Stenhagen, E. and Waern, K. (1970). *Arkiv Kemi*, **31**, 533
89. Odham, G. (1967). *Arkiv Kemi*, **27**, 231
90. Odham, G. and Waern, K. (1969). *Arkiv Kemi*, **29**, 563
91. Chong, R., King, R. R. and Whalley, W. B. (1971). *J. Chem. Soc. C*, 3566
92. Trost, B. M. (1970). *Accounts Chem. Res.*, **3**, 120
93. Tsizin, Yu. S. and Drabkina, A. A. (1970). *Russ. Chem. Rev.*, **39**, 498
94. Meyer, A. S. and Hanzmann, E. (1970). *Biochem. Biophys. Res. Commun.*, **41**, 891
95. Loew, P. and Johnson, W. S. (1971). *J. Amer. Chem. Soc.*, **93**, 3765
96. Loew, P., Siddall, J. B., Spain, V. L. and Werthemann, L. (1970). *Proc. Nat. Acad. Sci. U.S.*, **67**, 1462
97. Faulkner, D. J. and Petersen, M. R. (1971). *J. Amer. Chem. Soc.*, **93**, 3766
98. Slama, K. (1971). *Ann. Rev. Biochem.*, **40**, 1079
99. Schreurs, P. H. M., Montijn, P. P. and Hoff, S. (1971). *Rec. Trav. Chim.*, **90**, 1331
100. Keller-Schierlein, W., Brufani, M. and Muntwyler, R. (1971). *Helv. Chim. Acta*, **54**, 44
101. Brufani, M. and Fedeli, W. (1971). *Helv. Chim. Acta*, **54**, 51
102. Klitgaard, N. A., Hansen, K. P., Hjeds, H. and Jerslev, B. (1970). *Acta Chem. Scand.*, **24**, 33
103. Pattenden, G., Way, J. E. and Weedon, B. C. L. (1970). *J. Chem. Soc. C*, 235
104. Lijmbach, G. W. M., Cox, H. C. and Berends, W. (1970). *Tetrahedron*, **26**, 5993
105. Lijmbach, G. W. M., Cox, H. C. and Berends, W. (1971). *Tetrahedron*, **27**, 1839
106. Taylor, E. C., Altland, H. W. and McGillivray, G. (1970). *Tetrahedron Letters*, 5285
107. Moersch, G. W. and Burkett, A. R. (1971). *J. Org. Chem.*, **36**, 1149
108. Vesonder, R. F., Stodola, F. H., Rohwedder, W. K. and Scott, D. B. (1970). *Canad. J. Chem.*, **48**, 1985
109. Yano, I., Furukawa, Y. and Kusunose, M. (1970). *Biochim. Biophys. Acta*, **202**, 189
110. Yano, I., Furukawa, Y. and Kusunose, M. (1970). *Biochim. Biophys. Acta*, **210**, 105

111. Schildknecht, H. and Koob, K. (1971). *Angew. Chem., Int. Ed. Engl.,* **10,** 124
112. Phillips, B. E., Smith, C. R., Jr. and Tjarks, L. W. (1970). *Biochim. Biophys. Acta,* **210,** 353
113. Albro, P. W. and Fishbein, L. (1971). *Phytochemistry,* **10,** 631
114. Nozaki, H., Noyori, R. and Mori, T. *Japan Pat.,* 70 32 684 (*Chem. Abstr.,* **74,** 76 021f)
115. Crundwell, E. and Cripps, A. L. (1971). *Chem. Ind. (London),* 767
116. Siouffi, A. M. and Naudet, M. (1970). *Ann. Fac. Sci. Marseilles,* **43A,** 137 (*Chem. Abstr.,* **75,** 5167t)
117. Ames, D. E., Goodburn, T. G., Jevans, A. W. and McGhie, J. F. (1968). *J. Chem. Soc. C,* 268
118. McGhie, J. F., Ross, W. A., Spence, J. W. and (in part) James, F. J. (1971). *Chem. Ind. (London),* 1074
119. Ansell, M. F., Shepperd, I. S. and Weedon, B. C. L. (1971). *J. Chem. Soc. C,* 1840
120. Ansell, M. F., Redshaw, D. J. and Weedon, B. C. L. (1971). *J. Chem. Soc. C,* 1846
121. Ansell, M. F., Shepherd, I. S. and Weedon, B. C. L. (1971). *J. Chem. Soc. C,* 1857
122. Blackbourn, G. P. and Tighe, B. J. (1971). *J. Chem. Soc. C,* 257
123. Ucciani, E., Vantillard, A. and Naudet, M. (1970). *Chem. Phys. Lipids,* **4,** 225
124. Gunstone, F. D. and Said, A. I. (1971). *Chem. Phys. Lipids,* **7,** 121
125. Gunstone, F. D. and Inglis, R. P. (1970). *Chem. Commun.,* 877
126. Abbot, G. G. and Gunstone, F. D. (1971). *Chem. Phys. Lipids,* **7,** 290
127. Abbot, G. G. and Gunstone, F. D. (1971). *Chem. Phys. Lipids,* **7,** 303
128. Abbot, G. G. and Gunstone, F. D. (1971). *Chem. Phys. Lipids,* **7,** 279
129. Cornforth, J. W. and Cornforth, R. H. (1970). *Biochem. Soc. Symp.,* No. 29, 5
130. Blattmann, P. and Retey, J. (1970). *Chem. Commun.,* 1393
131. Blattmann, P. and Retey, J. (1970). *Chem. Commun.,* 1394
132. Cornforth, J. W. and Ross, F. P. (1970). *Chem. Commun.,* 1395
133. Scott, A. I., Phillips, G. T., Reichardt, P. B. and Sweeny, J. G. (1970). *Chem. Commun.,* 1396
134. Cornforth, J. W. (1969). *Quart. Rev. Chem. Soc.,* **23,** 125
135. Cornforth, R. H. (1971). *Tetrahedron Letters,* 2003
136. Corey, E. J., Vlattas, I. and Harding, K. (1969). *J. Amer. Chem. Soc.,* **91,** 535
137. Corey, E. J., Weinshenker, N. M., Schaaf, T. K. and Huber, W. (1969). *J. Amer. Chem. Soc.,* **91,** 5675
138. Corey, E. J., Schaaf, T. K., Huber, W., Koelliker, U. and Weinshenker, N. M. (1970). *J. Amer. Chem. Soc.,* **92,** 397
139. Corey, E. J., Noyori, R. and Schaaf, T. K. (1970). *J. Amer. Chem. Soc.,* **92,** 2586
140. Corey, E. J. (1971). *Ann. N.Y. Acad. Sci.,* **180,** 24
141. Corey, E. J., Shirahama, H., Yamamoto, H., Terashima, S., Venkateswarlu, A. and Schaaf, T. K. (1971). *J. Amer. Chem. Soc.,* **93,** 1490
142. Pace-Asciak, C. and Wolfe, L. S. (1971). *Biochemistry,* **10,** 3657
143. Crossley, N. S. (1971). *Tetrahedron Letters,* 3327
144. Welby-Gieusse, M., Laneelle, M. A. and Asselineau, J. (1970). *European J. Biochem.,* **13,** 164.
145. Maurice, M. T., Vacheron, M. J. and Michel, G. (1971). *Chem. Phys. Lipids,* **7,** 9
146. Brennan, P. J., Lehane, D. P. and Thomas, D. W. (1970). *European J. Biochem.,* **13,** 117
147. Ioneda, T., Lederer, E. and Rozanis, J. (1970). *Chem. Phys. Lipids,* **4,** 375
148. Laneelle, M. A. and Laneelle, G. (1970). *European J. Biochem.,* **12,** 296
149. Etemadi, A. H. and Gasche, J. (1965). *Bull. Soc. Chim. Biol.,* **47,** 2095
150. Asselineau, C., Tocanne, G. and Tocanne, J.-F. (1970). *Bull. Soc. Chim. Fr.,* 1455
151. Kondo, E., Kanai, K., Nishimura, K. and Tsumita, T. (1970). *Jap. J. Med. Sci. Biol.,* **23,** 315
152. Kanai, K., Wiegeshaus, E. and Smith, D. W. (1970). *Japan. J. Med. Sci. Biol.,* **23,** 327
153. Lederer, E. (1971). *Pure Appl. Chem.,* **25,** 135
154. Aldridge, D. C., Giles, D. and Turner, W. B. (1970). *Chem. Commun.,* 639
155. Barnett, W. E. and McKenna, J. C. (1971). *Tetrahedron Letters,* 2595
156. Danilyak, N. I., Kotovich, Kh.Z., Pokotilo, M. G., Nizovtseva, A. A., Cherevko, N. G. Mokryi, E. N. and Tolopko, D. K. (1971). *Prikl. Biokhim. Microbiol.,* **7,** 200 (*Chem. Abstr.,* **75,** 16546s)
157. Masuda, M. and Nishimura, K. (1971). *Phytochemistry,* **10,** 1401
158. Kato, M., Tanaka, R. and Yoshikoshi, A. (1971). *Chem. Commun.,* 1561
159. Martin, J., Watts, P. C. and Johnson, F. (1970). *Chem. Commun.,* 27
160. Oehler, E., Reiniger, K. and Schmidt, U. (1970). *Angew. Chem., Int. Ed. Engl.,* **9,** 457

161. Matsuda, Tsutomu; Kond o, Hiroshi and Nakamura, Noriyuki. (1971). *Kogyo Kagaku Zasshi*, **74**, 1135 (*Chem. Abstr.*, **75**, 63052n)
162. Heintz, M., Druilhe, A. and Lefort, D. (1971). *Chromatographia*, **4**, 167
163. Bohlmann, F. and Grenz, M. (1971). *Tetrahedron Letters*, 3623
164. Löffler, A., Norris, F., Taub, W., Svanholt, K. L. and Dreiding, A. S. (1970). *Helv. Chim. Acta*, **53**, 403
165. Klaren-De Wit, M., Frost, D. J. and Ward, J. P. (1971). *Rec. Trav. Chim.*, **90**, 1207
166. Kraatz, U., Wamhoff, H. and Korte, F. (1971). *Justus Liebig's Ann. Chem.*, **744**, 33
167. Jones, J. B. and Barker, J. N. (1970). *Can. J. Chem.*, **48**, 1574
168. Ducher, S. and Michet, A. (1970). *Bull. Soc. Chim. Fr.*, 4353
169. Daremon, C., Rambaud, R. and Verniette, M. (1971). *Compt. Rend. Acad. Sci., Ser. C*, **272**, 1734
170. Daremon, C. and Rambaud, R. (1971). *Bull. Soc. Chim. Fr.*, 294
171. Ijima, A., Mizuno, H. and Takahashi, K. (1971). *Chem. Pharm. Bull.*, **19**, 1053
172. Sarkisyan, P. A., Zalinyan, M. G. and Dangyan, M. T. (1971). *Arm. Khim. Zh.*, **24**, 245 (*Chem. Abstr.*, **75**, 63 050k)
173. Phillips, B. E., Smith, C. R., Jr. and Tjarks, L. W. (1970). *J. Org. Chem.*, **35**, 1916
174. Vesonder, R. F., Stodola, F. H., Wickerham, L. J., Ellis, J. J. and Rohwedder, W. K. (1971). *Can. J. Chem.*, **49**, 2029
175. Sheehan, J. C. and Mehdi Nafissi-V, M. (1970). *J. Org. Chem.*, **35**, 4246
176. Barton, D. H. R., Greig, D. G. T., Sammes, P. G. and Taylor, M. W. (1971). *Chem. Commun.*, 845
177. Bose, A. K., Spiegelman, G. and Manhas, M. S. (1971). *Tetrahedron Letters*, 3167
178. Kampe, K.-D. (1971). *Justus Liebig's Ann. Chem.*, **752**, 142
179. Kametani, T. and Ihara, M. (1971). *J. Chem. Soc. C*, 999
180. Igbal, A. F. M. (1971). *Tetrahedron Letters*, 3381
181. Stevens, R. V. and Kaplan, M. (1970). *Chem. Commun.*, 822
182. Fujii, T. and Yoshifuji, S. (1970). *Tetrahedron*, **26**, 5953
183. Fujii, T., Yoshifuji, S. and Tamai, A. (1971). *Chem. Pharm. Bull.*, **19**, 369
184. Fujii, T. and Yoshifuji, S. (1971). *Chem. Pharm. Bull.*, **19**, 1051
185. Black, D. St. C., Brown, R. F. C. and Wade, A. M. (1971). *Tetrahedron Letters*, 4519
186. Raber, D. J. and Gariano, P. (1971). *Tetrahedron Letters*, 4741
187. Cornforth, J. W., Redmond, J. W., Eggerer, H., Buckel, W. and Gutschow, C. (1970). *European J. Biochem.*, **14**, 1
188. Bordwell, F. G. and Knipe, A. C. (1970). *J. Org. Chem.*, **35**, 2956
189. Leffler, J. E. and Zepp, R. G. (1970). *J. Amer. Chem. Soc.*, **92**, 3713
190. Wheland, R. and Bartlett, P. D. (1970). *J. Amer. Chem. Soc.*, **92**, 6057
191. Adam, W. and Rucktäschel, R. (1971). *J. Amer. Chem. Soc.*, **93**, 557

9
Boron Compounds

K. J. TOYNE
University of Hull

9.1 INTRODUCTION

This chapter considers papers published during 1970 and 1971 which are directly relevant to aliphatic boron chemistry but it excludes important topics such as carboranes, boron hydrides and borazines[1]. Even with this limitation more than 500 references were considered and this fact demonstrates the development in organoboron chemistry since its effective beginning in 1956, and provides ample justification for Professor H. C. Brown's use of the latter part of the following couplet as the moral of his important work on hydroboration[2]: 'Large streams from little fountains flow, tall oaks from little acorns grow'.

During the last two years the most significant studies have been on radical reactions and nucleophilically-induced reactions of boron compounds, with particular emphasis on their uses in synthesis. In addition, the general methods of preparation of mono-, di- and mixed tri-alkylboranes and new *B*-alkyl-boracyclanes now available will enable a greater variety of carbon structures to be used in synthesis, and produce greater efficiency in the use of specific alkyl groups; in all of these areas, H. C. Brown and his colleagues have made a major contribution. However, it is important to emphasise that a review of this type reflects current activity and does not present boron chemistry in perspective. Many examples of procedures established previously can be found in recent publications on various topics as follows: the fundamentals of hydroboration[2], hydration of alkenes and alkynes via hydroboration[3], recent progress (1962–1968) in the hydroboration reaction[4]; selective reductions by diborane[5] and disiamylborane[6] of a large variety of organic compounds; synthetic applications of organoboranes[7–10], organoborane–carbon monoxide reactions[11], the reactions of isocyanides with boranes[12]; boron–carbon compounds[13]; cationic boron complexes[14]; diene synthesis via boronate fragmentation[15]; boronic ester neighbouring groups[16]; bimolecular homolytic substitution at boron[17]; a general survey of redistribution reactions which includes information on boron-containing systems[18]. Annual reviews[19–22] and continuing series[23, 24] on boron chemistry and less accessible reviews[25–27] on hydroboration and the reactions of organoboranes have also appeared.

9.2   SYNTHESIS OF ORGANOBORANES. HYDROBORATION

Important general methods of preparing mono-[28] and di-alkylboranes[29] and, from these, mixed trialkylboranes are now available and this permits a greater variation of the carbon structures used in established and new syntheses with organoboranes. The first general synthesis of monoalkylboranes (1,2-dialkyldiboranes)[28] used lithium aluminium hydride or aluminium

(1)
(a) X = R
(b) X = H

hydride reduction of 2-alkylbenzo[1.3.2]dioxaboroles (1a) which are readily available by reaction of an alkene with (1b) (see boronates, p. 369). The solution of the monoalkylborane can then be treated with an alkene to give a mixed trialkylborane, or kept as a pyridine derivative from which the monoalkylborane can be regenerated with boron trifluoride etherate.

Reduction[30] of methyl dialkylborinates (*B*-methoxydialkylboranes, $R_2^1B \cdot OMe$) with lithium aluminium hydride in the presence of an alkene gives mixed trialkylboranes ($R_2^1R^2B$), presumably because the dialkylborane initially formed reacts further with the alkene, although the dialkylborane itself could not be obtained. Moreover, reduction[31] of methyl dialkylborinates with aluminium hydride permits reducible functional groups to be present and a complex, $[(MeO)_3Al\cdot 3R_2BH]$, is precipitated. Formation of this complex stabilises the dialkylborane against disproportionation, and it is an active hydroborating agent, suitable for use in the formation of mixed trialkylboranes. The dialkylboranes can also be kept as the pyridine addition compounds. A further method[29], which gives excellent yields of almost pure dialkylboranes, uses the lithium aluminium hydride reduction of aryl dialkylborinates ($R_2B\cdot OAr$), themselves prepared from trialkylboranes and aryl borates (see borinates, p. 369).

Dimethyl sulphide–borane[32] may be useful as a convenient and less hazardous method of storing and handling diborane; it can be used for hydroboration and reduction[33] and to produce a distillable monoalkylborane[34], (S B)3-(methylthio)propylborane (2), which is reactive in hydroboration.

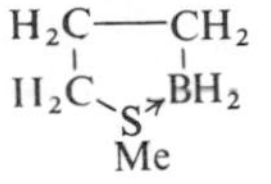

$$H_2C \text{---} CH_2$$
$$H_2C\underset{S}{\diagdown}{}_{\nearrow}BH_2$$
$$Me$$

(2)

The reactions of buta-1,3-diene with borane in tetrahydrofuran have now been clarified and have revealed a fascinating area of organoboron chemistry. When the ratio of buta-1,3-diene to borane is 3:2, the products from the reaction at 0 °C are (3) and (4)[35], but when the ratio is 1:1 the extra borane

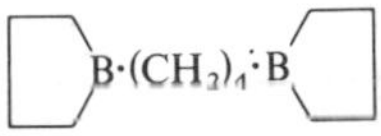

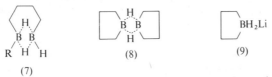

(3) 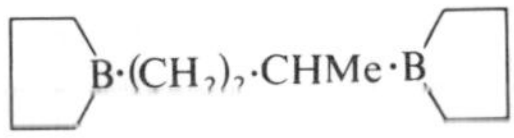   (4)

opens the borolane ring of the initial products to give polymeric 1,2-tetra-methylenediborane (5)[36] which on being heated gives 1,6-diboracyclodecane

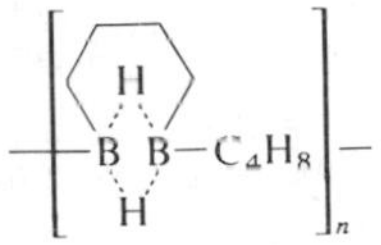

$$\left[\begin{array}{c} H \\ \diagup \diagdown \\ -B \quad B-C_4H_8- \\ \diagdown \diagup \\ H \end{array}\right]_n$$

(5)

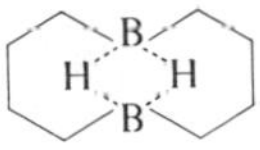

(6)

(6)[37]. This facile ring opening with borane is a general reaction of B-alkyl-borolanes[38] to give 1-alkyl-1,2-tetramethylenediboranes (7) whereas B-alkylborinanes do not give ring-opened products. The double hydrogen bridges in (7) react readily with water or methanol, whereas their reaction with alkenes is more difficult.

(7)

(8)

(9)

However, the reaction mixture containing the 1:1 butadiene–borane product reacts with alkenes in boiling tetrahydrofuran to give B-alkylborolanes in

55–65% yield[36]. Finally, biscorolane (8) has been prepared[37] by the reaction of lithium tetramethyleneborohydride (9) with methane sulphonic acid, and can be trapped by reaction with added alkene or methanol. In their absence (8) rapidly changes into (5) and (6), and so demonstrates the interrelationships of products (8), (3), (4), (5) and (6) derived from buta-1,3-diene.

Bisborinane (10a)[39], bis-(3,5-dimethyl)borinane (10b; 3,5DMB-6)[40] and bis-(3,6-dimethyl)borepane (10c; 3,6DMB-7)[40] (relatively unhindered dialkyl-

$$
\begin{array}{c}
X \qquad\qquad\qquad\qquad X \\
(CH_2)_n \quad B \underset{H}{\overset{H}{\cdots}} B \quad (CH_2)_n \\
X \qquad\qquad\qquad X
\end{array}
$$

(10)

(a) X = H, n = 1
(b) X = Me, n = 1
(c) X = Me, n = 2

boranes with two primary alkyl groups attached to a boron atom) have been prepared with the intention of studying their utility in hydroboration, in selective reductions, and in reactions in which 9-borabicyclo[3.3.1]nonane (9-BBN) (two secondary alkyl groups attached to boron) is not applicable (see Section 9.3.1 and Refs. 41 and 42). Penta-1,4-diene and borane (ratio 3:2) give a product which isomerises to form (11), which then reacts with borane

$$
\bigcirc\!\!-\!B\cdot(CH_2)_5\cdot B\!-\!\!\bigcirc
$$

(11)

to give (10a) and not the ring-opened product, as was formed in the reaction of borane with B-alkylborolanes (see p. 359).

Similarly, hydroboration[40] of 2,4-dimethylpenta-1,4-diene or 2,5-dimethyl-hexa-1,5-diene with borane in tetrahydrofuran at 0 °C followed by heating under reflux for 1 h leads to (10b) and (10c) respectively; (10a), (10b) and (10c) hydroborate alkenes to give B-alkylboracyclanes in excellent yield.

Directive effects in hydroboration of alkynes[43, 44], enynes[43], and conjugated diynes[45] have been studied and the preparation of the following compounds is possible; aldehydes, ketones, alkenes[43]; α,β-unsaturated esters, α-keto-esters[44]; conjugated cis-enynes, cis,cis-dienes, and α,β-acetylenic ketones[45]. Although diborane is essentially non-selective in addition to disubstituted acetylenes, the direction of addition of dicyclohexylborane or disiamylborane is significantly affected by the size of the substituents, with boron being directed to the less hindered position. Thexylborane generally gives a mixture of products (still useful for preparing cis-alkenes by protonolysis) but the dialkylboranes are more regiospecific and are therefore useful for preparing dialkyl(disubstituted-vinyl)boranes which can be converted into highly substituted ketones by oxidation, or into trisubstituted ethylenes by treatment with aqueous sodium hydroxide–iodine. Terminal alkynes and enynes (e.g., 3-methylbut-3-en-1-yne) react with dicyclohexylborane or thexylborane to give compounds with boron almost exclusively at the terminal position of the triple bond.

The hydroboration[44] of alkyl alk-2-ynoates with disiamylborane gives products with boron predominantly at the 2-position (equation (9.1)), and then protonolysis yields pure *cis*-α,β-unsaturated esters (13) when R^1 is primary or secondary alkyl, or a mixture of *cis*- and *trans*-esters (13) when R^1

$$R^1C : C \cdot CO_2R^2 \longrightarrow \underset{(12)}{\overset{R^1}{\underset{H}{}} C = C \overset{CO_2R^2}{\underset{BR_2^3}{}}} \xrightarrow{Me \cdot CO_2H} \underset{(13)}{R^1CH : CH \cdot CO_2R^2} \qquad (9.1)$$

is tertiary. Oxidation of (12) with hydrogen peroxide in the presence of a sodium dihydrogen phosphate buffer (pH *c.* 8) gives $>60\%$ yield of the α-keto-ester, but uncontrolled alkaline hydrogen peroxide oxidation gives a mixture of the *cis*- and the *trans*-ester (13), probably arising from rapid hydrolysis of the carbon–boron (vinyl) bond followed by protonation of the intermediate carbanion (equation (9.2)).

$$(12) \xrightarrow{HO} \underset{H}{\overset{R^1}{}} C = C \overset{CO_2R^2}{\underset{\underset{R^3 \quad OH}{B}}{} R^3} \longrightarrow [R^1CH : \bar{C} \cdot CO_2R^2] \xrightarrow{H_2O} R^1CH : CH \cdot CO_2 \qquad (9.2)$$

9.3 REACTIONS OF ORGANOBORANES. SYNTHETIC USES

The mechanisms of organoborane reactions can be broadly classified either as homolytic reactions, or as heterolytic reactions arising from coordination of a nucleophile with an electron-deficient boron atom. In some reaction sequences both processes may operate and an initial radical process may be followed by a nucleophilically-induced reaction. However, the division into these two general processes is used here to review the major synthetic uses of organoboranes revealed during the last two years, and it is convenient to consider the protonolysis and metallation reactions under separate headings.

9.3.1 Homolytic reactions

During the last two years the usefulness of clean radical reactions of organo-boranes in many general syntheses has become clear and mechanistic studies of homolytic reactions of organoboron compounds have also been published (see Section 9.4).

The fast reaction of organoboranes with α,β-unsaturated carbonyl compounds (equation (9.3)) was originally thought to proceed by a cyclic mechanism but has now been shown[46] to involve radical addition to give

$$R_3B + CH_2 : \overset{|}{C} - \overset{|}{C} : O \xrightarrow{H_2O} RCH_2 \cdot \overset{|}{CH} - \overset{|}{C} : O + R_2BOH \qquad (9.3)$$

the enol borinate (vinyloxyborane) which is then hydrolysed to the carbonyl compound. By using *B*-alkylboracyclanes[41] (see pp. 360 and 362), t-alkyl groups can now be added to α,β-unsaturated carbonyl systems to give, for example, highly branched ketones. The α,β-unsaturated carbonyl compounds (e.g., those with β-substituents and quinones) which had previously been regarded as inert can also react on addition of diacyl peroxides, or by photo-chemical initiation[47], or more conveniently on the introduction of small quantities of oxygen (air)[48, 49] and this affords new syntheses of a variety of aldehydes and ketones via hydroboration. With but-3-yn-2-one[50] the first product of radical addition is (14) which will react further to give (15), and

$$R^1CH : CH \cdot COMe \qquad R^1R^2CH \cdot CH_2 \cdot COMe \qquad R^1CH_2 \cdot CH{:}CH \cdot CH_2OH$$

$$\text{(14)} \qquad\qquad\qquad \text{(15)} \qquad\qquad\qquad \text{(16)}$$

with buta-1,3-diene monoxide the radical addition of organoboranes leads to 4-alkylbut-2-en-1-ols (16) and so provides a four-carbon atom homologation method[51]. The reactions with oxygen can be controlled to provide either an alternative method[52, 53] of producing alcohols from organoboranes (although the reaction is less stereospecific than the established method of oxidation with alkaline hydrogen peroxide) or, by oxidation at low temperatures, a general method[53] for preparing alkyl hydroperoxides.

Iodine inhibits[54, 55] the autoxidation of organoboranes by competing with oxygen for radicals (producing alkyl iodides), and the long induction periods caused by the presence of iodine suggest that the stage shown in equation (9.4) in the following scheme is slow. With sufficient iodine present, stage

$$\begin{array}{llll}
\text{Initiation} & R_3B + O_2 & \longrightarrow & R_2BO_2 \cdot + R \cdot & \text{(9.4)} \\
\text{Propagation} & R \cdot + O_2 & \longrightarrow & RO_2 \cdot & \text{(9.5)} \\
 & RO_2 \cdot + \rangle BR & \longrightarrow & RO_2B \langle + R \cdot & \text{(9.6)} \\
\text{Termination} & 2RO_2 \cdot & \longrightarrow & \text{inactive products} & \text{(9.7)}
\end{array}$$

(9.5) is completely inhibited: under these conditions the effect of the structure of the organoborane on the initiation stage can be studied and it is found[54] that the specific rate of initiation decreases with an increase in the steric crowding about the boron atom. The presence of alkyl iodides (R^2I) in the oxygen-induced reactions of organoboranes (R_3^1B) can also be used[56] to

$$\begin{array}{lll}
R^1 \cdot + R^2I & \longrightarrow & R^1I + R^2 \cdot & \text{(9.8)} \\
2R^2 \cdot & \longrightarrow & R^2R^2 & \text{(9.9)}
\end{array}$$

prepare alkyl iodides ($R'I$) (equation (9.8)) or coupled alkyl groups, e.g., biallyls, bibenzyls (equation (9.9)).

The alkyl radicals produced in stages (9.4) and (9.6) by photochemical initiation or with air, react with disulphides[57] (equation (9.10)) to give mixed dialkyl sulphides, and so B-alkyl-3,5-dimethylborinanes (see pp. 360 and

$$R^1 \cdot + R^2S \cdot SR^2 \longrightarrow R^1SR^2 + R^2S \cdot \qquad \text{(9.10)}$$

361) have proved useful for forming the corresponding alkyl methyl sulphides in good yield.

The reaction of N-chlorodimethylamine with tributylborane has been shown[58] to give butyldimethylamine and dibutylchloroborane by a polar process, (nucleophilically induced migration of an alkyl group from boron to nitrogen, see Section 9.3.2) and butyl chloride and dibutyl(dimethylamino)-borane by a simultaneous radical process (S_H2 attack of the dimethylamino radical at the metallic centre[59], see p. 368). The inhibition of the latter process may thus provide a useful method for preparing tertiary amines and dialkyl-chloroboranes.

In contrast to the reactions involving oxygen-, sulphur- and nitrogen-centred radicals discussed above, which generally involve bimolecular homolytic substitution at the boron atom with displacement of an alkyl radical (see Section 9.4.1), the reactions of organoboron compounds with bromine atoms leads to selective substitution at the position α to the boron atom[60], and the rearrangements and further reactions of the α-bromo-compounds promise to prove of considerable use in developing new syntheses. For the reaction of bromine with organoboranes in solution, which proceeds rapidly even in the dark, stages (9.11) and (9.12) are the important ones[61].

$$\text{Br}^{\cdot} + \text{R}_2\text{B}-\overset{|}{\underset{H}{\text{C}}}- \longrightarrow \text{R}_2\text{B}-\overset{|}{\underset{\cdot}{\text{C}}}- + \text{HBr} \qquad (9.11)$$

$$\text{R}_2\text{B}-\overset{|}{\underset{\cdot}{\text{C}}}- + \text{Br}_2 \longrightarrow \text{R}_2\text{B}-\overset{|}{\underset{\text{Br}}{\text{C}}}- + \text{Br}^{\cdot} \qquad (9.12)$$

$$\text{R}_2\text{B}-\overset{|}{\underset{\text{Br}}{\text{C}}}- + \text{HBr} \longrightarrow \text{R}_2\text{BBr} + \text{H}-\overset{|}{\underset{\text{Br}}{\text{C}}}- \qquad (9.13)$$

The hydrogen bromide produced in stage (9.11) can react in a slow stage (equation (9.13)) with the α-bromo-organoborane to give an alkyl bromide; the anti-Markownikov hydrobromination (see also Section 9.3.4) of internal alkenes is possible by using B-s-alkyl-9-borabicyclo[3.3.1]nonanes[42].

If water is present[62] the hydrogen bromide is absorbed and protonolysis of the α-bromo-organoborane (equation (9.13)) is avoided, but a nucleophilically induced shift[62, 63] of an alkyl group from boron to carbon takes place instead, for example, as in equation (9.14) (see Section 9.3.2). A second bromination and rearrangement of the borinic acid (17) is possible and subsequent oxida-

$$(\text{R}^1\text{R}^2\text{CH})_3\text{B} \xrightarrow{\text{Br}_2} (\text{R}^1\text{R}^2\text{CH})_2\text{B}\cdot\text{CR}^1\text{R}^2\text{Br} \xrightarrow{\text{H}_2\text{O}} \text{R}^1\text{R}^2\text{CH}-\overset{\overset{\displaystyle \text{OH}}{|}}{\underset{\underset{\displaystyle \text{CHR}^1\text{R}^2}{|}}{\text{B}}}-\text{CR}^1\text{R}^2 \qquad (9.14)$$

$$(17)$$

tion with alkaline hydrogen peroxide leads to highly substituted alcohols as a result of the combination of two or three molecules of an alkene.

A potentially important method[64] of preparing α-bromo-organoboranes without the simultaneous formation of hydrogen bromide employs the photochemically induced reaction of bromotrichloromethane with the trialkylborane (equations (9.15) and (9.16)), and demonstrates that the

$$\text{Cl}_3\text{C}^{\cdot} + \text{R}_2\text{B}-\overset{|}{\underset{H}{\text{C}}}- \longrightarrow \text{CHCl}_3 + \text{R}_2\text{B}-\overset{|}{\underset{\cdot}{\text{C}}}- \qquad (9.15)$$

$$\text{R}_2\text{B}-\overset{|}{\underset{\cdot}{\text{C}}}- + \text{BrCCl}_3 \longrightarrow \text{R}_2\text{B}-\overset{|}{\underset{\text{Br}}{\text{C}}}- + \text{Cl}_3\text{C}^{\cdot} \qquad (9.16)$$

trichloromethyl radical resembles the bromine atom in giving selective attack at the α-position.

Organoboranes react with neutral aqueous hydrogen peroxide by a radical process to give mixtures of alcohols, 'dimeric' and 'monomeric' alkanes and, in the presence of carbon tetrachloride, alkyl chlorides[65, 66].

9.3.2 Nucleophilically induced reactions. Migration of alkyl groups from boron to carbon, nitrogen, phosphorus or sulphur

The coordination of organoboranes with nucleophiles (see, for example, equation (9.17)) gives a product in which the transfer of an alkyl group as a carbanion to the neighbouring atom and release of the leaving group (L)

$$\underset{/}{\overset{R}{\underset{|}{\overset{|}{B}}}} + \bar{X}-L \longrightarrow \underset{/}{\overset{R}{\underset{|}{\overset{|}{B}}}}\!-\!X\!-\!\overset{\frown}{L} \longrightarrow \underset{/}{\overset{\backslash}{B}}\!-\!X\!-\!R + L^- \quad (9.17)$$

may occur, and the organoboron compound formed can then undergo further reaction.

The reactions of organoboranes with α-halogenocarbanions produced under the influence of a strong base have been used to prepare α-halogeno-boron compounds (see p. 363 for α-bromo-compounds) in which an alkyl group is then transferred from boron to carbon with displacement of halide ion. Protonolysis of the rearranged material gives the alkylated product.

$$\underset{/}{\overset{R}{\underset{|}{\overset{|}{B}}}} + \underset{\underset{\text{Hal}}{|}}{\overset{|}{\overset{-}{C}}}\!-\! \longrightarrow \underset{/}{\overset{R}{\underset{|}{\overset{|}{B}}}}\!-\!\underset{\underset{\text{Hal}}{\overset{|}{C}}}{\overset{/}{C}} \longrightarrow \underset{/}{\overset{R}{\underset{|}{\overset{|}{B}}}}\!-\!\overset{\overset{\centerdot}{}}{\underset{\backslash}{C}} \longrightarrow R\overset{/}{\underset{\backslash}{C}}H \quad (9.18)$$

These stages can be generalised as shown in equation (9.18). With ethyl bromocyanoacetate or bromomalononitrile[67] as the source of carbanions (which are less sterically hindered than that from diethyl chloromalonate) excellent yields of the alkylated malonic acid derivatives were obtained. By using dichloroacetonitrile[68] as the source of carbanions, α-chloronitriles and mixed dialkylacetonitriles can be prepared, so that by hydrolysis of the latter, a route is available to the dialkylacetic acids which allows the introduction of two secondary alkyl groups and is therefore more general than the familiar malonic ester method.

The analogous reaction[69] with ethyl 4-bromocrotonate (BrCH$_2$·CH:CH·CO$_2$Et) provides a four-carbon atom homologation, and illustrates allylic rearrangement in the protonolysis of the intermediate organoborane (see Section 9.3.3), so that the double bond moves out of conjugation with the ester group. In all these reactions the B-alkylboracyclanes offer the possibility of more efficient use of the alkyl group.

The alkoxide-induced reactions of chlorodifluoromethane[70] with tri-n-butylborane give three alkyl group migrations and provide a room temperature method for preparing trialkylcarbinols from trialkylboranes, a valuable alternative to the carbonylation method[11], which requires more vigorous conditions. Excellent yields were obtained with chlorodifluoromethane; the use of chloroform provides a more convenient procedure but with somewhat lower yields.

The reactions of trialkylboranes with diazocarbonyl compounds have been used to give aldehydes[71] by reaction with diazoacetaldehyde, or α-alkylcycloalkanones[72] by reaction with cyclic α-diazoketones; the former reaction provides a method for the two-carbon atom extension (ethanalation) of alkenes. These reactions of organoboranes with diazocarbonyl compounds ($N_2CH \cdot CO \cdot X$) possibly give initially a carbonylborane ($R_2B \cdot CHR \cdot CO \cdot X$; by nucleophilic attack on boron followed by loss of nitrogen and migration of an alkyl group), which rearranges to the vinyloxyborane [$RCH:CX(OBR_2)$] and is subsequently protonolysed[73]. The vinyloxyboranes [also formed either by the radical addition of trialkylboranes to α,β-unsaturated carbonyl compounds (see p. 361) or by anolysis of trialkylboranes (see p. 366)] react with alkyl lithium compounds to give the lithium enolates which are easily alkylated in a position-specific reaction, and in this way α,α- and α,β-dialkylated ketones have been prepared in good yield[74] from diazoketones and methyl vinyl ketone respectively.

The reaction of trialkylboranes with α-lithiofuran also involves transfers of alkyl groups from boron to carbon and provides a synthesis for 1,1-dialkyl-*cis*-but-2-en-1,4-diols[75].

A promising synthesis of secondary amines[76] arises from the probable coordination of an organic azide with triethylborane followed by loss of nitrogen and migration of the alkyl group; protonolysis of the intermediate ($REtN \cdot BEt_2$) gives the secondary amine.

Trialkylcyanoborates [$R_3\bar{B} \cdot CN\ Na^+$ (18), formed by the coordination of trialkylboranes with cyanide ion] are protonated on the nitrogen atom, which promotes an alkyl migration from boron to carbon (equation (9.19)).

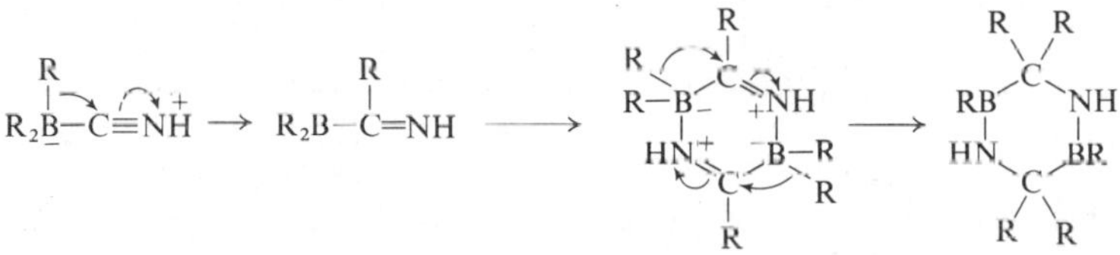

$$\tag{9.19}$$

A second migration in the dimerised intermediate gives a molecule which, after further stages, can be oxidised to give a dialkyl ketone[77], Acid chlorides or trifluoroacetic anhydride also react as electrophiles with (18) to give the reaction shown in equation (9.20) and provide mild, one-flask syntheses of dialkyl ketones in high yield[77, 78].

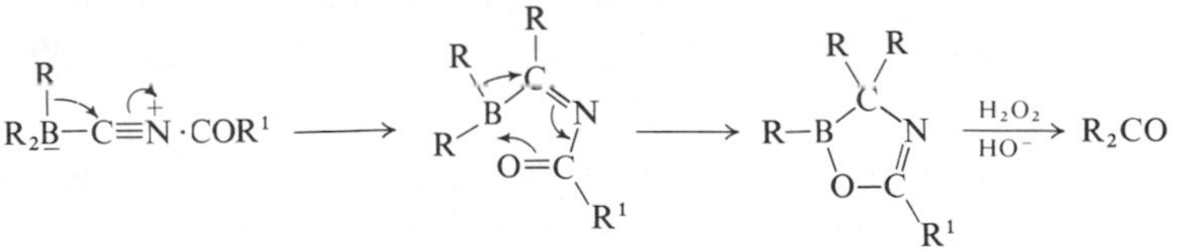

$$R_2B-C{\equiv}N \cdot COR^1 \longrightarrow \quad \longrightarrow R-B \xrightarrow[HO^-]{H_2O_2} R_2CO \tag{9.20}$$

(19)

When dialkylthexylcyanoborates (dialkyl-1,1,2-trimethylpropylcyanoborates from dialkyl-1,1,2-trimethylpropylborane) are used in these reactions, migration of the thexyl group does not occur and an excellent general synthesis of symmetrical, unsymmetrical, functionalised, and cyclic ketones is possible[79]. A third migration of an alkyl group from boron to carbon can be achieved by treatment of (19) with an excess of warm trifluoroacetic anhydride; subsequent oxidation gives high yields of trialkylcarbinols[80] (equation (9.21)).

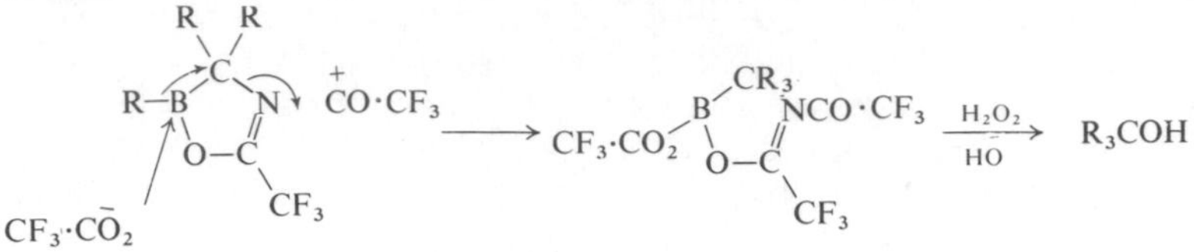

$$\text{(9.21)}$$

Alkyl transfer from trialkylboranes to phosphorus or sulphur with displacement of chlorine has been achieved with chlorophosphines or sulphenyl chlorides to give phosphines or sulphides respectively[81].

9.3.3 Protonolysis. Synthesis of hydrocarbons

The protonolysis of unsymmetrical trialkylboranes with propionic acid removes primary and secondary alkyl groups with only slight selectivity but, for steric reasons, t-alkyl groups react much more slowly[82]. However, the boron–carbon bond in vinyl compounds of boron is cleaved more easily by alcohols than a boron–alkyl bond because of the greater polarity of the boron–(sp^2)carbon bond[83] and this difference has been used in syntheses described below. Pivalic acid or diethylboryl pivalate $(Me_3C \cdot CO_2BEt_2)$ has a considerable catalytic effect on the protonolysis of triethylborane, a rapid and quantitative reaction with water or alcohols occurs even at room temperature[84], and so the reaction can be used for the determination of the water content of hydrates of salts[84] and for the enolysis of trialkylboranes with certain enolisable carbonyl compounds to give, for example, diethylvinyloxyboranes (see p. 365) in good yield[85]. Allylorganoboranes are protonolysed with acetic acid[86] or alcohols[69] and give an allylic rearrangement (equation (9.22)),

$$\text{(9.22)}$$

probably by a 6-membered cyclic process, so that if X were a carbonyl function the product would be unconjugated[69].

The allyl- and benzyl-boration of acetylenic and alkenic compounds has received considerable attention[83, 87-92] as reflected by the prodigious output

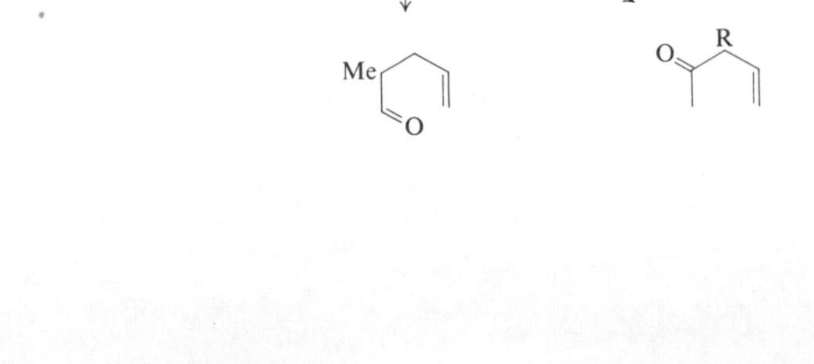

$$\text{(9.23)}$$

of Mikhailov and his co-workers. The main feature of these reactions has been their use in the synthesis of substituted penta-1,4-dienes[87, 93] and unsaturated carbonyl compounds[87], as illustrated in scheme (9.23). The terminal double bond of (20) can be hydrogenated before protonolysis to give the corresponding reduced products[87]. Vinyl ethers react with allylboranes in an analogous fashion to give an intermediate which eliminates a B·OR fragment thermally and produces the 1,4-diene[94], whereas the product (20, R = H) from the first stage of the reaction of triallylborane with alk-l-ynes will cyclise[88–91, 95] thermally to (21) and then to (22); alcoholysis of (21) (cleavage of the B-vinyl bond, see p. 366) and subsequent oxidation leads to substituted 1,6-dienes.

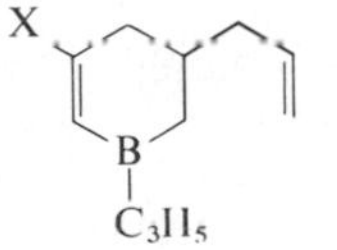

(21) 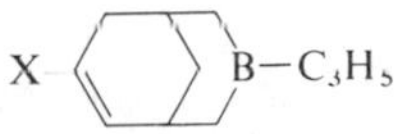   (22)

Other methods of preparing hydrocarbons include diene synthesis (particularly 1,5-dienes) by fragmentation of boronates[15], synthesis of terminal allenes from propargyl chlorides[96], and synthesis of alkenes by the reaction of organoboranes (with at least one β-CH group) with 2-methyl-2-nitrosopropane to give the stereospecific cis-elimination of one B-alkyl group as an alkene[97, 98], and the formation of O-dialkylboryl-t-butylhydroxylamine $(Bu^tNII·O·BR_2)$ by a cyclic process.

9.3.4   Metallation. Synthesis of alkyl halides

The transmetallation of organoboranes with certain mercury compounds gives alkylmercuric derivatives which react with bromine to form alkyl bromides (equation (9.24)) so that this sequence leads to the efficient anti-Markownikov hydrobromination of alkenes[99, 100]. The organoboranes derived

$$RCH : CH_2 \longrightarrow (RCH_2·CH_2)_3B \xrightarrow{Hg(OAc)_2} RCH_2·CH_2·Hg·OAc \xrightarrow{Br_2} RCH_2·CH_2Br$$

$$(9.24)$$

from terminal alkenes react with mercuric acetate at room temperature to give the alkylmercuric acetate and, by further reaction, a range of alkylmercuric salts[101]. Secondary alkyl groups, however, do not react so readily[102] and selective mercuration is possible when using mixed trialkylboranes[100, 101].

An alternative approach to the preparation of alkyl halides by anti-Markownikov hydrobromination is by the reaction of the trialkylborane derived from a terminal alkene with bromine and sodium methoxide in methanol[103]. The mechanism of the bromination has been investigated by using tri-exo-norbornylborane and unexpectedly reveals that the electrophilic substitution by bromine occurs with inversion of configuration[104] to give predominantly $endo$-bromonorbornane.

Alkyl halides have also been prepared by the reaction of trialkylboranes with aqueous cupric halides[105] and by radical reactions (see p. 363).

9.4 MECHANISTIC STUDIES

9.4.1 Homolytic reactions

Several investigations of the kinetics and mechanism of the homolytic re-actions of organoboron compounds have complemented the development of the synthetic applications of these reactions (see Section 9.3.1), and Davies, Roberts and their co-workers have made a major contribution in this area (see below and Ref. 106).

The kinetics of the autoxidation of organoboron compounds in solution are compatible with the reaction stages (9.4–9.7) shown on p. 363[17, 107, 108]. Absolute rate coefficients for equation (9.6) have been obtained and the factors controlling the magnitude of these rate coefficients have been discussed[17, 107]. The absolute rate coefficients have also been obtained for S_H2 reactions of alkoxy radicals at boron centres (equation (9.25); X = RO) by using independent competition techniques involving the analysis of products

$$X\cdot + \;\overset{\backslash}{\underset{/}{B}}{-}R \longrightarrow X\overset{/}{\underset{\backslash}{B}} + R\cdot \qquad (9.25)$$

by g.l.c., or by determination of the radical intermediate by e.s.r.[109, 110]. The substitutions by oxygen-centred radicals are very fast (typically $k = 10^5–10^7$ $M^{-1} s^{-1}$) and have low activation energies (0–5 kcal mol^{-1}).

Photochemical irradiation of ketones produces a triplet state which be-haves like an alkoxy radical and gives an S_H2 reaction at boron with a similar rate coefficient to the oxygen-centred radical reaction mentioned above[111, 112].

S_H2 reactions at boron have also been noted for nitrogen-centred radicals[59] (equation (9.25); X = Me_2N; see also p. 362), sulphur-centred radicals[113] (equation (9.25); X = RS; similar absolute rate coefficients to those for oxygen-centred radicals), iodine atoms[114] (equation (9.25); X = I) and the reaction of methyl radicals with triethylborane in the gas phase has been studied[115].

9.4.2 Heterolytic reactions

A detailed examination of the similarities in mechanism of nucleophilic substitutions at tetrahedral boron and tetrahedral carbon atoms has been reported[116–118]. The reactions of amines or phosphines as uncharged nucleo-philes with amine–borane substrates ($R_3^1\bar{B}{-}\overset{+}{N}R_3^2$) have been studied and orders of reaction, activation parameters, steric and electronic effects, deuterium kinetic isotope effects and the effect of solvent polarity have been used to show that two mechanisms of displacement similar to those in reactions at carbon centres can be distinguished, and appear in many cases to be con-current processes. Trimethylamine–borane and triethylamine–borane react with tri-n-butylphosphine through a dominant S_N2–B mechanism (S_N2 displacement on boron), which should result in inversion of configuration at the reaction site[116], and the reaction of trimethylamine–t-butylborane shows predominantly S_N1–B behaviour arising both from the neopentyl-like steric effect at the reaction site which would inhibit the S_N2–B process and from the internal steric repulsion which would favour the S_N1–B process[117]. Methyl-amine–diarylborane substrates react by an S_N1–B process because of the electronic stabilisation of the electron deficient S_N1–B transition state[118].

The reaction between trimethylphosphine–trimethylborane and an excess of trimethylphosphine[119] and the reactions of amine complexes of triallyl-borane with trimethylamine and with pyridine derivatives[120] have also been studied by n.m.r. spectroscopy.

Trimethylamine–tribromoborane reacts with trichloro- or trifluoro-borane to give mixed species arising from halogen exchange and, if the reaction occurs by ionisation of a boron–bromine bond to give $[Me_3N \cdot BBr_2]^+$, it may be possible to study the kinetics for the exchange of halogens on 4-coordinate boron[121].

9.5 ORGANOBORON ACIDS AND ESTERS

Simple, general syntheses are now available for alkaneboronic and dialkyl-borinic esters and acids to support the use of these compounds in syn-thesis[16, 30, 31]. The boronic esters can be prepared in nearly quantitative yield by the diborane-catalysed reaction of trialkylboranes with trimethylene borate[122], $[CH_2 \cdot (CH_2O)_2 \cdot BO \cdot CH_2]_2CH_2$, or o-phenylene borate[123] or alternatively by hydroborating an alkene with benzo[1.3.2]dioxaborole[123] (1b). A similarly catalysed redistribution of trialkylboranes with o-tolyl borate gives dialkylborinic esters[124]. Syntheses of diborinic[125], diboronic[126–128] and triboronic acid derivatives[128] have also been reported.

9.6 STRUCTURAL AND PHYSICAL INVESTIGATIONS

Numerous investigations of the structure and physical properties of organo-boron compounds have been made, including the following: the effect of alkyl substitution on the ^{11}B chemical shifts in aminoboranes and borates[129]; n.m.r. studies of addition compounds of boron- and phosphorus-containing molecules[130–132], of rotation about a carbon–nitrogen bond[133] and of the rotational barrier of the nitrogen—boron bond[134, 135]; x-ray crystallographic evidence for a boron–nitrogen analogue of an allene[136] and for a trans-annular boron—nitrogen bond in triethanolamine borate[137]; and photo-electron spectra[138–140].

9.7 MISCELLANEOUS TOPICS

Information about other topics in organoboron chemistry on which reports appeared during the last two years can be obtained by consulting the papers cited in references as follows: iminoboranes[141]; amide preparations involving boron reagents[142]; preparation of thioenol ethers[143], unsymmetrical disul-phides[144] and thioacetals[145] using thioboranes; the thermal decomposition of peroxyboron compounds[146].

9.8 SUMMARY OF TYPES OF COMPOUND SYNTHESISED

The following list gives the types of compound for which synthetic procedures are reported in papers included in this review, but the general references given in the Introduction are excluded.

α,β-Acetylenic ketones[45]; alcohols[52,62,64-66,88,91] and trialkylcarbinols[30,70,80]; aldehydes[43,48,71,88,91]; alkanes[65]; alkenes[43,97,98]; alkyl bromides[42,61,99,100,103-105]; alkyl chlorides[65,66,105]; alkyl hydroperoxides[53]; alkyl iodides[56,66]; alkyl-malonic acid derivatives[67]; alkylmercuric salts[99,101,102]; 2-alkylnaphthalene-1,4-diols[49]; allenes[96]; allyl alcohols[51,96]; amides[142]; amines (secondary[76], tertiary[58]); biallyls, bibenzyls[56]; α-chloronitriles[68]; dialkylacetic acids[68]; dialkylacetonitriles[68]; dialkylmercury[100]; cis,cis-dienes[45]; 1,4-dienes[87,88,91,93,94] 1,6-dienes[91]; 1,4-diols[75]; cis-enynes[45]; α-keto-esters[44]; ketones[41,43,47,48,50,60,72,74,77,79,87,92]; phosphines[81]; sulphides[57,81,144]; thioacetals[145]; thioenol ethers[143]; trialkylcarbinols — see alcohols; α,β-unsaturated esters[44]; β,γ-unsaturated esters[69]; α,β-unsaturated ketones[50]; vinyl ethers[83,87,92].

References

1. See (1972) 'Inorganic Chemistry Series One', MTP International Review of Science, Vol. 1 and 4 (London: Butterworths)
2. Brown, H. C. (1962). Hydroboration, (New York: Benjamin)
3. Zweifel, G. and Brown, H. C. (1963). Org. Reactions, 13, 1
4. Devaprabhakara, D. and Sethi, D. S. (1970). J. Sci. Ind. Res., 29, 280
5. Brown, H. C., Heim, P. and Yoon, N. M. (1970). J. Amer. Chem. Soc., 92, 1637
6. Brown, H. C., Bigley, D. B., Arora, S. K. and Yoon, N. M. (1970). J. Amer. Chem. Soc., 92, 7161
7. Cragg, G. M. L. (1969). J. Chem. Educ., 46, 794
8. Carruthers, W. (1971). Some Modern Methods of Organic Synthesis, 100–107, 207–238 (London: Cambridge University Press)
9. Brown, H. C. (1971). Chem. Brit., 7, 458
10. Brown, H. C. (probably June 1972). Boranes in Organic Chemistry, (Ithaca, New York: Cornell University Press)
11. Brown, H. C. (1969). Accounts Chem. Res., 2, 65
12. Casanova, J. (1971). Isonitrile Chemistry, 109 (The Reaction of Isonitriles with Boranes) (I. Ugi, editor) (New York: Academic Press)
13. Lappert, M. F. (1967). The Chemistry of Boron and its Compounds, 443 (Boron–Carbon Compounds) (E. L. Muetterties, editor) (New York: Wiley)
14. Shitov, O. P., Ioffe, S. L., Tartakovskii, V. A. and Novikov, S. S. (1970). Russ. Chem. Rev., 39, 905 [(1970). Usp. Khim., 39, 1913]
15. Marshall, J. A. (1971). Synthesis, 229
16. Matteson, D. S. (1970). Accounts Chem. Res., 3, 186
17. Ingold, K. U. and Roberts, B. P. (1971). Free Radical Substitution Reactions. Bimolecular Homolytic Substitutions (S_H2 Reactions) at Saturated Multivalent Atoms, 46 (New York: Wiley-Interscience)
18. Moedritzer, K. (1971). Organometallic Reactions, Vol. 2, 1 (E. I. Becker and M. Tsutsui, editors) (New York: Wiley-Interscience)
19a. Lappert, M. F., Smith, J. D. and Walton, D. R. M. (1969). Ann. Rept. Progr. Chem. B, 66, 271
19b. Cardin, D. J., Lappert, M. F., Smith, J. D. and Walton, D. R. M. (1970). Ann. Rept. Progr. Chem. B, 67, 271
20. Matteson, D. S. (1970). Organometal. Chem. Rev. B, 6, 323
21. Matteson, D. S. (1971). Organometal. Chem. Rev. B, 8, 1
22. Niedenzu, K. (1971). Organometal. Chem. Rev. B, 8, 45
23. Steinberg, H. and Brotherton, R. J. (1966). Organoboron Chemistry, Vol. 2 (New York: Wiley-Interscience)

24. (1970). *Progress in Boron Chemistry*, Vol. 2 and 3 (R. J. Brotherton and H. Steinberg, editors) (New York: Pergamon)
25. Suzuki, A. (1970). *Kagaku No Ryoiki, Zokan*, **89**, 213
26. Suzuki, A. (1970). *Yuki Gosei Kagaku Kyokai Shi*, **28**, 288
27. Uzarewicz, I., Zaidalewicz, M. and Uzarewicz, A. (1970). *Wiad. Chem.*, **24**, 1
28. Brown, H. C. and Gupta, S. K. (1971). *J. Amer. Chem. Soc.*, **93**, 4062
29. Brown, H. C. and Gupta, S. K. (1971). *J. Organometal. Chem.*, **32**, C1
30. Brown, H. C., Negishi, E. and Gupta, S. K. (1970). *J. Amer. Chem. Soc.*, **92**, 6648
31. Brown, H. C. and Gupta, S. K. (1971). *J. Amer. Chem. Soc.*, **93**, 1818
32. Beres, J., Dodds, A., Morabito, A. J. and Adams, R. M. (1971). *Inorg. Chem.*, **10**, 2072
33. Braun, L. M., Braun, R. A., Crissman, H. R., Opperman, M. and Adams, R. M. (1971). *J. Org. Chem.*, **36**, 2388
34. Braun, R. A., Brown, D. C. and Adams, R. M. (1971). *J. Amer. Chem. Soc.*, **93**, 2823
35. Brown, H. C., Negishi, E. and Gupta, S. K. (1970). *J. Amer. Chem. Soc.*, **92**, 2460
36. Brown, H. C., Negishi, E. and Burke, P. L. (1971). *J. Amer. Chem. Soc.*, **93**, 3400
37. Brown, H. C. and Negishi, E. (1971). *J. Amer. Chem. Soc.*, **93**, 6682
38. Brown, H. C., Negishi, E. and Burke, P. L. (1970). *J. Amer. Chem. Soc.*, **92**, 6649
39. Brown, H. C. and Negishi, E. (1971). *J. Organometal. Chem.*, **26**, C67
40. Brown, H. C. and Negishi, E. (1971). *J. Organometal. Chem.*, **28**, C1
41. Brown, H. C. and Negishi, E. (1971). *J. Amer. Chem. Soc.*, **93**, 3777
42. Lane, C. F. and Brown, H. C. (1971). *J. Organometal. Chem.*, **26**, C51
43. Zweifel, G., Clark, G. M. and Polston, N. L. (1971). *J. Amer. Chem. Soc.*, **93**, 3395
44. Plamondon, J., Snow, J. T. and Zweifel, G. (1971). *Organometal. Chem. Syn.*, **1**, 249
45. Zweifel, G. and Polston, N. L. (1970). *J. Amer. Chem. Soc.*, **92**, 4068
46. Kabalka, G. W., Brown, H. C., Suzuki, A., Honma, S., Arase, A. and Itoh, M. (1970). *J. Amer. Chem. Soc.*, **92**, 710
47. Brown, H. C. and Kabalka, G. W. (1970). *J. Amer. Chem. Soc.*, **92**, 712
48. Brown, H. C. and Kabalka, G. W. (1970). *J. Amer. Chem. Soc.*, **92**, 714
49. Kabalka, G. W. (1971). *J. Organometal. Chem.*, **33**, C25
50. Suzuki, A., Nozawa, S., Itoh, M., Brown, H. C., Kabalka, G. W. and Holland, G. W. (1970). *J. Amer. Chem. Soc.*, **92**, 3503
51. Suzuki, A., Miyaura, N., Itoh, M., Brown, H. C., Holland, G. W. and Negishi, E. (1971). *J. Amer. Chem. Soc.*, **93**, 2792
52. Brown, H. C., Midland, M. M. and Kabalka, G. W. (1971). *J. Amer. Chem. Soc.*, **93**, 1024
53. Brown, H. C. and Midland, M. M. (1971). *J. Amer. Chem. Soc.*, **93**, 4078
54. Brown, H. C. and Midland, M. M. (1971). *Chem. Commun.*, 699
55. Midland, M. M. and Brown, H. C. (1971). *J. Amer. Chem. Soc.*, **93**, 1506
56. Suzuki, A., Nozawa, S., Harada, M., Itoh, M., Brown, H. C. and Midland, M. M. (1971). *J. Amer. Chem. Soc.*, **93**, 1508
57. Brown, H. C. and Midland, M. M. (1971). *J. Amer. Chem. Soc.*, **93**, 3291
58. Davies, A. G., Hook, S. C. W. and Roberts, B. P. (1970). *J. Organometal. Chem.*, **23**, C11
59. Davies, A. G., Hook, S. C. W. and Roberts, B. P. (1970). *J. Organometal. Chem.*, **22**, C37
60. Pasto, D. J. and McReynolds, K. (1971). *Tetrahedron Letters*, 801
61. Lane, C. F. and Brown, H. C. (1970). *J. Amer. Chem. Soc.*, **92**, 7212
62. Lane, C. F. and Brown, H. C. (1971). *J. Amer. Chem. Soc.*, **93**, 1025
63. Brown, H. C. and Yamamoto, Y. (1971). *J. Amer. Chem. Soc.*, **93**, 2796
64. Brown, H. C. and Yamamoto, Y. (1971). *Chem. Commun.*, 1535
65. Bigley, D. B. and Payling, D. W. (1970). *J. Chem. Soc. B*, 1811
66. Davies, A. G. and Tudor, R. (1970). *J. Chem. Soc. B*, 1815
67. Nambu, H. and Brown, H. C. (1970). *Organometal. Chem. Syn.*, **1**, 95
68. Nambu, H. and Brown, H. C. (1970). *J. Amer. Chem. Soc.*, **92**, 5790
69. Brown, H. C. and Nambu, H. (1970). *J. Amer. Chem. Soc.*, **92**, 1761
70. Brown, H. C., Carlson, B. A. and Prager, R. H. (1971). *J. Amer. Chem. Soc.*, **93**, 2070
71. Hooz, J. and Morrison, G. F. (1970). *Can. J. Chem.*, **48**, 868
72. Hooz, J., Gunn, D. M. and Kono, H. (1971). *Can. J. Chem.*, **49**, 2371
73. Pasto, D. J. and Wojtkowski, P. W. (1970). *Tetrahedron Letters*, 215
74. Pasto, D. J. and Wojtkowski, P. W. (1971). *J. Org. Chem.*, **36**, 1790
75. Suzuki, A., Miyaura, N. and Itoh, M. (1971). *Tetrahedron*, **27**, 2775
76. Suzuki, A., Sono, S., Itoh, M., Brown, H. C. and Midland, M. M. (1971). *J. Amer. Chem. Soc.*, **93**, 4329

77. Pelter, A., Hutchings, M. G. and Smith, K. (1970). *Chem. Commun.*, 1529
78. Brehm, E., Haag, A., Hesse, G. and Witte, H. (1970). *Justus Liebigs Ann. Chem.*, **737,** 70
79. Pelter, A., Hutchings, M. G. and Smith, K. (1971). *Chem. Commun.*, 1048–1050
80. Pelter, A., Hutchings, M. G. and Smith, K. (1971). *Chem. Commun.*, 1048 (first paper)
81. Draper, P. M., Chan, T. H. and Harpp, D. N. (1970). *Tetrahedron Letters.*, 1687
82. Bigley, D. B. and Payling, D. W. (1971). *J. Inorg. Nucl. Chem.*, **33,** 1157
83. Bubnov, Yu. N., Korobeinikova, S. A., Isagulyants, G. V. and Mikhailov, B. M. (1970). *Izv. Akad. Nauk SSSR, Ser. Khim.*, 2023
84. Köster, R., Amen, K-L., Bellut, H. and Fenzl, W. (1971). *Angew. Chem. Internat. Ed.*, **10,** 748
85. Fenzsl, W. and Köster, R. (1971). *Angew. Chem. Internat. Ed.*, **10,** 750
86. Mehrotra, I. and Devaprabhakara, D. (1971). *J. Organometal. Chem.*, **33,** 287
87. Mikhailov, B. M., Bubnov, Yu. N., Korobeinikova, S. A. and Frolov, S. I. (1971). *J. Organometal. Chem.*, **27,** 165
88. Bubnov, Yu. N., Frolov, S. I., Kiselev, V. G., Bogdanov, V. S. and Mikhailov, B. M. (1970). *Zh. Obshch. Khim.*, **40,** 1311
89. Bubnov, Yu. N., Frolov, S. I., Kiselev, V. G. and Mikhailov, B. M. (1970). *Zh. Obshch. Khim.*, **40,** 1316
90. Mikhailov, B. M., Bubnov, Yu. N., Korobeinikova, S. A. and Bogdanov, V. S. (1970). *Zh. Obshch. Khim.*, **40,** 1321
91. Bubnov, Yu. N., Frolov, S. I., Kiselev, V. G., Bogdanov, V. S. and Mikhailov, B. M. (1970). *Organometal. Chem. Syn.*, **1,** 37
92. Mikhailov, B. M., Bubnov, Yu. N. and Korobeinikova, S. A. (1970). *J. Prakt. Chem.*, **312,** 998
93. Bubnov, Yu. N. and Mikhailov, B. M. (1971). *USSR Pat.* 304,246
94. Mikhailov, B. M. and Bubnov, Yu. N. (1971). *Tetrahedron Letters*, 2127
95. Mikhailov, B. M. and Cherkasova, K. L. (1971). *Izv. Akad. Nauk SSSR, Ser. Khim.*, 1244
96. Zweifel, G., Horng, A. and Snow, J. T. (1970). *J. Amer. Chem. Soc.*, **92,** 1427
97. Davies, A. G., Foot, K. G., Roberts, B. P. and Scaiano, J. C. (1971). *J. Organometal. Chem.*, **31,** C1
98. Foot, K. G. and Roberts, B. P. (1971). *J. Chem. Soc. C*, 3475
99. Tufariello, J. J. and Hovey, M. M. (1970). *J. Amer. Chem. Soc.*, **92,** 3221
100. Tufariello, J. J. and Hovey, M. M. (1970). *Chem. Commun.*, 372
101. Larock, R. C. and Brown, H. C. (1970). *J. Amer. Chem. Soc.*, **92,** 2467
102. Larock, R. C. and Brown, H. C. (1971). *J. Organometal. Chem.*, **26,** 35
103. Brown, H. C. and Lane, C. F. (1970). *J. Amer. Chem. Soc.*, **92,** 6660
104. Brown, H. C. and Lane, C. F. (1971). *Chem. Commun.*, 521
105. Lane, C. F. (1971). *J. Organometal. Chem.*, **31,** 421
106. Davies, A. G. and Roberts, B. P. (1971). *Nature Phys. Sci.*, **229,** 221 and refs. therein
107. Davies, A. G., Ingold, K. U., Roberts, B. P. and Tudor, R. (1971). *J. Chem. Soc. B*, 698
108. Grotewold, J., Hernandez, J. and Lissi, E. A. (1971). *J. Chem. Soc. B*, 182
109. Davies, A. G., Griller, D. and Roberts, B. P. (1971). *J. Chem. Soc. B*, 1823
110. Davies, A. G., Griller, D., Roberts, B. P. and Tudor, R. (1970). *Chem. Commun.*, 640
111. Davies, A. G., Griller, D., Roberts, B. P. and Scaiano, J. C. (1971). *Chem. Commun.*, 196
112. Davies, A. G., Roberts, B. P. and Scaiano, J. C. (1971). *J. Chem. Soc. B*, 2171
113. Davies, A. G. and Roberts, B. P. (1971). *J. Chem. Soc. B*, 1830
114. Lissi, E. A. and Sanhueza, E. (1971). *J. Organometal. Chem.*, **32,** 285
115. Grotewold, J., Lissi, E. A. and Scaiano, J. C. (1971). *J. Chem. Soc. B*, 1187
116. Budde, W. L. and Hawthorne, M. F. (1971). *J. Amer. Chem. Soc.*, **93,** 3147
117. Walmsley, D. E., Budde, W. L. and Hawthorne, M. F. (1971). *J. Amer. Chem. Soc.*, **93,** 3150
118. Lalor, F. J., Paxson, T. and Hawthorne, M. F. (1971). *J. Amer. Chem. Soc.*, **93,** 3156
119. Alford, K. J., Bishop, E. O., Carey, P. R. and Smith, J. D. (1971). *J. Chem. Soc. A*, 2574
120. Bogdanov, V. S., Baryshnikova, T. K., Kiselev, V. G. and Mikhailov, B. M. (1971). *Zh. Obshch. Khim.*, **41,** 1533
121. Krishnamurthy, S. S. and Lappert, M. F. (1971). *Inorg. Nucl. Chem. Letters*, **7,** 919
122. Brown, H. C. and Gupta, S. K. (1970). *J. Amer. Chem. Soc.*, **92,** 6983
123. Brown, H. C. and Gupta, S. K. (1971). *J. Amer. Chem. Soc.*, **93,** 1816
124. Brown, H. C. and Gupta, S. K. (1971). *J. Amer. Chem. Soc.*, **93,** 2802
125. Coutts, I. G. C. and Musgrave, O. C. (1970). *J. Chem. Soc. C*, 2225

126. Coutts, I. G. C., Goldschmid, H. R. and Musgrave, O. C. (1970). *J. Chem. Soc. C*, 488
127. Matteson, D. S. and Tripathy, P. B. (1970). *J. Organometal. Chem.*, **21**, P6
128. Matteson, D. S. and Thomas, J. R. (1970). *J. Organometal. Chem.*, **24**, 263
129. Davis, F. A., Turchi, I. J. and Greeley, D. N. (1971). *J. Org. Chem.*, **36**, 1300
130. Tuchagues, J-P. and Laurent, J-P. (1971). *Bull. Soc. Chim. Fr.*, 4246 and refs. therein
131. Jugie, G., Laussac, J-P. and Laurent, J-P. (1970). *Bull. Soc. Chim. Fr.*, 4238
132. Rudolph, R. W. and Schultz, C. W. (1971). *J. Amer. Chem. Soc.*, **93**, 6821 and refs. therein
133. Bushweller, C. H., Dewkett, W. J., O'Neil, J. W. and Beall, H. (1971). *J. Org. Chem.*, **36**, 3782 and refs. therein
134. Wells, R. L., Paige, H. L. and Moreland, C. G. (1971). *Inorg. Nucl. Chem. Letters*, **7**, 177
135. Totani, T., Tori, K., Murakami, J. and Watanabe, H. (1971). *Org. Magn. Resonance*, **3**, 627 and refs. therein
136. Bullen, G. J. and Wade, K. (1971). *Chem. Commun.*, 1122
137. Taira, Z. and Osaki, K. (1971). *Inorg. Nucl. Chem. Lett.*, **7**, 509
138. Bock, H. and Fuss, W. (1971). *Chem. Ber.*, **104**, 1687
139. Holliday, A. K., Reade, W., Johnstone, R. A. W. and Neville, A. F. (1971). *Chem. Commun.*, 51
140. Cetinkaya, B., King, G. H., Krishnamurthy, S. S., Lappert, M. F. and Pedley, J. B. (1971). *Chem. Commun.*, 1370
141. Meller, A. and Ossko, A. (1971). *Monatsh. Chem.*, **102**, 131
142. Pelter, A. and Levitt, T. E. (1970). *Tetrahedron*, **26**, 1899
143. Cragg, R. H. and Husband, J. P. N. (1971). *Inorg. Nucl. Chem. Lett.*, **7**, 221
144. Cragg, R. H., Husband, J. P. N. and Weston, A. F. (1970). *Chem. Commun.*, 1701
145. Cragg, R. H. and Husband, J. P. N. (1970). *Inorg. Nucl. Chem. Lett.*, **6**, 773
146. Maslennikov, V. P., Gerbert, G. P., Khodalev, G. F. and Shushunov, V. A. (1971). *Zh. Obshch. Khim.*, **41**, 592